T0261811

R Programming for Actuarial Science

R Programming for Actuarial Science

Peter McQuire
University of Kent
Canterbury
UK

Alfred Kume
University of Kent
Canterbury
UK

This edition first published 2024
© 2024 John Wiley & Sons Ltd

All rights reserved. No part of this publication may be reproduced, stored in a retrieval system, or transmitted, in any form or by any means, electronic, mechanical, photocopying, recording or otherwise, except as permitted by law. Advice on how to obtain permission to reuse material from this title is available at http://www.wiley.com/go/permissions.

The right of Peter McQuire and Alfred Kume to be identified as the authors of this work has been asserted in accordance with law.

Registered Offices
John Wiley & Sons, Inc., 111 River Street, Hoboken, NJ 07030, USA
John Wiley & Sons Ltd, The Atrium, Southern Gate, Chichester, West Sussex, PO19 8SQ, UK

For details of our global editorial offices, customer services, and more information about Wiley products visit us at www.wiley.com.

Wiley also publishes its books in a variety of electronic formats and by print-on-demand. Some content that appears in standard print versions of this book may not be available in other formats.

Trademarks: Wiley and the Wiley logo are trademarks or registered trademarks of John Wiley & Sons, Inc. and/or its affiliates in the United States and other countries and may not be used without written permission. All other trademarks are the property of their respective owners. John Wiley & Sons, Inc. is not associated with any product or vendor mentioned in this book.

Limit of Liability/Disclaimer of Warranty
While the publisher and authors have used their best efforts in preparing this work, they make no representations or warranties with respect to the accuracy or completeness of the contents of this work and specifically disclaim all warranties, including without limitation any implied warranties of merchantability or fitness for a particular purpose. No warranty may be created or extended by sales representatives, written sales materials or promotional statements for this work. This work is sold with the understanding that the publisher is not engaged in rendering professional services. The advice and strategies contained herein may not be suitable for your situation. You should consult with a specialist where appropriate. The fact that an organization, website, or product is referred to in this work as a citation and/or potential source of further information does not mean that the publisher and authors endorse the information or services the organization, website, or product may provide or recommendations it may make. Further, readers should be aware that websites listed in this work may have changed or disappeared between when this work was written and when it is read. Neither the publisher nor authors shall be liable for any loss of profit or any other commercial damages, including but not limited to special, incidental, consequential, or other damages.

A catalogue record for this book is available from the Library of Congress

Hardback ISBN: 9781119754978; ePub ISBN: 9781119754992; ePDF ISBN: 9781119754985; oBook ISBN: 9781119755005

Cover Design: Wiley
Cover Image: © Peter McQuire

Set in 9.5/12.5pt STIXTwoText by Integra Software Services Pvt. Ltd, Pondicherry, India
Printed and bound by CPI Group (UK) Ltd, Croydon, CR0 4YY

C9781119754978_091023

To my wife, Jenny, and daughter, Lauren, for their constant support and encouragement. (Peter McQuire)

To my wife Ortenca, for her support throughout the process. (Alfred Kume)

Contents

About the Companion Website

This book is accompanied by a companion website:

www.wiley.com/go/rprogramming.com

The website contains much of the R code used in this book, allowing copying of the suggested code, thus saving the reader significant time.

The website also includes numerous data files, such as investment data and mortality data. These data files will be analyzed using R code in several chapters of the book.

Introduction

1 Main Objectives of This Book

The overriding objective of this book is to help students of actuarial mathematics and related disciplines such as financial mathematics, develop programming skills which will enhance their understanding of actuarial, financial, and statistical concepts, enabling them to solve real-world problems encountered in these fields. Breaking this down further, the purposes of the book is two-fold:

1. To provide an introduction to the programming language, R. This is achieved using worked examples and undertaking exercises commonly seen in the fields of actuarial and financial mathematics.
2. Secondly, to improve the reader's level of understanding of actuarial and financial topics by using these programming skills. We believe that most students can develop a deeper understanding of mathematical material by solving problems using a programming language. From our experience of teaching actuarial mathematics and statistics, students often confirm that their understanding of a topic has vastly improved following the completion of a computer-based exercise or project.

   A similar effect is noted in students who opt to take a year out from their studies to work in the financial industry, often applying extensive programming skills to solve real-world problems. Such students invariably notice a similar level of improvement in their understanding of concepts. It is hoped, to some extent, that this learning experience can be mirrored throughout this book.

   The authors have significant teaching experience at both undergraduate and post-graduate levels, enhanced with experience in assessment processes for universities and the actuarial profession. This has given insights into the typical issues students experience with actuarial mathematics – problems often arise from a fundamental mis-understanding of introductory material. For example, a final year undergraduate may only fully understand a concept introduced in their first year whilst undertaking pro-gramming coursework in their final year on a specific application of the material taught in the first year, experiencing that "Eureka" moment.

   The reader should not underestimate the extent to which learning a programming language, such as R, to a level such that most exercises in this book can be completed,

R Programming for Actuarial Science, First Edition. Peter McQuire and Alfred Kume.
© 2024 John Wiley & Sons Ltd. Published 2024 by John Wiley & Sons Ltd.
Companion Website: www.wiley.com/go/rprogramming.com.

will help the reader in the employment market. Having a good working knowledge of R or similar language should improve the career prospects of the graduate.

A further motivating factor for writing this book originates from the decision taken by the Institute and Faculty of Actuaries (IFoA) in 2018 to choose the R programming language as an integral part of its syllabus. Indeed, much of the IFoA's syllabi for subjects CM1, CM2, CS1, and, in particular, CS2 are covered in the book.

2   Who Is This Book For?

This book is aimed at two main groups:

1. It has been written principally for university level actuarial and financial maths students, together with graduates undertaking professional actuarial exams (e.g. with the IFoA and SOA), and more generally to anyone aspiring to careers in actuarial mathematics and finance. The book should be useful to the student throughout their studies, whether first-year undergraduate or postgraduate, spanning topics from fundamentals of financial mathematics and Brownian motion, to a variety of mortality models and analysing investment strategies such as asset–liability matching and hedging.
2. Secondly, we hope the book appeals to more experienced professionals in related disciplines wishing to develop skills in a programming language, who may have had limited opportunities to do so earlier in their career. By undertaking examples and exercises related to material with which they are already familiar, this book provides an efficient journey to acquiring such programming skills. Such users of this book may therefore wish to review Chapters 3, 4, 23, and 25, which include traditional material most actuaries would be familiar with.

In writing this book, we have attempted to cater for a wide range of experiences and abilities. The overall style of the book aims to ensure that the basics of each topic are covered, with appropriate text, examples, and exercises, whilst including several more advanced tasks. As noted elsewhere, the reader should aim at expanding on the tasks included in this book.

It is assumed that the reader will have a knowledge of statistics and mathematics at a level expected from that of a first year undergraduate in a maths-based university degree.

The book includes the majority of topics covered in a typical undergraduate course in actuarial science. There is also perhaps a greater emphasis placed on a number of actuarial concepts which may not directly be assessed in traditional university courses; indeed, several examples involve addressing practical problems which the student will see in the workplace. For example, we introduce models which may help in improving how correlations are dealt with by insurance companies, and develop an understanding of fundamental risk management techniques such as hedging, asset-liability matching, and diversification. Ultimately, we hope the reader develops a good understanding of the problem-solving approaches used in the workplace.

3 How to Use This Book

To get the most from this book it is anticipated that during each study session the user will simultaneously:

- study the material in this book,
- access the book's website (code and data), and
- write and run code on their computer.

It would be expected that the user proceeds to write their own code and duplicate the results. The suggested code for each example/exercise is one of many possible solutions; it may be quite reasonable, depending on the scenario, for your code to be quite different to that set out in this book. It is important that the user practises writing their own, independent code, and does not try to learn, by rote, the code in the book. As noted in Chapter 1, the reader may wish to save a script file in respect of each chapter. Indeed the reader may wish to write functions incorporating and combining several sections of code from the website, improving the efficiency of their code.

We would expect most users to have had some prior exposure to, and knowledge of, the material in a chapter before embarking on it, either following an initial period of independent study, or attendance at related university lectures or tutorials; it is anticipated that readers will have access to alternative study material for each topic.

The website contains the majority of the R code included in the book, together with suggested code relating to the exercises. It is intended that the reader will treat the book and website as companions; it is not expected that most users use the book and website separately (for the most part at least). Note that a small amount of code is not included on the website (the missing code can simply be copied from the book) – this is to encourage more active learning of the material.

The vast majority of students will gain most benefit from frequent practise of writing code; occasional engagement is likely to end in less satisfactory results. It is hoped that the style of the book will lend itself to encouraging a greater level of creativity from the student, developing their own examples and exercises as their skills and knowledge increase.

4 Book Structure

We start by covering the fundamentals of R in Chapter 1, "R: What you need to know to get started", and Chapter 2, "Functions in R". If you are new to R we recommend that you first read these two chapters, and revisit them when required. Chapter 1 explains the key aspects of R, e.g. writing your first code in R, how objects are used etc. From experience, most students find it beneficial to initially read this chapter relatively quickly, referring back to it frequently. Readers new to R should benefit from spending some time digesting the examples in Chapter 2 to get a feel for writing basic R code and applying existing functions.

The typical actuarial and financial mathematics student is then likely to cover Chapters 3 and 4 – "Financial Mathematics I" (and "II"); the material included in these two chapters is usually covered in the first year of actuarial mathematics programmes at university.

As noted above, we think most readers will benefit from only a relatively brief study of Chapters 1 and 2, and to move onto the main chapters and start practising! It is unlikely to be beneficial to spend days memorising the material in these introductory chapters.

Most chapters are largely self-contained, with a few obvious exceptions, e.g. Financial Mathematics I and II, Contingencies I and II, the chapters on copulas, and Markov mortality models. There is a certain amount of grouping of chapters where the material is strongly related, and it is likely that most readers will tend to read a group of topics together.

A number of chapters lend themselves particularly to a relatively brief initial study, subsequently re-visiting them when studying a later chapter which uses that material. For example, application of the material in Chapters 5 and 6 is used in several later chapters of the book.

5 Chapter Style

Most chapters begin with setting key objectives and a broad discussion of the main ideas behind the topic of the chapter. This is usually followed with a certain amount of theory, the length of which is based on our experience of how well students generally tend to grasp the concepts. Compared to other texts, there will, in general, be less theory included in this book. Many topics covered in this book already have a wealth of excellent texts – repetition of the same theory is not warranted here. The importance of mathematical rigour should be stressed at this point; the student will benefit greatly over the long-term by developing a deeper understanding of the material which can be adapted to various scenarios (such comparisons are highlighted in several chapters of the book). Each chapter ends with a Recommended Reading list.

As noted in Section 3, most readers will require additional principal reading material on each topic to supplement the material in this book. This book focuses on solving actuarial problems by using the R programming language, and is not intended to be used as a student's sole source of learning for each subject.

The reader may also find it beneficial to own a copy of the "Formulae and Tables" issued by the Institute and Faculty of Actuaries (2002) (also freely available online at the time of publishing).

6 Examples and Exercises

There are over 400 examples and exercises included in this book which readers can use to develop their programming skills and understanding of the mathematical concepts. The book includes two main types of tasks:

1. Analysis of datasets (such as claims data, investment data, mortality data etc.), fitting various models to data, and testing the results. You will find these data sets on the book's website.
2. Other tasks do not require data sets. Code is used to develop a better understanding of actuarial concepts, often with the use of simulations. The book includes, for the most

part, the use of relatively simple code, aimed at communicating the fundamental ideas of the mathematics involved – it is the intention that the reader will develop their coding skills through self-study.

It is hoped that readers will also combine code from various parts of the book, developing their own more advanced models. For example, by combining code from various chapters on asset modelling, claims models, and mortality models, one could develop a model for an insurance company.

Ultimately, actuaries are involved in the management of risk – much of this book relates to measuring risk and uncertainty, and how to manage the risks identified. Indeed, inadequate risk management has contributed to many corporate failures, both on the macro or global level, and also within firms and industries. Many of the examples and exercises aim to develop these analytical skills. We mainly discuss risk in the context of financial risk (such as interest rate risk and market price risk), and demographic risk (such as mortality risk), although many of the principles could be applied to operational-type risks. Most of these discussions will relate to the fields of finance and insurance (both life and non-life).

A word of warning – the material in this book mainly relates to the *quantitative* management of risk, that is, analysing data and proposing statistical distributions and models to predict financial outcomes. It is important when analysing real-world risk that a qualitative approach is taken alongside such a quantitative approach – the relative weights assigned to the two approaches depending on the particular scenario. A risk in itself is an over-reliance on quantitative financial models, at the expense of any qualitative analysis and exercise of judgement.

The student is likely to benefit from a review of case study material which relate to risk management cases. Study of such cases will provide a more rounded education and knowledge-base of risk management, rather than solely understanding the mathematical approach discussed in this book. Examples of such case studies include: Robert Maxwell and the Mirror Group newspapers, Barings Bank, Equitable Life Assurance Company, Long-Term Capital Management, GFC 2008, Northern Rock, Lehmann Brothers, UK pension schemes/LDI crisis 2022, Silicon Valley Bank; and relevant regulations, such as: Basel Accords, Solvency 2, The Dodd-Frank Act, The Sarbanes-Oxley Act.

7 Verification of Code and Calculations – Best Practice

A key skill of the actuary is to verify complex calculations efficiently. For example, actuarial valuations of insurance companies and company pension schemes typically involve millions of calculations; clearly it is not sensible to check all of them. The actuary must be able to check calculations in an appropriate, cost-effective manner such that they, and other stakeholders, have sufficient confidence in them and can rely on their accuracy. Errors in these calculations may result in advice which has a significant impact on company balance sheets, solvency levels, profits, amount of additional funding required, dividend payouts, and even future career prospects. We will often provide more than one coded solution to a problem e.g. by performing an alternative, approximate calculation. This is a skill the authors believe most undergraduates would benefit from improving prior to entering the workplace.

8   Website: www.wiley.com/go/rprogramming.com

The book's website includes code from each chapter of the book, together with all data files used. It will also, periodically, be updated with extra coding exercises and solutions. We welcome feedback from our readers regarding areas which require further examples.

As referred to earlier, the website is fundamental in using this book efficiently. As well as providing solutions to exercises in the book, it allows copying of the suggested R code thus saving significant time.

9   R or Microsoft Excel?

It is expected that many readers will have some level of experience in using Microsoft Excel. Excel is a fantastic calculation tool. A significant benefit of Excel is its intuitive nature, making it relatively straightforward to learn the basics and quickly reach a reasonable level of competency. Indeed many financial institutions use Excel as a principal piece of software. Programming languages such as R have a significantly steeper learning curve than Excel; most new users, particularly those with no programming language background, will take several days to familiarise themselves with the basic workings of the R language.

In general, Excel is likely to be preferred to R for simpler tasks, or for more involved calculations which are unlikely to require numerous re-runs with various adjustments; the extra cost and time involved in writing R code may not be justified in such cases given the relatively small savings over the long term. In the same way that there are occasions when a calculator is a preferable tool to Excel, there are many occasions when Excel will be preferable to R.

It is important for the reader who has experience with Excel to develop an understanding of whether a programming language such as R will be more suited at solving a particular problem, or set of problems, than Excel. This should be achieved as the reader makes progress with this book. The reader is encouraged to tackle exercises in Excel (where possible) and to compare the process with R. An obvious example is that of the Loan Schedule discussed in Chapter 4 (where the reader is specifically encouraged to reproduce the calculations in Excel); it may well be the case here that using R code is not warranted. A number of calculation schedules involving Life Contingency examples (Chapters 23 and 24) may also prove more user-friendly with Excel. However, as these models become more complex (e.g. incorporating stochastic interest rate models) at some point it is likely that R will become more efficient.

For example, a pension scheme's valuation calculations may take several minutes to run in Excel compared to a few seconds in R; such calculations may be re-run hundreds of times throughout the analysis and verification process of the valuation, thus benefiting from faster running speeds. A decision is often required therefore regarding programming and running times and related costs when comparing Excel and a programming language such as R.

For a basic level of statistical analysis, Excel may be the preferred choice; however R will be the preferred route involving tasks which require anything more than basic analysis. To understand the benefits of using a programming language such as R requires a certain amount of practice and application; all the above should become clearer with experience.

For the casual data user, Excel is better, given the steeper learning curve required to learn most programming languages such as R. Excel is indeed used in several of our actuarial maths classes to demonstrate small-scale, simplified calculations; however, these often tend not to be particularly realistic, and ultimately lead to a better learning and teaching experience when carried out in R. Indeed, with students migrating towards languages like R and Python, a greater proportion of our assessments now involve R programming.

There are tasks which are particularly unsuited to, or just not possible in Excel. For example, it is not a simple task to calculate eigenvectors in Excel, but these can be calculated almost instantly with one line of R code; similarly, running large numbers of simulations using complex models, large matrix calculations, complex regression analysis etc. are problematic. Many student projects require the use of R (or similar language), and are simply not possible using Excel. R has many statistical functions which run significantly quicker, or are unavailable in Excel. We will see many examples of tasks in this book where using Excel is extremely slow and impractical, such as when running large tasks, and does not deal particularly well with huge datasets – often causing it to slow down and crash. Simple tasks in R such as analysing millions of rows of data across several databases can be extremely time-consuming in Excel, and prone to calculation errors. Several R script file (the file which contains the code) can be used to combine various tasks; with Excel the solution is significantly less elegant, and more difficult to verify and audit.

Also we are frequently required to solve problems numerically e.g. an exact solution may not be possible or easy to obtain. Such a numerical approach is usually more suited to R than Excel.

The use of R is also likely to reduce the risk of data corruption and other errors being made. Physically manipulating data and formulae (cutting, copying, deleting, pasting) in Excel is generally quick and easy, but not particularly robust. Human error introduces mouse-slips, moving to incorrect cells etc. Such processes may be required to be performed several times on many similar data sets – with R we run the same program requiring no manual interaction with the data. Less experienced users may suggest applying care is required when handling data; eventually, however, errors will be made. The experienced Excel user will only be too aware of such problems and the potential for calculation disaster. Ultimately, R is likely to be more robust, and less prone to manual errors.

A similar comparison could be made with the requirement for database systems. For small-scale data scenarios a spreadsheet may perform the required tasks adequately; similar to the comparison with R, Excel will tend to be initially more user-friendly than a database programme. However, with added complexity a robust, dedicated database system is required. Readers may wish to review the recent high-profile case relating to COVID-19 data held on a UK Government (Public Health England) spreadsheet "database" where a spreadsheet of data reached its maximum size resulting in data errors.

In short, if you are intending to become a serious user of data, or data analyst, learning R or a similar computer programme is a priority. At the risk of repeating a message from earlier, it is important for the reader to experience the advantages and disadvantages of R compared to Excel for themselves – please experiment! We would advise the reader to become competent with both tools.

Remark 1.1 For most readers of this book, learning R should really be about the process of learning a programming language; as is the case with a foreign language, once one language has been learnt to a reasonable level the hurdle to learning a second language is somewhat reduced. Thus, a further objective of the reader may be to subsequently learn other programming languages.

10   Caveats

All code, results, and analyses included in this book are provided following significant review by peers and colleagues. However the authors cannot guarantee their accuracy, and material from this book should not be used without further appropriate checking and review by their subsequent users.

11   Acknowledgements

Much of the inspiration for this book has come from teaching students of actuarial science and statistics, and discussing material with interested students, without which this book would not have been possible. We would like to thank all those students who have indirectly contributed to this book.

We would like to thank all the reviewers of chapters of this book for their their important and significant feedback: Shaun Parsley, Kevin Yuen, Dr Eduard Campillo-Funollet, Dhruv Gavde, Dr Daniel Bearup, Dr Pradip Tapadar, Dr James Bentham, Dr Peng Liu, John Millett, Professor Martin Readout, Professor Enkelejd Hashorva and Professor Malcolm Brown.

We would also like to acknowledge those involved with writing and developing the R programming language, and its predecessor, S. This includes all the authors of the packages used in this book of which there are too many to mention.

1

R : What You Need to Know to Get Started

```
library(MASS)
library(readxl)
```

1.1 Introduction

The purpose of this chapter is to introduce the fundamentals of the R programming language, and the basic tools you will need to use this book; it is therefore an important chapter for readers new to R. The reader is advised, following reading this chapter, to proceed to Chapter 2 which, together with this chapter, forms our introduction to programming in R.

R is an exceptional statistical computing tool which is increasingly used by researchers and students in many disciplines. R is an open-source language which one can use without purchasing a licence, and contribute to solving problems within the vast R community.

This chapter includes most of the topics we feel are crucial for new users to understand, although there are several important ideas not covered which are gradually introduced throughout the book. Hence the relatively short length of this chapter.

From our experience of teaching R programming, we find it is usually beneficial for the typical student new to R to be made aware of the key points (such as those included in this chapter), proceeding relatively quickly onto practising writing R code in realistic settings, returning to this chapter as a reference guide. We would expect most readers to gain less from an extended period of solely studying the basics of R.

This chapter will show you how to start using the R/RStudio package for basic tasks. A particularly good introduction to R can be found in its documentation page:

https://cran.r-project.org/doc/manuals/R-intro.pdf

which the reader may wish to read concurrently with this chapter. For a more extended introduction we recommend Crawley (2013) Chapters 1–6. (Alternative suggestions are included in the Recommended Reading section at the end of this chapter.) However, perhaps the best source of information is the number of excellent resources on the web devoted to R programming; it would be anticipated that many readers will use this alternative once the basics included in this chapter have been understood.

R Programming for Actuarial Science, First Edition. Peter McQuire and Alfred Kume.
© 2024 John Wiley & Sons Ltd. Published 2024 by John Wiley & Sons Ltd.
Companion Website: www.wiley.com/go/rprogramming.com.

1.2 Getting Started: Installation of R and RStudio

1.2.1 Installing R

You can download R onto your PC using the download link which can be found at:

https://cran.rstudio.com/

making the appropriate choice based on your operating system, for example, Windows, MacOS, Linux/Unix (see Figure 1.1). You must also select a CRAN mirror – simply follow the *CRAN* link on the homepage (most users should choose the closest location, which minimises download times).

Remark 1.1 What are CRAN mirrors?

The Comprehensive R Archive Network ("CRAN") is a collection of websites located globally which contain the latest versions of R software. CRAN mirrors is simply a reference to all the websites available globally.

1.2.2 What Is RStudio?

RStudio provides an excellent user interface for R which considerably improves the user's experience compared to using R directly. Once R has been installed (as detailed above), RStudio can be installed from:

https://www.rstudio.com/

following the appropriate links.

Figure 1.1 Downloading R.

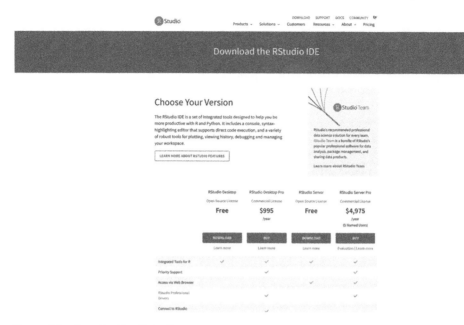

Figure 1.2 Downloading R-studio.

Figure 1.3 Starting R-studio, script window is closed.

The free desktop version would usually be a good first choice (see Figure 1.2). When you open RStudio, it should automatically link with the R programming environment. The standard layout of RStudio is shown in Figures 1.3 and 1.4 and contains four windows:

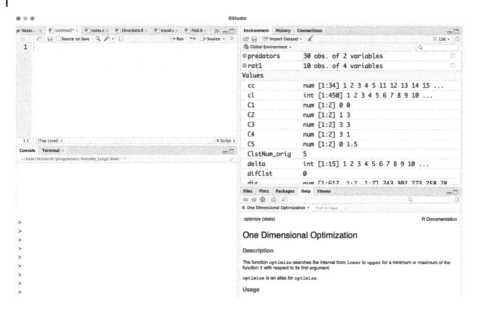

Figure 1.4 Script window is opened, by starting File>NewFile>R-Script.

- The top left window is the **script editor** where you will type your code in script files which can then be saved. If this does not appear automatically you may need to open a new script via File>New File>R Script. This is where you will spend most of your time when programming in R. See Section 1.12 for a further discussion on script files.
- The bottom left window is the console, where R commands are executed, and where output from commands can be viewed.
- The top right window provides information about variables created, and other coding history, etc.
- In the bottom right window you can select tabs for various options, such as packages loaded, help pages, files in the local/working directory and recently run plots.

After some initial familiarization, the reader may wish to customise these windows.

Remark 1.2 The distinction between RStudio and the language, R, should be noted. R is the underlying programming language and consists of code and objects which are discussed later in this chapter; RStudio is the convenient environment in which we will use R. Note that when we refer to R we may be referring to RStudio or the language R itself, it should be clear from the context which is being referred to. Where the distinction is important we will specify either R or RStudio.

RStudio incorporates many mouse-click actions to carry out tasks, improving efficiency – each click carries out specific R code which is produced automatically in the console. Some examples are introduced later in this chapter, for example, when downloading data to R.

1.2.3 Inputting R Commands

As noted above, commands will usually be entered in the script editor in RStudio; it is straightforward to re-run code from the editor and the script can be saved. Alternatively, commands may be entered directly at the command line in the console window (the bottom left panel), although this will usually result in a worse user experience.

As noted earlier, you can open a new file in the script editor with File>New File>R Script>. To get started, type the following in this currently empty script editor file, followed by Ctrl+Enter (Windows), or Cmd+Return (Mac) as appropriate, or using the Run button on the top right margin of the script window – note that the cursor must be on the same line as the piece of code you wish to run:

```
2+3

[1] 5
```

and save the file as testscript. (Note that the file will be recognised as a script R-file, with suffix .R or .r.)

Remark 1.3 This script can, of course, be used to include all the R coding covered in this chapter.

It is worth noting at this point that comments can be included with your R code by adding #. This is considered good protocol, which can improve the clarity of your code, for example, when revisiting your code after a period of time, or for others to use. Similarly, any line of code beginning with # is ignored by R. For example:

```
#my first code in R : 2+3
#here is an example of an easy sum
2+3 #easy sum

[1] 5
```

There are a number of alternative keyboard shortcut options available to run code (in addition to CTRL+Enter or CMD+Return noted above):

- To run the entire code in the current script window, press CTRL+A (or CMD+Return) (to highlight the whole script) then CTRL+Enter or CMD+Return accordingly.
- You can also select particular lines of code manually before applying CTRL+Enter or CMD+Return.
- To re-run sections of code, CTRL+Shift+P or CMD+Shift+P is very useful, for example, when repeatedly making small code changes and re-running large selections of code.

(Please see Section 1.12 for further details on Script files.)

1.3 Assigning Values

When writing code we will usually *assign* values to a variable, using '<-' ; to see the contents or value of a variable we type the name of the variable at the command prompt. For example:

```
x <- 2 #assigning the value 2 to x
x

[1] 2
```

The instruction x <- 2 can be read as "assign the value of 2 to x". This is a similar operation to using '=':

```
xx = 5 # x equals 5
xx

[1] 5
```

Many users prefer the '=' sign, although it can be considered misleading – it does *not* mean "equals" in a mathematical sense. For example:

```
xxx = 5
xxx = xxx + 3
xxx

[1] 8
```

Most of the instructions in R are via functions. They are the R commands which are typically input by a name followed by a pair of round brackets, for example, exp(), sqrt(), sin().

```
sin(pi/6) #sine of 30 degrees
[1] 0.5
```

Writing functions is perhaps the key skill you will need to acquire; after reading this chapter most readers would benefit from next referring to Chapter 2, which provides a brief introduction to writing functions in R.

Remark 1.4 (A comment on syntax used specifically for the purposes of this book.) Note that throughout this book, rather than assigning additional lines for each individual instruction, we will often use a sequence of commands in one line using the ';' as a separator. We also occasionally enclose the command within brackets to obtain the output automatically.

```
y<- sin(pi/6)   ;        yy<-y+7
(y<-sin(pi/6)) ;        (yy<-y+7)

[1] 0.5
[1] 7.5
```

However we would not envisage that the reader adopts the same usage in writing code (we adopt this usage for environmental reasons, that is, to reduce paper usage).

1.4 Help in R

It is worth noting at this early stage that a help facility exists within R. For example to get help, and an example, on the plot function, type:

```
?plot           #provides basic guidance
example(plot)   #provides examples of its usage
```

Alternatively for help, press f1 whilst the cursor is over your function of choice. In addition, many packages provide guidance on applications and the use of functions; one can click on the required downloaded package under the packages tab in the bottom-right window.

1.5 Data Objects in R

In the R programming environment there are four main types of data structures:

1. Vectors
2. Matrices
3. Dataframes
4. Lists

For the reader to develop their R skills it is important that these objects, and the differences between them, are understood. These are discussed in Sections 1.6–1.9.

1.6 Vectors

Vectors are the fundamental data structure in R. Most of the functions you will write are likely to involve vectors. They are essentially a string of data entries where each entry is of the same type (compare *Lists* and *Dataframes*). Vectors can be of four types:

1. numeric
2. logical
3. character
4. factor

1.6.1 Numeric Vectors

Numeric vectors are the most commonly used. We can create a numeric vector using the concatenate function, c() (the same principles apply for the other vector types, as will be seen later.) For example, the following code creates numeric vectors for the age (years), height (cm), and weight (kg) of six individuals:

```
age <- c(54, 64, 49, 37, 24, 16)
ht <- c(179.7, 167.0, 186.7, 181.6, 181.0, 186.1)   #units - cm
wt <- c(68.7, 72.6, 69.3, 109.4, 83.6, 78.0)
```

One can perform arithmetic operations, such as addition or division with vectors; the following calculates the body mass index, BMI, for each individual:

$$BMI = \frac{weight}{(height)^2}$$ (1.1)

where weight is in kg and height is in metres.

```
ht_m <- ht / 100        # convert height from cm to m
(BMI <- wt / ht_m^2)
```

```
[1] 21.27456 26.03177 19.88129 33.17307 25.51815 22.52174
```

Concatenation can also be used to create a new vector from existing vectors. Consider for example two vectors x and y:

```
x<-c(2,3,5,7,11);   y<-c(3,6,10)
(z<-c(x,y))
```

```
[1]  2  3  5  7 11  3  6 10
```

However, care must be taken that the groups of vectors are of the correct length. The output of z is showing the concatenation of x and y. However, this output is interesting:

```
x/z
```

```
Warning in x/z: longer object length is not a multiple of shorter object length
```

```
[1] 1.0000000 1.0000000 1.0000000 1.0000000 1.0000000 0.6666667 0.5000000
[8] 0.5000000
```

Check the warning message. The vectors x and z are of different lengths, but there is still a vector output. This suggests the calculation is proceeding in R such that the shortest vector is (partially) repeated as many times until the lengths match. This is called the "recycling rule" in R – it is important to be aware of this in practice as it could lead to the misinterpretation of output.

Another set of useful vectors of numerical type are those containing numbers obtained from equally spaced values or from some regular pattern. For example:

```
1:10; 10:1
```

```
[1]  1  2  3  4  5  6  7  8  9 10
[1] 10  9  8  7  6  5  4  3  2  1
```

The simplest use of the seq() function achieves the same result, but also offers signifi-cant functionality:

```
seq(1,10); seq(10,1)
  [1]  1  2  3  4  5  6  7  8  9 10
  [1] 10  9  8  7  6  5  4  3  2  1
seq(0, 10, by=2) #sets the interval
  [1]  0  2  4  6  8 10
seq(0, 10, length=5) #sets the length of the resultant vector
  [1]  0.0  2.5  5.0  7.5 10.0
```

Example 1.1

Plot the function x^2 over the interval $[-2, 2]$. This can be achieved with a sequence of 100 evenly spaced points between -2 and $+2$ and plotting x^2 using the plot function, joining the points by straight lines, type="l", to give the appearance of a smooth curve. (We will look at plots in more detail later in this chapter.)

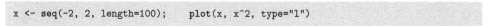

```
x <- seq(-2, 2, length=100);    plot(x, x^2, type="l")
```

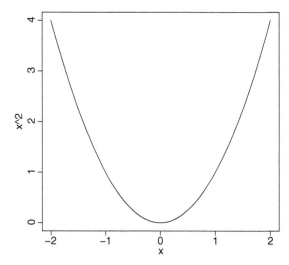

Simple code such as this can be instructive to undergraduate students developing their understanding of various functions and statistical distributions.

Exercise 1.1

1. *Experiment with different values for the* length *argument to see how this affects the smoothness of the plot.*
2. *Plot the function* $\cos(x)$ *between* $(0, 4\pi)$
3. *Check the length of the vector* x *as above using* length(x).

1.6.2 Logical Vectors

We will often want to extract data based on chosen criteria. The instruction below tests whether each element of our vector z (from earlier) is greater than 4:

```
z; z>4

[1]   2  3  5  7 11  3  6 10
[1] FALSE FALSE  TRUE  TRUE  TRUE FALSE  TRUE  TRUE
```

This command tests whether each element of z meets two conditions:

```
z>4 & z<8

[1] FALSE FALSE  TRUE  TRUE FALSE FALSE  TRUE FALSE
```

In both of these examples the output is a logical vector, that is, a vector with entries TRUE and FALSE; you can check the type of the variable using mode:

```
mode(z);        mode(z>4)

[1] "numeric"
[1] "logical"
```

The *logical operators*, <, >, <=, >=, == (logical equals), and != (logical not equals) are used for generating logical vectors. In Boolean Algebra the True/False entries are equivalently treated as 1/0 respectively. Therefore, using the sum function on logical vectors converts them into numeric values (1 for True, and 0 for False) producing an output which is the number of cases which meet the conditions:

```
sum(z>4 & z<8)

[1] 3
```

It is important to recognise the usage of ==, and how it differs to =:

```
aa=5     #means ''aa is equal to 5''
aa==5    #means ''Is aa equal to 5?''

[1] TRUE

aa==4    #means ''Is aa equal to 4?''

[1] FALSE
```

One could create sub-vectors using the **square bracket operator** [ ] which should contain information about the position of the sub-vector entries:

```
z[c(1,3,1,8)]

[1]   2  5  2 10
```

Often we will want to extract the entries which meet chosen criteria:

```
z[(z>4)]        #list all entries greater than 4
[1]  5  7 11  6 10
```

In the following lines we obtain complementing outputs:

```
z[2:6];   z[-(2:6)]
[1]  3  5  7 11  3
[1]  2  6 10
```

A similar bracket structure is used for matrices and more advanced structures such as lists and dataframes (see later).

Example 1.2

The following code generates a random sample of length 1,000 from a standard Normal Distribution, stored as x. The objective is to calculate how many values in the output exceed 0.

```
set.seed(123);       x<-rnorm(1000)
sum(x>0)    #number in sample greater than zero (expect 500)
[1] 505
```

Exercise 1.2 *Determine the number of entries in x between (-1,2), and compare with the expected number.*

Remark 1.5 The function, `rnorm`, simulates numbers from a chosen Normal Distribution. This, and many similar, important functions, are discussed in more detail in Chapter 5, and are used extensively throughout the book. In the meantime, the reader is invited to apply the code to the following: `x<-rnorm(1000,5,1)`, and `x<-rnorm(1000,0,5)` and perform calculations similar to Exercise 1.2.

1.6.3 Character Vectors

Up to this point we have discussed numerical and logical vector types. However, our data will frequently include data such as sex, post code, category of policyholder, smoker/non-smoker? To represent such qualitative data, we introduce *character* vectors, such as C1 below:

```
C1<-(c("Yes","Maybe","No", "No","No", "Yes","Maybe","Yes","Maybe","Maybe"))
```

C1 is character vector with three possible entries ("Yes","Maybe","No"); to carry out statistical analysis we must convert such qualitative data into numeric data. The `as.factor` command does this for us, converting a character vector into a factor vector, which is the subject of the next section.

1.6.4 Factor Vectors

Factor vectors are used to represent categorical variables (as opposed to quantitative variables) – fundamentally, we cannot use words in equations. The specified categories held in a data set (e.g. male/female, very fit/fit/unfit) are held in the factor vector as integer values – this is to allow many types of statistical analysis to be carried out. When we download a .csv file, all columns of data which are held as strings will, by default, be converted into factor type vectors.

We will encounter factors in several chapters later in the book. However, you may need to convert, or *coerce*, a character vector to a factor vector to undertake analysis on the data:

```
C1

 [1] "Yes"    "Maybe" "No"     "No"     "No"     "Yes"    "Maybe" "Yes"     "Maybe"
[10] "Maybe"

class(C1)

[1] "character"

(FC1<-as.factor(C1))

 [1] Yes    Maybe No     No     No     Yes    Maybe Yes     Maybe Maybe
Levels: Maybe No Yes

class(FC1)

[1] "factor"
```

Here we are "coercing" a character vector into a factor vector (see Appendix for more details on coercing).

Remark 1.6 See Section 1.19 for a general discussion on converting data types. Converting, or coercing data into factor type objects is perhaps the most important example in R where we must ensure the data is held in a particular format.

As noted above, factors in R contain integer values; they are converted back to text for the benefit of the programmer. We can see this by further coercing the data to numeric type:

```
as.numeric(FC1)
 [1] 3 1 2 2 2 3 1 3 1 1
```

`levels` outputs the different levels of the factor (note that R assigns these alphabetically):

```
levels(FC1)
[1] "Maybe" "No"     "Yes"
```

The names of the levels of the factor can be changed using a character string of the appropriate length:

```
(levels(FC1)<-c("M","N","Y"))

[1] "M" "N" "Y"
```

The function `ordered()` can be used for ordinal data so that one can assign the ordered levels:

```
(OrderedFC1<-ordered(FC1,c("N","M","Y")))

 [1] Y M N N N Y M Y M M
Levels: N < M < Y
```

This re-ordering is often important when analysing data which has a natural rank, as above where it is natural to hold 0=No, 1=Maybe, 2=Yes.

1.7 Matrices

Matrices store data of a single data type, arranged by rows and columns, usually containing only numeric values. (We shall shortly discuss dataframes – these are similar to matrices in R but, in general, contain various data types.)

We can set up a matrix by presenting its values as a vector:

```
(x <- matrix(1:12, nrow=4, ncol=3))

     [,1] [,2] [,3]
[1,]    1    5    9
[2,]    2    6   10
[3,]    3    7   11
[4,]    4    8   12
```

Note that the matrix is filled column by column; this is the default. If we want it filled up row by row, we use `byrow=TRUE`; we can also omit the `nrow=` and `ncol=`:

```
(y <- matrix(1:12, 4, 3, byrow=TRUE))

     [,1] [,2] [,3]
[1,]    1    2    3
[2,]    4    5    6
[3,]    7    8    9
[4,]   10   11   12
```

Exercise 1.3

1. *Apply the following functions to* y: `nrow()`, `ncol()`, *and* `dim()`
2. *What do the following commands produce?*

   `diag(x)`

   `diag(4)`

   `matrix(0, 4, 4)`

3. *Check that* x+y *and* x-y *do what you expect. What does* x*y *produce?*
4. *Set up a* matrix z *with different dimensions and see what happens when you type* x+z.

Matrix multiplication %*%

You should have discovered above that x*y multiplies each element of x by the corresponding element of y. To perform standard matrix multiplication, we use the special operator %*%. Using the above examples we can multiply x by the *transpose* of y, using the t() function:

```
x %*% t(y)

      [,1] [,2] [,3] [,4]
[1,]    38   83  128  173
[2,]    44   98  152  206
[3,]    50  113  176  239
[4,]    56  128  200  272
```

We will use matrix multiplication frequently in this book to improve the efficiency of a number of calculations.

Exercise 1.4 *One will often want to manipulate data in a matrix. Review the output of the following code (similar data manipulation methods to (1) and (2) can be used on dataframes – see Section 1.8):*

1. *Access individual rows, individual columns, or individual elements of a matrix:*

```
x[3,];  x[,2];  x[2,3];  x[1:3,];  x[1:3,c(1,3)]
```

2. *Label the rows and columns of matrix* x *using* rownames() *and* colnames() *for example,*
   colnames(x) = c("Col1", "Col2", "Col3")
3. *Review the output of the following instructions in* R *:*

```
v = matrix(c(5,4,2,2,3,1,1,3,4), 3, 3, byrow=TRUE) # Set up a 3x3 matrix
t(v)                                     # Transpose
det(v)                                   # Determinant
solve(v)                                 # Inverse
eigen(v)                                 # Eigenvalues and eigenvectors provided in a list
```

1.8   Dataframes

As noted above, dataframes are similar objects to matrices; the key difference being that data held in each column of a dataframe can be of different types. For example, insurance company data may consist of columns, each containing only names, id number, amount insured, policy dates etc., with each individual row containing the data in respect of one policyholder. Thus a dataframe may be thought of as broadly analogous to a database.

It is straightforward to create a dataframe; using the BMI data from earlier in the chapter and adding persons' names:

```
names_x<-c("Ann", "John", "James", "Lauren", "Britney", "Colin")
(employ.data <- data.frame(names_x,age, ht, wt, BMI))

  names_x age    ht    wt      BMI
1     Ann  54 179.7  68.7 21.27456
2    John  64 167.0  72.6 26.03177
3   James  49 186.7  69.3 19.88129
4  Lauren  37 181.6 109.4 33.17307
5 Britney  24 181.0  83.6 25.51815
6   Colin  16 186.1  78.0 22.52174
```

To further illustrate dataframes, we look at the larger `painters` dataframe, found in the MASS library.

Extracting the first 3 rows as an example:

```
head(painters, 3)
           Composition Drawing Colour Expression School
Da Udine            10       8     16          3      A
Da Vinci            15      16      4         14      A
Del Piombo           8      13     16          7      A
```

In a similar way to matrices, we can extract data using [ ]:

```
painters$Expression[1:2]

[1]  3 14

painters[1:3,4]

[1]  3 14  7

painters[1:2,4:5]

         Expression School
Da Udine          3      A
Da Vinci         14      A
```

The following code extracts all painters from school "C".

```
painters[which(painters[,5]=="C"),]
            Composition Drawing Colour Expression School
Barocci              14      15      6         10      C
Cortona              16      14     12          6      C
Josepin              10      10      6          2      C
L. Jordaens          13      12      9          6      C
Testa                11      15      0          6      C
Vanius               15      15     12         13      C
```

Exercise 1.5 *Extract all painters who produced 10 or more compositions of each type.*

1.9 Lists

All the elements of a vector must be of the same type (e.g. numeric, logical, character). Internally, matrices are stored as vectors with some additional information about the number of rows and columns. So their elements must also all be of the same type and length.

On the other hand, much statistical analyses produce output of various types and lengths; R uses *lists* to store these results in a single object. It would be anticipated that you will encounter lists when viewing output from functions involved in statistical analysis, rather than creating your own list objects.

Example 1.3

As a trivial example, we generate data (*x* and *y*) and perform linear regression of *y* on *t* (regression is covered in more detail in Chapter 5).

```
set.seed(191020);set.seed(200); t <- seq(1:1000);  y <- 1.2+ 0.5 * t + rnorm(1000);
regmod <- lm(y ~ t); names(regmod)
```

Note that the command `regmod <- lm(y ~ t)` doesn't produce any output; instead it stores the output in the variable `regmod`, which is a *list* of 12 objects. We will frequently be required to identify a particular output from a function with which we may not be familiar with – we can use `names` to help pick the output we require:

```
names(regmod)

[1] "coefficients"  "residuals"    "effects"    "rank"
[5] "fitted.values" "assign"       "qr"         "df.residual"
[9] "xlevels"       "call"         "terms"      "model"
```

These objects are not all of the same type (e.g. some contain numerical information, others have textual details about the R code that was used to fit model). We can obtain the first vector in the list using either of the following:

```
regmod$coefficients; regmod[[1]]    #2 equivalent methods to obtain the output

(Intercept)          t
  1.1805591    0.5000703
(Intercept)          t
  1.1805591    0.5000703
```

The reader is encouraged to look through all the vectors in the output list, i.e. `regmod[[2]]`, `regmod[[3]]`, `regmod[[4]]`...`regmod[[12]]`.

1.10 Simple Plots and Histograms

The `plot()` function produces a plot in a new window. Using the `age` and `height` data from earlier (repeated below):

```
        [,1] [,2]  [,3]   [,4] [,5]   [,6]
age    54.0    64  49.0   37.0   24   16.0
ht    179.7   167 186.7  181.6  181  186.1
```

```
plot(age, ht,cex=2)     #cex - size parameter
```

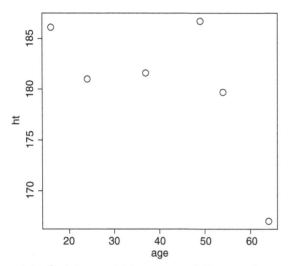

The first two variables represent the *x* and *y* coordinates of the dots plotted in the window; cex makes the dots bigger. It is important that the length of the two vectors is the same. One can customise these plots by adding several settings within the plot command such as titles, axes labels, colours, size of font, whether a line or points are required, type and size of point, thickness of line, and many more. Adding data to the plot is straightforward, using points:

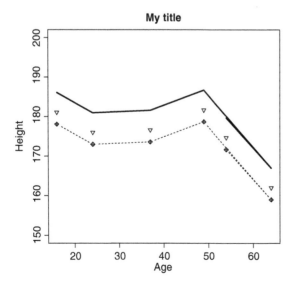

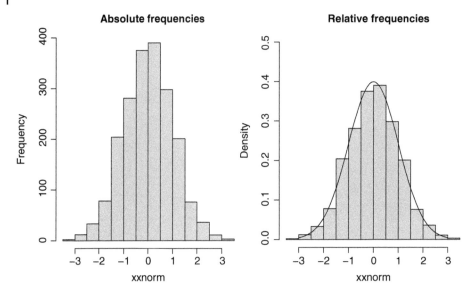

Figure 1.5 Histograms : 2,000 simulations from N(0,1).

```
plot(age, ht,xlab="Age",ylab="Height", lwd=3,main="My title",
type="l",pch=4,cex.lab=1.5,ylim=c(150,200))
points(age, ht-5,type="p",pch=6)
points(age, ht-8,type="b",lty=2,pch=9)    #type = b produces both points and lines
```

For more information see ?plot.

Histograms can be obtained using the command hist. Figure 1.5 (left) shows a histogram of 2,000 simulated data points from a standard Normal distribution (the function rnorm() is covered in a later chapter). The second histogram shows the relative frequency of simulated points, together with the underlying distribution:

```
set.seed(100);    xxnorm<-NULL
xxnorm=rnorm(2000)
hist(xxnorm,,main="absolute frequencies")
hist(xxnorm,main="relative frequencies",freq = FALSE,ylim = c(0,0.5))
lines(seq(-3,3,by=0.01),dnorm(seq(-3,3,by=0.01)))
```

Both the plot and histogram functions are used extensively throughout this book.

1.11 Packages

A package is a collection of functions and data which have a common theme developed by R users. For example the tseries package includes many functions which are useful in the analysis of time series.

There are thousands of packages in R. To use many of the functions in this book you will first need to download a particular package. However, a number of packages are automatically downloaded when you first install R – these packages are generally the most commonly used, and are used to perform particularly common commands.

Below is listed a small selection of the packages that you will use extensively throughout this book:

`lifecontingencies, demography, copula, MASS, survival, tseries`

The reader may wish to peruse the vignettes on some of these packages e.g. `browse Vignettes("survival")`, `browseVignettes("lifecontingencies")`. Indeed, the trainee actuary may find these vignettes an extremely useful addition to their study material for various actuarial topics.

Another particularly useful package, although less utilised in this book, is `dplyr`, a data manipulation package, allowing an advanced level of manipulating large datasets. This package is highly recommended to the reader whose work includes large-scale data manipulation.

To use functions and data included in a package we first *install* the package, and then *load* it. A loose analogy with installing and loading is with applications on a smart phone; to use an app you must first download it onto your phone – this is analogous to installing the package. When we want to use the app you need to touch the app icon – this is analogous to loading the package.

There are several ways to install a package. The simplest is via `Tools/Install packages...`, entering the required package name in the `Packages` field. Alternatively, click on `Install` under the `Packages` tab in the bottom-right window.

Once installed it will appear under this `Packages` tab, with the checkbox unticked; to load the package simply click on the checkbox. You will see the code in the console.

You can view the names of all installed packages on your hard drive using `library()`. Note that once a package is installed it is saved onto the hard drive – it does not need to be installed again (unless you remove it or need to update the package version). When you close R however, you must re-load the package on reopening R (by ticking the checkbox) – it is likely that you may forget this step initially, and will receive the message:

Error in *code* : could not find function *function_name*

Note you can remove a package from your hard drive using `remove.packages ("package\_name")`:

1.12 Script Files

The reader has perhaps already concluded that the most efficient way of running a list of R commands for a session is to use script files, which were discussed briefly in Section 1.2.3. A script file is a plain text file that contains R commands. Script files usually have the extension `.r` or `.R`.

As noted in Section 1.2.2, to create a script file, click File → New File → R script. The majority of users will simply open the required script file they have previously been working on, run through code, and proceed to change/update code as required.

It is good practice to use different script files for different tasks. For example, the reader may eventually have a script file in respect of each chapter of this book.

Sourcing script files

The reader may find using the source function leads to more efficient coding in the future, once the volume of code and functions used has grown sufficiently; this option is highlighted at this point purely to make the reader aware of this functionality.

Rather than repeatedly copying potentially large amounts of code for functions and/or commands to numerous script files we are working on, we can simply run the source code from our current script file. The command source runs code without opening the file which contains the required functions. The files from which code is sourced are ordinary script files which include functions, objects etc. The reader may therefore wish to create such files which include functions from this book as they progress through it, rather than repeatedly copying code for these functions to several script files.

You can run the sequence of functions/commands saved in *testsourcescript.R* using source("testsourcescript.R"). (If the file being sourced is not in the working directory (see Section 1.14), you would need to specify the complete file path and not just the file name.) Note also the shortcuts available in RStudio.

1.13   Workspace, Saving Objects, and Miscellany

As you progress through a specific task/exercise, writing more R code, you will generate an increasing amount of plots, new functions, data objects etc. These can all be saved in the current "workspace", managed by R in *.RData* type files. When quitting R / RStudio you are asked to save the latest version of the workspace file; by choosing Yes the workspace file will be saved in the current working directory (see Section 1.14). You can save the current workspace you are working on at any time with save.image. Other save options are available, for example, you can choose only to save data.

Thus, when you re-start R, you can access the objects from a previous workspace by re-loading it, avoiding the need to re-run the code to obtain the objects (which can be time-consuming). Use load("Workspace1.RData") to call the objects you saved from a previous workspace, here called "Workspace1" (or alternatively via the Session > Load Workspace... tab on the top menu within the RStudio interface).

Note that in the early days of writing R code you will have less need to use workspaces. They are particularly useful when undertaking large projects which include many tasks.

The list of objects created from an R session can be obtained using objects() or ls(); alternatively the list is available in the top right window under the Environment tab. On the same menu you can click the Clear objects from the workspace button; this is similar to closing R and restarting, so don't do it unless you have already saved any work you need (you can't undo).

The `history()` function sets out a list in the top right window in `RStudio` containing a complete list of commands from the current session.

1.14 Setting Your Working Directory

It is often useful to set a default directory from which data files are read from (or written to). This is known as the working directory, and can be set with:

```
setwd("~/directoryforthisproject")          #setting the working directory
getwd()                                      #outputs the working directory
```

For Windows users, note that pathnames include back slashes, \, which must be changed to forward slashes, /.

To read a file from that directory (now the working directory), one is required to enter only the filename, and not the entire path. For example, all the data files from this book could be downloaded and saved to one directory, which could then be accessed simply with a command which includes the filename, but not the pathname. If the file is in a sub-directory relative to the working directory simply include the additional path details. Also, if you share script files between users, filenames will work for other users, whereas exact pathnames are likely to differ. Further advantages of setting an appropriate working directory will become obvious with further use.

1.15 Importing and Exporting Data

1.15.1 Importing Data

Data will usually be saved in either `.csv` or Excel files; there are three equivalent ways to import such data files into R. The reader can download the files *"dataframe_example.csv"* or *"excel_import_example.xlsx"* using the methods described below:

(1) To import a `.csv` data file into R use the commands `read.csv()` or `read.table()`. (These have different defaults for headings in the data; it is best to include `header=TRUE`, or `header=FALSE` as required. Note that `read.csv()` calls `read.table()`, but with options set which assume values are separated with commas rather than the usual "white space".)

```
dataframe_example <- read.csv("~/dataframe_example.csv", header=FALSE)
```

One can read Excel files using the `read_excel` function from the `readxl` package:

```
excel_example <- read_excel("~/excel_import_example.xlsx")
```

(2) Alternatively we can import data from `.csv` files or Excel by clicking the `Import Dataset` button under the `Environment` tab in the top right window. Once this

option is chosen you will be given the opportunity to choose your data file and format options via an interactive menu, for example, for .csv files choose the From Text option. Once downloaded the name of the data file appears in the console window.

(3)Alternatively the data can be downloaded into R using the menu options:

```
File -> Import Dataset -> From ....
```

If a .csv file is required, choose From Text, select the required file, and choose separator = comma amongst the options on the subsequent screen. The code and name of the object will be shown in the console window. Options 2 and 3 provide a range of options to download data of various other types.

If data files are saved in the working directory (see Section 1.14), only the filename is required (no pathname is required).

Note you can always get data into R (quickly) by copying the data (e.g. from Excel) onto the clipboard (<Control+C>) and running the following:

```
new_data <- read.table("clipboard")
```

In general this would not be recommended as it is prone to human error, for example, missing data when copying. It should be noted that this method can be operating-system specific.

If you need to enter new data into R which does not exist elsewhere (e.g. such as a large covariance matrix), it is often convenient to do this first in Excel rather than directly using R code. The file containing the data can be saved in Excel as a .csv file and then imported into R as described above (or use the "clipboard" method).

1.15.2 Exporting Data

Occasionally we may wish to review results from calculations carried out in R in a spreadsheet. It is straightforward to export data from R using write.csv (R_data_object,path/name). For example:

```
data_export<-matrix(1:9,3,3)
write.csv(data_export,"~/newfile.csv") #first create a file with this name
```

The file will be saved in your assigned working directory. Note that the file must not be open when writing data to it.

1.16 Common Errors Made in Coding

The following list includes some of the most common errors made by those new to the R programming language:

- misspelling of functions and variables
- text is case sensitive

- using the incorrect type of bracket, or missing brackets. For example:
    ( )are used for functions
    [ ] are used for vectors, matrices, dataframes
    {} are used when writing new functions
- poor naming conventions, for example, using the same name several times to identify different entities
- using back-slash and forward-slash symbols incorrectly
- missing/wrong quotation marks
- forgetting to load a previously installed package
- failing to coerce data correctly

1.17 Next Steps

As noted earlier, we recommend reading Chapter 2 following completion of this chapter.

1.18 Recommended Reading

Included below are a list of books which include introductory chapters on R.

- Crawley M.J. (2013). *The R Book*, 2e. Chichester, West Sussex, UK: John Wiley and Sons.
- Everitt, B.S. and Hothorn, T. (2006). *A Handbook of Statistical Analyses using R*. Boca Raton, Florida, US: Chapman and Hall/CRC.
- Rizzo, M.L. (2019). *Statistical Computing with R, Second Edition*, 2e. Boca Raton, Florida, US: Chapman and Hall/CRC.
- Verzani, J. (2014). *Using R for Introductory Statistics*, 2e. Boca Raton, Florida, US: Chapman and Hall/CRC.

1.19 Appendix: Coercion

Remark 1.7 It is debatable whether such a topic should be included in an "R basics" chapter such as this; the reader should at least be aware of issues which may arise due to holding data in an incorrect format.

In Section 1.6.4 we saw an example where our initial data, was converted, or coerced, from a character vector to a factor vector. This was crucial in order to undertake statistical analysis. In this section we demonstrate the general concept of coercion in R, by looking at a number of further examples.

Example 1.4 Automatic coercion to character type object.

Vectors must consist of only one type of data. R must therefore coerce vectors where they consist of different types, using a set of rules (not discussed in detail here). Take the following vector, consisting of numeric, character, and logical types, where all entries are automatically converted in characters:

```
ww<-c(2,"peter",TRUE);   ww
[1] "2"     "peter" "TRUE"
class(ww)
[1] "character"
```

Exercise 1.6 *Automatic coercion to a numeric vector. Try running the following code:*

```
www<-c(0,TRUE,1);     class(www);   sum(www);   www
www<-c(FALSE,TRUE,TRUE);   class(www);     sum(www)
```

For now, it is important for the reader to note that some functions in R require data objects to be in a particular class, thus one type may need to be changed to another, for example, using as.data.frame, as.matrix etc. It is likely that the reader will encounter issues which require coercion of data; the reader is therefore encouraged to read further into this topic at some point.

Data wrongly held as a factor type object

It's worth noting that data may be held as factors where it is preferable not to be. We may, therefore, need to convert a factor to a numeric vector such that a particular function will work; this can initially cause problems as the error may not be obvious to a new user. For example, the same of our data below is 17; however, the following does not work:

```
factor_example #this may be how our data is presented to us
[1] 4 5 7 1
Levels: 1 4 5 7
sum(factor_example)
Error in Summary.factor(structure(c(2L, 3L, 4L, 1L), .Label = c("1", "4", :  'sum'
not meaningful for factors
class(factor_example)
[1] "factor"
levels(factor_example);     as.numeric(factor_example) #problems!
[1] "1" "4" "5" "7"
[1] 2 3 4 1
```

This is not our original data. The 1,4,5,7 are actually held in R as 1,2,3,4. To correct this we should coerce the factor into a numeric, as follows:

```
data_numeric<-as.numeric(levels(factor_example)[factor_example])
data_numeric;     sum(data_numeric)
[1] 4 5 7 1
[1] 17
```

Thus it is often worth checking the class of object if the output from R is not as expected.

2

Functions in R

Peter McQuire

2.1 Introduction

2.1.1 Objectives

The objective of this chapter is to introduce the reader to using functions in R. We will, briefly, look at using some of the most important basic R functions, then proceed to write our own functions. It is expected that this chapter, together with Chapter 1, will be one of the first read by most users of this book. It provides a simple introduction to using R for the first-time user and provides material which is likely to be fundamental in future R programming.

There are generally three types of functions we encounter in R:

- Core functions,
- Package functions, and
- User Defined functions

This chapter is split into two main sections: the first introduces pre-defined functions (Core functions and Package functions), and the second part introduces user-defined functions. We complete the chapter with a section which covers two commonly used functions: for and integrate - used to write loops and perform integration in R.

Please note that it is assumed throughout this chapter that the reader is new to programming. There may also be a small amount of repetition of some material from Chapter 1.

2.1.2 Core and Package Functions

Core functions in R are immediately available to use when the R programming language is first downloaded onto your computer. In addition there is a huge number of functions written by other R users which exist in packages. To use these functions we must first install and load the relevant package (as discussed in Chapter 1). Most chapters in this book require the use of such functions; any key packages required are detailed at the start of each chapter.

R Programming for Actuarial Science, First Edition. Peter McQuire and Alfred Kume.
© 2024 John Wiley & Sons Ltd. Published 2024 by John Wiley & Sons Ltd.
Companion Website: www.wiley.com/go/rprogramming.com.

Only a relatively small number of these functions is discussed here; many more ready-to-use functions will be introduced gradually throughout the book. Please note that key *statistical* functions, such as rnorm, lm, fitdistr, are not included here - they are discussed in Chapter 5.

Many "R textbooks" include a chapter with extensive lists and descriptions of Core and Package functions; however, we do not think this is particularly useful in our context. Instead, we gradually introduce functions throughout the book by way of examples, hopefully making their application clearer. We would argue it is of little use to set out hundreds of important functions in an introductory chapter. Indeed we would suggest that the most efficient way to learn the correct syntax of functions is by using internet search engines. There are several excellent websites (and some less good) which the reader should identify. The reader should find that, as they progress through the book, they can incorporate previously used functions in tackling new, more complex problems, effectively building up a substantial library of code.

Note also, that much of the code included in this chapter utilises only a basic application of the functions; the reader should be aware that many of these functions have extra functionality which can allow for significant flexibility. As noted elsewhere, help on these functions can be obtained within R using <F1> or search engines.

2.1.3   User-Defined Functions

The second part of this chapter deals with User-Defined Functions. When undertaking a particular task in R we often find that no "ready-to-use" function exists which is appropriate for our specific problem or situation. In these cases we must write our own function. We call these "user defined functions" or "udf"s.

Udfs tend to be written when the same operation is required several times; writing a function can save significant time over the longer term. With experience the analyst can judge when it is beneficial to write a particular function.

Writing udfs is a core skill in learning any computer language. As such this section is indispensable.

2.2   An Introduction to Applying Core and Package Functions

2.2.1   Examples of Simple, Common Functions

Here are 12 examples of commonly used functions which require little or no explanation:

```
seq  sum  prod  length  max  min  sort  round  rep  sort  mean  var
```

For example:

```
seq(3,8)
[1] 3 4 5 6 7 8
sum(seq(3,5));     prod(seq(3,5))
```

```
[1] 12
[1] 60
```

```
rep(2,5);           length(rep(3,121))
```

```
[1] 2 2 2 2 2
[1] 121
```

```
max(sum(seq(3,5)),prod(seq(3,5)),length(rep(3,121)))
```

```
[1] 121
```

```
sort(c(4,3,8,1,17,2),decreasing = FALSE)
```

```
[1]  1  2  3  4  8 17
```

```
mean(seq(1,10));     var(seq(1,10))
```

```
[1] 5.5
[1] 9.166667
```

Many additional options are often available which the reader should explore, for example

```
rep(2:5, each = 2)
```

```
[1] 2 2 3 3 4 4 5 5
```

```
seq(18,58,by=10)
```

```
[1] 18 28 38 48 58
```

The following commonly used functions need a little more explanation:
which apply rowSums cbind order
table pmin pmax set.seed

Let's look at each of these in turn:

1. which (and related functions, which.max, which.min)

   This function returns the position in a vector which equals a value of interest (i.e. "which position").

   ```
   aa<-c(1,0,5,74,1,0,-2,53)
   which(aa==0)  #returns the position
   ```

   ```
   [1] 2 6
   ```

   ```
   which(aa==0 | aa<0)  # 'or'
   ```

   ```
   [1] 2 6 7
   ```

   ```
   aa[which(aa!=0)]    #returns all items not equal to 0
   ```

   ```
   [1]  1  5 74  1 -2 53
   ```

   ```
   which.max(aa); which.min(aa)  #returns the positions of the max and min
   ```

   ```
   [1] 4
   [1] 7
   ```

2. `apply`: this function applies a chosen function (in this example, `sum`) to a set of data; the 2nd argument determines whether it is applied by rows (1) or by columns (2):

```
(sides1<-matrix(c(3,5,4,12),nrow=2))

     [,1] [,2]
[1,]   3    4
[2,]   5   12

apply(sides1,1,sum);    apply(sides1,2,sum)

[1]  7 17
[1]  8 16
```

(See Example 2.1 where `apply` is used with a udf.)

The reader should also investigate the closely related, and important, functions `tapply, sapply , lapply` etc.

3. `rowSums`: however, summing row or column entries is easier using:

```
rowSums(sides1)

[1]  7 17

colSums(sides1)

[1]  8 16

sum(sides1[1,])#just the first row

[1] 7
```

4. `cbind`: this is an invaluable function for combining columns of data. Here we combine exam scores for 3 students:

```
math_s<-c(67,78,45)
english_s<-c(70,77,52)
student_id<-c(1,2,3)
(all_scores<-cbind(student_id,math_s,english_s))

      student_id math_s english_s
[1,]           1     67        70
[2,]           2     78        77
[3,]           3     45        52
```

5. `order`: this function re-orders data according to particular criteria - this is similar to `sort` but should be used when dealing with multivariate data, for example, dataframes. Using the height datafile, we re-order our data from smallest to tallest:

```
height_data<-read.csv("~/heights_mf.csv")
```

```
head(order(height_data$height)) #positions of the smallest data

[1]  81   5 196 105 131 119
```

```
tail(order(height_data$height)) #positions of the tallest data
[1]   8 121   4  11  10  94
```

`order` assigns a value to each entry in the ordered list corresponding to its position in the original list i.e. it returns the position of the shortest person (81), then the second shortest person (5) etc. The tallest person is in position 94. Setting out the shortest and tallest people:

```
head(height_data[order(height_data$height),],2)   #2 shortest people

    id   height gender
81 81 148.6166      2
5   5 151.4652      2

tail(height_data[order(height_data$height),],2)   #2 tallest people

    id   height gender
10 10 192.1432      1
94 94 197.6158      1

head(height_data[order(-height_data$height),],2) # -ve commonly used

    id   height gender
94 94 197.6158      1
10 10 192.1432      1
```

6. `table`: this function returns the numbers of each variable value in a dataset, and can provide a useful summary of the data, here returning the numbers of apples (A) and melons (M) in the data (53 apples and 47 melons):

```
fruit <- read.csv("~/KNNpmacsimple.csv")
table(fruit$type)
```

```
 A  M
53 47
```

7. `pmin` ; `pmax` these functions look at a vector of values and compares each entry in that vector with a number, returning another vector of the same length. It is best explained by way of an example:

```
simple_data<-c(5,10,15,20)   #our data
pmax(12,simple_data);     pmin(12,simple_data)

[1] 12 12 15 20
[1]  5 10 12 12
```

8. `set.seed`

The reader should be aware of the `set.seed` function which is used in many examples and exercises throughout the book. Effectively, it fixes the random number generator in R such that the same series of random numbers can be reproduced. Many examples in the book use a series of randomly generated values on which analysis is carried out and conclusions are determined. It would not make for very convincing reading if the reader was unable to reproduce the same set of random numbers and subsequent results! Here we randomly generate three numbers between 0 and 1 using the the Uniform distribution, and then rerun the code:

```
runif(3)
[1] 0.8021878 0.5365745 0.3749753
set.seed(200); runif(3); runif(3)
[1] 0.5337724 0.5837650 0.5895783
[1] 0.6910399 0.6673315 0.8392937
set.seed(200); runif(6)
[1] 0.5337724 0.5837650 0.5895783 0.6910399 0.6673315 0.8392937
```

Care should therefore be taken to reset the seed before re-running the code.

This has provided the briefest of introductions to a number of functions you are likely to use frequently. Now the reader has some exposure to using functions in R we proceed to write a few simple functions of our own.

2.3 User-Defined Functions

2.3.1 What does a "udf" consist of?

Every new function we write in R contains the following four parts:

1. The chosen name of the function.
2. The word "function".
3. The function's required arguments (in round brackets).
4. The operation, or body, usually written within curly brackets, which sets out the details of the calculation. (If this code is written entirely on one line these brackets are not required, although it is best practice still to use them.)

Then to call, or apply, your new function we type its name followed with the values of the arguments defined in 3 above.

2.3.2 Naming Conventions

Do ensure the name given to a new function is not currently in use for e.g. do not call it "sum" or "which". (This should be obvious as you type its name, as R tries to anticipate the function you may be requiring.) When naming your new function, use a short, meaningful name. The name must not contain spaces; for clarity therefore most programmers either choose to split words with "." or "_". We tend to use the latter in this book.

2.3.3 Examples and Exercises

Example 2.1

Write a function which calculates the length of the hypotenuse of a right angled triangle. First we need the equation:

$$z = (x^2 + y^2)^{\frac{1}{2}}$$

This is the "operation" (part 4 above). The length of the hypotenuse depends on the lengths of the two shorter sides, say x and y (our arguments). Writing the function, which we'll call *hyp*:

```
hyp<-function(x,y){(x^2+y^2)^0.5}
```

Testing our new code and assigning the answer, using the example we learnt in our first Pythagorean class in school:

```
first_ans<-hyp(3,4)
first_ans

[1] 5
```

Using the `apply` function introduced in Section 2.2:

```
hyp1<-function(x){(x[1]^2+x[2]^2)^0.5}
(sides1<-matrix(c(3,5,4,12),nrow=2))

     [,1] [,2]
[1,]    3    4
[2,]    5   12

apply(sides1,1,hyp1)

[1]  5 13
```

Exercise 2.1 *Using the following expression for y:*

$$y = a^2 + b^2 - c$$

calculate y for each of the following 500 sets of of a, b, and c (in z below), and calculate the mean value of y:

```
z<-matrix(seq(1,1500),nrow=500, ncol=3);  colnames(z)<-c("a","b","c")
head(z,3)

     a   b    c
[1,] 1 501 1001
[2,] 2 502 1002
[3,] 3 503 1003
```

Exercise 2.2 *Write a function which returns the volume of a cuboid.*

Exercise 2.3 *Write a function to calculate the temperature in Fahrenheit, given the temperature in Celsius.*

Exercise 2.4 *Write a function to calculate the area under a cosine curve with period T and amplitude of A, between the limits t_0 and t_1. Use your function to calculate the area under a cosine curve with period $= 4\pi$ and amplitude $= 2$, between $-\pi$ and $+\pi$.*

Exercise 2.5 *Plot a graph of the height of an object falling due to Earth's gravity, against time (at one second intervals), from an initial height, $h_0 = 2,000\,m$, with an initial vertical speed $u = 0\,ms^{-1}$. The height of the ball, h_t, at time, t, is given by the following expression:*

$$h_t = h_0 + ut - \frac{1}{2}g\,t^2$$

Recalculate h_t for the case where $u = 80\,ms^{-1}$. Assume that the rate of acceleration due to gravity, g, equals $10\,ms^{-2}$.

The results are shown in Figure 2.1 (left).

Example 2.2

How long does the object in Exercise 2.5 take to reach the ground if we are standing at a height of 20 m and throw the ball upwards at $u = 30\,ms^{-1}$.

We can easily solve this equation exactly; the correct answer is 6.60555s. However, we may encounter situations (we will see several later in the book) where a closed-form solution is not available, and need to use a method similar to that below, to solve our equations. (For example, a similar method is applied to determine a bond's gross redemption yield.)

```
height2<-function(ho,u,t){ho+u*t-5*t^2}
step_size<-0.000001
h2<-height2(20,30,seq(0,7,step_size))
which(abs(h2-0)==min(abs(h2-0)))*step_size

[1] 6.605552
```

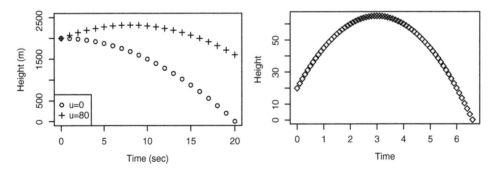

Figure 2.1 Effect of gravity

2.4 Using Loops in R - the "for" Function

Running loops is a method used extensively in this book. The basic code structure is as follows:

```
x<-NULL
x[1]<-0
for (i in 1:5) {
x[i+1]<-2*x[i]+1}
x

[1]  0  1  3  7 15 31
```

The reader should note that typing `<for>` helpfully prompts the suggested syntax:

```
for (variable in vector) {.
```

We frequently use nested loops throughout the book (a loop within a loop):

```
array_ex<-matrix(0,3,4)    #set up a blank matrix
for(i in 1:3) {for(j in 1:4)
{array_ex[i,j] = 100*i+j }
}
array_ex

     [,1] [,2] [,3] [,4]
[1,]  101  102  103  104
[2,]  201  202  203  204
[3,]  301  302  303  304
```

Note that loops can be quite memory intensive and thus slow to run; alternative coding is often preferable. However, the use of loops lends itself to very clear and easy-to-follow code.

`replicate` can be a better choice than using a loop, here simulating two sets of 4 realisations from a standard Normal distribution:

```
replicate(2,rnorm(4,0,1))

            [,1]        [,2]
[1,]  0.55806524 -0.29705130
[2,]  0.05975527  0.16815003
[3,] -0.11464087  1.41987233
[4,] -1.02057835 -0.09952507
```

Loops tend to be used when we need to record a value for the next calculation.

2.5   Integral Calculus in R

2.5.1   The "Integrate" Function

We will be required to carry out integration on several occasions throughout this book. Simple integration is straightforward in R by using the `integrate` function.

Example 2.3

We wish to integrate the following simple function, with respect to x:

$$y = x^2$$

First we need to write our function, and give it a name; let's call it y:

```
y<-function(x){x^2}
y(13)        #test

[1] 169
```

We use the `integrate` function as follows:

```
integrate(function,lower_limit,upper_limit)
```

The `integrate` function produces the following outputs:

```
[1] "value"        "abs.error"    "subdivisions" "message"       "call"
```

Performing the following integration: $\int_2^{10} x^2 \, dx$

```
(area2_10<-integrate(y,2,10)$value)

[1] 330.6667
```

The exact answer is $\frac{1}{3}(10^3 - 2^3) = 330.66666...$

Example 2.4

Consider the following function:

$$f(x) = 3\sin^2(0.5x)$$

Plot a graph of this function and find the area under the curve between $x = 0$ and $x = 10\pi$.

```
my_sine_fn<-function(x)3*(sin(0.5*x))^2
integrate(my_sine_fn,0,10*pi)$value

[1] 47.12389
```

Alternatively, the above can be written:

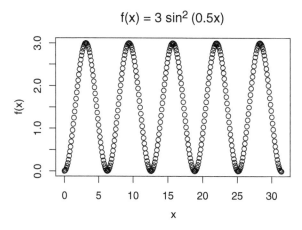

Figure 2.2 Plotting a function

```
my_sine_fn_int<-function(a,b)integrate(my_sine_fn,a,b)
my_sine_fn_int(0,10*pi)$value
```

```
[1] 47.12389
```

Figure 2.2 is obtained from the following code:

```
range1<-c(seq(0,10*pi, 0.1))
plot(range1,my_sine_fn(range1),xlab="x", ylab="f(x)",
main=bquote("f(x) = 3" ~ {sin}^2 ~"(0.5x)"))
```

2.5.2 Numerical Integration

Although not a function, this is an obvious point at which to introduce numerical integration. Often an exact solution to an integral is not obtainable (we shall see several examples later in the book), and we need to use numerical integration techniques to solve our problems. It can also be very useful to perform a check on our calculations.

For our purposes the following simple methods will usually be adequate. We may simply split the area in which we are interested into a large number of rectangles of equal width, h, calculate the average of the function at the two end-points of each sub-interval (alternatively calculate the function at the mid-point of each sub-interval, see Equation 2.2), calculate the area of each rectangle, and add together the areas:

$$\int_a^b f(x)dx = \frac{h}{2}f(a) + h\sum_{i=1}^{n-1} f(x_i) + \frac{h}{2}f(b) \tag{2.1}$$

where $x_0, x_1, x_2, ..., x_n$ are the set of equally spaced points in the interval $[a, b]$ and $h = x_{i+1} - x_i$. Alternatively:

$$\int_a^b f(x)dx = h\sum_{i=0}^{n-1} f(x_i + \frac{h}{2}) \tag{2.2}$$

Care should be taken by the reader in understanding the accuracy of the numerical method applied. A brief introduction to this topic can be found in Rizzo (2019) Section 13.3.

Applying Equation 2.2 to Examples 2.3 and 2.4:

```
z1<-function(x)x^2;        z2<-function(x)3*(sin(0.5*x))^2

own_int_fn<-function(stepx,lowxx,highxx,int_fn){
xx<-seq(lowxx+stepx/2,highxx-stepx/2,stepx)   #mid-points
yy<-int_fn(xx)
int_soln<-sum(yy)*stepx
}

(fn_x_sq<-own_int_fn(0.01,2,10,z1));   (fn_exmp<-own_int_fn(0.01,0,10*pi ,z2))

[1] 330.6666
[1] 47.12389
```

Thus we have written our own function which carries out numerical integration.

Exercise 2.6 *Carry out numerical integration on* $y = x^2$, *between* $x = 10$ *and* $x = 11$.

2.6 Recommended Reading

Although we would primarily recommend search engines to understand the required syntax for various functions in R, the following texts provide valuable guidance on applying functions in R:

- Charpentier, A. et al. (2016). *Computational Actuarial Science with R*. Boca Raton, Florida, US: Chapman and Hall/CRC.
- Crawley, M.J. (2013). *The R Book, (2e)*. Chichester, UK. Wiley and Sons Ltd.
- Rizzo, M.L. (2019). *Statistical Computing with R, (2e)*. Boca Raton, Florida, US: Chapman and Hall/CRC.
- Thisted, R.A. (1988) *Elements of Statistical Computing*. Boca Raton, Florida, US: Chapman and Hall/CRC.
- Verzani, J. (2014) *Using R for Introductory Statistics, (2e)*. Boca Raton, Florida, US: Chapman and Hall/CRC.

3

Financial Mathematics (1): Interest Rates and Valuing Cashflows

Peter McQuire

3.1 Introduction

One of the most fundamental tasks of the actuary is to estimate, using particular assumptions, how much money is required at a point in time such that we can expect to be able to make contractual payments in the future. To estimate this amount we will generally need to make three assumptions – the level of interest rates (or investment returns) expected in the future, the probability that the payment or payments will be made, and an assumption regarding the amount of the payment. In this chapter we simplify the situation and concern ourselves only with future payments which are certain and for a known amount; only an interest rate assumption is required. We will introduce the other factors in later chapters.

The principal objective of this chapter is therefore to develop methods to calculate the amount required in respect of a series of future cashflows, based on a particular set of interest rates. These interest rates may reflect the expected future investment return on the assets we hold; often, however, they may be adjusted to allow for a degree of prudence, other commercial factors, or simply be prescribed by legislation.

There are various types of interest rates used in practice, such as the force of interest, effective interest rates, and nominal interest rates. The majority of this chapter will discuss the force of interest; however we will also compare the concepts of effective and nominal rates of interest and discuss why various types of interest rates are used in practice.

It is expected that many readers will have at least some familiarity with the concept of effective interest rates. The approach taken in this chapter is however different to many texts covering this topic; unlike most texts on this important, often misunderstood topic, we will start with a discussion on the force of interest.

Throughout the chapter we will undertake calculations by coding numerical calculations of various complexities, which should provide the reader with a good, fundamental understanding of the key concepts. The chapter includes formulae and corresponding coding to allow students to solve problems requiring the calculation of the present value of a series of known cashflows, typical of those found on first-year actuarial and financial mathematics undergraduate courses and exams.

R Programming for Actuarial Science, First Edition. Peter McQuire and Alfred Kume.
© 2024 John Wiley & Sons Ltd. Published 2024 by John Wiley & Sons Ltd.
Companion Website: www.wiley.com/go/rprogramming.com.

The following chapter contains discussions on a variety of topics also fundamental to financial mathematics, together with an extensive number of exercises, some of which are similar to those included in this chapter. The reader may therefore wish to study these two chapters together.

The reader may wish to have to hand one of the following texts: McCutcheon and Scott (1986) is the classic text, or Garrett (2016) which provides a more modern writing approach.

3.2 The Force of Interest

Understanding the force of interest is a critical part of the student actuary's training. It is perhaps best understood from the following statements.

The amount of interest earned in a small time interval Δt on a fund of value X_t is approximately equal to:

$$X_t \, \delta_t \, \Delta t \tag{3.1}$$

where δ_t is the force of interest which applies at time t. As $\Delta t \to 0$ the increase in the fund value, dX_t, over the "infinitesimally small" time period, dt, is therefore given by:

$$dX_t = X_t \, \delta_t \, dt \tag{3.2}$$

Thus our formal definition of δ_t is:

$$\delta_t = \frac{1}{X_t} \frac{dX_t}{dt} \tag{3.3}$$

Note that X_t is a continuous function of t.

Remark 3.1 Equation 3.3 is often expressed in the following equivalent form, where $A(t, t+h)$ is the fund value at time $t + h$ following an investment of 1 at time t (h represents a small period of time):

$$\delta_t = \lim_{h \to 0} \frac{A(t, t + h) - 1}{h}$$

Let's insert values into Equation 3.2 to get a better understanding of δ_t:

- $X_0 = 100$ (fund value at $t = 0$), and
- $\delta_0 = 0.06$ per annum

Of course we cannot look at an infinitesimally small period of time, dt, in practice; we will look at a small period of time Δt which should give us an answer close to the correct answer – we will choose one day, so $\Delta t = \frac{1}{365}$ years. The approximate amount of interest earned over Δt is given by Equation 3.1:

$$£100 \times 0.06 \times \frac{1}{365} = £0.0164384$$

Hence the fund value at the end of day one will be, approximately, £100.0164384. (It is important to note that the units of δ must be "per unit time" e.g. "per year" as in this case.) However this is clearly only approximate as interest will have been earned, and added to our fund, after say one hour, one second, etc. To arrive at the exact result we must look at infinitesimally small time intervals; applying elementary calculus to Equation 3.2 we can show that:

$$X_{t_1} = X_{t_0} e^{\int_{t_0}^{t_1} \delta_t \, dt} \tag{3.4}$$

where X_{t_0} and X_{t_1} are the fund values at times t_0 and t_1 respectively.

This is one of the most important equations in finance. (We will see similar equations in Chapter 16 which we can use to calculate the probability a life will survive a particular period of time.)

Where the force of interest is constant throughout time ($\delta_t = \delta$) we obtain the following important, and exact, equation from Equation 3.4:

$$X_{t_1} = X_{t_0} e^{\delta \, (t_1 - t_0)} \tag{3.5}$$

Let's see how close our approximations, using Equation 3.2, compare with the exact values using Equation 3.5.

Example 3.1

Given $X_0 = 100$ and $\delta = 6\%$ per annum, calculate X_1. The exact answer is given by applying Equation 3.5:

```
delta<-0.06;   100*exp(delta*1)

[1] 106.1837
```

And approximately (using Equation 3.2) by recalculating the fund value every day:

```
dt<- 1/365;   fund_value<-100
for(a in 1:365) {fund_value<-fund_value*delta*dt+fund_value}
fund_value

[1] 106.1831
```

Exercise 3.1 *Why is the approximate value lower than the correct value? Re-calculate the above using hourly calculations, and then every second.*

The results are shown below; we get very close to the correct value using very small Δt:

```
fund_value_hr;   fund_value_sec

[1] 106.1836
[1] 106.1837
```

Exercise 3.2 *Using the fund values at various times from the above exercise, demonstrate that the gradient of the plot of fund value against time equals δX, as per Equation 3.3.*

It is important to note at this point that the force of interest is a true rate. By this we mean that a force of interest, equal to say 12% per annum, is *exactly the same* as a force of interest of 1% per month (assuming every month is the same length). This is analogous to the statement that travelling at a speed of 60 km per hour is exactly the same as travelling at a speed of 1 km per minute.

Example 3.2

Calculate X_1 using a monthly force of interest, $\delta = 0.5\%$ per month, and $X_0 = 100$.

The exact answer is given by:

```
delta_month<-0.06/12
100*exp(delta_month*12)

[1] 106.1837
```

It's trivial to see that we will arrive at the same answer as above (Example 3.1). However this is an important point which we will revisit later in the chapter when we discuss effective rates of interest (*which are not true rates*).

3.3 Present Value of Future Cashflows

Up to this point we have concerned ourselves with the accumulated value of our fund at a future time. However, actuaries will mainly be concerned with estimating the amount of money which is required now (or other specified time) such that, allowing for future investment returns (i.e. accrued interest), we will be likely to make any promised payments in the future. Dividing both sides of Equation 3.4 by:

$$e^{\int_{t_0}^{t_1} \delta_t \, dt}$$

we obtain one of the most important equations in actuarial science:

$$X_{t_0} = X_{t_1} e^{-\int_{t_0}^{t_1} \delta_t \, dt} \tag{3.6}$$

where X_{t_1} is the amount of money due at time t_1 and X_{t_0} is commonly referred to as the present value ("PV") at time t_0 of the payment X_{t_1} i.e. X_{t_0} is the amount of money we estimate that we need at t_0 so we can pay X_{t_1} at time t_1.

More generally, for n future payments $X_1, X_2, ..., X_n$, payable at times $t_1, t_2, ..., t_n$, the present value of these n payments at $t = 0$, PV, can be calculated using Equation 3.6, and is given by:

$$PV = \sum_n X_n e^{-\int_0^{t_n} \delta_t \, dt} \tag{3.7}$$

This is our most general equation when calculating the present value of a series of future cashflows which are certain to happen; provided we know δ_t at all times, t, we can calculate the present value of any series of known cashflows.

In Section 3.5 we develop a function in R in respect of Equation 3.7. Before we do this we first develop a series of simplifying equations that may help us solve common problems more quickly.

If δ_t is constant, Equation 3.7 simplifies to:

$$PV = \sum_n X_n e^{-\delta\, t_n} \tag{3.8}$$

Example 3.3

Calculate the PV of the following cashflows using a constant force of interest of 0.5% per month:

1) £40 in 9 months, followed by
2) £70 in 3 years

Solution : $PV = 40e^{-0.005\times9} + 70e^{-0.005\times36} = 96.71$

Writing code in respect of the above:

```
cashflow_12<-matrix(c(40,70,9,36),ncol=2)
deltax<-0.005
x<-cashflow_12[,1]*exp(-deltax*cashflow_12[,2])
sum(x)
```
```
[1] 96.70881
```

Often the payments will be of equal amounts, X. Equation 3.8 simplifies to:

$$PV = X \sum_n e^{-\delta\, t_n} \tag{3.9}$$

Finally, the simplest, common scenario we encounter in finance is where each of our n equal payments are paid at regular time intervals i.e. at $t_1 = 1, t_2 = 2, t_3 = 3, ..., t_n = n$. This is a geometric sum:

$$PV = X(e^{-\delta} + e^{-2\delta} + e^{-3\delta}... + e^{-n\delta})$$

giving the following important equation for an annuity certain, $a_{\overline{n}|}$:

$$PV = X a_{\overline{n}|} = X \frac{1 - e^{-\delta n}}{e^{\delta} - 1} \tag{3.10}$$

where X is the constant payment made at the end of each time period (known as "in arrears"); where payments are made at the start of each time period ("in advance") this equation should be multiplied by e^{δ}.

Therefore to solve any scenario which involves a series of constant payments made at regular intervals, and where δ_t is constant, we simply determine n (the total number of payments), and apply the appropriate δ which corresponds to the frequency of these payments.

For example, if we have 4 years of monthly payments where $\delta = 6\%$ per annum, then $n = 48$ and $\delta = 0.5\%$ per month. Or if we have 10 years of twice yearly payments (i.e. payable every six months) where $\delta = 7\%$ per annum, then $n = 20$ and $\delta = 3.5\%$ per six months.

Exercise 3.3 *Write a function in respect of Equation 3.10, and find the PV of the two series of cashflows in the above paragraph where each payment is equal to £1 and is paid in arrears.*

Remark 3.2 The presentValue function, which is included in the lifecontingencies package, calculates the present value of a series of simple cashflows; it is discussed in Chapter 4.

The remaining common scenario is where payments are *continuous* for a time, t. Clearly this is a theoretical scenario but can be very useful and is often a good approximation of reality e.g. daily payments for 30 years. In this scenario the appropriate formula is:

$$\bar{a}_{\overline{t}|} = \frac{1 - e^{-\delta t}}{\delta} \tag{3.11}$$

This is easily derived from Equation 3.10 as follows: as the payment period becomes smaller δ becomes smaller. As $\delta \to 0$, $e^{\delta} \to 1 + \delta$ (using Maclaurin series expansion), leaving the denominator as required.

Exercise 3.4 *Demonstrate the equivalence of Equations 3.10 and 3.11 by using the 2 equations to calculate the present value of £1m per annum payable continuously for 10 years, with $\delta = 5\%$ p.a. (Equation 3.10 approximately by using a suitably small time step, and Equation 3.11 exactly).*

Equation 3.10 deals with scenarios where there is a series of regular, equal cashflows and a constant force of interest. These scenarios are quite common in practice so will be of some practical use to the reader; it is also useful when checking calculations under more complex scenarios.

3.4   Instantaneous Forward Rates and Spot Rates

In Section 3.5 we will address problems which involve non-constant interest rates. This is a good juncture at which to discuss forward and spot rates.

The force of interest at some future point in time from now, δ_t, is commonly referred to as the *instantaneous forward interest rate* at time, t. Forward rates in general refer to the expected rate of interest which will apply over some future period of time (e.g. between $t = 3$ years and $t = 5$ years), and are mainly determined by prices on the bond and interest rate swap markets. Loosely, we can think of this instantaneous forward rate as the interest rate which the market is expecting over a short period of time, say one day, from time t. It was formally defined in Equation 3.3.

We will use the following example of a forward rate curve in many of the problems in this chapter (the same example is used in Section 3.5 and is plotted in Figure 3.1):

$$\delta_t = 0.03 - 0.01e^{-0.5t} \tag{3.12}$$

Note that this is an example of how the forward rate may stand today; the forward rate curve will in general change continuously with time.

δ_t can alternatively be defined in terms of the present value of £1 payable at some future time, t, here denoted by P_t:

$$\delta_t = -\frac{d}{dt} \log P_t \qquad (3.13)$$

This is demonstrated below using the rate curve specified in Equation 3.12, by calculating the present value of £1 payable at $t = 2$ and $t = 2.001$, and comparing with δ_2:

```
delta<-function(t){0.03-0.01*exp(-0.5*t)}          #forward rate
dt_x<-0.00001                                      #dt
p1<-exp(-integrate(delta,0,2)$value); p2<-exp(-integrate(delta,0,2+dt_x)$value)
(log(p1)-log(p2))/dt_x

[1] 0.02632121

delta(2)

[1] 0.02632121
```

The reader should compare Equations 3.3 and 3.13; Equation 3.3 describes the evolution of a fund value over a small period of time, whereas Equation 3.13 calculates the change in the present value of a payment if that payment is deferred by a small amount of time.

Then we can define the *spot rate*, R_T, which can be thought of as the average of all the δ_t's between $t = 0$ and some future time, T:

$$R_T = \frac{\int_0^T \delta_t \, dt}{T} \qquad (3.14)$$

Thus the spot rate is a convenient way to express how a fund may be expected, today, to increase between $t = 0$ and $t = T$ using a single interest rate. For example, a 10-year spot rate of 8% per annum means that we expect to receive an average force of interest of 8% per annum for the next 10 years. From Equations 3.4 and 3.14 we arrive at:

$$X_{t_1} = X_0 e^{R_{t_1} t_1} \qquad (3.15)$$

Note that the spot rate curve, as with the forward rate curve, provides a snapshot of expected future interest rates, and will change with time as markets reappraise expectations.

3.5 Non-Constant Force of Interest

3.5.1 Discrete Cashflows

As promised, we now discuss PV calculations in situations where the force of interest is not constant; in reality this will often be the case. We will use the following time-dependent δ_t in the examples which follow:

$$\delta_t = 0.03 - 0.01e^{-0.5t} \qquad (3.16)$$

where t is in years.

Exercise 3.5 *Calculate the approximate amount of interest due on a sum of £100 m to be borrowed for one day in 10 days' time, assuming the (instantaneous) forward rates in Equation 3.16. Ignore expenses. (We will carry out exact calculations shortly.)*

Exercise 3.6 *Determine, using Equation 3.14, the spot rate yield curve from this forward rate curve.*

Functions for the forward rate and spot rates are set out below, with plots of the rates shown in Figure 3.1:

```
delta<-function(t){0.03-0.01*exp(-0.5*t)}          #forward rate
spot<-function(t){(0.03*t+0.02*exp(-0.5*t)-0.02)/t} #spot rate
range1<-seq(0.1,20,0.1)
forw_curve<-delta(range1); spot_curve<-spot(range1)
```

We can use a number of methods to calculate the present value of future cashflows where the interest rate is not constant.

Example 3.4

Calculate, using Equation 3.7, the present value of a single payment of £100 payable in 2 years' time using the above interest rate curve. The equation is given by:

$$PV = 100e^{-\int_0^2 (0.03-0.01e^{-0.5t})\, dt}$$

We can use the `integrate` function to calculate this:

```
100*exp(-integrate(delta,0,2)$value)
[1] 95.37463
```

Alternatively using the spot rate function:

```
100*exp(-spot(2)*2)
[1] 95.37463
```

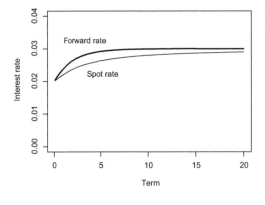

Figure 3.1 Non-constant forward rates and spot rates.

Remark 3.3 If the reader has not carried out numerical integration before, they may find it instructive to write code to carry out a numerical integration of the above calculation, as follows:

```
time_step1<-seq(0.05,1.95,0.1)
(force_int1<-delta(time_step1))
```

```
 [1] 0.02024690 0.02072257 0.02117503 0.02160543 0.02201484 0.02240428
 [7] 0.02277473 0.02312711 0.02346230 0.02378115 0.02408445 0.02437295
[13] 0.02464739 0.02490844 0.02515675 0.02539296 0.02561765 0.02583138
[19] 0.02603469 0.02622808
```

```
100*exp(-sum(force_int1)*0.1)
```

```
[1] 95.3745
```

This is consistent with our exact calculation above. We can of course get more accurate answers by shortening the dt's. This is effectively how the `integrate` function in R works.

Thus we have calculated the present value of this single cashflow using both the forward rates and the applicable spot rate.

Exercise 3.7 *By using only the forward rates and Equation 3.2 write code to show the (approximate) growth of £1 from $t = 0$ to $t = 10$ years using daily time intervals. Verify using an exact calculation.*

We now calculate the PV of a series of cashflows – this is a simple extension of the above.

```
pv_x<-function(z){
delta<-function(t){0.03-.01*exp(-0.5*t)}
x<-z[1]*exp(-integrate(delta,0,z[2])$value)
x}
```

where z[1] and z[2] are the size and payment time of each cashflow respectively.

Example 3.5

Let's say we are required to pay £10, £20, and £50 in one, two, and three years' time respectively:

```
cashflow_123<-matrix(c(10,20,50,1,2,3),ncol=2)
colnames(cashflow_123)=c("payment","time");  cashflow_123
```

```
     payment time
[1,]      10    1
[2,]      20    2
[3,]      50    3
```

To calculate the present value of these cashflows it is easiest to use the `apply` function:

```
sum(apply(cashflow_123,1,pv_x))
```

```
[1] 75.26816
```

Exercise 3.8 *Consolidate the above code within one function.*

Hence, the above code can be used to value any discrete set of cashflows using a particular forward yield curve (defined as δ_t above).

Before we move on the reader may wish to attempt the following two exercises:

Exercise 3.9 *Calculate exactly the amount of interest due on a sum of £100 m to be borrowed for one day in 10 days' time (using the forward rate set out above). This is a repeat of Exercise 3.5 above, but here calculated exactly.*

Exercise 3.10 *Calculate the projected value at $t = 10$ years of the following two payments:*
£100 received at $t = 0$ years
£50 received at $t = 7$ years.
You should assume the following instantaneous forward rates:

```
delta<-function(t){pmax(0.1-0.005*t,0.07)}
```

3.5.2 Cashflows Which Are Continuous

There is only one main scenario left to deal with; to value a continuous, changing rate of payment ρ_t, with a changing rate of interest. This situation may arise for example where we have varying daily payments over 40 years and we can very closely approximate these payments with a parametric curve. The present value of these payments at $t = 0$, made between $t = s_1$ and $t = s_2$, at the rate of $\rho(s)$ per unit time is given by:

$$PV = \int_{s_1}^{s_2} \rho(s)e^{-\int_0^s \delta_t \, dt} ds \tag{3.17}$$

This equation, together with Equation 3.7, will allow us to deal with any scenario with known cashflows. The key to understanding this equation is to note that $\rho(s)$ is a rate of payment, and that $\rho(s)ds$ is the infinitesimally small cashflow amount in the time period ds. We then discount each cashflow in a similar way to Equation 3.6.

We first look at a continuous but constant payment stream, of 1,000 per year. We can think of this, approximately, as 365 payments, each of 2.74, each discounted for the corresponding number of days. Using the original forward rates from above and discretizing:

```
delta<-function(t){0.03-.01*exp(-0.5*t)}
x<-NULL; rate<-NULL
dt<- 1/365; total_time<-1
steps<-total_time/dt
for(t in seq(1,steps)){
x[t]<-1000/365*exp(-integrate(delta,0,(t-0.5)*dt)$value)
}

sum(x)

[1] 989.3389
```

Of course, we can do better by calculating the present value exactly, using the derived spot rates:

```
pv_spot<-function(t){exp(-(0.03*t+0.02*exp(-0.5*t)-0.02))}
1000*(integrate(pv_spot,0,1)$value)
```

```
[1] 989.3389
```

But what if our cashflows are changing?

Example 3.6

Our payments are now payable for 5 years at the rate of $10 - t$ per annum from $t = 0$ to 5:

$$\rho_t = 10 - t$$

By using the derived spot rates we remove the need for the second integration:

```
pv_paymentrate<-function(t){(10-t)*exp(-(0.03*t+0.02*exp(-0.5*t)-0.02))}
(pv_var<-integrate(pv_paymentrate,0,5)$value)
```

```
[1] 35.52032
```

Alternatively, using the original forward rates and applying Equation 3.17 by discretizing:

```
delta<-function(t){0.03-.01*exp(-0.5*t)}
t=1
x<-NULL; rate<-NULL
dt<-.01; total_time<-5
steps<-total_time/dt
for(t in seq(1,steps)){
rate[t]<-10-(t-1)*dt-0.5*dt
x[t]<-rate[t]*dt*exp(-integrate(delta,0,(t-0.5)*dt)$value)
}
sum(x)
```

```
[1] 35.52032
```

To verify this calculation approximately, we can first simplify the position by using an equivalent, constant δ of say 2.5% per annum, and an average annual payment of 7.5:

$$PV \approx \int_0^5 7.5e^{-0.025s} ds = 35.2509292 \tag{3.18}$$

```
pv_approx_fn<-function(t){(7.5)*exp(-(0.025*t))}
(pv_approx<-integrate(pv_approx_fn,0,5)$value)
```

```
[1] 35.25093
```

Exercise 3.11 *Repeat the calculation using a similar method to that used in Example 3.5.*

Example 3.7

A company, which provides services to the seasonal UK tourist industry, is committed to make daily payments over the next year to a particular group of employees. The payments can be represented with the following formula (where t is in years):

Annual rate of payment $= £60\sin(\pi t)$ million

A plot of the cashflows is shown in Figure 3.2. Calculate the amount of money at $t = 0$ which will be sufficient to make these payments, assuming $\delta = 5\%$ per annum.

So, for example, on the 50th day it will be paying at the annual rate of £25.03 m per annum i.e. it will pay approximately £68580 on that day. Proceeding to calculate the present value:

```
delta<-0.05
cflowdiscount<-function(t)60*sin(pi*t)*exp(-delta*t)
(annual_pv<-integrate(cflowdiscount,0,1)$value)

[1] 37.2563
```

We estimate that the company would need £37.3 million at $t = 0$ to pay these benefits over the next year.

Exercise 3.12 *The company wants some understanding of how interest rate risk may affect the ability to make the payments. Set out a table of results based on interest rates falling prior to the project e.g. a range from 1% to 5% (constant rates).*

This concludes our study of the various cashflow / discount rate combination problems we may encounter. We will now move onto a discussion on effective and nominal rates of interest, and to what extent they are used in practice.

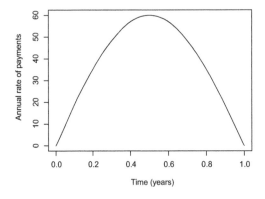

Figure 3.2 Cashflow payments.

3.6 Effective and Nominal Rates of Interest

Many readers may be concerned about where the discussion on (1) effective rates of interest, and (2) nominal rates of interest, is included in this book; most courses will start with this discussion. The short answer to this is that the force of interest conveys the most information and is generally the preferred rate used in mathematical settings. Effective rates can be useful in particular circumstances, as we shall see shortly. However, nominal rates are less commonly used, although remain useful in a limited number of situations and must still be understood.

3.6.1 Effective Rates of Interest

Let's start with the definition of i, the effective rate of interest:

Given an effective rate of interest of i per unit time (e.g. 7% per annum), a fund value of X_0 at $t = 0$ will equal $X_0(1 + i)$ at $t = 1$. That is:

$$X_1 = X_0(1 + i) \tag{3.19}$$

It is important to note that without further information we cannot calculate the fund value at $0 < t < 1$. The idea that an effective rate equal to i necessarily means that interest is received annually in arrears is flawed. It is correct to say that if you hold £100 at the start ($t = 0$) and hold £107 at the end ($t = 1$), this is equivalent to an effective rate of 7% per annum, and that one possibility is that we received £7 interest at the end of the period. But what fund would you have at $t = 0.5$? If interest is received annually in arrears you would only have £100 as you haven't received the interest yet. This may or may not be correct – we don't know. The "effective rate of interest" refers to "what effectively do I have after one time period?" What happens up to that point is not considered.

However if we make the assumption that our force of interest is constant between $t = 0$ and $t = 1$ then we can calculate fund values at any time between $t = 0$ and $t = 1$, by applying Equation 3.5. If $\delta_t = \delta$ (i.e. constant) then the fund value at $t = 1$ is given by:

$$X_1 = X_0 e^{\delta} = X_0(1 + i)$$

as per Equations 3.5 and 3.19, giving the following important relationship:

$$\delta = \log(1 + i) \tag{3.20}$$

It is important to note that knowing i provides little information about δ_t, other than its average value.

Exercise 3.13 *Using Equation 3.2 and a timestep of 0.01 years, show that a fund growing at $\delta = \log(1.06)$ per annum (constant) results in the same fund value at $t = 1$ as using an effective interest rate of 6% per annum.*

This exercise demonstrates one of the key points of the chapter – starting with a fund value of £1, applying δ over very short time intervals, and calculating the interest earned over these time intervals, results in a fund of £$(1 + i)$ at the end of the period.

Using Equation 3.20 we can re-write Equation 3.10 using effective rates, giving us the present value of a series of n equal payments of amount 1:

$$PV = a_{\overline{n}|} = \frac{1 - v^n}{i} \tag{3.21}$$

where $v = \frac{1}{1+i}$.

The following should be noted with respect to Equation 3.21:

- n is the number of equal payments, each made at the end of n equal time periods.
- i is the effective rate of interest per time period which is consistent with the payment frequency e.g. if payments are made monthly, the effective rate of interest should also be monthly.

Exercise 3.14 *Write a function in R in respect of Equation 3.21.*

Exercise 3.15 *Show that the equation of an annuity with term n can be considered as the difference between two infinite annuities, with one deferred for n years.*

It is worth considering which particular calculations can be made using Equation 3.21. For example, can we calculate the present value of 12 monthly payments (in arrears), each of £10, if we only know the annual effective rate of interest? The answer is no. We don't know how much interest we will receive, say, in the first month; we only know what the total rate is over the entire year. We could estimate the value if we assume that the force of interest is constant; without making this assumption we would either need details of the monthly effective rates, or δ_t for $0 \leq t \leq 1$.

Exercise 3.16 *Calculate the annual effective rate of interest between $t = 0$ and $t = 1$ given $\delta_t = (t+2)\%$. Now calculate the present value of a series of 12 equal monthly payments if $\delta_t = (t+2)\%$. Recalculate the present value of these 12 monthly payments assuming $\delta_t = 2.5\%$ per annum (constant).*

Non-financial example – a growing tree

It may also help to think of examples in the physical world, rather than finance. (Indeed, areas of physics and engineering use analogous concepts to the force of interest; however they will tend not to use the equivalent of effective interest rates.)

Consider the following statement: a 10 m tree grows at the rate of 6% per annum. This is an ambiguous statement. However, to the layperson this statement is likely to be quite clear, and be interpreted to mean that the expected tree height at the end of the year is 10.6 m. However it could be interpreted as a constant annual "force of growth". There is the added complication that it is likely to grow faster in the summer.

Let's say that the actual annual "force of growth" is as follows:

$$\delta_t = \cos((t - 0.5)/2) * 2 - 1.920963 \tag{3.22}$$

Writing a function for δ_t (plotted in Figure 3.3 (left)):

```
growth_fn<-function(t)(cos((t-0.5)/2)*2-1.920963)
time_step<-seq((0),1,.001)
tree_height<-growth_fn(time_step)
```

where the tree grows at its fastest rate in the summer, around $t = 0.5$. The height of the tree at the end of the year, h_1, is given by the following expression:

$$h_1 = 10 \, e^{\int_0^1 \delta_t dt}$$

Calculating the height of the tree at $t = 1$:

```
10*exp(integrate(growth_fn,0,1)$value)

[1] 10.6
```

Thus our tree has grown by 6% over the year (or very close to it); *the effective annual rate of growth is 6% per annum*. It's certainly not true to say that "the force of growth was 5.83% per annum" – this is the average force of growth.

Stating that the tree grows at an effective rate of 6% per annum is true but incomplete – it doesn't tell us what happened over the year. The best description of growth is given by Equation 3.22.

However if we are only interested in the height of the tree every year-end then this effective rate is useful. Thus effective rates can and probably should be used if we are only interested in quantities at particular regular points in time, which is often the case in finance.

Exercise 3.17 *Calculate and plot the height of the tree between $t = 0$ and $t = 1$ using (1) the force of growth rate, δ_t, and (2) the misleading effective growth rate assuming a constant force of growth [see Figure 3.3 (right)].*

Hence the force of interest is the more natural way of expressing these rates. (Any reader who has studied physics or engineering may be more comfortable with the force of interest

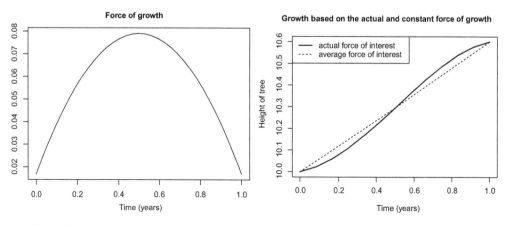

Figure 3.3 Tree growth.

than effective rates of interest; the equivalent of the force of interest is used in these subjects as it is a better way of describing nature.)

Exercise 3.18 *Calculate the present value of 10 payments, each of £1, paid at t = 1, 2, ..., 10 years, using the yield curve shown by Equation 3.16. Recalculate the present value using an effective rate of 2.5% per annum and 3.0% per annum.*

3.6.2 Why Do We Use Effective Rates?

Effective rates are often used in finance due to both the regular, discrete nature of cashflows we frequently encounter in the financial and insurance sectors, and also their ease of communication. As cashflows are often received or paid out at fixed, regular intervals (such as monthly pension payments or insurance premiums), using an effective rate of interest with a period consistent with the frequency of payments may accurately, and succinctly, reflect the real-world position. For these calculations we do not particularly require a force of interest. Indeed, in Chapter 23 (Contingencies) we almost exclusively use effective rates of interest as most cashflows can only occur at specific, regular points in time. Where the force of interest can be considered to be constant (i.e. not a function of time), the use of effective rates is usually preferred.

Regarding communications, the use of effective rates of interest is much clearer for the layperson. If we start the year with £100 and end it with £108, it is intuitive that we have received 8% interest (compared with receiving interest continuously at the average rate of 7.7% per annum).

However in situations where interest rates are changing and/or we have irregular payment patterns we are likely to require applying the force of interest in our calculations.

Remark 3.4 A note on numerical integration and forces of interest; when present values are calculated numerically using forces of interest, as in our examples earlier in the chapter, we are using small effective rates of interest over small periods of time.

3.6.3 Nominal Interest Rates

We leave much of the discussion on nominal rates to other texts which discuss them in some detail. Below is a brief discussion together with a possible explanation for their usage. Stated below for completeness is the relationship between nominal rates, $i^{(p)}$, and effective rates, i:

$$i^{(p)} = p((1 + i)^{\frac{1}{p}} - 1) \tag{3.23}$$

From discussions earlier in this chapter, we can loosely interpret effective rates of interest as receiving the interest at the end of the period. In the same way nominal rates can be interpreted as receiving interest payments more frequently than once per unit period. For example a nominal rate of "12% per annum payable (or convertible) monthly" can be interpreted as receiving 1% at the end of each month, and is exactly equivalent to an effective rate of 1% per month.

Perhaps the main reason why nominal rates are used is to aid communication; the most convenient way to state interest rates which apply to cashflows which involve regular payments is often by using nominal rates. For example, bonds typically have 6-monthly cashflows; the most natural interest rate to use for such bonds is a 6-monthly effective rate of interest, which is most easily communicated using the nominal rate, $i^{(2)}$p.a. (as there is no convenient expression for "a 6-monthly effective rate of interest"). Or if a monthly effective rate of 1% applies, it may be more usual to define this as a nominal rate of 12% per annum payable monthly.

McCutcheon and Scott (1986) also imply the potential redundancy of nominal rates – "The general rule to be used in conjunction with nominal rates is very simple. Choose as the basic unit of time the period corresponding to the frequency with which the nominal rate of interest is convertible and use $\frac{i^{(p)}}{p}$ as the effective rate of interest per unit time" e.g. change 10% per annum nominal rate convertible quarterly to an effective rate of 2.5% per quarter.

The use of nominal rates is partly historical; before the advent of computers, analysts used actuarial tables which included nominal rates to aid calculations which involved several regular payments each year (e.g. monthly). If we have a table of annuity rates we can quickly calculate the rates for annuities which involve payments more frequently simply by applying Equations 3.21, 3.23, and a ratio of interest rates:

$$a_{\overline{n}|}^{(p)} = a_{\overline{n}|} \frac{i}{i^{(p)}} \tag{3.24}$$

It is important to note that nominal rates are not true rates in the mathematical sense.

Example 3.8

Calculate the present value of a 3-year annuity contract, with total annual payments of £120 each year, payable monthly in arrears. The nominal interest rate of 8% per annum payable quarterly.

This is an effective interest rate of 2% per 3 months. This is exactly equivalent to an quarterly force of interest $\delta = \ln(1.02) = 0.0198026$.

We can calculate our 36 payment annuity using a monthly force of interest and Equation 3.10. Note that in doing so we must assume that each monthly effective rate, or the force of interest, is constant (as noted earlier in the chapter):

$$a_{\overline{36}|} = \frac{1 - e^{-0.0198026/3 \times 36}}{(e^{0.0198026/3} - 1)} = 31.9365997$$

Each payment is for £10 so the present value of the payments is 319.3659974.

Exercise 3.19 *We hold £319.37 at the start of the above contract in an interest-bearing account with $i^{(4)} = 8\%$ p.a. Write code to show the amount held in the account after each payment, and that we will have £0 at $t = 3$ years at the completion of the contract.*

3.7 Appendix: Force of Interest – An Analogy with Mortality Rates

Note: Readers who have not yet studied contingencies and survival analysis may wish to defer reading this appendix.

The comparison of effective rates and forces of interest is directly analogous to that of the mortality rate terms q_x and μ_x; q_x tends to be intuitive ("what is the probability a life, aged x, dies in the next year?"), whereas force of mortality perhaps requires a little more thought. The same conclusions made in this chapter can be reached; if we simply need to estimate survival probabilities between regular discrete ages then q_x is likely to adequately meet our objectives. However if we need to calculate the probabilities over non-integer age ranges (e.g. "a life bought a policy at age 53 and 2 months, what is the probability she will survive 10 years and 1 month?") it is more intuitive and accurate to apply the force of mortality μ_x in our calculations. A similar argument can be made when comparing Markov chain models (in discrete time) with continuous time jump models (discussed in Chapter 22).

3.8 Recommended Reading

- Garrett S.J. (2016) *An Introduction to the Mathematics of Finance: A Deterministic Approach.* Oxford, UK: Butterworth Heinemnan.
- McCutcheon and Scott. (1986) *An Introduction to the Mathematics of Finance.* Oxford, UK: Butterworth Heinemnan.
- Wilmott, P. (2007) *Introduces Quantitative Finance.* Chichester, West Sussex, UK: Wiley.

4

Financial Mathematics (2): Miscellaneous Examples

Peter McQuire

```
library(lifecontingencies)
```

4.1 Introduction

This chapter does not follow the typical style of most chapters in this book; there is no specific theme which it covers, consisting mainly of a series of examples and exercises based around topics included in a typical first year of an undergraduate programme for actuarial or financial mathematics. As such, this chapter should provide good practice to those readers new to writing code, and for consolidating one's understanding of fundamental actuarial concepts.

Students may also wish to try the examples and exercises included in this chapter in Microsoft Excel, and compare the advantages and disadvantages with using R. Many tasks in this chapter are relatively simple compared with those in later chapters, such that Excel may be preferable in some cases. Throughout an actuary's career decisions will often be required regarding the optimal software to use for a particular task.

The aims of this chapter are to:

1. cover miscellaneous topics typically studied in the first year of an undergraduate programme in actuarial science;
2. provide further practice in writing functions, and other code, in R;
3. apply functions, including those which exist in current R packages, to solve simple problems.

There is also an emphasis throughout the chapter (and in the book in general) on the importance of testing any code you write, and indeed checking any calculations. The skills learnt in this chapter will aid the reader when it comes to writing more complex functions which will be introduced throughout the book.

R Programming for Actuarial Science, First Edition. Peter McQuire and Alfred Kume.
© 2024 John Wiley & Sons Ltd. Published 2024 by John Wiley & Sons Ltd.
Companion Website: www.wiley.com/go/rprogramming.com.

4.2 Writing Annuity Functions

4.2.1 Writing a function for an annuity certain

In the previous chapter we noted the following fundamental equation to calculate the present value of n payments, for amounts X_n, payable at times t_n, discounted at the constant force of interest, δ:

$$PV = \sum_{1}^{n} X_n e^{-\delta t_n} \tag{4.1}$$

From Equation 4.1 we derived one of the first equations that students of actuarial science use in their studies – that for an *immediate annuity certain* where $X_n = 1$ for all n:

$$a_{\overline{n}|} = \frac{1 - v^n}{i} \tag{4.2}$$

where $v = \frac{1}{1+i}$ and $i = e^\delta - 1$.

This important equation was discussed in the previous chapter; the reader may already have written a function for this basic expression. Our task is to develop a function in R for an annuity certain which incorporates functionality to allow for payments which increase, various frequencies of payments, and for payments in advance rather than arrears.

(The annuity and presentValue functions are discussed later in the chapter – the reader may subsequently wish to use these to verify/simplify the functions developed below.)

Example 4.1 Calculate $a_{\overline{10}|}^{@4\%}$ and verify the calculation.

```
annuity_certain<-function(i,term){x<-(1-(1+i)^-term)/i
return(x)
}

annuity_certain(0.04,10)

[1] 8.110896

v<-1/1.04
sum(v^(1:10)) # a check

[1] 8.110896
```

Note that the return command is not required; however it can make the code easier to read, especially where the code is more complex and there are multiple outputs. The following alternative code is perhaps a more transparent way to write the function (not using return here):

```
ann_z<-function(i,n){v<-1/(1+i)
        x<-seq(1,n)
        vv<-v^x
        pres_val1<-sum(vv)
        pres_val1
}
ann_z(0.04,10)
```

```
[1] 8.110896
```

We should allow for payments in advance or arrears, and for various payment frequencies. The exact code which you write is likely to be quite different to that shown below. Here we have calculated $\ddot{a}^{(12)}_{\overline{20}|}$ at $i = 10\%$ p.a.:

```
ann_z<-function(i,n,p,advance){i_req<-(1+i)^(1/p)-1
v_req<-1/(1+i_req)
x<-seq(1,n*p)
vv<-v_req^x
pres_val1<-sum(vv)/p/ ifelse(advance == 1,v_req,1)
pres_val1
}
ann_z(0.1,20,12,1)
```

```
[1] 8.968001
```

```
ann_z(0.04,10,1,0)   #as per above
```

```
[1] 8.110896
```

Note that most programmers will develop this type of code by initially splitting it into several parts, and then simplifying it. Your final code is unlikely to be written with your first attempt.

Of course you can check your answers with published actuarial tables, for example the IFoA's Formulae and Tables book.

Exercise 4.1 *Extend your function to allow for regular payments where each payment increases by one unit, that is $Ia_{\overline{n}|}$, and use it to calculate the present value of the following eight payments: £100 at $t = 1$ years, £200 at $t = 2$ years, £300 at $t = 3$ years, ...£800 at $t = 8$ years, at a constant force of interest equal to 5% per annum.*

Exercise 4.2 *Extend your function further to allow for regular payments where each payment increases by a fixed percentage amount. Apply the function to the cashflows in Exercise 4.1, except the payments increase at 6% per annum compound, and not by £100.*

Exercise 4.3 *Calculate the present value of an annuity which has the following cashflow pattern and a term of 10 years:*

£100 payable monthly in advance in year 1. In year 2, the monthly payments are increased by 3%, that is £103 per month. For year 3 the monthly payments are further increased by 3%, that is £106.09, and so on, that is the monthly payments in year 10 are for £130.48. Assume a monthly effective interest rate of 0.5%. (Also perform reasonableness checks on your calculations.)

The reader may also wish to write this in the form of a function.

We would recommend the reader to write various interest rate conversion functions, for example between $i, i^{(p)}, d, d^{(p)}, \delta$.

Example 4.2 Convert an annual effective rate, $i = 4\%$ p.a., to a p-thly rate $i^{(12)}$, and a p-thly interest rate, $i^{(12)} = 6\%$ p.a., to an annual force of interest.

```
i_to_ip<-function(i,p) ((1+i)^(1/p)-1)*p
i_to_ip(0.04,12)
```

```
[1] 0.03928488
```

```
ip_to_delta<-function(ip,p) p*log(1+ip/p)
ip_to_delta(0.06,12)
```

```
[1] 0.0598505
```

We now approach solving the above exercises similar to those above, but using a function which currently exists in R.

4.3 The 'presentValue' Function

The presentValue function, which is included in the lifecontingencies package in R, calculates the present value of a series of cashflows. This serves a similar purpose to the functions developed earlier in this chapter and in previous chapters, but allows for more flexibility with the timing of payments and changing interest rates; this function can prove very useful.

Remark 4.1 Note that the type of interest rate required for this function is the effective interest rate.

Example 4.3

Calculate how much money we may need now in order to pay £2 at the end of year 1 and £3 at the end of year 4, assuming a constant effective interest rate of 8% per annum.
     Using the presentValue function:

```
presentValue(c(2,3),c(1,4),0.08)
```

```
[1] 4.056941
```

```
2/1.08+3/1.08^4   # a check
```

```
[1] 4.056941
```

Example 4.4

Calculate the present value of the following cashflows using a constant force of interest of 0.5% per month (repeating Example 3 from the previous chapter):

1. £40 in 9 months, followed by
2. £70 in 3 years.

```
presentValue(c(40,70),c(0.75,3),exp(0.06)-1)     #OR

[1] 96.70881

presentValue(c(40,70),c(9,36),exp(0.005)-1)

[1] 96.70881
```

Exercise 4.4 *Write a function incorporating the* `presentValue` *function which uses the force of interest as an input.*

Example 4.5 Calculate the present value of the 300 payments set out in the *'pv_example.csv'* file, using a constant effective annual interest rate of 4.5% p.a. (all times are in years), and perform a broad check.

```
data_ex<-read.csv("~/pv_example.csv")
```

```
presentValue(data_ex[,2],data_ex[,1],0.045)

[1] 132780.5

presentValue(sum(data_ex[,2]),mean(data_ex[,1]),0.045) # a broad check

[1] 131450.9
```

Example 4.6 Using the `presentValue` function, calculate the present value of £1 m per annum payable continuously for 10 years, with $\delta = 5\%$ p.a., using a suitably small time step (compare Exercise 4 from the previous chapter).

```
annual_pay<-1000000
steps_pa<-100000;  n_years<-10;  deltaz<-0.05
amount1<-annual_pay/steps_pa
c_flow<-rep(amount1,steps_pa*n_years)
timeszz<-seq(1/steps_pa,10,by=1/steps_pa)
presentValue(c_flow,timeszz,exp(deltaz)-1)

[1] 7869385

1000000*(1-exp(-deltaz*n_years))/deltaz  #theoretical value

[1] 7869387
```

We can apply different spot rates of interest to different cashflows. For example, here we apply 8% p.a. for 1 year, and 9% p.a. for 4 years:

```
presentValue(c(2,3),c(1,4),c(0.08,0.09))
[1] 3.977127
2/1.08+3/1.09^4    # a check
[1] 3.977127
```

Example 4.7

Here we repeat Example 5 from the previous chapter, using the same forward rate yield curve (Equation 4.3) and derived spot rates, but instead using the `presentValue` function:

$$\delta_t = 0.03 - 0.01e^{-0.5t} \tag{4.3}$$

```
spot1<-function(t){(0.03*t+0.02*exp(-0.5*t)-0.02)/t}
presentValue(c(10,20,50),c(1,2,3),exp(spot1(c(1,2,3)))-1)
[1] 75.26816
```

Exercise 4.5 *Calculate the present value of the cashflows in the 'pv_example.csv' file, using the forward rate curve (Equation 4.3).*

Perhaps a payment at a future time is dependent, or contingent, on a particular event. We can include a probability in the function to allow for this. (This functionality is useful when tackling problems typical in survival analysis and life contingencies, where future payments are not guaranteed to be made – see Chapters 16 and 23.)

Example 4.8

Repeat Example 4.3, with the additional constraint that payments are made to the owner of the policy only if the life is alive at that time the payment is due (assume the probability of the life being at alive at $t = 1$ and $t = 4$ are 0.98 and 0.91 respectively).

```
presentValue(c(2,3),c(1,4),0.08,c(0.98,0.91))
[1] 3.821446
```

We'll finish this section calculating various standard cashflows using the `presentValue` function (with $i = 3\%$ p.a., assuming constant δ).

1. Annual payments in arrears: $a_{\overline{3}|}$

```
payments<-5    #also used below
presentValue(rep(1,payments),1:payments,0.03)
[1] 4.579707
```

2. Annual payments in advance: $\ddot{a}_{\overline{5}|}$

```
presentValue(rep(1,payments),1:payments-1,0.03)

[1] 4.717098
```

3. Monthly level payments, in arrears: $a_{\overline{5}|}^{(12)}$

```
freq<-12
presentValue(rep(1/freq,payments*freq),seq(1/freq,payments,by=1/freq),0.03)

[1] 4.642342

#or
presentValue(rep(1/freq,payments*freq),seq(1,payments*freq),1.03^(1/freq)-1)

[1] 4.642342
```

4. Monthly payments which increase annually at a simple rate: $Ia_{\overline{5}|}^{(12)}$

```
years<-5;  freq<-12;  x<-1:years
y<-rep(x/freq, each = freq); head(y,15)

 [1] 0.08333333 0.08333333 0.08333333 0.08333333 0.08333333 0.08333333
 [7] 0.08333333 0.08333333 0.08333333 0.08333333 0.08333333 0.08333333
[13] 0.16666667 0.16666667 0.16666667

timez<-1:(freq*years)
presentValue(y,timez/12,0.03)

[1] 13.65269
```

5. Monthly payments which increase annually (at a compound rate of 2% p.a.):

```
years<-5;    freq<-12;    x<-1:years-1;    inc<-0.02;  disc<-0.03
y<-rep(x, each = freq)
bb<-(1+inc)^y/freq
presentValue(bb,timez/freq,disc)

[1] 4.826135

df<-(1+inc)^(0:(years-1))/12   #alternative code
sum(rep(df,each=freq)*(1+disc)^-(seq(1:(years*freq))/freq))

[1] 4.826135
```

And try the following:
6. $I\bar{a}_{\overline{5}|}$, i.e. a continuously payable annuity, with stepped increases.
7. $\bar{I}\bar{a}_{\overline{5}|}$, i.e. a continuously increasing annuity, payable continuously.
   Note that an approximation is required for the last two annuities when using the presentValue function.

4.4   Annuity Function

Alternatively we can calculate the present value of standard annuity cashflows using the annuity function in the lifecontingencies package. The examples above are repeated here:

```
annuity(0.03,5)                          #the default - payment in arrears
[1] 4.579707
annuity(0.03,5,m=0,k=1,type="arrears") #m refers to deferral, and k is the frequency
[1] 4.579707
annuity(0.03,5,m=1,k=1,type="advance") #equivalent to above
[1] 4.579707
annuity(0.03,5,m=0,k=1,type="advance")
[1] 4.717098
annuity(0.03,5,m=0,k=12,type="arrears")
[1] 4.642342
increasingAnnuity(0.03,5,type="advance")*annuity(0.03,n=1,k=12,type="arrears")
[1] 13.65269
```

The monthly annuity, payable in arrears, increasing annually (here at 2% p.a.), is easily calculated as follows:

```
net_ratex<-1.03/1.02-1
(annuity(net_ratex,5,type="advance"))*annuity(0.03,1,k=12,type="arrears")
[1] 4.826135
```

Of course, the above should be written as a function if any of these compound expressions were regularly required.

Exercise 4.6 *For readers who have a book of actuarial tables issued by the Institute and Faculty of Actuaries, recreate the tables of compound interest calculations using the various functions covered in the preceding sections.*

4.5   Bonds – Pricing and Yield Calculations

It is a straightforward extension of the above to value a coupon-bearing bond. (In this section it is assumed that the reader has a basic knowledge of what constitutes a coupon-bearing bond.)

Example 4.9

Calculate the price of £100 nominal of a newly issued 12-year bond, level annual coupons equal to £4 paid at the end of each year, with the next coupon payable in 12 months' time, at δ=9% per annum. The bond is redeemable at par. The value is given by:

$$4 \, \frac{(1 - e^{-0.09 \times 12})}{e^{0.09} - 1} + 100 \, e^{-0.09 \times 12}$$

```
term<-12;  forcex<-0.09;   coup<-4;   red<-100
time<-1:term
(payments<-c(rep(coup,term-1),red+coup))
```

```
[1]   4   4   4   4   4   4   4   4   4   4   4 104
```

```
presentValue(payments,time,exp(forcex)-1)
```

```
[1] 62.00986
```

```
#or
```

```
4*annuity(exp(0.09)-1,12,m=0,k=1,type="arrears")+100*exp(-0.09*12)
```

```
[1] 62.00986
```

Exercise 4.7 *Calculate the price of the bond in Example 4.9 where, instead of 12 annual coupons equal to £4, the coupon payments are 6-monthly, that is 24 payments of £2.*

Exercise 4.8 *Write a function to calculate the price of a bond.*

As noted earlier, we can incorporate the probability of payments being received, for example for a corporate bond. If you believe that every coupon payment has an 80% chance of being paid, and the final payments at redemption (final coupon plus redemption payment) are only 50% likely:

```
prob<-c(rep(0.8,11),0.5)
presentValue(payments,time,0.09,prob)
```

```
[1] 40.26442
```

It is important to note the effect which interest rates have on bond prices and other cashflows. The following loop calculates the price of the bond described in Example 4.9 at various effective interest rates, ranging from 1% per annum to 9% per annum:

```
pv<-0;   counter<-1
for(rate in seq(0.01,0.09,0.001)){pv[counter]<-presentValue(payments,time,rate)
counter<-counter+1}
```

(See Figure 4.1.) This inverse relationship should be studied alongside the material in Section 4.7; here we are describing the immediate change in the price of the 12-year bond following a change in interest rates, for example if interest rates increased from 3% p.a. to 7% p.a. overnight, the price of the 12-year bond would fall from £110 to £76 (highlighted). (In Section 4.7 we monitor the price changes of a bond throughout its lifetime, allowing for changes in interest rates.)

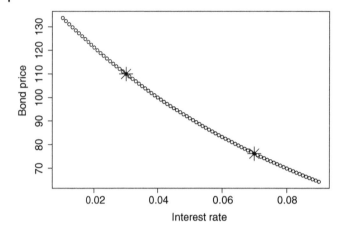

Figure 4.1 Effect of interest rates on bond prices.

A common requirement is to calculate the gross redemption yield (GRY) of a bond given its price and known cashflows. Given the price, P, of a bond, the GRY is defined as follows (as an effective rate of interest):

$$\sum_{n=1}^{Y} C_n(1 + GRY)^{-t_n} + R(1 + GRY)^{-t_Y} = P \tag{4.4}$$

where C_n is the amount of each coupon payment, Y is the number of coupon payments and R is the amount of the redemption payment at time t_Y. Usually the general, first term in Equation 4.4 can simply be denoted by $a_{\overline{Y}|}^{@GRY}$, where all the coupons are for an equal amount. 'Gross' refers to the situation where no tax is liable either on the coupon payments, or on any potential capital gain arising from the redemption payment. The Net Redemption Yield is appropriate where the investor is subject to tax on the receipts from the bond.

Remark 4.2 The Gross Redemption Yield is an example of the Internal Rate of Return (IRR), which is discussed in more general terms below.

We can most easily determine the GRY by calculating several prices based on a series of GRYs:

Example 4.10

Calculate the GRY of the following bond: £100 nominal of a newly issued 12-year bond with coupons of £3 p.a. payable every 6 months and redemption at par (with the first coupon payable in 6 months) if the price is £91.

```
price<-91;    term<-12;    times<-seq(0.5,term,0.5)
payments<-c(rep(3*0.5,2*term-1),101.5)

price_xx<-NULL;   counter<-1;   steps<-0.000001
for(i in seq(0.01,0.1,steps)){
price_xx [counter]<-presentValue(payments,times,i)
```

```
counter<-counter+1
}
```

```
(position_x<-(which(min(abs(price_xx -price))==abs(price_xx -price)))))
```

[1] 29880

```
(gry_x<-0.01+position_x*steps)          #answer - GRY
```

[1] 0.03988

```
presentValue(payments,times,gry_x)      #verifying
```

[1] 90.99924

Thus GRY $= 3.988\%$ p.a. If a more accurate yield is required we can simply lower the step value. Alternative code is included below, which iteratively recalculates the price using a 'while-loop', and also calculates the force of interest:

```
yprice<-91;    term<-12;    times<-seq(0.5,term,0.5)
payments<-c(rep(3*0.5,2*term-1),101.5)

minz<-0;        maxz<-0.2    #set yield range, care to set range wide enough
price_est<-100               #set initial estimate
cc<-1                        #counter to break while loop

while(abs(price_est -yprice)>0.00001){
        if(cc==100){break};
        gryx<-(minz+maxz)/2
        price_est<-presentValue(payments,times,exp(gryx)-1)
        if(price_est<yprice){maxz<-gryx}else{minz<-gryx}
        cc<-cc+1
}
cc;    price_est;    gryx
```

[1] 24
[1] 91.00001
[1] 0.03910449

As noted above, GRY is a particular example of determining the internal rate of return (IRR) from a series of cashflows. From a given series of cashflows we can, using Equation 4.1, determine the *IRR*. We have, using the effective rate of interest equivalent:

$$\sum_n X_n(1 + IRR)^{-t_n} = 0 \qquad (4.5)$$

where X_n is the amount of the nth cashflow, including both income and outgo, and t_n is the time of the nth payment. This has many applications, such as to determine a bond's GRY (as shown above), or to determine the expected or achieved return on a capital project.

Exercise 4.9 *An investor is assessing whether she should proceed with a property investment which involves renting out the property for 20 years, at which time she would sell the property. Calculate the IRR assuming the following series of cashflows:*

- *Purchase price of property* = £4,000,000
- *Rent collected at t* = 1: £100,000
- *Rent increases each subsequent year by 3%*
- *Final rent received at t* = 19
- *Expected sale price at t* = 20: £8,000,000

The cashflows are set out below:

```
years<-20 ;   x<-0:(years-2);     intz<-0.03
rent<-(1+intz)^x
(cflow<-c(-4000000,rent*100000,8000000))
```

```
[1] -4000000.0   100000.0   103000.0   106090.0   109272.7   112550.9
[7]    115927.4   119405.2   122987.4   126677.0   130477.3   134391.6
[13]   138423.4   142576.1   146853.4   151259.0   155796.7   160470.6
[19]   165284.8   170243.3  8000000.0
```

Also determine the purchase price at t = 0 *if the required IRR*= 10% *per annum.*

Example 4.11 Pricing an equity.

An analyst has estimated that an equity will payout the following dividends on each share:

- nil dividends at $t = 1$ and $t = 2$ years
- 40p at $t = 3$ years
- followed by increases of 10% p.a. for the following 5 years ($t = 4$ to 8 years)
- followed by increases of 3% p.a. thereafter (from $t = 9$ years).

Calculate the price of the share assuming a required return of 7% p.a.

```
divx1<-c(0,0,1.10^(0:5),1.10^5*1.03^(1:1000))*0.4
presentValue(divx1,1:length(divx1),0.07)
```

[1] 11.75619

```
divx1<-c(0,0,1.10^(0:5),1.10^5*1.03^(1:10000))*0.4
presentValue(divx1,1:length(divx1),0.07)
```

[1] 11.75619

Remark 4.3 This is perhaps a good example of a calculation which may be simpler to carry out in Microsoft Excel. However if many similar calculations are required, the application in R may still prove superior.

Reverting back to our discussion on bonds, a useful exercise is to understand how changes in the price of a bond, resulting from changes to interest rates, depends on the term of the bond. (This important concept is related to the 'duration' of the bond and is discussed in more detail later in the book.) For example does the price of a 2-year bond tend to vary as much as a 40-year bond given a particular parallel movement in the yield curve?

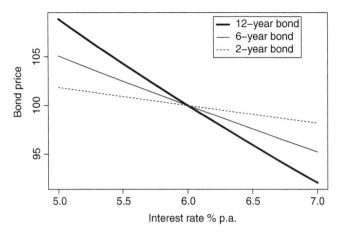

Figure 4.2 Effect of term on bond price variability.

Exercise 4.10 *The task is to calculate the prices of 2-year, 6-year and 12-year bonds, each paying coupons of 6% each year, using 21 flat yield curves – the reader should calculate these prices with yields at 5.0%, 5.1%, ...6.9% and 7.0% p.a.*

The results from Exercise 4.10 are shown in Figure 4.2. It is clear that the price sensitivity to changes in interest rates is greatest for the 12-year bond, with the 6-year bond being more sensitive than the 2-year bond. This is an extremely important result, and one we will be returning to in later chapters.

4.6 Bond Pricing: Non-Constant Interest Rates

In this section we calculate the current price of a bond using the instantaneous forward rate from earlier in the chapter (Equation 4.3, repeated below for convenience):

$$\delta_t = 0.03 - 0.01e^{-0.5t}$$

Hence, functions for the instantaneous forward rate and spot rates are:

```
delta1<-function(t){0.03-0.01*exp(-0.5*t)}
spot1<-function(t){(0.03*t+0.02*exp(-0.5*t)-0.02)/t}
```

We shall also estimate the future price of a bond, and the future returns from bonds.

In Exercise 4.5 we calculated the present value of a series of future payments using these functions; it is straightforward to extend this to calculate the price of a bond, and how this price may evolve in the future. More generally, we can use the equation for forward rates, between times t_1 and t_2 :

```
forw_1_2<-function(t1,t2){(0.03*t2+0.02*exp(-0.5*t2)-0.03*t1-0.02*exp(-0.5*t1))/
(t2-t1)}
```

Example 4.12

Calculate the price of £100 nominal of a newly issued 12-year bond, annual coupons of £4 paid at the end of each year, with the next coupon payable in 12 months' time, using the instantaneous forward rate set out in Equation 4.3:

```
term<-12;     timexy<-1:term
y3<-forw_1_2(0,timexy)     #spot rates, t1=0
payments<-c(rep(4,term-1),104)
(pv_bondat0<-presentValue(payments,timexy,exp(y3)-1))   #price of bond at t=0

[1] 111.5677

#alternatively
y4<-spot1(timexy)   #spot rates
(pv_bondat0_alt<-presentValue(payments,timexy,exp(y4)-1))   #price of bond at t=0

[1] 111.5677
```

We may wish to estimate the fund value at $t = 1$ following the purchase of the bond at $t = 0$ for £111.57, assuming no changes to our expectations of forward rates:

```
term2<-11;     timexy2<-1:term2
y2<-forw_1_2(1,timexy2+1)          #at t=1
payments2<-c(rep(4,term2-1),104)      #only 11 payments
(pv_bondat1<-presentValue(payments2,timexy2,exp(y2)-1)) #ignoring first coupon

[1] 110.0643
```

Including the coupon due at $t = 1$, the estimated fund value is 114.0643, that is a (continuous) return over the first year of 2.21%, consistent with the average instantaneous forward rate over this period, and one-year spot rate:

```
mean(delta1(seq(0,1,by=0.00001)))

[1] 0.02213061

spot1(1)

[1] 0.02213061
```

Extending this calculation to the redemption date, allowing for the reinvestment of coupons received at times $t = 1, 2, 3, ..., 11$ (again with the assumption that they achieve returns in line with the forward rates), the estimated fund value at redemption and corresponding investment return are as follows:

```
coupon_val<-sum(4*exp(forw_1_2(1:11,12)*(11:1)))
(end_fund_value<-coupon_val+104)

[1] 156.7546
```

```
#or
exp(spot1(12)*12)*pv_bondat0
```

```
[1] 156.7546
```

```
(ret_xyz<-log(end_fund_value/pv_bondat0)/12)
```

```
[1] 0.02833746
```

```
#or
spot1(12)
```

```
[1] 0.02833746
```

Thus, if we invest £1 m in this bond at $t = 0$, the estimated fund value at redemption, $t = 12$ years, is £1.405 m ($e^{12 \times 0.0283375} = 1.405$).

Of course, the actual bond fund value at redemption would in general be different to this, depending on changes to the yield curve and whether there have been any payment defaults (see Exercise 4.15 which calculates the fund value following a change in yields).

Exercise 4.11 *Verify the calculations in the above section with alternative methods.*

Exercise 4.12 *Calculate the GRY of the bond in Example 4.12, given the forward rates used in the example.*

4.7 The Effect of Future Yield Changes on Bond Prices Throughout the Term of the Bond

This section provides a brief introduction to Asset–Liability matching.

A frequently quoted rule is: 'bond prices will fall if interest rates rise' (see Figure 4.1). Although a useful statement, it can be misleading. The following example traces the price of a zero-coupon bond (ZCB) throughout its lifetime, and provides an introduction to the concept of managing interest rate risk by matching liabilities with particular assets.

Example 4.13

A company has a contractual payment equal to £100 m due in 10 years. The yield curve at $t = 0$ is flat at $i = 3\%$ p.a. Based on this data the Finance Director has proposed purchasing ZCBs to the value of £74.4 m which he has calculated will be sufficient to meet this liability in 10 years, given an effective return of 3% p.a.

Three different bond strategies are proposed: (1) to buy a 10-year ZCB, (2) to buy a 20-year ZCB and sell this at $t = 10$ and (3) to repeatedly buy one-year ZCB each year, for ten years. A colleague has suggested it does not matter which strategy is adopted as 'returns from bonds are guaranteed (if we ignore defaults)'.

Demonstrate the path of the bond fund value between $t = 0$ and $t = 10$ years under each of these strategies, assuming the flat yield curve rises during this period from 3% p.a. to 8% p.a. (uniformly, at 0.5% each year). Also consider the opposite scenario where the flat yield curve decreases over the period (from 8% p.a. to 3% p.a.).

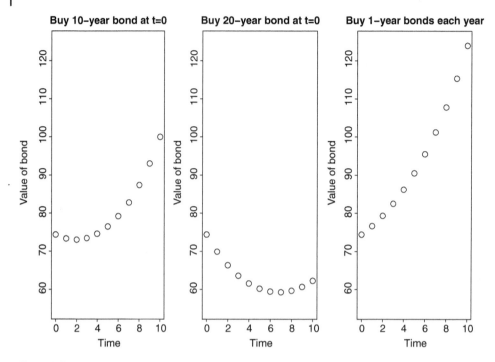

Figure 4.3 Effect on bond price from increasing interest rates over 10 years.

The results are plotted in Figures 4.3 and 4.4. We can see that the main risks are buying long bonds (20-year bonds) with rates subsequently increasing, and buying short bonds (1-year bonds) with rates subsequently falling in the future; in both cases we have insufficient assets at $t = 10$ ($A < 100$ m). If we buy the 10-year bond we are immune to interest rate changes. Hence, our colleague is mistaken; it does matter which bond is purchased.

It is important to note that if the bonds involve coupon payments the results are slightly altered as the coupons received throughout the lifetime of the bond receive different rates of return. If we invest in a 10-year coupon-bearing bond and rates subsequently fall, the value of the asset at $t = 10$ will be (slightly) less than the liability as the coupons will have been reinvested at lower rates – this is known as 'reinvestment risk'.

This important topic is discussed in significantly more detail in later chapters.

Exercise 4.13 *A pension scheme invests £60 m in UK gilts by purchasing £30 m in each of two UK gilts, each paying 6% annual coupons, with redemption at par in 3 and 10 years' time. The current yield curve is flat and is assumed to remain so (only parallel movements in the yield curve are permitted under this model). The effective rate of interest is currently 5% p.a., and in one year's time is assumed to have the following distribution: $i \sim N(0.05, 0.015^2)$.*

Simulate bond fund valuations in one year's time, and estimate the probability that the fund will have fallen in value over the year. Assume the next coupons received will be at $t = 1$ year. A plot of the simulated fund values at $t = 1$ is shown in Figure 4.5. Comment on the appropriateness of this model.

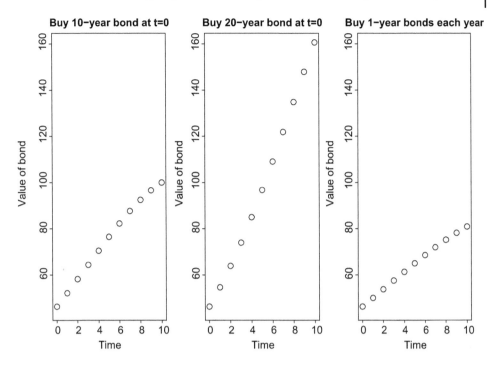

Figure 4.4 Effect on bond price from decreasing interest rates over 10 years.

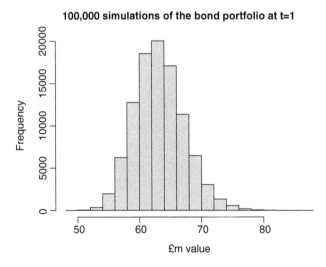

Figure 4.5 Simple bond valuation model: interest rate risk.

Exercise 4.14 *(Continuation of Exercise 4.13.) Instead of purchasing the 10-year bond the manager can alternatively buy a longer term bond, but is committed to purchase the 3-year bond. Determine the longest term bond he is permitted to buy such that there is less than 1% probability that the total fund value is less than £52 m at t = 1 year.*

Exercise 4.15 *Use the coupon-bearing bond in Example 4.12 for this exercise. This bond has been issued on 1 January 2022. An institution has invested £10 m in this bond at issue (i.e. 1 January 2022). The yield curve on 1 January 2022 is the same as that set out in Section 4.6, and changes as expected over the following month until 1 February 2022, when it increases by 1% at all terms, that is, to:*

$$\delta_t = 0.04 - 0.01e^{-0.5t}$$

Using these yield curves, estimate the value of this £10 m investment between issue and redemption on 1 January 2034, and plot a graph showing the change in the estimated bond fund value between 2022 and 2034. Assume coupons achieve returns in line with the yield curve.

4.8   Loan Schedules

4.8.1   Introduction

This section sets out an example which provides further practice for readers new to coding.

When a bank lends money to an individual or company, there will follow a series of repayments of the loan from the borrower to the bank. These repayments may take many forms. For example, the repayments may be for a fixed amount for the entirety of the loan period; alternatively each repayment instalment may be in respect of the interest due together with a final payment at the end of the loan period equal to repayment of the capital amount. Another example would involve a single repayment in the future equal to the initial loan amount plus accrued interest.

In this section we will look at the first of the above examples – fixed, level payments for an agreed period of time, where the repayments are calculated using a particular fixed rate of interest. As the loan is gradually repaid over the course of the loan the amount of loan outstanding will reduce until the final payment is made and no loan is outstanding. The bank will regularly monitor its total loans outstanding.

The details of the payments (both past and future) together with loan outstanding are recorded in the loan schedule. Our aim in this section is to write code which calculates all the items in a typical loan schedule. We adopt two approaches: Method 1 takes us through each step in the loan schedule methodically, whilst Method 2 involves much less code.

Example 4.14

Mary borrows £1,000,000 from the bank to buy a house. The effective interest rate charged by the bank is 8% per annum and the entire loan is to be repaid over 25 years with equal, annual instalments paid at the end of each year.

First define these parameters:

```
loan<-1000000;  interest_rate<-0.08;  term<-25
```

To calculate the annual repayments the bank must equate the £1,000,000 loan with the present value of the annual future repayments from the borrower, X. Note that in this example we shall ignore any expenses that the bank will incur (e.g. employee salaries) and profit requirements. Thus we need to solve:

$$X \, a_{\overline{25}|} = £1,000,000 \quad @i = 8\% \, \text{p.a.}$$

Making use of our earlier `annuity_certain` function (from Section 4.2), we can calculate the annual payment required under this loan agreement:

```
(payment_amount<-round(loan/annuity_certain(interest_rate,term),2))
[1] 93678.78
```

Thus Mary will be required to pay £93,678.78 to the bank at the end of each year.

Exercise 4.16 *Perform a reasonableness check on this calculation.*

4.8.2 Method 1

Before we write the code using Method 1 we will first set out clearly what the entries for the first year of our schedule should be. Prior to the first interest payment at $t = 1$ the loan will have grown to £1,080,000, due to the effective interest rate of 8% p.a.

If only £80,000 was paid at $t = 1$ there would remain £1,000,000 outstanding at $t = 1$. These annual payments could be continued, such that the outstanding loan amount at the end of every year would always be £1,000,000, and the loan may never be paid off. (This type of loan contract is known as an interest-only loan, where only the interest amount is paid of periodically; the capital is paid off at the end of the loan period.) However, the amount paid at the end of the year is £93,678.78 (as required from our equation of value above). The idea is that this excess yearly payment reduces the amount of loan outstanding each year. The outstanding loan amount at the end of year 1 is therefore:

$$£1,000,000 - (£93,678.78 - £80,000) = £986,321.22$$

Now let's write the code to calculate the entries for all 25 years in our loan schedule, first defining the loan amount outstanding at the beginning of each year, the interest due on that loan and the amount of the loan which has been paid off during the year.

Remark 4.4 Please note that simpler code than the following can be written to obtain a full Loan Schedule – see Method 2. The objective of this chapter is to practise writing code, and not, for the time being at least, to write the most efficient code.

```
loan_outstanding<-NULL;   capital_paid_off<-NULL;   interest_payment<-NULL

for (t in 1:term) {
loan_outstanding[1]<-loan #set the initial loan
interest_payment[t]<-interest_rate*loan_outstanding[t]
capital_paid_off[t]<-payment_amount-interest_payment[t]
loan_outstanding[t+1]<-loan_outstanding[t]-capital_paid_off[t]
```

```
interest_payment[term+1]<-0
capital_paid_off[term+1]<-0
}
#end of loop
loan_outstanding<- round(loan_outstanding,2)
(cum_cap_paid<-cumsum(capital_paid_off)[-26]) #plotted

 [1]    13678.78    28451.86    44406.79    61638.11    80247.94   100346.56
 [7]   122053.06   145496.09   170814.56   198158.50   227689.96   259583.94
[13]   294029.43   331230.57   371407.79   414799.20   461661.91   512273.65
[19]   566934.32   625967.84   689724.05   758580.75   832945.99   913260.45
[25]  1000000.07
```

From Figure 4.6 we can see how the payments relate to paying off the loan over 25 years, and that the total initial loan of £1 m has been paid off after 25 years (as expected).

Finally it's best practice to write the above code for the Loan Schedule as a function.

Exercise 4.17 *Write a Loan Schedule function, and apply it to the above scenario.*

Applying the function, schedule_function (code not shown here), it is a simple matter of running it with the required parameter values:

```
payments_x<-schedule_function(0.08,25,1000000)
head(payments_x,3)    #extract of output

        t loan_outstanding interest_payment capital_paid_off payment_amount
[1,] 0         1000000.0          80000.00         13678.78       93678.78
[2,] 1          986321.2          78905.70         14773.08       93678.78
[3,] 2          971548.1          77723.85         15954.93       93678.78

tail(payments_x,3)    #extract of output

         t loan_outstanding interest_payment capital_paid_off payment_amount
[24,] 23        167054.06         13364.325         80314.45       93678.78
[25,] 24         86739.61          6939.169         86739.61       93678.78
[26,] 25             0.00             0.000             0.00           0.00
```

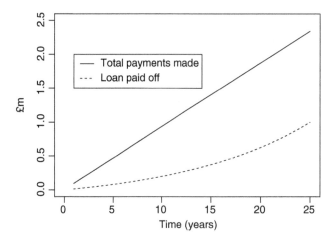

Figure 4.6 Effect of payments made under Loan Schedule.

If we wish, it is straightforward to export the output to a `.csv` file or Excel, using `write.csv(payments_x,path/file.csv)`. (See Chapter 1.15.2.)

Writing a function is just neater and easier to use for multiple cases. For example a bank may have thousands of loans, each with different interest rates and time left to repayment completion, and want to know the total loan capital outstanding today. If the above loan had been written five years ago we could easily find the outstanding capital:

```
payments_x[6,2]   #picks out the 6th row and 2nd column

loan_outstanding
        919752.1
```

Exercise 4.18 *(challenging) A bank has written 100 loans over the last four years (see data set 'Loans100forexercise.csv') and wishes to calculate the total loan outstanding on this book of loans as at 1 January 2021.*

Calculate the total loan amount outstanding on the 100 loans on 1 January 2021. Also calculate the estimated amounts outstanding at the end of each of the next 25 years.

(Tip: using the apply function can simplify your code.)

The total loans outstanding on 1 January 2021 were calculated to be £46,933,798. The predicted changes in loans outstanding over the next 25 years are plotted in Figure 4.7.

Exercise 4.19 *Re-write the above code such that it is applicable for loans with monthly payments.*

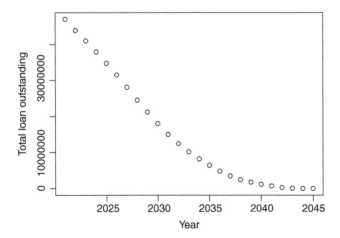

Figure 4.7 Solution from Exercise (outstanding amount on 100 loans).

With more examples the reader will hopefully see the potential advantages of a programming language like R over Excel:

- With R (or similar language) we only need a few lines of code with the necessary inputs. Indeed, calling the function requires only one line of code.
- There is no danger of large spreadsheets being corrupted with incorrect entries, deleted formulae, errors resulting from incorrect copying of cells etc. Experienced Excel users will only be too aware of cell corruption issues.
- The spreadsheet may require adjustments to allow for different lengths of loan period.
- If you have hundreds of data sheets you may need hundreds of calculation spreadsheets.
- Running large programs in R will be significantly quicker than running them in Excel.

4.8.3 Method 2

Of course, with the particular example of a simple Loan Schedule we can easily obtain exact solutions without compiling a full schedule.

It is straightforward to show that, for a loan of term n with level payments of X:

1. the capital outstanding at time t (immediately after the payment at time t), L_t, is equal to $X\, a_{\overline{n-t}|}$;
2. capital repaid in the payment at time t (immediately after payment at that time) is equal to $L_t - L_{t-1}$;
3. the interest paid equals the excess amount of payment, that is $X - (L_t - L_{t-1})$.

For example, the loan outstanding at the start of the first three years, immediately after any payment made, is shown below:

```
payment_amount*annuity_certain(interest_rate,c(25,24,23))

[1] 1000000.0  986321.2  971548.1
```

The capital repaid in the second payment at $t = 2$:

```
payment_amount*annuity_certain(interest_rate,24)-
payment_amount*annuity_certain(interest_rate,23)

[1] 14773.08
```

And the interest paid in the second payment at $t = 2$ must be:

```
payment_amount - (payment_amount*annuity_certain(interest_rate,24)-
payment_amount*annuity_certain(interest_rate,23))

[1] 78905.7

# alternative calculation
payment_amount*annuity_certain(interest_rate,24)*interest_rate

[1] 78905.7
```

Exercise 4.20 *Using the above expressions, write a simple function to calculate the amount of loan outstanding at time t, together with the capital and interest elements included in each annual payment.*

The reader is encouraged to carry out the calculations in Section 4.8 in Excel, and compare the advantages of using R and Excel (and both methods) to solve these problems. As will often be the case in the workplace, a decision will be required regarding which software is most suitable in any particular situation.

4.9 Recommended Reading

- Booth et al. (2005) *Modern Actuarial Theory and Practice.* (2nd edition). Boca Raton, Florida, US: Chapman and Hall/CRC.
- Garrett SJ. (2016) *An Introduction to the Mathematics of Finance: A Deterministic Approach.* Oxford, UK: Butterworth Heinemnan.
- McCutcheon and Scott. (1986) *An Introduction to the Mathematics of Finance.* Oxford, UK: Butterworth Heinemnan.

5

Fundamental Statistics: A Selection of Key Topics

Dr Alfred Kume

```
library(moments)
library(psych)
library(tseries)
library(nortest)
library(MASS)
```

5.1   Introduction

This chapter covers many of the key statistical techniques used in this book, and loosely follows some of the key concepts set out in various introductory actuarial and statistical resources. In particular, a number of topics discussed are included in statistics modules in the first year of a typical undergraduate maths-related degree. Most topics are covered relatively briefly, with the use of examples highlighting where common misunderstandings often arise. As with several chapters in this book, we recommend that the reader consults a standard statistical textbook. A number of recommended texts are included at the end of this chapter. Several references are made in this chapter which will be relevant to other parts of the book. A certain amount of background knowledge is assumed however, such as probability definitions and basic statistical theory regarding concepts of the mean and variance of a random variable.

5.2   Basic Distributions in Statistics

In general, the random variables used in probability and statistics are of two types: discrete and continuous. A random variable X is discrete if it takes only a finite number of possible values with positive probability, as in the case of the Binomial distribution with index n and probability p:

R Programming for Actuarial Science, First Edition. Peter McQuire and Alfred Kume.
© 2024 John Wiley & Sons Ltd. Published 2024 by John Wiley & Sons Ltd.
Companion Website: www.wiley.com/go/rprogramming.com.

$$P(X = k) = \binom{n}{k}p^k(1 - p)^{n-k} \quad k = 0, 1, 2, \cdots, n$$

or countable number of values as in the case of the Poisson distribution:

$$P(X = k) = e^{-\lambda}\frac{\lambda^k}{k!} \quad k = 0, 1, 2, 3, \cdots$$

Other commonly used discrete distributions include Negative Binomial, Geometric, Hyper-Geometric. Examples of continuously distributed random variables include Normal, t, F, Gamma, and subfamilies such as Chi-square, Exponential, Weibull, log-Normal, Beta, etc. Be aware that some distributions (e.g. Exponential, Gamma) have more than one type of parameterisation and their implementation requires some care. For example, the pdf of Gamma distribution is shown below in both versions:

$$f(x) = \frac{1}{s^\alpha \Gamma(\alpha)}x^{\alpha-1}e^{-x/s} = \frac{\beta^\alpha}{\Gamma(\alpha)}x^{\alpha-1}e^{-\beta x} \quad x > 0, \quad \beta = \frac{1}{s}$$

There also exist random variables of a mixed type, partly continuous and partly discrete. The Compound Poisson distribution, which is discussed extensively in Chapter 25, is such an example.

The main standard distributions are set out in Table 5.1. For any particular distribution, for example gamma, norm, pois we simply prefix one of the letters r,p,d,q to produce any of the following quantities:

r Random sample generator,
p Cumulative distribution functions (cdf) i.e. $F(x) = P(X \leq x)$,
d Density function $f(x)$,
q Quantile function, the inverse of the cdf, F^{-1}.

Some examples, with their output are set out below:

```
rgamma(5,shape=4,scale=0.6)
[1] 1.290986 2.554305 2.126349 3.196666 2.907644
rexp(3,rate=6)
[1] 0.2899345 0.1253313 0.1214136
pnorm(c(-1.96,-1.64,0,2.33),0,1)
[1] 0.02499790 0.05050258 0.50000000 0.99009692
dgamma(1:5,shape=4,scale=0.6)
[1] 0.24289558 0.36701639 0.23395649 0.10474352 0.03863964
qnorm(seq(0.05,0.95,by=.1),mean=0,sd=1)
[1] -1.6448536 -1.0364334 -0.6744898 -0.3853205 -0.1256613  0.1256613
[7]  0.3853205  0.6744898  1.0364334  1.6448536
```

Table 5.1 Some distributions and their use in R.

	Distribution name	R code	Parameters	Sampling 10 for some parameters
1	Binomial	binom	size,probability	rbinom(10,size=3,prob=0.6)
2	Poisson	pois	rate: lambda	rpois(10,lambda=6)
3	Geometric	geom	prob	rgeom(10,prob=0.6)
4	Uniform	unif	min,max	runif(10,min=0,max=1)
5	Normal	norm	mean,sd	rnorm(10,mean=0,sd=1)
6	F	f	df1,df2	rf(10,df1=1,df2=1)
7	Chi-Square	chisq	df	rchisq(10,df=1)
8	Exponential	exp	rate	rexp(10,rate=1)
9	Student-t	t	df	rt(10,df=1)
10	Weibull	weibull	shape,scale	rweibull(10,shape=1,scale=1)
11	LogNormal	lnorm	meanlog,sdlog	rlnorm(10,meanlog=0,sdlog=1)
12	Negative Binomial	binom	size,probability	rnbinom(10,size=4,prob=0.6)
13	Gamma	gamma	shape,size	rgamma(10,shape=4,scale=0.6)
14	Beta	beta	shape1,shape2	rbeta(10,shape1=4,shape2=1.62)
15	Cauchy	cauchy	location,sscale	rcauchy(10,location=4,scale=1/2)

Such coding syntax is used extensively throughout the book. Please note that the parameter values used above are for illustrative purposes only and the user can vary them according to the distribution of interest. These examples are easily extendable to any standard distributions; some of which are shown in the Table 5.1.

Note that parameter names would not usually be included in the code, for example rgamma(10,4,0.6) – it is only shown above for completeness.

One can explore the shape of these probability density functions by either generating histograms of many simulated observations or by plotting the density functions. In Example 5.1 we plot a histogram of simulated values from a Gamma(2, 0.5), and also its density function:

```
Example 5.1 x=rgamma(1000,2,0.5)        #simulate 1000 observations
xval=seq(0,15,length=50)  #specify the range of x values
hist(x);                  #plot histogram
plot(xval,dgamma(xval,2,0.5),type="l",lwd=2,main="Gamma density
function",xlab="x")
```

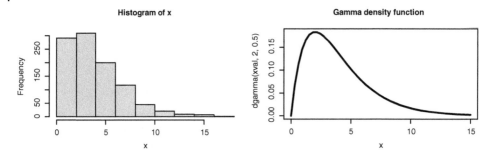

Figure 5.1   Histogram of simulated values from a gamma distribution and its density function.

Alternatively, we can obtain the respective empirical (using the simulated data above) and the theoretical cumulative distribution functions:

```
Example 5.2 plot.ecdf(x,main="From simulations")
#ecdf of the simulated data
xval=seq(0,15,length=50);       plot(xval,pgamma(xval,2,0.5),
type="l",lwd=2,main="Theoretical values")
```

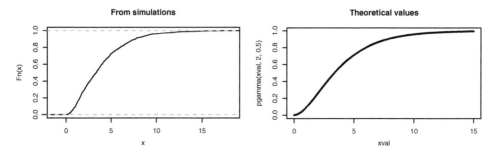

Figure 5.2   Empirical and theoretical cumulative distribution functions from a gamma distribution.

Such tools can also be used if one wants to compare/fit the appropriate distributions to certain data.

Exercise 5.1

- *Perform a similar graphical output for the normal distribution $N(\mu = 2, \sigma^2 = 4)$.*
- *Perform a similar graphical output for the t distribution with degrees of freedom $n = 7$.*
- *Perform a similar graphical output for the Poisson distribution with rate $\lambda = 5$.*

Example 5.3 (Poisson Process) We can simulate a realisation from a Poisson process by simply using the exponential arrival times:

```
set.seed(230);    x=rexp(n=30,rate=2)
plot(cumsum(x),1:30,type = 's',xlab="",ylab="",xlim=c(0,15),
ylim=c(0,35))
points(cumsum(x),1:30,pch=1)
```

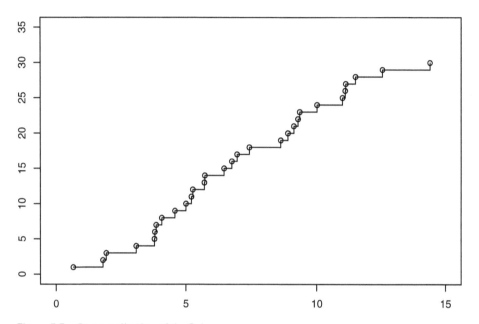

Figure 5.3 Some realisation of the Poisson process.

Example 5.4 (Normal approximation of t distributions) It is known that t-distribution approaches the standard normal for the degrees of freedom $n > 30$. Confirm this by plotting the standard Normal density together with the t_n-density for $n = 2, 10$, that is 3 curves.

```
x=seq(-4,4,by=0.1);              plot(x,dnorm(x),type="l",lwd=2)
lines(x,dt(x,df=2),lty=2);       lines(x,dt(x,df=10),lty=3)
legend(x="topright",legend=c("Normal","t10","t2"),lty=c(1,3,2),
lwd=c(2,1,1))
```

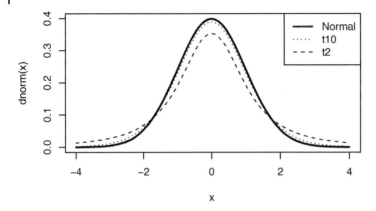

Figure 5.4 Normal and t distributions.

Remark 5.1 Note that we will, in general, not show code in relation to plotting figures; the code tends to be similar and its inclusion leads to a less favourable reading experience. All code is however included in the book's website (www.wiley.com/go/rprogramming.com).

Example 5.5 (A demonstration of the Central Limit Theorem (CLT)) As a standard derivation of the CLT, the distribution of the sample mean estimator $\bar{x} = \frac{\sum_{i=1}^{n} x_i}{n}$ follows the normal distribution with mean μ and variance $\frac{\sigma^2}{n}$, where μ and σ^2 are the corresponding population quantities, namely, $\sqrt{n}(\bar{x} - \mu) \sim N(0, \sigma^2)$ or $\sum_{i=1}^{n} x_i \sim N(n\mu, n^2\sigma^2)$.

Here we can demonstrate this for the standard Uniform distribution $U(0,1)$ ($\mu = 1/2$ and $\sigma^2 = 1/12 = 0.08333333$), by fixing $n = 500$ and generating $m = 3000$ random values of $\bar{x}$, each generated from $n = 500$ realisations x, and showing their histogram jointly with the pdf of the asymptotic distribution $N(0, \frac{1}{12})$.

```
set.seed(123);n=500;m=3000;dd=vector(length=m)
for (i in 1:m){dd[i]=mean(runif(n))-1/2;dd[i]=dd[i]* sqrt(n)}
```

Exercise 5.2 *Confirm the CLT as above for the Binomial distribution with parameters: size $= 10$ and probability $p = 0.4$.*

Note that the CLT holds generally for cases where the population variance σ^2 is finite.

Exercise 5.3 *Cauchey distribution which is used for the modelling of very dispersed data has tail probabilities fatter than that of the normal resulting in an undefined variance.*

- *Use* rchauchy *to simulate large numbers of observations and confirm that their sample variance is not converging to any fixed value.*
- *Confirm therefore that the CLT does not hold for this distribution.*

There are several examples in this book where we apply the CLT in approximating the sum of random variables, and comparing the results with simulations, for example adding claim amounts across several departments of an insurance company and approximating the results with an appropriate Normal distribution.

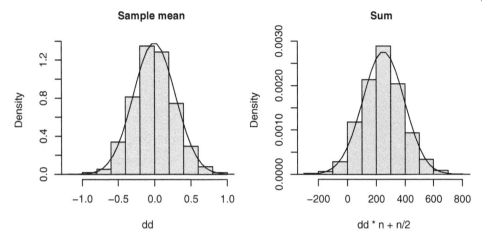

Figure 5.5   CLT for sample mean of U(0,1).

5.3   Some Useful Functions for Descriptive Statistics

5.3.1   Introduction

Consider the dataframe trees from the datasets package. First we change the variable name Girth to Diameter:

```
colnames(trees)[1]<-"Diameter"   #preferred name
```

A simple summary can be obtained:

```
summary(trees)

    Diameter        Height        Volume
Min.   : 8.30   Min.   :63   Min.   :10.20
1st Qu.:11.05   1st Qu.:72   1st Qu.:19.40
Median :12.90   Median :76   Median :24.20
Mean   :13.25   Mean   :76   Mean   :30.17
3rd Qu.:15.25   3rd Qu.:80   3rd Qu.:37.30
Max.   :20.60   Max.   :87   Max.   :77.00
```

One could obtain the particular quantities for only one variable using mean (trees[,1]), var(trees[,1]) and quantile(trees[,1]) (in this case for Diameter). Alternatively we can use mean(trees$Diameter). The following summary plots can be useful tools for initial exploration of the data:

```
par(mfrow=c(1,3));hist(trees$Diameter);hist(trees$Height);boxplot(trees)
```

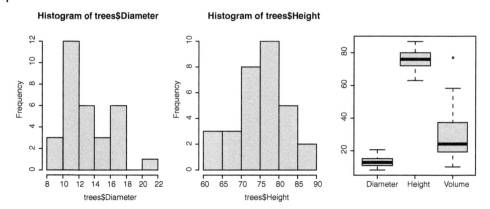

Figure 5.6 Histograms and a box plot.

5.3.2 Bivariate or Higher Order Data Structure

Analysing and understanding the dependence between variables is crucial in developing accurate and appropriate models. Many examples throughout the book will analyse such dependencies.

The dependence between the three variables in the trees dataset can be explored using pairwise plots such as plot(trees$Diameter,trees$Height), or a more complete set of pairwise plots using plot(trees) (see Figure 5.7, left). However, a more informative graphical output, including fitted regression lines, histograms, and pairwise correlation values (see Figure 5.7, right) is obtained using pairs.panels from the psych package:

```
plot(trees);pairs.panels(trees,show.points=TRUE, lm=TRUE)
```

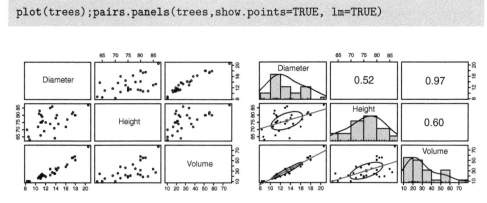

Figure 5.7 Two options for pairwise plots.

As it can be seen graphically, the variables Diameter and Volume have a strong linear relationship. This pairwise dependence is also confirmed from the (Pearson) correlation value using cor(trees$Diameter,trees[,3],method="pearson") which confirms the value 0.9671 shown at the top right corner of Figure 5.7 (right).

Note that for a pair of data vectors, x and y, the Pearson correlation coefficient, r, is defined as:

$$r = \frac{S_{XY}}{\sqrt{S_{XX}S_{YY}}} \tag{5.1}$$

where

$$S_{XY} = \sum (x_i - \bar{x})(y_i - \bar{y}) \quad S_{XX} = \sum (x_i - \bar{x})(x_i - \bar{x}) \quad S_{YY} = \sum (y_i - \bar{y})(y_i - \bar{y}) \tag{5.2}$$

Such a measure of dependence is strongly related to the slope parameter in the linear regression model *Diameter* $= \alpha + \beta \times Volume$, where $\hat{\beta} = 0.1846321 =$ lm(trees$Diameter~trees[,3])$coeff[2] (see later for a more detailed description of such a model and the use of lm function).

Exercise 5.4 *Run the following code:*

```
Sxx=sum((trees$V-mean(trees$V))^2);        Syy=sum((trees$D-mean(trees$D))^2)
Sxy=sum((trees$V-mean(trees$V))*(trees$D-mean(trees$D)));     Sxy/(Sxx*Syy)^0.5
lm(trees$Diameter~trees[,3])$coeff[2]*(Sxx/Syy)^0.5
```

and convince yourself that

- *the Pearson correlation is the same as* Sxy/(Sx*Sy)^0.5.
- *the regression coefficient, β, and the Pearson correlation measure differ by the factor* $\sqrt{\frac{S_{xx}}{S_{yy}}}$.

An alternative measure of dependency is discussed in Chapter 13.

5.4 Statistical Tests

In this section we outline a number of key statistical tests.

5.4.1 Exploring for Normality or Any Other Distribution in the Data

Testing for normal distribution
While there are many methods for checking whether some data is generated or best described by some particular distribution (e.g. Normal), initially a visual check using qqnorm and qqline is appropriate:

```
qqnorm(trees$Diameter);       qqline(trees$Diameter)
```

Normal Q–Q

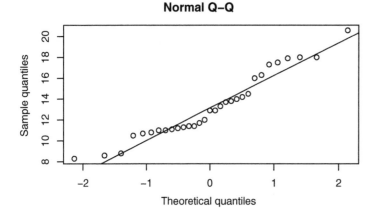

Figure 5.8 Exploring sample quantiles with those of the standard Normal distribution.

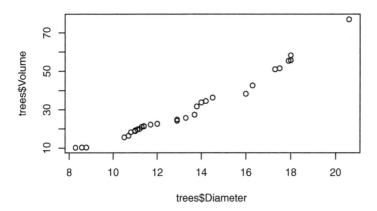

Figure 5.9 Exploring sample quantiles for each variables.

The closer these dots to the line the more evidence of normality is present. Note that this visual approach is appropriate for any type of known distribution (using qqplot). In particular, we can compare whether two particular data sets are of the same distribution:

```
qqplot(trees$Diameter,trees$Volume)
```

It seems that Diameter and Volume data are similarly distributed as the qqplot seems close to linear. One can also formally test for normality using various tests, such as the Shapiro test:

```
set.seed(14);  x=rnorm(300);  shapiro.test(x)  #test 1, a random dataset


Shapiro-Wilk normality test

data:  x
W = 0.99626, p-value = 0.702

shapiro.test(trees$Diameter)                   #test 2, tree diameters data


Shapiro-Wilk normality test

data:  trees$Diameter
W = 0.94117, p-value = 0.08893
```

In the first test, the p-value is 0.702, so we do not have evidence against the normality assumption (as expected). In the second run for the data `trees$Diameter`, the p-value is 0.089. This suggests that there is some evidence against the normality assumption (at 10% significance level we reject the Null assumption of normality). The Shapiro test is preferred in the situations where there are not any parameters of the normal distribution specified.

Additional tests for normality are available – some of them can be found in the `nortest` library. For example, one can use Anderson-Darling test as:

```
ad.test(x); ad.test(trees$Diameter);


Anderson-Darling normality test

data:  x
A = 0.26585, p-value = 0.6895

Anderson-Darling normality test

data:  trees$Diameter
A = 0.7455, p-value = 0.04668
```

This test is adapted from the Kolmogorov-Smirnov test in order to account for the tail behaviour of the data. The Kolmogorov-Smirnov test (`ks.test`) is also applicable for any type of distribution provided it is clearly specified in its arguments. In particular, the adjustments for Binomial and Poisson distributions are made using `ks.test(y,pbinom,size = 100, prob= 0.4)` or `ks.test(y,ppois,lambda=4)` for some count data y.

Testing for normality of our simulated data x we use:

```
ks.test(x,"pnorm",0,1)          #compare x with the N(0,1)
ks.test(trees$Volume,"pnorm",mean(trees$Volume),var(trees$Volume))
#compare the data in Volume with the standard normal with the matching moments of
the data
```

Exercise 5.5 *Run the above code and interpret the output. Also run the following code:* `ks.test(x, pgamma, shape=4, rate=2)`.

Exercise 5.6 *For some count data* `xpois<-c(3,4,6,12,14,3,7,5,4,7,9,12)`, *perform the Kolmogorov-Smirnov test using* `ks.test`, *for testing whether the Poisson distribution with mean* `mean(xpois)` *is appropriate.*

One can also perform normality testing using `jarque.bera.test` from the `tseries` package. For example:

```
jarque.bera.test(x);jarque.bera.test(trees$Diameter)
```

Each of the tests mentioned have specific strengths in testing for normality. For example, `jarque.bera.test` uses the skewness and kurtosis measures of deviations from normality. The same test can be run using `jarque.test` from the `moments` package. We recommend that you use more than one of the available tests for your practical problem and pay special attention to the underlying assumptions for each of the tests employed.

5.4.2 Goodness-of-fit Testing for Fitted Distributions to Data

These tests measure the closeness of the data to some target distribution of interest. (The Normal distribution was the specific target distribution in Section 5.4.1.)

5.4.2.1 Continuous distributions

For continuous data one can use the `ks.test` to check whether certain data can be appropriately explained by some theoretical distribution. In the following, we simulate some data form the t_3-distribution and compare it with both the t_3 and the standard normal distribution:

```
set.seed(123); x=rt(n=100,df=3); ks.test(x,"pt",3); ks.test(x,"pnorm");

Asymptotic one-sample Kolmogorov-Smirnov test

data:   x
D = 0.051736, p-value = 0.9517
alternative hypothesis: two-sided

Asymptotic one-sample Kolmogorov-Smirnov test

data:   x
D = 0.10002, p-value = 0.2698
alternative hypothesis: two-sided
```

Note that the null hypothesis for both tests of these are not rejected. This is not surprising since t and normal distributions are not far from each other. (The reader should re-run the test with $n = 1,000$.) However, for the gamma distribution we have a different conclusion as in `ks.test(x,"pgamma", shape=1)`; with p-value: 0.

5.4.2.2 Discrete distributions

The examples above relate to inference with continuous distributions. For cases when we need to test whether data are from a particular discrete distribution, we can apply the chi-squared test:

$$\chi^2 = \sum_{i=1}^{k} \frac{(E_i - O_i)^2}{E_i} \sim \chi^2_{k-p} \tag{5.3}$$

where E_i and O_i are the estimated and observed frequencies of some particular range of possible values, and p is the number of estimated parameters. For example, we may wish to assess whether mortality rates derived from data are similar to rates from a recognised, previously tested, table of mortality rates (see Chapter 18). Below we generate 120 data points from a Poisson ($\lambda = 2$) distribution:

```
set.seed(234);xx<-rpois(120,lambda=2);freq_table=table(xx) #generates the table of
frequencies
mean(xx)

[1] 1.95
```

Let us assume that the true rate is not known but we will fit to this data a Poisson distribution with $\lambda = \bar{x} = 1.95$. We calculate the observed and expected frequencies as vectors in TT of the fitted Poisson distribution with $\lambda = 1.95$:

```
SS=t(as.matrix(freq_table))              #generate the frequency table
y=as.numeric(dimnames(freq_table)[[1]])  #extract the possible drawn values
Ex=dpois(y,lambda = mean(xx))*length(xx) #calculate expected
Ex[7]=(1-ppois(5,lambda = mean(xx)))*length(xx);(TT=rbind(SS,Ex)) #combine
obs and exp

          0        1        2        3        4        5        6
    11.00000 36.00000 36.00000 25.00000 10.00000 1.000000 1.000000
Ex  17.07289 33.29213 32.45983 21.09889 10.28571 4.011426 1.779125
```

Note that the χ^2 test is advised to run only the expected frequencies that are greater than 5, therefore we will group together entries which correspond to values 4, 5, and 6:

```
TT1=cbind( TT[,1:4],TT[,5]+TT[,6]+TT[,7]);TT1  #adjusted table for test

          0        1        2        3        4
    11.00000 36.00000 36.00000 25.00000 12.00000
Ex  17.07289 33.29213 32.45983 21.09889 16.07626
```

Now we generate the chi-square statistic for $k = 5$ and $p = 1$ and compare it with χ^2_4:

```
chisq.test(TT1[1,],p=TT1[2,]/120)      #TT1[2,]/120 generates the vector of
probabilities


Chi-squared test for given probabilities

data:  TT1[1, ]
X-squared = 4.5214, df = 4, p-value = 0.34
```

(Note that the value of the statistic agrees with sum((TT1[1,]-TT1[2,])^2/TT1[2,])= 4.5214 using Equation 5.3.) The p-value suggests that the Poisson distribution is appropriate.

Exercise 5.7 *Change entries of the original vector* SS *such that H_0 could be rejected.*

See Chapter 18 for a practical application of this test relating to mortality rates.

5.4.3 T-tests

In statistical theory the basic tests are those related to the presence of some non-zero parameter θ in the model. Generally speaking the standard hypothesis testing $H_0 : \theta = 0$ versus $H_1 : \theta \neq 0$ is based on the statistics

$$t\text{-value} := \frac{\hat{\theta}}{s.e(\hat{\theta})}$$

where $s.e(\hat{\theta})$ represents the standard error of the estimator $\hat{\theta}$, evaluated in the estimation process (see the likelihood section for example). The smaller the relative value of $s.e(\hat{\theta})$ with respect to $\hat{\theta}$, the higher the t-value and the weaker is the evidence for $H_0 : \theta = 0$. We compare the resulting t-value against t_1, the t-distribution with 1 degrees of freedom. Normally, the t-value above 2 is considered as sufficient evidence against the null hypothesis H_0. Note that the t_n distribution has a parameter n, which represents the degrees of freedom. This distribution has fatter tails than the standard normal but as n increases this distribution gets closer to that of the standard normal (see Example 5.4 in page 91). If $n \geq 30$ there is practically no difference with the standard normal distribution.

5.4.3.1 One sample test for the mean

If one wanted to test the hypothesis $H_0 : \mu = \mu_0$ against $H_1 : \mu \neq \mu_0$ we can run the t-test using the statistics

$$t = \frac{\bar{x} - \mu_0}{\hat{\sigma}/\sqrt{n}} \quad \text{where} \quad \hat{\sigma}^2 = \frac{S_{xx}}{n-1}$$

and compare it with t_{n-1}, the t-distribution with $n-1$ degrees of freedom. Note that $\hat{\sigma}^2$ is an estimate of the sample variance. We perform this test in R:

```
set.seed(23);    x=rnorm(30,mean=0);    t.test(x,mu=3)


One Sample t-test

data:  x
t = -18.545, df = 29, p-value < 2.2e-16
alternative hypothesis: true mean is not equal to 3
95 percent confidence interval:
 -0.1562716  0.4707449
sample estimates:
mean of x
0.1572366
```

The p-value for $H_0 : \mu = 3$ is obtained from the output above, or more specifically, from t.test(x,mu=3)$p.v= 1.258×10^{-17}. This is extremely small; we can therefore be very confident in rejecting H_0.

If, however, $H_0 : \mu = 0$, we have a different outcome with a p-value equal to 0.3134841 (running t.test(x,mu=0)).

The $1 - \alpha$ confidence intervals for the mean are defined as

$$(\bar{x} - \frac{\hat{\sigma}}{\sqrt{n}} t_{n-1,\alpha/2}, \bar{x} + \frac{\hat{\sigma}}{\sqrt{n}} t_{n-1,\alpha/2})$$

Note that this t.test has many settings, for example the default confidence interval is conf.level = 0.95, but can be set as required, e.g. t.test(x,mu=3, conf.level = 0.90). The corresponding p-value here is two-sided, but we can run the one-sided test for $H_1 : \mu < 3$ as t.test(x,mu=3, alternative = "less"). The reader should verify all the above output.

5.4.3.2 Two sample tests for the mean

There are situations when the analyst wants to test equality in the means for two popula-tions, for example analyse whether claim sizes are distributed with the same mean from two particular geographical regions. For two samples from these populations with sam-ple means $\bar{x}_1$ and $\bar{x}_2$, sample variances $\hat{\sigma}_1^2$ and $\hat{\sigma}_2^2$ and number of observations n_1 and n_2 respectively, the t-statistic is defined as

$$t = \frac{\bar{x}_2 - \bar{x}_1}{\hat{\sigma}\sqrt{\frac{1}{n_1} + \frac{1}{n_2}}}$$

where $\hat{\sigma}^2$ is the pooled sample variance estimator

$$\hat{\sigma}^2 = \frac{(n_1 - 1)\hat{\sigma}_1^2 + (n_2 - 1)\hat{\sigma}_2^2}{n_1 + n_2 - 2}$$

Under the null hypothesis that the means of the populations from which the samples are drawn are equal, the t-statistic follows a t-distribution with $n_1 + n_2 - 2$ degrees of freedom.

For two sample data vectors, say x and y, one can test whether they come from distri-butions with the same mean as follows (note the implicit assumption of common variance estimated as $\hat{\sigma}^2$ above):

```
y=rnorm(30,mean=3);     t.test(x,y)          #R output is removed here
```

Here the test gives a p-value $= 2.364 \times 10^{-17}$, so we can reject H_0, and conclude that we are confident that x and y have different means. This test is in fact equivalent to the difference between the corresponding means is zero. It is this difference whose confidence interval, $(-3.3039606, -2.3653258)$, is provided in the output from t.test(x,y).

Exercise 5.8 *Re-run the above test but with $y \sim N(1,1)$, $y \sim N(0.2,1)$, and $y \sim N(0,1)$.*

If the one-sided tests are needed you could try `alternative = "less"` or `alternative = "greater"` if required.

5.4.4 F-test for Equal Variances

The assumption for equal variances is useful in the general ANOVA analysis in the data. Even in the two sample t-test for the mean this is a common assumption which requires initial investigation. Such tests are based on the F-statistics:

$$F = \frac{\hat{\sigma}_1^{\,2}}{\hat{\sigma}_2^{\,2}}$$

and compared to the critical values of the F-distribution F_{n_1-1,n_2-1}.

For example, in order to test whether `trees$Volume` and `trees$Diameter` have the same variance one can use:

```
var.test(trees$Volume, trees$Diameter,  alternative = "two.sided")


	F test to compare two variances

data:   trees$Volume and trees$Diameter
F = 27.438, num df = 30, denom df = 30, p-value = 1.51e-14
alternative hypothesis: true ratio of variances is not equal to 1
95 percent confidence interval:
 13.22966 56.90397
sample estimates:
ratio of variances
          27.43757
```

The p-value is practically zero providing strong evidence to reject $H_0 : \sigma_1 = \sigma_2$. The box plot in Figure 5.6 confirms visually that these data vectors do not have an equal variance.

Exercise 5.9 *Using the output of* `dim(trees)`, *confirm that the degrees of freedom for the F-test above are both* 30.

5.5   Main Principles of Maximum Likelihood Estimation

Remark 5.2 The method of Maximum Likelihood Estimation is used extensively throughout this book.

5.5.1   Introduction

Suppose that we have observed the value, y, of some continuous random variable, Y, whose probability density function, $f(y;\theta)$, is known in terms of some parameter, θ. If Y is discrete then $f(y;\theta)$ is naturally replaced by the distribution function, the probabilities associated at the data point y.

In general, y and θ can be both vectors as in linear regression models (see 5.7 in page 118) where data is the collection of observations of both dependent (response) variable Y and predictors(covariates) $X_1,...X_p$; and the vector parameter θ contains $\theta_0,\theta_1...\theta_p$. Note that we normally use the bold notation for vectors and matrix objects. The likelihood function

$$L(\theta) = f(Y;\theta) \quad \theta \in \Theta$$

is considered as a function of θ while data are considered fixed and Θ is the space of possible parameter values.

If we have a collection of independent observations $(y_1, y_2, ..., y_n)$ the likelihood function is defined

$$L(\theta) = \prod_{i=1}^{n} f(y_i;\theta) \quad \theta \in \Theta \tag{5.4}$$

The Maximum Likelihood Estimate, MLE, of θ is the particular value, $\hat{\theta}$, for which the probability of obtaining the observed data is the highest. Formally we can write:

$$\hat{\theta} = \arg\max_{\theta \in \Theta} L(\theta) = \arg\max_{\theta \in \Theta} \log L(\theta)$$

This maximisation can be done analytically in some cases, but also numerically if the analytical method is difficult. In the following we will consider the maximisation of the likelihood function related to the exponential distribution. This is illustrated both numerically and theoretically to demonstrate their equivalence.

Remark 5.3 (Estimate/Estimator) Note that the symbol ^ as in $\hat{\theta}$ corresponds to the numerical value, called estimate that is generated from some data points y_i. However, if $\hat{\theta}$ is seen as a function of some collection of random variables y_i, then $\hat{\theta}$ is also a random variable called estimator.

5.5.2 MLE of the Exponential Distribution

Let $y = (y_1, ..., y_n)$ be observations from the exponential distribution, with density function:

$$f(y;\theta) = \theta^{-1}e^{-\frac{y}{\theta}}, y > 0, \theta > 0$$

The resulting likelihood function is therefore:

$$L(\theta) = \prod_{i=1}^{n} \theta^{-1}e^{\frac{y_i}{\theta}} = \theta^{-n}e^{-\frac{\sum_{i=1}^{n} y_i}{\theta}} = \theta^{-n}e^{-n\frac{\bar{y}}{\theta}} \quad \theta > 0 \tag{5.5}$$

Example 5.6 Spring Failure Times (Davidson 2009)

The strengths of a number of mechanical springs were tested with a stress of 950 N/mm2; the spring failure times (in seconds) are:

$$y = (225, 171, 198, 189, 189, 135, 162, 135, 117, 162); \quad \bar{y} = 168.3, n = 10$$

We now fit the exponential distribution to this data, determining the MLE both numerically and analytically.

5.5.2.1 Obtaining the MLE numerically using R

In the following code we define the likelihood function, $\mathcal{L}$, as per Equation (5.5) and plot $\mathcal{L}$ against θ over the range $1 < L < 500$. We repeat the calculations replacing L with log L. Both of these functions indicate a mode at around 168 (see Figure 5.10). Note also the efficient use of the apply function below:

```
y=c(225, 171, 198, 189, 189, 135, 162, 135, 117, 162)
L=function(th)  {Y=prod(exp(-y/th)/th);Y}
theta_x=1:500;theta1<-matrix(theta_x,length(theta_x),1);ydat=apply(theta1,1,L)
theta_xx=100:300;theta2=matrix(theta_xx,length(theta_xx),1);lydat<-log(apply
(theta2,1,L))
theta1[which.max(ydat)]    #the answer

[1] 168
```

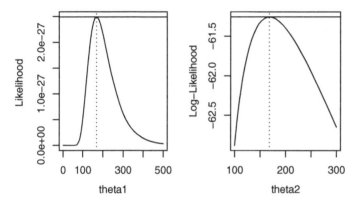

Figure 5.10 Exploring the Likelihood and log-Likelihood numerically.

The value of θ which corresponds to the maximum evaluated likelihood value is 168. Clearly such a value can be calculated at various degrees of accuracy, depending on the increments and range of θ's explored.

Remark 5.4 (Numerical optimisation) Rather than performing a visual search for finding the mode of the likelihood function which could only work in simple univariate cases, a more robust and simpler numerical approach is to use the standard numerical procedures for maximizing/minimising functions in R\Rstudio. Such methods are borrowed from the numerical analysis part of the mathematics which, in general, optimise the multivariate functions numerically within some pre-specified absolute tolerance. The tolerance values can be refined depending on the accuracy required for the solution. The standard values in R are sufficiently small and the commands which are appropriate for such maximisation/minimisation procedures are optimize (adjusted for univariate functions like L here), or optim, nlm for multivariate functions. Application of the optimize function below yields a consistent result:

```
(a=optimize(L,c(1,500),maximum = TRUE))

$maximum
[1] 168.3

$objective
[1] 2.490084e-27
```

the default tolerance option `tol = .Machine$double.eps^0.25` indicates that the solution is within the tolerance `.Machine$double.eps`$^{\frac{1}{4}}$ and `.Machine$double.eps` = 2.220446×10^{-16} is the practical zero value in R. Note that `optimize` produced a list that we called a, one of its components is the optimal value `a$max`$=168.2999997$. There are many options that this function can take when called, such as whether the maximum or minimum value is sought, for example `maximum = TRUE`.

5.5.2.2 Obtaining the MLE analytically

Maximizing $\mathcal{L}(\theta)$ on θ is the same as finding that $\hat{\theta}$ for which $\frac{\partial \log \mathcal{L}(\theta)}{\partial \theta} = 0$. This produces

$$\hat{\theta} = \frac{\sum_{i=1}^{n} y_i}{n}$$

i.e. the mean value of the data. The function $\log \mathcal{L}(\theta)$ in this case is a convex function around $\hat{\theta}$ as the second order derivative of $\log \mathcal{L}(\theta)$ is

$$\frac{\partial^2 \log \mathcal{L}(\theta)}{\partial^2 \theta} = \frac{n}{\theta^2}\left(1 - \frac{2\hat{\theta}}{\theta}\right)$$

which at $\theta = \hat{\theta}$, is

$$\frac{\partial^2 \log \mathcal{L}(\theta)}{\partial^2 \theta}\Big|_{\hat{\theta}} = \frac{n}{\hat{\theta}^2}(1 - 2) = -\frac{n}{\hat{\theta}^2} < 0$$

so $\hat{\theta} = \frac{\sum_{i=1}^{n} y_i}{n}$ maximises $\log \mathcal{L}(\theta)$ as well as $\mathcal{L}(\theta)$. The mean value for our data is:

```
mean(y)

[1] 168.3
```

This corresponds to $\hat{\theta}$, which is consistent with the MLE determined using `optimize`. Please note that, whenever possible, it is preferable to use close form expressions for the MLE estimation – the numerical methods are not always reliable. For verification and educational purposes, it is best, if possible, for the reader to adopt both approaches.

Exercise 5.10 (Poisson distribution) *The following simulated data represent the number of monthly motor claims incurred by an insurance company, over 36 consecutive months:*

```
set.seed(100);    y<-rpois(36,100)
```

Show, analytically, that the MLE for the Poisson rate is $\hat{\theta} = mean(y) = 99.5$. *Re-run with simulated data using different argument values in* set.seed.

Example 5.7 (We now explore the problem in Exercise 5.10 numerically) In the following code, we define two functions L and logL for the likelihood and log-likelihood, respectively; we then evaluate these functions at various values $\theta = 1 : 500$ to plot and visually identify the MLE point.

```
L=function(lambda)      {Y=exp(-lambda)*lambda^(y)/gamma(y+1);Y=prod(Y);Y}
logL=function(lambda)   {Y=exp(-lambda)*lambda^(y)/gamma(y+1);Y=sum(log(Y));Y}
```

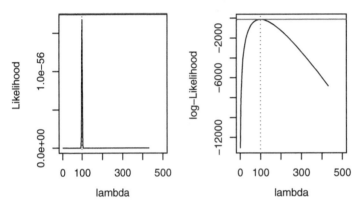

Figure 5.11 Likelihood (and Log) function for accident number data, max at $\hat{\theta} = 99.5$.

Note that log-likelihood function is more convenient both theoretically, as we can evaluate the log-likelihood derivatives easily and numerically, see, for example Figure 5.11 where the mode of the function around $\hat{\theta} = 99.5$ is more clearly displayed.

Exercise 5.11

- *Apply the function* which.max *to identify that value of* θ *from* 1:500 *which maximises the function values at* apply(th,1,L) *above.*
- *Confirm that the* optimize *function here produces a value* $\hat{\theta} = 99.5$.

5.5.3 Large Sample (Asymptotic) Properties of MLE

The parameter estimates output in R/Rstudio are almost always associated with the corresponding standard errors. These indicate the associated confidence of each estimated parameter and are also used to produce the corresponding t-values. In the following, we show the relevant theoretical justification for these numerical quantities.

The log-likelihood function, $\ell(\theta)$, is not only preferred over the likelihood function, $\mathcal{L}(\theta)$, due to the analytical and numerical convenience (as seen in the examples above); its main use is in understanding the limiting behaviour of the MLE estimators. Let's assume that for $y_1, y_2, ..., y_n$ sample points:

$$\ell(\theta) = \log \mathcal{L}(\theta) = \log \prod_{i=1}^{n} f(y_i; \theta) = \sum_{i=1}^{n} \ell_i(\theta) \quad \theta \in \Theta$$

where $\ell_i(\theta) := \ell_i(\theta; y_i) = \log f(y_i; \theta)$. We call $J(\theta) = -\frac{\partial^2 \ell(\theta)}{\partial^2 \theta}$ the observed information. From the discussion in Section 5.5.2 related to the Exponential distribution:

$$\ell(\theta) = \mathcal{L}(\theta) = -n \log \theta - \frac{1}{\theta} \sum_{i=1}^{n} y_i = n \left(\log \theta - \frac{\bar{y}}{\theta} \right)$$

it can be shown that the observed information is given by $\frac{\partial^2 \ell(\theta)}{\partial^2 \theta} = n \frac{1}{\bar{y}^2}$. Note that as n increases the second derivative of $\ell(\theta)$ at its maximum at $\theta = \bar{y}$ increases linearly as in $n \frac{1}{\bar{y}^2}$. Therefore $\ell(\theta)$ becomes more spiky around the optimal point $\hat{\theta}$. This intuition is in fact consistent with the following facts; under some regularity conditions on the true distribution $f(y; \theta)$:

- The consistency of the MLE:

  $\hat{\theta} \rightarrow \theta_0$, as $n \rightarrow \infty$ where θ_0 is the true parameter.

- Asymptotic normality:

  $\hat{\theta} \sim N(\theta_0, I(\theta_0)^{-1})$

  where $I(\theta_0) = -E_{\theta_0}(\frac{\partial^2 \ell(\theta)}{\partial^2 \theta}) \approx J(\hat{\theta})$ ($\approx$ is sometimes strictly $=$)

  $I(\theta_0)$ is called the Fisher or expected information.
  For the vector case $\theta = (\theta_1, \theta_2, ..., \theta_p)$ then $I(\theta_0)$ is a matrix such that

  $$I(\theta_0)_{r,s} = -E_{\theta_0}\left(\frac{\partial^2 \ell(\theta)}{\partial \theta_r \partial \theta_s}\right) \quad \text{and} \quad J(\hat{\theta})_{r,s} = -\left(\frac{\partial^2 \ell(\theta)}{\partial \theta_r \partial \theta_s}\right)_{\hat{\theta}}$$

In practice we use the fact that $I(\theta_0) \approx J(\hat{\theta})$, that is the expected information $I(\theta_0)$ is estimated by the observed information $J(\hat{\theta})$ at the maximum likelihood estimate, $\hat{\theta}$. In many practical situations, it is easier to calculate $J(\hat{\theta})$ and use $\frac{1}{\sqrt{J(\hat{\theta})}}$ as the standard error for our MLE estimator θ. Thus using the observed information $J(\hat{\theta})$ we can construct $(1 - \alpha)\%$ confidence intervals as

$$\left(\hat{\theta} - z_{\alpha/2}J(\hat{\theta})^{-1/2}, \ \hat{\theta} + z_{\alpha/2}J(\hat{\theta})^{-1/2}\right)$$

$$P(Z > z_\alpha) = \alpha \quad Z \sim N(0, 1)$$

When θ is a vector, the corresponding standard error for the θ_rth component is the square root of the rth diagonal element of $J(\hat{\theta})$. In the following cases, θ is simply a vector of length 1, and $J(\hat{\theta})^{-1}$ is a scalar.

Example 5.8 (CI for Exponential parameter) In the Spring failure data (using the exponential distribution), $n = 10$, $\hat{\theta} = \bar{y} = 168.3$, and $J(\hat{\theta}) = I(\hat{\theta}) = 10/168.3^2$. A 95%

confidence interval for the true θ_0 is then (where $z_{0.025} = 1.96$):

$$\left(\bar{y} - z_{0.025}/\sqrt{10/168.3^2}, \bar{y} + z_{0.025}/\sqrt{10/168.3^2}\right) = (63.99, 272.61)$$

This is a 95% confidence interval for the parameter θ using the MLE. We use the function `fitdistr` (see below) to obtain these standard errors directly. We shall see similar applications of this result several times throughout the book.

Exercise 5.12 (CI for the Poisson rate) *For the Poisson data example, we had $n = 36$, $\bar{y} = \frac{\sum_{i=1}^{36} y_i}{36} = 99.5$. Show that the 95% CI for the true rate of the Poisson distribution θ is*

$$\left(\bar{y} - 1.96\sqrt{\frac{\bar{y}}{36}}, \bar{y} + 1.96\sqrt{\frac{\bar{y}}{36}}\right) = (96.24, 102.76)$$

Example 5.9 From Government statistics the total number of births for the period of 2011–2015 is 3, 537, 963 of which 1, 86, 2738 of them (or 52.65%) are boys. (See (https://www.gov.uk/government/statistics/gender-ratios-at-birth-in-great-britain-2011-to-2015).)

• Derive the likelihood function for this data assuming that the probability that a baby is born a boy is $p = 0.5$.
• Derive the MLE for p.

Can you numerically evaluate the likelihood function for $p = 0.5$ above? In fact it is numerically not possible (in R at least) to calculate $\binom{n}{k}$ for such values $n = 3, 537, 963$, $k = 1, 862, 738$. However, one can show that:

$$\frac{\partial \ell(p)}{\partial p} = \frac{k}{p} - \frac{n-k}{1-p} = 0 \quad \text{implies} \quad \hat{p} = \frac{k}{n} = 0.5265 = \frac{1, 862, 738}{3, 537, 963}$$

Exercise 5.13

• Derive the second-order derivative of $\log \mathcal{L}(p)$ for the example above and show that $\hat{p} = \frac{k}{n}$ maximises the likelihood.
• Find the corresponding standard error using the observed information $\frac{\partial^2 \ell(p)}{\partial^2 p}|_{\hat{p}}$.

5.5.4 Fitting Distributions to Data in R Using MLE

`fitdistr` is a useful function which fits a range of distributions, determining the MLEs of the appropriate parameters. We show below how to fit the Gamma distribution (including the calculation of standard errors) to random data, x, generated from a Gamma(5,2) distribution:

```
set.seed(12);    x=rgamma(1000,5,2)
fitdistr(x, "gamma")$est;    fitdistr(x, "gamma")$sd

   shape      rate
5.096861 2.035381
      shape         rate
0.22088173 0.09269981
```

Exercise 5.14 *Re-run the above code but with 100,000 simulations, and compare the estimates.*

Exercise 5.15 *Fit a normal distribution to the data x generated above. Is there enough evidence to reject the hypothesis that x comes from a normal distribution?*

Exercise 5.16 *Use the data, y, from Exercise 5.10 to fit the Poisson distribution and confirm that the standard error for the rate parameter is $\sqrt{\bar{y}/36}$.*

(Please see Chapter 26 for a more detailed example.)

5.5.5 Likelihood Ratio Test, LRT

The LRT is a particularly useful test for nested models and uses a χ^2 penalty for any unnecessary additional parameters. Let's assume that we have models `Full` and `Part` with parameter vectors (θ_1, θ_2) and θ_1, respectively. Note that `Part` can be seen as a special case of `Full` when $\theta_2 = 0$. Note also that θ_1 and θ_2 can be vector parameters of lengths p_1 and p_2 say. As a result, the number of parameters for the respective models are: $\#param(Full) = p_1 + p_2 = p$; $\#param(Part) = p_1$, that is $dim(\theta_1) = p_1$ and $dim(\theta_2) = p_2$.

Our objective is to determine whether the addition of p_2 additional (non-zero) parameters results in a more suitable model. The hypothesis test of interest is

$$H_0 : \theta_2 = 0 \quad \text{vs} \quad H_1 : \theta_2 \neq 0$$

Note that H_0 points to the reduced model `Part` while H_1 points to the extended model `Full`.

Result 5.1 (LRT for nested models) If $H_0 : \theta_2 = 0$, that is model `Part` is true,

$$2\ell(Full) - 2\ell(Part) \sim \chi^2_{p_2} \tag{5.6}$$

Example 5.10

Consider for example the data set `Weib.csv` which represents the claim size distribution of some 2,000 insurance claims (the data is plotted in Figure 5.12). We want to determine whether this data is best explained by an Exponential or Weibull distribution. Note that Weibull is a generalisation of the Exponential distribution by introducing a power transform determined by the additional scale parameter, that is they are nested models, Weibull with 2 parameters and Exponential with 1 parameter. Therefore, $p_2 = 1$ and an appropriate critical region is defined as $P(\chi^2_1 \leq 3.841) = 0.95$.

```
weib=read.csv(Weib.csv,header=TRUE)[,2]
fitW=fitdistr(weib,"weibull");fitE=fitdistr(weib,"exponential")
```

The resulting log-likelihood values of these fitted MLE models are `fitE$log=` $-11,013$ and `fitW$log=` $-10,783$, an increase of 230, giving an LRT statistic of

2(fitW\$log-fitE\$log)= 461.1, which is highly significant in the scale of χ_1^2 (the chi-square distribution with 1 degree of freedom). Therefore there is extreme evidence against H_0. This is not surprising since the shape parameter is highly significant with the corresponding t-value=1.51316/0.02634 = 57.4 (see the appropriate R output for fitW). This is consistent with a visual analysis (see Figure 5.12) – the Weibull model certainly appears more appropriate than the simpler Exponential model.

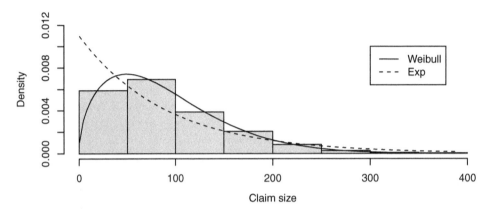

Figure 5.12 Fitting Exponential and Weibull distribution to claim data.

Exercise 5.17 *In the* Weib *data set we found that the Weibull shape parameter is 1.5131649. Perform a similar LRT as above after the corresponding power transformation is performed as* weib=weib^1.51. *Is your conclusion the same as before?*

5.6 Regression: Basic Principles

Regression theory is perhaps the main technique used in statistical modelling. The objective is to infer some function, f, such that the response variable, Y, is explained by some other p predictor variables, X_i:

$$Y = f(X_1, X_2, ..., X_p) + error$$

This model can be used in various forms. For example, f can be linear or polynomial, or the data variables X and Y can be discrete or continuous. In the following sections we will focus on the linear regression models.

For a full discussion of Linear Regression please see the Recommended Reading at the end of the chapter.

5.6.1 Simple Linear Regression

In the simple linear regression models the data are assumed to be in pairs $(y_i, x_i), i = 1, 2, \cdots n$ and the model is such that

$$y_i = \alpha + \beta x_i + e_i$$

where e_i is some error in observing the value y_i of the response (or dependent) variable y, x_i is an observation of the explanatory (or predictor) variable x. Additionally, e_i's are assumed to be independently normally distributed with mean $= 0$ and variance $= \sigma^2$. Such a model is based on the following conditional distribution assumption:

$$Y|(X = x) \sim N(\alpha + \beta x, \sigma^2)$$

The parameters α and β, called the intercept and slope, are fitted using the Least Sum of Squares principle, which minimises

$$RSS = \sum_{i=1}^{n} \hat{e}_i^{\,2} = \sum_{i=1}^{n} (y_i - (\alpha + \beta x_i))^2$$

After differentiating with respect to both α and β and equating them to zero one finds the optimal values which produce the following estimates:

$$\hat{\beta} = \frac{\sum_{i=1}^{n} x_i y_i - n\bar{x}\bar{y}}{\sum_{i=1}^{n} x_i^2 - n\bar{x}^2} = \frac{S_{XY}}{S_{XX}} \qquad \hat{\alpha} = \bar{y} - \hat{\beta}\bar{x}$$

For example, in the trees data set (Section 5.3.2), the pair of variables which shows some linear dependence is Diameter (x) and Volume (y). Carrying out a linear regression of Volume against Diameter, using the important lm function ("linear model"):

```
model=lm(trees$Volume~trees$Diameter)
```

The data and regression line are shown in Figure 5.13.

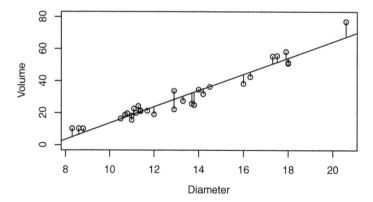

Figure 5.13 A bivariate plot with a fitted linear regression line and residuals as the vertical lines joining the observations with their fitted values.

The output available from the lm function is shown below:

```
names(model)

[1] "coefficients"   "residuals"   "effects"   "rank"
[5] "fitted.values"  "assign"      "qr"        "df.residual"
[9] "xlevels"        "call"        "terms"     "model"
```

The standard output is obtained from model; a more detailed summary is obtained using summary:

```
summary(model)


Call:
lm(formula = trees$Volume ~ trees$Diameter)

Residuals:
   Min    1Q Median    3Q    Max
-8.065 -3.107  0.152  3.495  9.587

Coefficients:
                Estimate Std. Error t value Pr(>|t|)
(Intercept)     -36.9435     3.3651  -10.98 7.62e-12 ***
trees$Diameter    5.0659     0.2474   20.48  < 2e-16 ***
---
Signif. codes:  0 '***' 0.001 '**' 0.01 '*' 0.05 '.' 0.1 ' ' 1

Residual standard error: 4.252 on 29 degrees of freedom
Multiple R-squared:  0.9353, Adjusted R-squared:  0.9331
F-statistic: 419.4 on 1 and 29 DF,  p-value: < 2.2e-16
```

From the list model, one can extract the intercept, α, and slope, β, using model$coeff[1] and model$coeff[2], giving the fitted linear model:

$$Volume = -36.943 + 5.066 \times Diameter$$

This implies that the Volume of the tree is predicted to increase by approximately five units for each additional increase of one unit in Diameter. We can obtain the fitted values using this regression line as model$fitted.values, or fitted(model). We used the latter option for adding the vertical lines on the regression plot of Figure 5.13.

Remark 5.5 Please note the asymmetry that exists between x_i and y_i variables in the simple regression model. The fitted line is that which passes closest to the data in terms of the vertical distances $\hat{e}_i$.

Exercise 5.18 *Perform the same analysis as above but swapping the axes, that is*
lm(trees$Diameter~trees$Volume) *and comment on any differences in the resulting fitted line.*

5.6.2 Quantifying Uncertainty on $\hat{\beta}$

Our predictions of y from the regression model are only as good as the quality of the estimates; it is vital, therefore, that we understand the accuracy of these parameter estimates.

From the model assumptions we assume that y_i are observations from the Normal distribution $N(\alpha + \beta x_i, \sigma^2)$, therefore an estimator for σ^2 could be obtained as a sample variance of the residuals:

$$\hat{e}_i = y_i - (\hat{\alpha} + \hat{\beta} x_i)$$

The unbiased estimator for σ is given by:

$$\hat{\sigma}^2 = \frac{RSS}{n-2} = \frac{1}{n-2} \sum_{i=1}^{n} \hat{e}_i^2. \tag{5.7}$$

Note that $n-2$ is due to the fact that we estimate two parameters in the process of generating residuals. Note also that $\hat{\beta} = \frac{S_{XY}}{S_{XX}}$ and while seen as a linear combination of the random quantities $y_i \sim N(\alpha + \beta x_i, \sigma^2)$, it follows that

$$E(\hat{\beta}) = \beta \qquad Var(\hat{\beta}) = (se(\hat{\beta}))^2 = \frac{\sigma^2}{S_{XX}}$$

where S_{XX} is defined in Equation (5.2), and

$$\frac{\beta - \hat{\beta}}{se(\hat{\beta})} = \frac{\beta - \hat{\beta}}{\hat{\sigma}/\sqrt{S_{XX}}} \sim t_{n-2}$$

One can verify the standard error for the slope obtained above from `summary(model)`:

```
n=length(trees$Volume)
sxx=sum((trees$Diameter-mean(trees$Diameter))^2)
shat=sum(model$residuals^2)/(n-2)
sqrt(shat/sxx)      #se estimate,  agrees!

[1] 0.247377
```

The corresponding t-value: $\frac{5.066}{0.247} = 20.48$ is highly significant (agrees the output in `model`). Similar arguments apply for the evaluation of the standard errors of the intercept. Confidence intervals for the parameters can be obtained as follows:

```
confint(model,level=0.95)

                 2.5 %      97.5 %
(Intercept)    -43.825953 -30.060965
trees$Diameter   4.559914   5.571799
```

If either of the intervals contain zero then there is evidence that the respective parameter is zero, that is not present in the model. Such a conclusion would be highly relevant for the slope parameter β which is related to the correlation between the variables.

Remark 5.6 Please refer to Faraway (2014) and Sheather (2009) for further technical details regarding linear regression theory.

One can easily generate the prediction values with pred.int:

```
pred.int <-predict(model,data.frame(x=trees$Diameter),interval='confidence',
level=0.99)
```

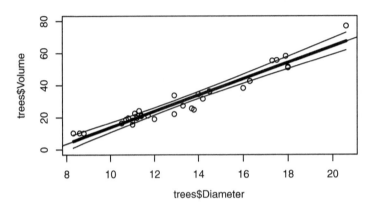

Figure 5.14 A bivariate plot with a fitted linear regression line and predicted intervals.

It is important to note that the confidence intervals are largest for those x values furthest from $\bar{x}$ (see later).

5.6.3 Analysis of Variance in Regression

We use Analysis of Variance (ANOVA) when the data variability of one variable, say Y, can be expressed in terms of some other variables, X. In the simple regression setting we can express S_{yy} as:

$$S_{yy} = RSS + \frac{S_{xy}^2}{S_{xx}} \quad \text{or} \quad TSS = RSS + RegSS$$

with $TSS = S_{yy}$, the total sum of Squares, $RSS = \sum_{i=1}^{n} \hat{e}_i^2$, the Residual Sum of Squares, and $RegSS = \frac{S_{xy}^2}{S_{xx}}$, the regression Sum of Squares. These quantities have the corresponding degrees of freedom $n - 1$, $n - 2$, and 1. In the ANOVA setting we test effects and in this case we can test whether the variable X has an effect on Y. We can test this using the F-test statistic:

$$F = \frac{RegSS}{RSS/(n-2)} \sim F_{1,n-2}$$

This F-test for the ANOVA effect of X on Y is equivalent to the t-test seen above for testing the slope hypothesis $H_0 : \beta = 0$. Applying the aov function to the trees data:

```
aovmodel=aov(Volume~Diameter,data=trees); summary(aovmodel)

            Df Sum Sq Mean Sq F value Pr(>F)
Diameter     1   7582    7582   419.4 <2e-16 ***
Residuals   29    524      18

---
Signif. codes:  0 '***' 0.001 '**' 0.01 '*' 0.05 '.' 0.1 ' ' 1
```

One can easily check the above F-value from the t-value calculated earlier ($20.48^2 = 419.36$). Therefore the F-test statistic is simply the square of the t-statistic for the slope in model output as in page 111. They appear to be two different tests but perform the same statistical inference.

5.6.3.1 R^2 and adjusted R^2 Coefficient of Determination

The following provides an indication of the goodness of fit of the model:

$$R^2 = \frac{RegSS}{TSS} = \frac{S_{xy}^2}{S_{xx}S_{yy}} = 1 - \frac{RSS}{TSS}.$$

The closer is the value of R^2 to 1, the better the model fit. The adjusted R^2 coefficient is adjusted for the degrees of freedom (n-2):

$$AdjR^2 = 1 - \frac{RSS/(n-2)}{TSS(n-1)}$$

Note that the roles of X and Y data are symmetric and R is simply the Pearson correlation coefficient as seen before.

Exercise 5.19 *Obtain the R^2 coefficient and its adjusted value for*
model=lm(trees$Diameter~trees$Volume) *that we studied earlier.*

5.6.4   Some Visual Diagnostics for the Proposed Simple Regression Model

It is worth reminding the reader that the key assumptions for fitting this model relate to the errors, $e_i = y_i - \alpha - \beta x_i$, such as:

- being independent,
- having a common variance σ^2,
- being normally distributed.

It is therefore important to investigate whether these requirements are jointly met when inferring from such a model; we do this here by running some visual diagnostic checks. In fact, more rigorous standard tests for white noise in Chapter 30 or simply testing for normality of the residuals are also relevant here.

However, a selection of visual tests can take us a long way in identifying the potential issues with the residuals for the whole model or some specific outlier data points. For example in Figure 5.15, three data examples are considered. They represent some constructive pathological cases. The corresponding data points and the fitted regression lines are shown in the first row; the second row includes appropriate displays of the corresponding residuals. The code used to generate this data (and plots shown in Figure 5.15) is as follows:

```
set.seed(123);x=1:40;y1=3+x+5*abs(rnorm(40,sd=0.1))*rep(c(1,-1),20)
y2=x^2+(x-2)^3;y3=3+x+rnorm(40,sd=1);   y3[35:40]=y3[10:15]*1.6
plot(y1~x);    m1=lm(y1~x);    abline(m1)  #for plot 1 in the first row
plot(y2~x);    m2=lm(y2~x);    abline(m2)
plot(y3~x);    m3=lm(y3~x);    abline(m3)
plot(sign(m1$res));plot(m2$res);hist(m3$res) #second row
```

The reader is encouraged to experiment with other pathological cases.

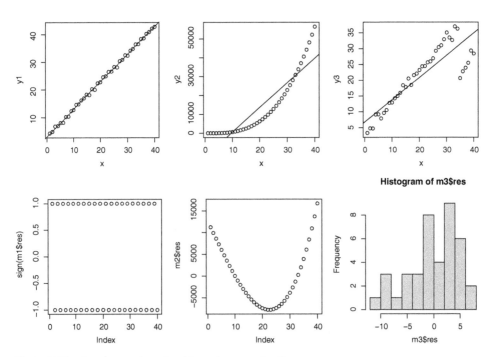

Figure 5.15 Some examples of problematic regression lines.

In the first example the residuals have alternating signs (note the residual signs plotted) and hence the independence assumption is violated. In the second example a quadratic function seems more appropriate here than a linear one, and the residuals are negatively valued only around the middle of the data points. Also the histogram of residuals (not shown here) indicates a departure from the desired symmetry around zero. Therefore the normal distribution assumption does not hold.

In the third example the histogram of the residuals is showing asymmetry around 0 especially in the negative range. This is in fact due to the cluster of the last six set of points which seem to be out of line with the rest. These data points are indeed outliers. A way to identify outliers is using the standardised residuals.

How do we use the standardised residuals?

Note that

$$var(\hat{e}_i) = var(y_i - \hat{\alpha} - \hat{\beta}x_i) = \sigma^2 \left(\frac{n-1}{n} - \frac{(x_i - \bar{x})^2}{S_{XX}} \right)$$

this implies that the variance of the residuals in the correct model depends on x_i values and therefore we need to count for that by standardising as

$$t_i = \frac{\hat{e}_i}{\hat{\sigma} \sqrt{\frac{n-1}{n} - \frac{(x_i - \bar{x})^2}{S_{XX}}}}$$

and we need to base our analysis based on these t-values. In fact it is advisable that the model is deemed appropriate if it agrees on these three visual checks:

- Residuals versus fitted values (for any apparent structural issues like the ones seen);
- QQ-plot of the standardised residuals (for checking the normality); and
- Square root of standardised residuals versus fitted values.

These relevant plots are easily identified in the standard output using `plot (lm-model)`. For example, the third set of observations in Figure 5.15 is generating model m3 and the command `plot(m3)` produces Figure 5.16. In the top left figure one can see that the residuals for the last 6 observations are indeed large values. The line is roughly linear until the last cluster of these observations. The QQ-plot seems reasonable except for the first negative values (corresponding to the same cluster of points).

The plots of the second row are used to check the equal variance assumption and for the presence of any influential observations. In particular, the `scale-location` plot should be close to a horizontal line if equal variance holds. The `Residual vs Leverage` is used to indicate any specific observation being unusually influential (high leverage) in the model. We encourage the keen reader to read more about these advanced measures independently.

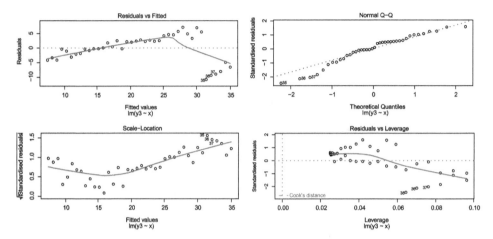

Figure 5.16   Model 3: Analysis of the residuals, observations 35 to 40 are outliers.

The observations 35 to 40 in m3 have unusually high residual values and are outliers. One can remove these from the model, explore if there is any error in the data generation process for these items, or even consider reviewing the regression model, that is quadratic function or additional covariates.

Exercise 5.20   *Carry out the same analysis on data sets* m1 *and* m2.

5.7   Multiple Regression

5.7.1   Introduction

In the multiple regression model we have more than one dependent variable in the model. Indeed, the number of variables could be several hundred, for example in the field of genetics a number of genes could be potential predictors for medical conditions.

Consider the following data set where some response data, Y, and four dependent, or predictor, variables $X1, X2, X3$, and $X4$ are generated:

```
n=200;set.seed(123);X1=rnorm(n)+35;X2=rnorm(n)+5;X3=rnorm(n)+15
X4=0.2*X1+0.2*X2+0.1*X3+rnorm(n)*0.1
Y=100+2*X1+3*X2+1.5*X3+1/3*X4+rnorm(n)*3
```

Note that the last line of code is generating data as follows:

$$Y_i = 100 + 2X1_i + 3X2_i + 1.5X3_i + \frac{1}{3}X4_i + e_i$$

It is important to note the following: 1)$X4$ is a linear combination of the other three plus some noise; 2)Y is generated as a linear combination of $X1, X2, X3$, and $X4$.

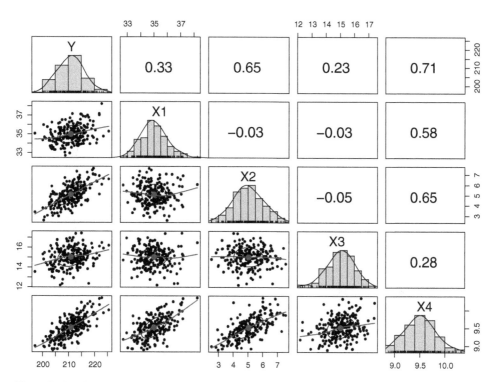

Figure 5.17 Pairwise plots for the generic multivariate data.

Alternatively, the data can be downloaded from:

```
Test=read.csv("~/TestMulReg.csv")
```

The graphical display of Figure 5.17 showing the pairwise plots indicates Y to be linearly dependent, to some degree, on each of the four variables X'i. For modelling this relationship jointly for all these predictors we use the multivariate regression model:

$$Y_i = \theta_0 + \theta_1 X1_i + \theta_2 X2_i + \theta_3 X3_i + \theta_4 X4_i + e_i$$

The predictive function here is now a hyperplane determined by θ_i's and $X1, X2, X3, X4$ variables. Note that such a model is also represented in the literature in terms of matrix and vector notation as:

$$\mathbf{Y} = \mathbf{X\theta} + \mathbf{e}$$

$$\mathbf{Y} = (Y_1, Y_2, \cdots, Y_n)^t \quad \mathbf{\theta} = (\theta_0, \theta_2, \cdots, \theta_4)^t \quad \mathbf{e} = (e_1, e_2, \cdots, e_n)^t$$

$$\mathbf{X} = \begin{pmatrix} 1 & X1_1 & X2_1 & \cdots & X4_1 \\ \cdots & \cdots & \cdots & \cdots & \cdots \\ 1 & X1_n & X2_n & \cdots & X4_n \end{pmatrix}$$

where the intercept term is seen as some additional predictor variable with entries of ones. Note that in general if we have $p - 1$ predictor variables the matrix $\mathbf{X}$ (called the design matrix) will be of dimension $n \times p$ and the vector θ will be of length p. In our example $p = 5$. Models similar to the one above could be used, for example, in predicting company share price (Y), based on potential relevant factors (X_i) such as revenue, costs, profits, employee turnover, level of gearing, recent volatility of share price, etc. The objective would be to determine θ, and hence a regression model, such that we can estimate the share price.

5.7.2 Regression and MLE

The standard approach to regression analysis is based on the method of Least Squares; this is closely related to that of MLE except for some minor adjustments for the estimation of the error variance. We state below the MLE arguments but the reader could easily adopt them for the Least Squares.

5.7.2.1 Multivariate Regression
It can be shown that, for the true model in the multivariate regression setting, $\mathbf{Y} = \mathbf{X\theta} + \mathbf{e}$

$$\log \mathcal{L}(\theta, \sigma) = -\frac{n}{2} \log \sigma^2 - \frac{n}{2} \log 2\pi - \frac{1}{2\sigma^2}(\mathbf{Y} - \mathbf{X\theta})^\top (\mathbf{Y} - \mathbf{X\theta})$$

If $\mathbf{X}$ has full rank the least squares estimator is also the MLE:

$$\hat{\theta} = (\mathbf{X}^\top \mathbf{X})^{-1} \mathbf{X}^\top \mathbf{Y}$$

and for the MLE of σ^2 we need to maximise

$$\log \mathcal{L}(\hat{\theta}, \sigma) = -\frac{n}{2} \log \sigma^2 - \frac{n}{2} \log 2\pi - \frac{1}{2\sigma^2}(\mathbf{Y} - \mathbf{X\hat{\theta}})^\top (\mathbf{Y} - \mathbf{X\hat{\theta}})$$

Hence,

$$\tilde{\sigma}^2 = \frac{(\mathbf{Y} - -\mathbf{X\hat{\theta}})^\top (\mathbf{Y} - \mathbf{X\hat{\theta}})}{n} = \frac{\text{RSS}}{n}$$

Again, note the difference with the standard unbiased estimator $\hat{\sigma}^2 = \dfrac{\text{RSS}}{n-p}$. We can fit a multivariate regression in R using the lm function:

```
model_full=lm(Y~X1+X2+X3+X4,data=Test)
(coeff=round(summary(model_full)$coeff,3))

            Estimate Std. Error t value Pr(>|t|)
(Intercept)  105.516      9.098  11.598    0.000
X1             2.257      0.482   4.679    0.000
X2             3.763      0.502   7.502    0.000
X3             1.644      0.320   5.133    0.000
X4            -1.813      2.160  -0.839    0.402
```

The values of $\hat{\theta}$ are shown in the first column ("Estimate"); the reader should compare these values with those used to simulate the data. Note that the term Y~X1+X2+X3+X4 is essentially a symbolic formula that R is interpreting as "Y versus X_1, X_2, X_3, and X_4". A similar syntax is used for many modelling instructions in R, and is seen in several of the chapters of the book. The parameter estimates θ_i of model_full, including the intercept, are shown together with the corresponding standard errors, t-values, and p-values. It is clear that the intercept parameter θ_0 is particularly significant (p-value $= 0$). The same applies for θ_1, θ_2, and θ_3, but not for θ_4, the coefficient related to the $X4$ variable . In particular, the corresponding p-value is 0.402 suggesting there is insufficient evidence against $H_0 : \theta_4 = 0$. On the other hand, one can easily see that a simple linear regression between Y and X_4 is indeed valid:

```
summary(lm(Y~X4, data=Test))$coeff

            Estimate Std. Error  t value      Pr(>|t|)
(Intercept) 99.47826  7.8601672 12.65600 2.752161e-27
X4          11.72307  0.8262977 14.18747 5.528375e-32
```

and the corresponding residual analysis: plot(lm(Y~X4, data=Test)) confirms the validity of this univariate (simple regression) model (see the bottom-left plot in Figure 5.17). How is it possible that the variable $X4$ is individually a strong predictor for Y but not in the joint multivariate model? The predictive power of $X4$ could also be explained indirectly by the presence of other variables in the multivariate model. In particular, as seen in the pairwise plots (Figure 5.17), the variable $X4$ seems to have some strong linear relation with $X1$ and $X2$ and not so much with $X3$. An extreme hypothetical scenario is that if $X4 = X2$ or even a linear combination of two or more variables, for example $X4 = 3X1 + 2X2 + X4$. This suggests that the most appropriate model is the one which ignores those unnecessary predictor variables – $X4$ is potentially one of them in this example. The reduction of the number of valid covariates addresses the so-called collinearity in the space of covariates. In a more general setting where more than one linear constraint is present, we need to find a certain subset among the variables $\{1, X1, X2, X3, X4\}$ which provides the most suitable model; however, there are in general $2^5 = 32$ sub-model possibilities, or 2^p in the case of p variables.

5.7.3 Tests

We can try to find the most suitable model by determining some measure of comparing the models based on quantities previously discussed: RSS, R^2, adjusted R^2, $F - test$, log-likelihood ratios. We will introduce two additional measures, AIC and BIC, later in this chapter.

F-test

In the case of p variables, we can run the following F-test to test whether some sub-model (Restricted) using only covariates $\mathbf{X}_1$ is more appropriate than a bigger (Full) model containing a wider list of covariates $\mathbf{X}_1$ and $\mathbf{X}_2$:

$$\text{Full model}: \quad \mathbf{Y} = \mathbf{X}_1\boldsymbol{\theta}_1 + \mathbf{X}_2\boldsymbol{\theta}_2 + \mathbf{e} \quad dim(\boldsymbol{\theta}_1, \boldsymbol{\theta}_2) = p_1 + p_2$$

$$\text{Restricted model}: \quad \mathbf{Y} = \mathbf{X}_1\boldsymbol{\theta}_1 + \mathbf{e} \quad dim(\boldsymbol{\theta}_1) = p_1 \tag{5.8}$$

For example we may try $Y_i = \theta_0 + \theta_1 X1_i + e_i$ and then $Y_i = \theta_0 + \theta_1 X1_i + \theta_3 X3_i + e_i$ where $p_1 = 2$, $p_2 = 1$, and $\mathbf{X}_1 = X1$, $\mathbf{X}_2 = (X1, X3)$. Let RSS_{12} and RSS_1 represent the corresponding quantities for the models Full model and Restricted model, clearly $\text{RSS}_{12} < \text{RSS}_1$.

Result 5.2 The F-statistic used for testing whether the p_2 parameters are not to be present in the model, that is $H_0 : \boldsymbol{\theta}_2 = 0$ is

$$F = \frac{(\text{RSS}_1 - \text{RSS}_{12})/p_2}{\text{RSS}_{12}/(n - p)} \qquad p = p_1 + p_2 \tag{5.9}$$

which is compared to the critical values $F_{p_2, n-p}$ distribution.

If the p-value is greater than $\alpha\%$, then we do not have enough evidence to reject H_0 at the $(100-\alpha)\%$ significance level. Note that this F-test is equivalent to that seen in the ANOVA for simple linear regression (Section 5.6.3) in page 114.

5.7.3.1 Likelihood Ratio Test in Regression

LRT is particularly useful for nested models. Similar to the general setting stated in Section 5.5.5 one can see that for our regression modelling:

Result 5.3 (LRT for nested regression models) If $H_0 : \boldsymbol{\theta}_2 = 0$ then

$$2\ell(\text{Full}) - 2\ell(\text{Rest.}) = n(\log \text{RSS}_1 - \log \text{RSS}_{12}) \sim \chi^2_{p_2} \tag{5.10}$$

Note the difference with the F-statistic (Equation 5.9) under the same null $H_0 : \theta_2 = 0$ hypothesis. They are different tests but generally have the similar p-values; see the numerical outputs when we run both of these tests later.

5.7.3.2 Akaike Information Criterion: AIC

The AIC and BIC tests are used frequently throughout this book. The absolute value of AIC does not provide information on the quality of the model; it is the relative value compared to other models, or the pairwise difference, which is relevant:

$$\Delta AIC = AIC_1 - AIC_2$$

For a given model with number of parameters p, the AIC is defined as follows:

$$AIC = -2 \log \mathcal{L}(\hat{\theta}) + 2p \qquad (5.11)$$

The smaller the AIC, the better the model, thus allowing us to compare two models; they do not need to be nested models. We can also use AIC as a cost for (stepwise) regression model selection (using the `leaps` package – see later).

5.7.3.3 AIC and Regression model selection

In the linear regression model (after removing some constant terms):

$$AIC = -2 \log \mathcal{L}(\hat{\theta}) + 2p = n \log \left(\frac{RSS}{n} \right) + 2p$$

and this sum gets smaller if the model prediction improves (smaller residuals obtained from larger models). The pairwise difference between two regression models, with p_1 and p_2 parameters, is

$$\Delta AIC = AIC_1 - AIC_2 = n \log \left(\frac{RSS_1}{RSS_2} \right) + 2(p_1 - p_2)$$

5.7.3.4 Bayesian Information Criterion: BIC

For a given model with number of parameters p, the BIC is defined as follows:

$$BIC = -2 \log \mathcal{L}(\hat{\theta}) + p \log n$$

Note the difference with AIC is due to the relative weight, $\log n$, on the number of parameters, p. As a result, if $n > 8$, $\log n > 2$, then BIC penalises more on p and so will tend to select simpler models than AIC. Generally speaking, AIC is better for prediction while BIC is better for simpler models.

5.7.4 Variable Selection, Finding the Most Appropriate Sub-Model

The standard variable selection procedures in the linear regression model involve adding ("forward selection") or removing ("backward elimination") one variable at a time, that is $p_2 = 1$ in the F-test statistics (Equation 5.9).

5.7.5 Backward Elimination

The **Backward elimination steps** are as follows:

1. fit the full model with all the predictors.
2. find the predictor with the largest associated p-value (or the smallest t-value or F-value statistics).
   -if its p-value is above some chosen threshold (e.g. 0.05) go to step 3.

-if not, keep the corresponding predictor and stop.

3. delete the predictor, re-fit the model, and go to step 2.

Note that $p > 0.05$ (which is commonly used) corresponds to $|t| < 2$ (or $F < 4$) approximately.

In backward elimination we drop only one variable at a time, that is $p_2 = 1$, and the remaining variables are those which have a greater impact on the regression, that is the smallest p-value in some appropriate list. Note that the model_full has originally four explanatory variables X1+X2+X3+X4. We use drop1 as:

```
drop1(model_full, test="F")

Single term deletions

Model:
Y ~ X1 + X2 + X3 + X4
         Df Sum of Sq    RSS    AIC F value    Pr(>F)
<none>                 1883.1 458.47
X1        1    211.41 2094.5 477.75  21.893 5.383e-06 ***
X2        1    543.55 2426.6 507.19  56.287 2.166e-12 ***
X3        1    254.47 2137.6 481.82  26.351 6.860e-07 ***
X4        1      6.80 1889.9 457.19   0.704    0.4025
---
Signif. codes:  0 '***' 0.001 '**' 0.01 '*' 0.05 '.' 0.1 ' ' 1
```

The function drop1 with test="F" uses the F-test criterion for dropping variables. The output provides the RSS, F-values, and p-values of each sub-model after dropping each variable from the starting model_full model. Thus the sub-models which drop1 compares here are:

$$Y \sim X2 + X3 + X4$$
$$Y \sim X1 + X3 + X4$$
$$Y \sim X1 + X2 + X4$$
$$Y \sim X1 + X2 + X3$$

The variable corresponding to the smallest RSS, the smallest F-value, or the largest p-value, is X4. Since the p-value here is 0.4025 (RSS = 1889.88, F-value = 0.704), very far from 0.05, we decide to drop the variable X4 and generate a new model, $Y \sim X1 + X2 + X3$, as newfit = update(model_full, .~.-X4).

This analysis implies that there is not enough evidence to support inclusion of $X4$ in the model. We now consider dropping additional variables of this new model newfit:

```
newfit = update(model_full, .~.-X4);
drop1(newfit,test="F")

Single term deletions

Model:
Y ~ X1 + X2 + X3
         Df Sum of Sq    RSS    AIC F value    Pr(>F)
<none>                 1889.9 457.19
X1        1    639.91 2529.8 513.51  66.365 4.323e-14 ***
```

```
X2        1    2254.57 4144.4 612.24 233.823 < 2.2e-16 ***
X3        1     391.03 2280.9 492.80  40.554 1.331e-09 ***
---
Signif. codes:  0 '***' 0.001 '**' 0.01 '*' 0.05 '.' 0.1 ' ' 1
```

The list of possible variables to be dropped here is one of the X1+X2+X3. Since none of those departures from the model generates p-values above 5% we stop here and propose this as the most appropriate model, that is $Y \sim X1 + X2 + X3$.

5.7.6 Forward Selection

In the forward selection procedure we successively add variables, using the add1 function; the variables to add are those which have a greatest impact on the regression, that is the highest p-value in some appropriate list. We start with the null model:

```
model_null=lm(Y~1,data=Test)
add1(model_null,test="F",scope =  .~.+X1+X2+X3+X4)

Single term additions

Model:
Y ~ 1
         Df Sum of Sq    RSS    AIC F value     Pr(>F)
<none>                 4991.9 645.45
X1        1     547.66 4444.3 624.21  24.399 1.661e-06 ***
X2        1    2101.83 2890.1 538.15 143.996 < 2.2e-16 ***
X3        1     275.55 4716.4 636.10  11.568 0.0008117 ***
X4        1    2516.49 2475.4 507.17 201.284 < 2.2e-16 ***
---
Signif. codes:  0 '***' 0.001 '**' 0.01 '*' 0.05 '.' 0.1 ' ' 1
```

The variable $X4$ appears to be most influential in that its addition produces the largest F-value (201.3) and the smallest p-value: 5.53×10^{-32}. Updating our model:

```
newfit_v1= update(model_null, .~.+X4)   #our latest regression model
```

We consider further additions (without showing the output), noting that the scope is +X1+X2+X3:

```
add1(newfit_v1,test="F",scope =  .~.+X1+X2+X3)    #output not shown here
```

From the output (not shown here) we notice that X2 is contributing the most, so we will include this variable (our model is now X2+X4). Considering adding the remaining variables:

```
newfit_v2= update(newfit_v1, .~.+X2)
add1(newfit_v2,test="F",scope =  .~.+X1+ X3)
newfit_v3= update(newfit_v2, .~.+X3);
add1(newfit_v3,test="F",scope =  .~.+X1)#our latest regression model
```

In the final stage the variable $X1$ is also added because the p-value of the corresponding test is less than 5%. Note that the output of the last line above coincides with the same test as that applied at the first step in the backward elimination method using drop1 above (with F-value $= 21.9$).

Exercise 5.21 *Verify the above process by analysing the output.*

From the above analysis the most appropriate model is determined to be the full model, $Y \sim X1+X2+X3+X4$, which is different to that proposed using the backward selection method. Hence it is not guaranteed that both of these approaches will converge to the same model (see later).

Exercise 5.22 (New York Restaurant data)

```
nyc = read.csv("http://www.math.smith.edu/~bbaumer/mth247/sheather/nyc.csv")
```

This data set is studied in Sheather (2009). It represents the average Price *of a dinner in 168 Italian restaurants in New York City, together with the customer ratings (measured on a scale of 0 to 30) of the* Food, Decor, *and* Service. *In addition, there is another variable indicating whether the restaurant is located to the east or west of 5th Avenue.*

1. *Justify the following lines of code:*

```
full <- lm(Price ~ Food + Decor + Service, data = nyc)
drop1(full, test="F")
newfit = update(full, .~.-Service)
stepAIC(full, test="F",scope=~Food + Decor + Service,direction ="backward")
```

2. *Perform the Forward selection procedure by starting with*

```
null <- lm(Price ~ 1, data = nyc)
```

5.7.7 Using AIC/BIC Criteria

In backward/forward implementations we started the search only from models full or null respectively, and determined the change based on the F-test. In fact we could have started from any other sub-model that we think is appropriate using an *alternative measure* of importance for including the variables in the model. They will not necessarily stop at the same model (according to some criterion such as highest R^2, AIC, etc.). The following line generates the AIC value for a particular model (say newfit):

```
extractAIC(newfit)[2]
```

```
[1] 457.1898
```

For the BIC criterion the additional term $k = \log(\dim(\text{Test})[1])$ (sample size=dim(Test)[1]) is needed:

```
extractAIC(newfit, k = log(dim(Test)[1]))[2] #n=dim(Test)[1] is the sample size

[1] 470.3831
```

If, in the Backward selection procedure, we wanted to use the AIC criterion then the process is carried out using stepAIC from the MASS library (the algorithm is explained below the following output):

```
stepAIC(model_full,scope=~X1+X2+X3+X4,direction ="backward")

Start:  AIC=458.47
Y ~ X1 + X2 + X3 + X4

        Df Sum of Sq    RSS    AIC
- X4     1      6.80 1889.9 457.19
<none>               1883.1 458.47
- X1     1    211.41 2094.5 477.75
- X3     1    254.47 2137.6 481.82
- X2     1    543.55 2426.6 507.19

Step:  AIC=457.19
Y ~ X1 + X2 + X3

        Df Sum of Sq    RSS    AIC
<none>               1889.9 457.19
- X3     1    391.03 2280.9 492.80
- X1     1    639.91 2529.8 513.51
- X2     1   2254.57 4144.4 612.24

Call:
lm(formula = Y ~ X1 + X2 + X3, data = Test)

Coefficients:
(Intercept)          X1          X2          X3
    105.399       1.903       3.386       1.456
```

In this output the search is done by dropping each of the variables in turn. Initially, the algorithm discards X4, since there is a problematic p-value there and that reduced model has the smallest AIC=457 (see the R-output above next to the -X4 line), and therefore this is used in the next step.

In the following step, the model Y~X1+X2+X3 is similarly analysed and the AIC values obtained by dropping the three variables in turn are 492, 513, and 612. Since these values are larger than the previously chosen model with AIC=457, the process stops. Therefore, based on the AIC criterion the same model Y~X1+X2+X3 is chosen as above. To use BIC the following code is required (no output shown here):

```
stepAIC(model_full,scope=~X1+X2+X3+X4,k = log(dim(Test)[1]),direction ="backward")
```

The default value, $k = 2$, is applicable for the AIC, while $k = n$ (as above) is applicable for the BIC values. Note that $n = 200 =$ dim(Test)[1].

Exercise 5.23 *By exploring the R-output of the line above, identify the chosen model whose BIC value is 470.38.*

5.7.8 LRT in Model Selection

We now perform a variable selection procedure based on the Likelihood ratio test explained above which uses the χ^2 distribution. The following line is the same as in backward selection with the change from test="F" to test="Chisq":

```
drop1(model_full, test="Chisq")

Single term deletions

Model:
Y ~ X1 + X2 + X3 + X4
        Df Sum of Sq    RSS    AIC  Pr(>Chi)
<none>                1883.1 458.47
X1       1   211.41 2094.5 477.75 3.967e-06 ***
X2       1   543.55 2426.6 507.19 1.066e-12 ***
X3       1   254.47 2137.6 481.82 4.781e-07 ***
X4       1     6.80 1889.9 457.19    0.3959
---
Signif. codes:  0 '***' 0.001 '**' 0.01 '*' 0.05 '.' 0.1 ' ' 1
```

The output is providing the RSS, *AIC* values, and p-values of the χ^2 tests of each sub-model after dropping each variable from the starting full model. Based on the output, we chose to drop the appropriate variable corresponding to the highest p-value ($X4$ in this case). In particular the test statistics relevant to the sub-model which discards $X4$ is

$$200 \log(RSS_{-X4}) - 200 \log(RSS_{model_full}) = 200 \log \left(\frac{1889.8759}{1883.0776} \right) = 0.7207$$

and $P(\chi^2_{df=1} > 0.7207) = pchisq(0.7207, df = 1, lower.tail = F) = 0.3959$ (in agreement with the output). Since this p-value is far from 5%, we accept the smaller model, that is the predictor $X4$ is dropped. In fact the same value of the χ^2 statistics could have been obtained from Equation 5.11: $457.19 - 458.47 + 2 = 0.72$.

Exercise 5.24 *Once you understand how to run the χ^2 test for hypothesis testing, perform the additional steps for the variable (Forward and Backward) selection for the New York data set using the χ^2 test.*

- Perform the variable (Forward and Backward) selection for the New York data set using the χ^2 test.
- Perform the variable selection for the (Forward and Backward) Bridge construction data set using the χ^2 test.

5.7.9 Automatic Search Using R-squared Criteria

We can also perform automatic model selection in R using the package `leaps` which is designed to perform an exhaustive search for the possible models based on R^2 and its Adjusted version, without the need for manual intervention. *Due to space efficiency*, we will run this analysis initially for the dataset `trees` (the same dataset used earlier in the chapter); the Test data is analysed as an exercise. Let us consider these lines of code:

```
library(leaps);    (M=leaps(x=trees[,1:2],y=trees[,3],method="r2"))

$which
      1     2
1  TRUE FALSE
1 FALSE  TRUE
2  TRUE  TRUE

$label
[1] "(Intercept)" "1"              "2"

$size
[1] 2 2 3

$r2
[1] 0.9353199 0.3579026 0.9479500
```

Note that the first argument, x, is a matrix containing the predictors as its columns, and the second variable y is the response. The search is involving *all* the possible model subsets and would be in general too large to show (hence the use of the `trees` data).

The starting model here is $y \sim X$ (`Volume ~ Diameter`). Note that in this interface the variable names are not used but just simply 1,2 (for Diameter and Height, respectively) as that is the order of the input x matrix.

One could provide the models in the correct order (i.e. in increasing R^2 order) by the criteria used:

```
M$which[ order(M$r2), ]

      1     2
1 FALSE  TRUE
1  TRUE FALSE
2  TRUE  TRUE
```

These results show the `Volume~Height` model has the lowest R^2 value, with the `Volume~Diameter+Height` model having the highest. A cleared table for these models ordered by the R^2 is obtained as

```
cbind(M$which[ order( M$r2 ), ],sort(M$r2))

  1 2
1 0 1 0.3579026
1 1 0 0.9353199
2 1 1 0.9479500
```

These results are consistent with the results obtained earlier.

Exercise 5.25

- *Run the following lines and try to understand its output*

```
M1=leaps(x=Test[,-1] ,y=Test[,1],method="r2")
cbind(M1$which[ order( M1$r2 ), ],sort(M1$r2))
```

5.7.10 Concluding Remarks on Test Data

The chosen model here is therefore `final_fit=lm(Y~X1+X2+X3,data=Test)`:

$$Y = 105.399 + 1.903 \times X1 + 3.386 \times X2 + 1.456 \times X3$$

Of all these three predictors, the largest effect in Y is that from $X2$. A drop or increase by one unit in $X2$ effects Y by about three units; this is about twice as much as that of $X3$. These variables here are chosen for illustrative purposes only but in practice such variables will represent real quantities of interest.

The stepwise backward/forward procedures that we implemented for the Test dataset lead to different models (one was the full model and the other was the model stated above). These procedures compare nested models. In general, the forward selection might stop too early. The chosen model from backward selection would appear to fit too well (overfit) as its choice would depend on the chosen criterion. It is therefore advisable to use other alternatives to the standard F-test. In our case, the AIC/BIC and LRT as well as the t-scores seems to steer our decision towards the model without the variable $X4$. In general, if any specific knowledge about some particular variable is present we can force that in the model prior to adding/deleting the rest of variables.

5.7.11 Modelling Beyond Linearity

If, however, some other polynomial term is needed in the modelling the linear regression could be easily adjusted as `=lm(Y~X1+X2+I(X2^2) +X3,data=Test)`, where we have added the predictor `X2^2`. The decision to add such power terms depends on the data structure but some care is needed if higher powers are used as it is harder to distinguish between the errors and the polynomial predicting functions. In finance applications, it is common to apply additional transformations to the variables so that the underlying assumptions of the fitted models are met. For example a log transform of the dependent variable Y as `lm(log(Y) ~X1+X2 +X3)` might improve on the validity of the model. Most of transformations considered are those of the Box-Cox type:

$$y_\lambda = \begin{cases} \frac{y^\lambda - 1}{\lambda} & \lambda \neq 0 \\ \log(y) & \lambda = 0 \end{cases}$$

One could consider fitting the linear regression to the `trees` data set for various values of λs using the `boxcox` function:

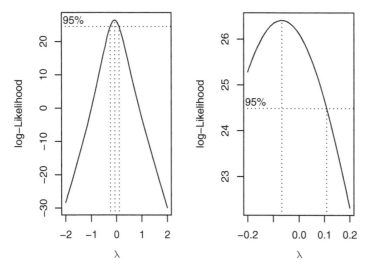

Figure 5.18 Plots for various lambda with log(covariates).

suggesting that $\lambda = 0$ is not a bad choice. It seems that the log transformation is appropriate here. This is not surprising as in biology when dealing with growth, just like in finance when dealing with constant force of interest, or mortality rates as in survival models, the exponential transformations appear naturally.

5.8 Dummy/Indicator Variable Regression

5.8.1 Introducing Categorical Variables

This section refers to situations where one or more of the explanatory variables in a standard regression model is "categorical". This is in contrast with "continuous" variables, which have been seen in the examples in this chapter up to this point, for example diameter of trees, height, etc. Common examples of categorical variables include sex, post code, car model, etc. Note that the number of categories may be chosen by the analyst designing the investigation (e.g. level of expertise could be categorised as beginner/competent/expert, or perhaps only as beginner/expert); other times there may be less, or no choice (e.g. post code, model of car). Also note that categorical variables can be either numeric or non-numeric, whereas continuous variables are always numeric. We will see further examples of categorical variables in Chapters 21 and 27.

In order to work with categorical variables we must assign what are referred to as "indicator", or "dummy" variables – we are simply identifying various categories with numbers rather than words. This will become clearer from the examples below (and those in Chapters 21 and 27).

For now, we consider the following simple example.

Example 5.11 In order to save on fuel, a farmer changed the vehicle that was used for delivering goods to local shops. The farmer wants to analyse the impact of the new vehicle (which uses the same type of fuel) and has collated relevant data – see `Farmer.csv`. (The first three and the last three entries of this data are shown in Table 5.2.)

The column "Amount" represents the amount of fuel used for each delivery, the column "Load" represents the weight of goods in each delivery, and "Vehicle" is a categorical variable, defined as 0 if the old vehicle is used, and 1 if the new one is used (see the box plot in Figure 5.19 which sets out an initial analysis of the data). We want to develop a method to test whether the new vehicle has indeed resulted in a reduction in fuel costs.

Table 5.2 A few entries from Farmer data.

	X	Amount	Load	Vehicle
1	1	74	38	0
2	2	90	30	0
3	3	64	32	0
921	921	61	34	1
922	922	68	35	1
923	923	64	35	1

We can set up a simple regression model, $Y = \theta_0 + \theta_1 X + e$, with:

- Y = Amount of fuel consumption.
- X = An indicator variable taking value 0 for Old vehicle, and 1 for New vehicle.

After loading the data as `Farmer=read.csv(''~/Farmer.csv'')` we run the ordinary regression model:

```
DM=lm(Amount~Vehicle,data=Farmer)
```

The output in R is summarised as follows:

	Estimate	Std. Error	t-value	Pr(> |t|)
(Intercept)	79.9300	0.4178	191.32	0.0000
Vehicle	-15.2436	0.5550	-27.47	0.0000

The model has two forms depending on the values of indicator variable X:

$$Y = \theta_0 + \theta_1 X + e = \begin{cases} \theta_0 + \theta_1 + e & X = 1 \\ \theta_0 + e & X = 0 \end{cases}$$

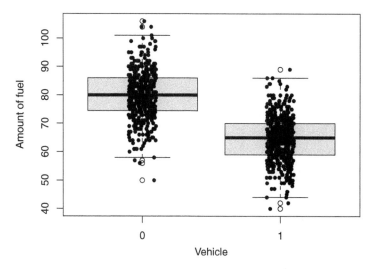

Figure 5.19 Box plots of Amount of fuel, 0-Old Vehicle, 1-New Vehicle.

The fitted regression line is therefore:

$$Amount = 79.93 - 15.24 \, Vehicle + e$$

The predicted Amounts of fuel for the Old and New vehicle (from this simple model) are therefore 79.93 and 64.69, respectively – the difference is the slope coefficient of the $Vehicle$ indicator variable.

The appropriate hypothesis testing in this case is:

$$H_0 : \theta_1 = 0 \quad \text{vs} \quad H_1 : \theta_1 < 0.$$

The corresponding t-statistic takes the value -27.466; there is therefore extreme evidence against H_0 and the assumption of $\theta_1 < 0$ holds. The 95% confidence interval for $\hat{\theta}_1$ is $\hat{\theta}_1 \pm t(\alpha/2, n-2)se(\hat{\theta}_1)$ and is obtained using:

```
confint(DM,level=0.95)

                2.5 %     97.5 %
(Intercept)   79.11009   80.74991
Vehicle      -16.33280  -14.15435
```

In fact what this regression model is doing is essentially a t-test for differences in means for the two vehicles (see Section 5.4.3.2):

```
t.test(Amount~Vehicle,data=Farmer,var.equal=TRUE)

Two Sample t-test

data:  Amount by Vehicle
t = 27.466, df = 921, p-value < 2.2e-16
alternative hypothesis: true difference in means between group 0 and group 1 is
not equal to 0
```

```
95 percent confidence interval:
 14.15435 16.33280
sample estimates:
mean in group 0 mean in group 1
       79.93000          64.68642
```

Note the sample estimates for each group in the `t-test` are the same as those predicted by the regression model (79.93 and 64.69).

5.8.2 Continuous and Indicator Variable Predictors – Including Load in the Model

In the above section we included only vehicle type in our model. We now consider the variable Load in our modelling which will, hopefully, result in significant improvements to our proposed model. Note that the indicator variable in the example above has affected the choice of intercept part of the models, this was equivalent to the means difference as the `t-test` confirmed. There are other ways that the indicator variable can alter the models. For example, the new vehicle could have tempted the farmer to carry less load so that could have also been a reason for less fuel consumption. This interaction between the values of Load and type of Vehicle is studied below. We now have:

- Y = Amount of fuel consumption.
- X = Load.
- C = An indicator variable taking value 0 for Old vehicle, and 1 for New vehicle.

We proceed to consider four possible models (1–4):

1. *Coincident regression lines* (see Figure 5.20(1)).

$$Amount = \theta_0 + \theta_1 Load + e$$

```
mod_c=lm(Amount~Load,data=Farmer)
```

Summary of results:

	Estimate	Std. error	t-value	Pr(> |t|)
(Intercept)	84.4285	2.2166	38.09	0.0000
Load	-0.3442	0.0573	-6.01	0.0000

The model predicts the following (a graphical representation of this fitted model is shown in Figure 5.20(1)):

$$Amount = 84.43 - 0.34 \times Load$$

Remark 5.7 Each model (1-4) produces one (as above), or two straight lines, that is of the form: $Amount = m \times (Load) + c$.

2. *Parallel regression lines* (see Figure 5.20(2)).

$$Amount = \theta_0 + \theta_1 Load + \theta_2 C + e = \begin{cases} \theta_0 + \theta_1 Load + e & Vehicle = 0 \\ (\theta_0 + \theta_2) + \theta_1 Load + e & Vehicle = 1 \end{cases}$$

```
mod_pr=lm(Amount~Load+Vehicle,data=Farmer)
```

	Estimate	Std. error	t-value	Pr(> |t|)
(Intercept)	80.3261	1.6834	47.72	0.0000
Load	-0.0110	0.0451	-0.24	0.8081
Vehicle	-15.2041	0.5786	-26.28	0.0000

The model predicts the following:

$$Amount = \begin{cases} 80.33 - 0.01 \times Load & Vehicle = 0 \\ 65.12 - 0.01 \times Load & Vehicle = 1 \end{cases}$$

Note that the difference in intercepts is simply 15.2.

3. *Regression lines with equal intercepts but different slopes (interaction)* (see Figure 5.20(3)).

$$Amount = \theta_0 + \theta_1 Load + \theta_3(Load \times C) + e = \begin{cases} \theta_0 + \theta_1 Load + e & Vehicle = 0 \\ \theta_0 + (\theta_1 + \theta_3)Load + e & Vehicle = 1 \end{cases}$$

```
mod_eq_in=lm(Amount~Load+Load:Vehicle,data=Farmer)
```

	Estimate	Std. error	t-value	Pr(> |t|)
(Intercept)	71.9252	1.7420	41.29	0.0000
Load	0.2186	0.0483	4.53	0.0000
Load:Vehicle	-0.3988	0.0152	-26.29	0.0000

$$Amount = \begin{cases} 71.9251789 + 0.2186416 \times Load & Vehicle = 0 \\ 71.9251789 - 0.1801729 \times Load & Vehicle = 1 \end{cases}$$

This model, and model (4), represent the scenario where the effect of the Load on the amount of Fuel consumed depends on whether the Vehicle is Old or New. We shall see various examples of such "interactions" later in the book.

4. *Unrelated regression lines* (see Figure 5.20(4)).

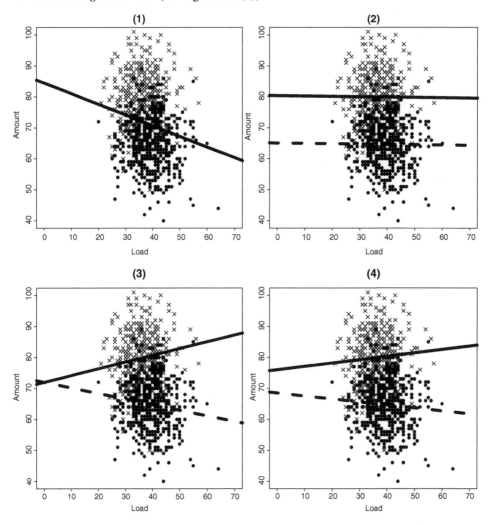

Figure 5.20 The fitted regression lines of the four models with data points as: Circles-New, Crosses-Old.

$$Amount = \theta_0 + \theta_1 Load + \theta_2 C + \theta_3 (Load \times C)$$

$$+e = \begin{cases} \theta_0 + \theta_1 Load + e & Vehicle = 0 \\ (\theta_0 + \theta_2) + (\theta_1 + \theta_3) Load + e & Vehicle = 1 \end{cases}$$

```
mod_un_rel=lm(Amount~Load+Vehicle+Load:Vehicle,data=Farmer)
```

$$Amount = \begin{cases} 75.97 + 0.11 \times Load & Vehicle = 0 \\ 68.49 - 0.1 \times Load & Vehicle = 1 \end{cases}$$

	Estimate	Std. Error	t value	Pr(> |t|)
(Intercept)	75.9731	2.5660	29.61	0.0000
Load	0.1095	0.0701	1.56	0.1184
Vehicle	-7.4830	3.4889	-2.14	0.0322
Load:Vehicle	-0.2053	0.0915	-2.24	0.0251

Note that in Model 4 both the intercept and slope differ between the vehicle types, namely the effect of Vehicle affects both of these parameters via θ_2 and θ_3.

We can test whether the type of vehicle has an effect on the amount of fuel spent by testing, whether

$$H_0 : \theta_2 = 0 = \theta_3 \quad H_1 : \text{at least one of them is not zero}$$

This is done by applying the ordinary F-test between the restricted (coincident lines) and unrestricted model (unrelated regression lines) as:

```
        anova(mod_c,mod_un_rel)

Analysis of Variance Table

Model 1: Amount ~ Load
Model 2: Amount ~ Load + Vehicle + Load:Vehicle
  Res.Df    RSS Df Sum of Sq      F    Pr(>F)
1    921 112556
2    919  63946  2     48610 349.3 < 2.2e-16 ***
---
Signif. codes:  0 '***' 0.001 '**' 0.01 '*' 0.05 '.' 0.1 ' ' 1
```

The extremely small p-value suggests that the Vehicle type has an effect on the amount of fuel used, and we should reject H_0.

Exercise 5.26 *Use the AIC criteria to determine the most suitable of the four models considered above.*

In these examples we have considered only indicator variables (such as C with 0-1 values) as real valued explanatory variable in R. We could have also considered instead regressing as categorical variables by using $as.factor(C)$ (which treats C as categorical) in regression. The results would have been the same as above. In general, explanatory variables can have multiple categories. We shall see such examples in Chapter 27 when we analyse the effect of different postcodes on claims.

5.9 Recommended Reading

- Davison, A. (2003). *Statistical Models*. Cambridge, UK: Cambridge University Press.
- Faraway, J.J. (2014). *Linear Models with R, 2e*. New York: Chapman and Hall/CRC.
- Sheather, S. (2009). *A Modern Approach to Regression with R*. New York, NY: Springer.
- Verzani, J. (2014). *Using R for Introductory Statistics, 2e*. Boca Raton, Florida, US: Chapman and Hall/CRC.

6

Multivariate Distributions, and Sums of Random Variables

Peter McQuire

```
library(MASS); library(psych); library(mvtnorm); library(mnormt)
```

6.1 Multivariate Distributions – Examples in Finance

Actuaries will frequently be interested in the joint behaviour of multiple variables, and in particular, the distributions of the sums of these variables. For example:

- we may wish to understand the distribution of investment returns from an asset portfolio, which consists of the individual returns from a number of individual assets, and to understand the benefits of diversification (see Chapters 7 and 8);
- in Chapter 10 Section 7, we look at the sum of asset values and liability values to determine the likelihood of an entity's future insolvency, and hence develop an appropriate investment strategy to manage this risk;
- similarly, in Chapter 11 we look at the sum of price changes of an asset and a hedging instrument with the aim of finding a combination such that the net price change is as stable as possible;
- in Chapter 15 we model price changes of a bond portfolio, allowing for correlations between credit ratings of the individual constituent bonds;
- we may wish to understand how likely it is that an insurer's total claims will exceed a particular level; the total claims will consist of claims incurred across individual policies (see Chapter 24 and 25), and also under various types of policies (e.g. motor, buildings, marine insurance).

  The study of multivariate distributions is therefore likely to be of particular importance to actuaries. McNeil et al. (2005) provides an excellent introduction to multivariate modelling.

R Programming for Actuarial Science, First Edition. Peter McQuire and Alfred Kume.
© 2024 John Wiley & Sons Ltd. Published 2024 by John Wiley & Sons Ltd.
Companion Website: www.wiley.com/go/rprogramming.com.

6.2   Simulating Multivariate Normal Variables

We will find it particularly useful to produce simulations from multivariate distributions when discussing a number of topics included in this book. We do this in some detail in Chapters 13, 14, and 15, and also in those chapters referred to in Section 6.1.

In this section we generate simulations from a simplified model – a multivariate Normal distribution. This will be useful in obtaining a basic understanding of scenarios where many variables are involved, such as those listed above. Of course, when modelling particular scenarios great care is required to ensure the multivariate model is an accurate reflection of the data; use of the multivariate Normal distribution is likely to be an approximation and judgement may be required to determine whether such approximations are acceptable. (The reader is directed towards Chapter 13 which describes a methodology for potentially more accurate statistical modelling of several variables. An exercise is also included below which introduces the reader to the multivariate t-distribution.)

Where $\mathbf{X} = (X_1, X_2, ..., X_d)$ has a multivariate Normal distribution with d dimensions, the density function is given by:

$$f(\mathbf{x}) = \frac{1}{(2\pi)^{d/2}(\det \Sigma)^{0.5}} \exp[-\frac{1}{2}(\mathbf{x} - \mu)'\Sigma^{-1}(\mathbf{x} - \mu)] \tag{6.1}$$

where μ is the vector containing the mean values i.e. $E(\mathbf{X})$, and Σ is the covariance matrix. (See Example 6.1 for a calculation using Equation 6.1.)

In the examples and exercises which follow, we will use a multivariate Normal distribution, as described by Equation 6.1, consisting of the following three Normal distributions ($d = 3$):

$A \sim N(2, 2^2)$, $B \sim N(5, 4^2)$, $C \sim N(8, 6^2)$, with correlation coefficients: $\rho_{AB} = 0.1$, $\rho_{AC} = 0.5$, $\rho_{BC} = 0.9$.

Setting up these parameter values – the mean vector (μ) and covariance matrix (Σ):

```
mean_vecx<-c(2,5,8)      # mu

varA<-4; varB<-16; varC<-36; corAB<-0.1; corAC<-0.5; corBC<-0.9
covAB<-corAB*(varA*varB)^0.5
covAC<-corAC*(varA*varC)^0.5
covBC<-corBC*(varB*varC)^0.5

(covar_matx<-matrix(c(varA,covAB,covAC,covAB,varB,covBC,covAC,covBC,varC),3,3))

      [,1] [,2] [,3]
[1,]  4.0  0.8  6.0
[2,]  0.8 16.0 21.6
[3,]  6.0 21.6 36.0
```

The covariance matrix includes the variance terms as diagonal entries, and covariance terms as the appropriate non-diagonal entries. For example $Cov(A, B) = \rho_{AB}\sigma_A\sigma_B = 0.1 \times \sqrt{4} \times \sqrt{16} = 0.8$. ($\sigma_i$ is the standard deviation of the ith random variable.)

Example 6.1

By writing suitable code in respect of Equation 6.1, calculate $f(0, 1, 1)$, and verify your answer using the dmnorm function from the mnormt package:

```
data_xyz<-c(0,1,1) # data point
fx_num<-exp(-(t(data_xyz-mean_vecx)%*%solve(covar_matx)%*%(data_xyz-mean_vecx))/2)
fx_den<-(2*pi)^(length(data_xyz)/2)*(det(covar_matx))^0.5
fx_num/fx_den

              [,1]
[1,] 0.002852655

dmnorm(data_xyz, mean_vecx, covar_matx, log = FALSE)      #alternative calc

[1] 0.002852655
```

To simulate realisations from this multivariate Normal distribution, we use the mvrnorm function from the MASS package; this function requires the following inputs: (1) the number of simulations required, (2) a vector, μ, containing the mean values of each univariate distribution, and (3) a matrix containing all covariance terms, that is, the covariance matrix, Σ. The following code produces 10,000 simulations, using the parameter values set out above:

```
set.seed(20000);   simsN<-10000
mvnTry<-mvrnorm(simsN,mean_vecx,covar_matx)           #10,000 simulations
apply(mvnTry,2,mean);   cov(mvnTry)      #checks on mean and cov matrix

[1] 2.001322 5.012909 8.020593
            [,1]          [,2]          [,3]
[1,] 3.9205044   0.8198129   5.906483
[2,] 0.8198129  16.0688023  21.768155
[3,] 5.9064828  21.7681552  36.133111
```

The simulated data is plotted in Figure 6.1 (using the pairs.panels function from the psych package); it also shows the histograms of the simulations from each univariate distribution, and the correlation coefficient between each pair (e.g. $\rho_{AB} = 0.1$). The statistics calculated above are consistent with the parameter values used in the simulations (allowing for stochastic error). A sample of the first 5 simulations is shown below:

```
round(mvnTry[1:5,],2)

        [,1]   [,2]   [,3]
[1,]   1.48   0.02   1.43
[2,]  -3.33   0.31  -4.34
[3,]   0.55  -0.88  -2.36
[4,]   0.98   4.80   8.21
[5,]   5.23   8.28  15.98
```

Plots of Bi-variate Normal distribution density functions: (AB) and (BC)

Figure 6.2 shows two 3-dimensional plots of a bivariate Normal distribution density function, using the same univariate distributions (1) A and B, and (2) B and C (described earlier), and Equation 6.1 (with $d = 2$):

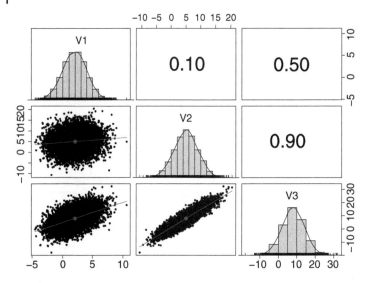

Figure 6.1 Simulations from a Multivariate Normal distribution.

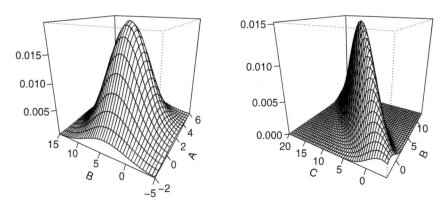

Figure 6.2 Bi-variate Normal distributions – AB and BC.

```
#showing only the code for AB.....
mean_vecAB<-c(2,5)                                    #X and Y only
(covar_vecAB<-matrix(c(varA,covAB,covAB,varB),2,2)) #X and Y only

        [,1] [,2]
[1,]    4.0  0.8
[2,]    0.8 16.0

x_bi<-seq(-2, 6, 0.5);    y_bi<-seq(-5, 15, 0.5)  #ranges for plot
f_bi<-function(x, y) dmnorm(cbind(x, y), mean_vecAB, covar_vecAB)
z_val<-outer(x_bi, y_bi, f_bi)
```

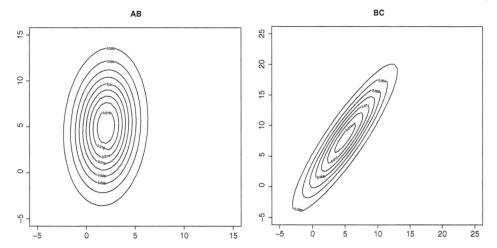

Figure 6.3 Bi-variate Normal distribution – Contour plots of density function.

Figure 6.3 shows contour plots of the equivalent density distributions.

Exercise 6.1 *Verify, using Equation 6.1, a number of values shown in Figure 6.3.*

Remark 6.1 The mvrnorm function uses the Singular Value Decomposition ("SVD") method to simulate data points from a multivariate Normal distribution. An alternative method is the Choleski decomposition of the relevant covariance matrix. The reader may wish to review and compare these two approaches to simulating such data.

Exercise 6.2 ***Multivariate t-distribution*** *The reader is encouraged to also experiment with the multivariate t-distribution. Included below is sample code which produces simulations from such a distribution, using the rmvt function from the mvtnorm package.*

```
mvnTry_t<-rmvt(simsN, sigma=covar_matx, df = 10, delta=mean_vecx, type="shifted")
```

6.3 The Summation of a Number of Random Variables

As noted at the start of Section 6.1 we will frequently be interested in the distribution of the sum of a number of random variables. Thus, where we have n variables, $X_1, X_2, X_3, X_4, ...X_n$:

$$X_T = X_1 + X_2 + ... + X_n = \sum_{i=1}^{n} X_i \tag{6.2}$$

where, for example, X_T is a random variable representing the insurer's total claim amount, and X_i is a random variable representing the claim amount from department i. Equation 6.3 gives the expression for the mean of the sum of n random variables:

$$E(X_T) = E(X_1 + X_2 + ... + X_n) = \sum_i^n E(X_i) \tag{6.3}$$

Equation 6.4 gives the important expression for the variance of the sum of a number of random variables, which will be applied several times in this book:

$$Var(X_T) = Var(X_1 + X_2 + ... + X_n) = \sum_i^n \sum_j^n Cov(X_i, X_j) \tag{6.4}$$

In words, "the variance of a sum equals the sum of all the covariances".

Below, we look at two relatively simple cases. First, that of two Poisson random variables; the 2nd example involves the sum of three Normal random variables. We look at examples where the variables are independent, and then where dependency exists to varying degrees.

Sum of random variables from two Poisson distributions

For the 2-variable case (Y and Z), Equation 6.4 becomes (where $Cov(Y,Z) = \rho_{YZ}\sigma_Y\sigma_Z$):

$$Var(Y + Z) = \sigma_Y^2 + \sigma_Z^2 + 2\rho_{YZ}\sigma_Y\sigma_Z \tag{6.5}$$

We will assume: $Y \sim Poisson(5)$, $Z \sim Poisson(5)$. If Y and Z are independent, $\rho_{YZ} = 0$ and $Var(Y + Z) = 5 + 5 = 10$. If Y and Z are completely dependent, $\rho_{YZ} = 1$, and $Var(Y+Z) = 5 + 5 + 2 \times 1 \times \sqrt{5}\sqrt{5} = 20$. This is demonstrated below by running simulations, first where $\rho = 0$:

```
set.seed(1110);   sims<-100000;  y<-rpois(sims,5); z<-rpois(sims,5)
x<-y+z
mean(x);     var(x)

[1] 9.99961
[1] 9.97285
```

Now if they are completely dependent ($\rho = 1$):

```
xx<-y+y #because z=y
mean(xx);    var(xx)

[1] 9.9889
[1] 19.941

head(x,8);   head(xx,8)

[1] 11 11 13  8 10  9 11 15
[1] 10 14 16 10  6  4 10 22
```

Even from a sample of only 8 values, one can see a greater spread where the values are dependent. Mean values are not affected by the degree of correlation – in both cases: $E(Y + Z) = 5 + 5 = 10$.

In reality, of course, the level of dependency is likely to fall between the two extremes demonstrated above. The above scenarios may reflect the position where an insurer writes two portfolios of marine insurance; if the portfolios are in similar geographical positions it may be that the number of claims from the portfolios are highly correlated. If they are from two areas in quite different parts of the world there may be little correlation between the number of claims.

Sum of random variables from a multivariate Normal distribution

We can easily extend the scope of our calculations using the mvrnorm function from Section 6.2. Here we look at the sum of the three random variables, A, B, and C, from Section 6.2.

Exercise 6.3 *The variance of the sum of 2 random variables can be calculated using Equation 6.5. Write down an expression for the variance of the sum of 3 random variables, and calculate the variance of $A + B + C$ (112.8 – see below).*

Analysing the simulations from Example 6.1 and the mvnTry object created:

```
sum_margx<-rowSums(mvnTry)
mean(sum_margx);        var(sum_margx)

[1] 15.03482
[1] 113.1113

sum(mean_vecx);         sum(covar_matx)   #check against theory

[1] 15
[1] 112.8
```

Exercise 6.4 *Re-run the simulations above assuming zero correlation ($\rho = 0$) between each of the 3 variables A, B, and C. The results are shown below:*

```
covar_matx2

      [,1] [,2] [,3]
[1,]    4    0    0
[2,]    0   16    0
[3,]    0    0   36

mean(sum_margx2);      var(sum_margx2)

[1] 14.97255
[1] 56.58274

sum(covar_matx2)       #check against theory

[1] 56
```

The variance of the sum of variables exhibiting little correlation is lower than that for highly correlated variables; this important concept is developed in many chapters of the book.

6.4 Conclusion

As noted in the Introduction, there are several examples throughout this book of applications of the simulations and variance calculations discussed above. The reader may wish to review this chapter before reading those chapters referred to in the Introduction.

Exercise 6.5 *The reader is encouraged to experiment with numerous combinations of parameter values relating to all calculations set out in this chapter.*

6.5 Recommended Reading

- Anderson, T.W. (2003). *An Introduction to Multivariate Statistical Analysis*, 3e. Hoboken, New Jersey, US: John Wiley and Sons.
- Giri, N.C. (2004). *Multivariate Statistical Analysis: Revised and Expanded*, 2e. Boca Raton, Florida, US: CRC Press.
- McNeil, A.J., Frey, R., and Embrechts, P. (2005). *Quantitative Risk Management*. Princeton, NJ: Princeton University Press.
- Ripley, B.D. (1987). *In Stochastic Simulation*. Wiley Online Library: John Wiley and Sons.
- Rizzo, M.L. (2019). *Statistical Computing with R, Second Edition*, 2e. Boca Raton, Florida, US: Chapman and Hall/CRC.
- Sweeting, P. (2017). *Financial Enterprise Risk Management*, 2e. Cambridge, UK: Cambridge University Press.
- Venables, W.N. and Ripley, B.D. (2002). *Modern Applied Statistics with S*, 4e. New York, NY: Springer.

7

Benefits of Diversification

Peter McQuire

```
library(MASS)
```

7.1 Introduction

In this chapter we will look at why diversification is one of the key tools in risk management. In particular, we shall look at:

- the importance of holding uncorrelated risks, or at least risks which can be considered not to be highly correlated;
- the effect of increasing the number of risks.

We shall also demonstrate these concepts by simulating returns from various asset portfolios. The chapter does contain less coding than most chapters in the book.

Remark 7.1 When we discuss "risks" in this chapter, we will typically be referring to insurance policies, loans or assets. For example the first bullet point could relate to insurance companies selling uncorrelated insurance policies, or to a pension scheme investing in several uncorrelated assets.

7.2 Background

Diversification is used by most financial institutions as a way of reducing uncertainty in future financial results; banks diversify by writing loans to many different individuals and institutions; insurance companies diversify by writing lots of largely uncorrelated insurance policies. Investment strategies adopted by insurers, occupational pension schemes and other institutions will be heavily influenced by the concept of diversification by investing in several different types of investment opportunities. Companies in general should be aware of the benefits which diversification can bring, for example by trading in various countries, selling a wide variety of products and using various trading

R Programming for Actuarial Science, First Edition. Peter McQuire and Alfred Kume.
© 2024 John Wiley & Sons Ltd. Published 2024 by John Wiley & Sons Ltd.
Companion Website: www.wiley.com/go/rprogramming.com.

partners. Through diversification institutions can make their financial outcomes more predictable – this is the key point.

For example, you are unlikely to be willing to insure a friend's £1 m house, as the outcomes are very uncertain. If it is destroyed in a storm or fire you would be liable for the costs of rebuilding the house. Or, more likely, you will incur no costs (it isn't destroyed). The possible cashflows in one year are, broadly speaking, £0 or say, £1 m; insuring only one house is an extremely poorly diversified position. However an insurer may well insure the house, largely due to the fact that they are already insuring thousands of houses, and insuring one more may make their results even more predictable. (They will also have more expertise and historic data so can probably measure the risk more accurately.)

In general, the more policies written, or assets invested in, the more diversified the position will be, leading to greater certainty in future financial results. However, writing lots of highly correlated risks will not result in a well-diversified portfolio; buying shares in lots of different supermarket companies will not lead to a particularly well-diversified position.

In this chapter we will compare asset portfolios which consist of both a small and large number of different assets, and also portfolios which exhibit high and low degrees of correlations between the assets. These numerical examples should demonstrate the above ideas. Before we do so we first need to formalise our ideas with some fundamental mathematics.

7.3 Key Mathematical Ideas

First note that the return on our portfolio of n assets is given by:

$$R_p = w_1 R_1 + w_2 R_2 + \ldots + w_n R_n = \sum_{i=1}^{n} w_i R_i \tag{7.1}$$

where w_i is the proportion of the portfolio invested in asset i, $\sum_{i=1}^{n} w_i = 1$, and R_i is the return on asset i. The key equation to understanding the effects of diversification is the following:

$$\text{Variance}(R_p) = \text{Variance}(w_1 R_1 + w_2 R_2 + \ldots + w_n R_n) = \sum_{i}^{n} \sum_{j}^{n} w_i w_j \text{Cov}(R_i, R_j)$$

$$= \sum_{i}^{n} w_i^2 \text{Var}(R_i) + \sum_{i}^{n} \sum_{j \neq i}^{n} w_i w_j \text{Cov}(R_i, R_j) \tag{7.2}$$

where $\text{Var}(R_i)$ is the variance of the returns on asset i, and $\text{Cov}(R_i, R_j)$ is the covariance of the returns from assets i and j. In words, the variance of a sum is the sum of the covariances. It is difficult to over-state the importance of this equation.

Equation 7.2 can be written in matrix form, which simplifies our calculations. It is not a quick exercise to write the full expression for the variance of the returns from a 100-asset portfolio. In matrix form:

$$\text{Var}(R_p) = w^T C w \tag{7.3}$$

where w is a column vector containing the weights of each asset (w^T is its transpose, a row vector) and C is the covariance matrix. The diagonal entries in C are the variances of the returns of the n assets and the non-diagonal entries are the covariances of each asset pair.

For n assets, *where we have invested equally in each asset*, such that $w_i = \frac{1}{n}$, we can write:

$$\text{Var}(\frac{R_1}{n} + \frac{R_2}{n} + ... + \frac{R_n}{n}) = \frac{\overline{\text{Var}(R_i)}}{n} + (n-1)\frac{\overline{\text{Cov}(R_i, R_j)}}{n} \tag{7.4}$$

where $\overline{\text{Cov}(R_i, R_j)}$ is defined as the average of the covariances between all the asset pairs.

This equation is very important and perhaps provides a better understanding of diversification. It is exact provided that we are invested equally in our n assets. Note that we do not need to assume that all variances and covariances are equal (see later examples).

Exercise 7.1 *Explain the steps involved in arriving at Equation 7.4, and verify that it is consistent with our 2-asset case.*

Exercise 7.2 *Write down, using Equation 7.2, an expression for the variance of the returns from a 4-asset portfolio with allocations 10%, 20%, 30% and 40%.*

How are we to interpret Equation 7.4? First note that as n gets large (e.g. 100 assets in our portfolio), the 1st term on the RHS approaches zero. This is often referred to as "diversifying away our specific (or diversifiable) risk."

As n gets large the 2nd term on the RHS approaches $\overline{\text{Cov}(R_i, R_j)}$, the average covariance between our asset pairs. It's also important to note that, if all the covariance terms are of similar size, this term gets a little bigger as we keep increasing n (see Figure 7.1). *We cannot remove this by increasing the number of assets*; we can only reduce it by reducing the degree of correlation between our assets.

Figure 7.1 illustrates the relationship between the variance of returns from our portfolio and n. For simplicity we'll assume that each asset return has the same standard deviation ($\sigma = 0.1$) and the correlation between each pair of asset returns is 0.2 ($\rho = 0.2$); note that it is the average variance and covariance which is important.

```
var_sum<-0;    covar_sum<-0;    total_sd<-0;    total_var<-0
rho<-0.2                                        #key parameter
sd<-0.1;        variance<-sd^2
covariance<-rho*variance                        #as all rho's are equal
for(n in 1:50){
var_sum[n]<-variance/n
covar_sum[n]<-(n-1)/n*covariance
total_var[n]<-var_sum[n]+  covar_sum[n]
total_sd[n]<-total_var[n]^0.5}
```

From Graph A we can see that the uncertainty relating to the variance term in Equation 7.4 has almost fallen to 0, with most of the total risk coming from the covariance term in the equation.

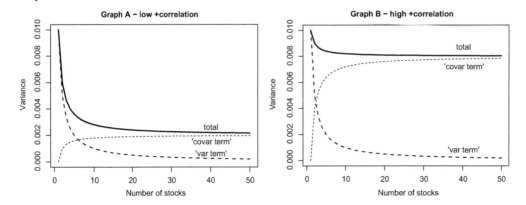

Figure 7.1 The effect of the number of assets, and the degree of correlation, on risk.

Re-running the above code with highly correlated assets, for example, $\rho = 0.8$ gives the results in Figure 7.1 Graph B. Although the variance of returns falls as we invest in more assets, it is reduced to a lesser extent as the assets are highly correlated.

Figure 7.2 compares the standard deviation of returns (rather than the variances) from our highly correlated portfolio with that of our largely uncorrelated portfolio. The standard deviation of the portfolio returns is reduced significantly (to 4.6% for 50 assets) if we invest in a large number of largely uncorrelated assets, compared to 10% if we invest in just one asset. The effect of investing in lots of assets which are highly correlated is much less (only reducing to 9%).

Another point of interest is that the diversification benefit from investing in one additional asset reduces somewhat as n gets large. It appears there is little benefit from investing in more than, say 20 assets, particularly when these marginal benefits are compared with additional administration and trading costs.

The key conclusion from this example is that diversification can be increased, and thus uncertainty reduced, by investing in (1) lots of assets, which (2) have little correlation with each other.

Exercise 7.3 *Re-run the above calculations with (1) perfectly correlated assets ($\rho = 1$) and (2) uncorrelated assets ($\rho = 0$).*

Figure 7.2 The effect of the degree of correlation on the standard deviation of returns.

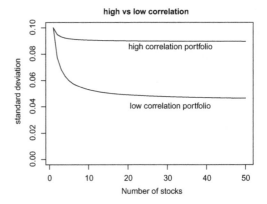

Exercise 7.4 *Plot a graph showing how the variance of a portfolio's return changes as the degree of correlation between each asset pair reduces from 1 to 0. Assume a portfolio consists of 10 assets.*

Exercise 7.5 *Verify that Equation 7.4 applies where the constituent assets have differing variances and covariances of returns, but equal weightings, by calculating the variance of returns using Equation 7.3.*

However Equation 7.4 does require equal asset allocations in the portfolio.

Exercise 7.6 *Verify that Equation 7.4 does not apply where the constituent assets have different weightings.*

The reader may have noted that we have not considered the situation where $\rho < 0$ i.e. asset returns are negatively correlated. This condition relates to the situation where the return from one asset will, in general, offset the return from another resulting in a "hedged position," and even more predictable results than above. We discuss this in Chapter 11.

7.4 Running Simulations

This section illustrates the above findings by running a series of simulations, and provides further coding practice. Figure 7.3 includes the simulated investment returns from two portfolios; one which consists of just 2 assets, with the other consisting of 20 assets. Investment returns from all stocks were assumed to be identically distributed ($\mu = 0$, $\sigma = 0.1$), and jointly Normally distributed with the same degree of correlation between all stocks ($\rho = 0.2$). Each portfolio consists of assets in equal proportions. Sample code for the 2-asset portfolio is set out below:

```
set.seed(500);      sims<-100000
no_of_stocks<-2;    rho<-0.2
var_each_stock<-0.01

sigma<-matrix(rho*var_each_stock,no_of_stocks,no_of_stocks)
diag(sigma)<-var_each_stock
sigma                          # covariance matrix

      [,1]   [,2]
[1,] 0.010 0.002
[2,] 0.002 0.010

mean<-c(rep(0.00,no_of_stocks))
sim1<-mvrnorm(sims,mean,sigma)   # simulate multivariate normal variables
sim_return1<-rowMeans(sim1)      # simulated 2-stock portfolio return
```

From running 100,000 simulations of future investment returns, there were 9875 simulations which resulted in returns < -0.1 from the 2-stock portfolio, compared with only 2003 simulations which resulted in such losses from the 20-stock portfolio. By investing in lots of stocks there is less chance of catastrophic performance from the portfolio. Of course the same principle applies if we were to look at a portfolio of bank loans or insurance policies.

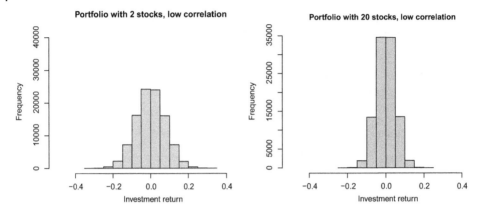

Figure 7.3 Simulated returns.

Exercise 7.7 *The standard deviations of the two portfolios' sets of simulations are:*

```
c(sd(sim_return1),sd(sim_return2))
```

```
[1] 0.07718841 0.04899649
```

Check that these standard deviations calculated from the simulations are consistent with the theoretical values.

Exercise 7.8 *Calculate the 99% VaR from these simulations, and compare with theory.*

Exercise 7.9 *Rewrite this code so it is more general and can be used for stocks with different characteristics, and for various number of stocks. For example, we may have a portfolio consisting of shares in 10 companies – you will need 10 variances and 45 covariances.*

Exercise 7.10 *Rerun the simulations with the much higher correlation of $\rho = 0.8$ and compare with the results from $\rho = 0.2$ above. Verify the variance of the returns using Equation 7.3.*

The reader is encouraged to reframe the above ideas to various contexts. The following exercise demonstrates the importance to property insurance:

Exercise 7.11 *General insurance application*

All houses in an insurance company's property book are valued at £1 m. Let's assume the same, simple annual claim distribution for each property:

- 95% probability that a policy has no claims,
- 4.9% probability that a policy has exactly one claim for £5,000,
- 0.09% probability that a policy has exactly one claim for £100,000,
- 0.01% probability that a policy has exactly one claim for £1 m,
- no policies make more than one claim, and
- all property claims are independent.

1) Obtain a simulated distribution of future total claims where the insurer sells (i) 10, (ii) 100 and (iii) 1,000 such policies.
2) Calculate 99% VaRs for books where the insurer sells (i) 10, (ii) 100 and (iii) 1,000 such policies.
3) How are the results affected if all property claims are completely dependent on each other, such that $\rho = 1$ (this may be appropriate for houses located in the same earthquake or flood zone)?

7.5   Recommended Reading

- Booth et al. (2005) *Modern Actuarial Theory and Practice.* (2nd edition). Boca Raton, Florida, US: Chapman and Hall/CRC.
- Hull, J. (2015) *Risk Management and Financial Institutions* (4th edition). Hoboken, New Jersey, US: Wiley.
- Wilmott, P. (2007) *Introduces Quantitative Finance.* (2nd edition). Chichester, West Sussex, UK: Wiley.

8

Modern Portfolio Theory

Peter McQuire

```
library(dplyr)
```

8.1   Introduction

In 1952 Harry Markovitz published his paper "Portfolio Selection" in the *Journal of Finance*. The paper proposed a methodology which allowed investors to analyse the balance between risk and returns between various asset portfolios. The theory became popularly known as "mean-variance portfolio theory," "Modern Portfolio Theory (MPT)," "mean-variance optimisation," or even "Markovitz Portfolio Theory".

In this chapter we discuss the key aspects of this theory. We proceed to calculate the expected returns and variance of returns from various combinations of assets, comparing our results from portfolios which consist of a range of risky and risk-free assets (we will define exactly what is meant by "risk" under MPT shortly). We will do this by writing our own code rather than use one of the many functions within R already available, which should aid the learning process. An Appendix is included which describes a method to determine efficient portfolios using Lagrange multipliers.

Remark 8.1 There are various packages in R which are related to the material in this chapter, such as fPortfolio and PortfolioAnalytics. The reader may wish to review these in due course.

MPT allows an investor, at least in theory, to choose a particular portfolio of assets which is expected to provide the greatest investment return subject to an acceptable level of risk which the investor is willing to take. Thus our calculations may provide a wide range of potential portfolios from which the investor can choose, each with a different risk/return profile; the portfolio chosen by the investor will depend on the level of risk which is most desirable to them.

R Programming for Actuarial Science, First Edition. Peter McQuire and Alfred Kume.
© 2024 John Wiley & Sons Ltd. Published 2024 by John Wiley & Sons Ltd.
Companion Website: www.wiley.com/go/rprogramming.com.

One of the key assumptions under MPT is that risk is measured in terms of the variance of investment returns. This seems to be a reasonable statement; a "risky" asset can certainly be deemed to be one which has significantly uncertain future returns. For example, equities are generally considered to be riskier than cash, gilts, or Treasury bills because returns from equities are, in general, more uncertain.

However there are several alternative measures of risk. For example, an institution may primarily be concerned with the ability to meet its liabilities when they are due to be paid – the risk here may be that of having insufficient liquidity to do so or holding assets whose market value does not move in line with its liabilities.

(Note that we will often calculate and plot the expected returns against the *standard deviation of returns* (i.e. the positive square root of the variance), which is a more intuitive way to consider the riskiness of an asset's returns.)

When we discuss "risky" assets in the context of MPT we are referring to assets with a non-zero variance (or standard deviation) of returns; under MPT the higher the variance the riskier the asset is considered to be. We will include non-risky assets in a number of portfolios in this chapter; this simply means that the returns on those assets are certain. Although in practice such assets do not exist (e.g. there will always be a chance of default) various assets may be considered to be close to risk-free, such as US or German short-term government bonds.

Exercise 8.1 *List three sovereign bonds which have defaulted in the last 30 years.*

8.2   2-Asset Portfolio

First we look at a simple portfolio, P, consisting of just two risky assets, A and B. The idea behind the examples in this chapter is to look at many combinations of asset allocations and to compare which portfolios give the most preferred combinations of return and risk i.e. is it preferable to invest everything in A, everything in B, or some combination of both?

The return on the portfolio, R_P, is given by:

$$R_P = w_A R_A + w_B R_B \tag{8.1}$$

where R_A and R_B are random variables representing the return on assets A and B, respectively, and w_A and w_B are the proportions, or weights, of the allocations in A and B, respectively.

The *expected* return from the 2-asset portfolio is given by:

$$E(R_P) = w_A E(R_A) + w_B E(R_B) \tag{8.2}$$

and the variance of returns:

$$Var(R_P) = w_A^2 Var(R_A) + w_B^2 Var(R_B) + 2w_A w_B Cov(R_A, R_B) \tag{8.3}$$

In matrix form we have:

$$Var(R_P) = w^T \Sigma w \tag{8.4}$$

where Σ is the covariance matrix and w is a vector containing w_A and w_B. (These expressions were also discussed in previous chapters.) We will now proceed to calculate the expected returns and variance of returns for various portfolios. Before we can do this we will need parameter values for:

1) the expected returns from the assets, A and B,
2) the variance of returns from A and B,
3) the correlation of the returns from A and B.

i.e. 5 parameters in total. We will set the parameter values as follows: $E(R_A) = 6, E(R_B) = 8, Var(R_A) = 4, Var(R_B) = 9$ and $Cov(R_A, R_B) = 1$. Therefore:

```
(r<-matrix(c(6,8),nrow=2,ncol=1))      #return vector

      [,1]
[1,]   6
[2,]   8

(sigmalow2<-matrix(c(4,1,1,9),2,2))    #covariance matrix

      [,1] [,2]
[1,]    4    1
[2,]    1    9
```

We can consider asset B to be riskier, and to provide higher expected returns, than asset A.

Exercise 8.2 *Given the above values, what is the value of the correlation coefficient between the returns on A and B?*

If we invested 80% in A and 20% in B the expected return and variance of returns would be respectively:

$$E(R_P) = 0.8 \times 6 + 0.2 \times 8 = 6.4$$

$$Var(R_P) = 0.8^2 \times 4 + 0.2^2 \times 9 + 2 \times 0.8 \times 0.2 \times 1.0 = 3.24$$

Now set up the various portfolios with appropriate allocation percentages of A and B. We need to decide how many to use – let's look at 5% steps i.e. 21 portfolios with the following splits: 100/0, 95/5, 90/10,...,0/100.

```
steps<-.05;     runs<-1/steps+1
```

Thus we are looking at all allocations with 5% steps only; the sum for each must add up to one:

```
a<-seq(0,1,steps);     b<-1-a
p<-as.matrix(data.frame(a,b))
head(p,2);    tail(p,2)

         a      b
[1,] 0.00   1.00
[2,] 0.05   0.95
         a      b
[20,] 0.95 0.05
[21,] 1.00 0.00
```

These are the 21 assets allocations we will look at. Of course you could, and probably should, look at finer splits e.g. every 1%. Calculating the expected returns, standard deviation of returns, and variances of returns from all the chosen portfolio splits, using Equation 8.4, is straightforward in R:

```
mean_mpt<-NULL;  var_mpt<-NULL

for(h in 1:runs){mean_mpt[h]<-(p[h,])%*%r}              #return
for(g in 1:runs){var_mpt[g]<-t(p[g,])%*%sigmalow2%*%p[g,]}  #variance
 sd_mptlow2<-sqrt(var_mpt)                               #sd
```

Collating and plotting the results:

```
output<-cbind(p,mean_mpt,sd_mptlow2)
head(output,2);  tail(output,2)    #a selection only

          a     b mean_mpt sd_mptlow2
[1,] 0.00 1.00      8.0   3.000000
[2,] 0.05 0.95      7.9   2.868362
          a     b mean_mpt sd_mptlow2
[20,] 0.95 0.05      6.1   1.930673
[21,] 1.00 0.00      6.0   2.000000
```

Figure 8.1 (left) is a classic plot obtained under MPT from a 2-asset portfolio. If we invest 100% in asset A the expected return is 6% and the standard deviation of the returns ("risk") is 2%, but if we invest in both A (70%) and B (30%) we get a higher expected return of 6.6% and lower uncertainty (standard deviation = 1.79%). (See rows 15 and 21 in *output*.) Thus, according to MPT, a rational investor should invest, at least partially, in the more risky asset (B) – by doing so they can expect to achieve higher returns with less risk compared to investing solely in a safer asset (A).

Exercise 8.3 *Under what circumstances may a knowledgeable investor invest 100% in asset A?*

How would the results change if the degree of correlation between R_A and R_B is higher?

Exercise 8.4 *Re-run the above calculations with a higher correlation of $\rho = 0.9166667$.*

A plot of the results with two highly correlated assets ($\rho = 0.9166667$) is shown in Figure 8.1 (right). None of the portfolios now have asset allocations which exhibit a higher return with lower risk (there is no "turning point"). The benefits of diversification do not appear here because assets A and B are highly correlated.

It is important to note that with just two assets all points in "mean – sd" space lie on one curve. What happens if we have three, or more, assets?

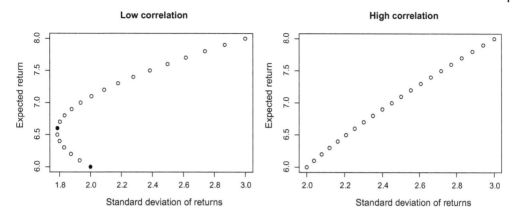

Figure 8.1 2-Asset portfolios.

8.3 3-Asset Portfolio

We proceed to analyse portfolios with more than two assets; we will consider a 3-asset portfolio. This will require us to re-write our code. Choosing the full range of portfolio allocations (still with a 5% gradation) will result in many more combinations than for the 2-asset case.

Exercise 8.5 *Write code which lists all the possible combinations of portfolios with exactly three assets, given a "step" size of 5% e.g. [0,0,1], [0,0.05,0.95], [0,0.1,0.9]...[1,0,0]. Call the 231 × 3 matrix containing these portfolio splits* port_split *– this is used in subsequent code below.*

Choosing values for the 3×3 covariance matrix, with a low degree of correlation:

```
(sigma3<-matrix(c(4,1,1.6,1,9,2.4,1.6,2.4,16),3,3))#3 risky, low corr

     [,1] [,2] [,3]
[1,]  4.0  1.0  1.6
[2,]  1.0  9.0  2.4
[3,]  1.6  2.4 16.0
```

and the vector of expected returns:

```
r_risky<-matrix(c(6,8,10),3,1)
```

Exercise 8.6 *Calculate the expected return and standard deviation of returns of each chosen portfolio split obtained from Exercise 8.5, using similar code to that used for the 2-asset case.*

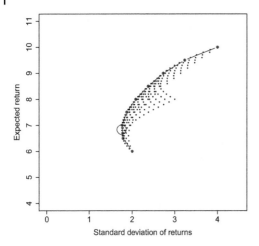

Figure 8.2 3-Asset portfolio.

The results to Exercise 8.6 are plotted in Figure 8.2.

How does this compare with the 2-asset portfolio? The key observation is there are clearly many *sub-optimal* portfolios. For example, there is a portfolio with an expected return of 8% and standard deviation of 2.1%, and a portfolio with an expected return of 8% and standard deviation of 3.0%; clearly the first portfolio is preferable under MPT as it is expected to produce the same return but is less "risky". Similarly there is a portfolio with an expected return of 9% and standard deviation of 3%, and a portfolio with an expected return of 7.8% and standard deviation of 3%; clearly the first portfolio is preferable – we can expect a better return whilst taking the same amount of risk.

Those portfolios which have the highest return for a given level of risk (i.e. standard deviation of returns) can be described as being "mean-variance" efficient; the curve joining all such portfolios is known as the efficient frontier. According to MPT, a rational investor should only consider those portfolios which sit on the efficient frontier.

The following code determines a number of portfolios which are the most "efficient," by taking regular standard deviation intervals and finding the portfolio with the highest expected return in each interval:

```
rangex<-0.02;     stepx<-0.5
min_req_ret<-6;    max_req_ret<-10
pfolio<-matrix(0,(max_req_ret-min_req_ret)/stepx+1,3)
s_dev1<-NULL
req_ret<-min_req_ret
ii<-1

#start loop
for (req_ret in seq(min_req_ret,max_req_ret,by=stepx)) {
pfolio[ii,]<-port_split[which(mean_mpt_risky > req_ret-rangex &
mean_mpt_risky < req_ret+rangex)[which.min(var_mpt[which(mean_mpt_risky >
req_ret-rangex & mean_mpt_risky < req_ret+rangex)])],]

s_dev1[ii]<-min(var_mpt[which(mean_mpt_risky > req_ret-rangex &
mean_mpt_risky < req_ret+rangex)])^0.5
```

```
ii<-ii+1
}
#end loop

res_summ<-as.data.frame(cbind(pfolio,seq(min_req_ret,max_req_ret,by=stepx),s_dev1))
colnames(res_summ) <- c("alloc A", "alloc B", "alloc C", "exp_ret","st_devn")
res_summ

  alloc A alloc B alloc C exp_ret  st_devn
1    1.00    0.00    0.00     6.0 2.000000
2    0.75    0.25    0.00     6.5 1.785357
3    0.60    0.30    0.10     7.0 1.762385
4    0.45    0.35    0.20     7.5 1.868556
5    0.30    0.40    0.30     8.0 2.084226
6    0.15    0.45    0.40     8.5 2.379811
7    0.00    0.50    0.50     9.0 2.729469
8    0.00    0.25    0.75     9.5 3.234579
9    0.00    0.00    1.00    10.0 4.000000
```

The selection of efficient portfolios are shown as joined-up bold points in Figure 8.2. Note that the first two rows in the table above are not efficient portfolios.

Remark 8.2 The Appendix to this chapter sets out an optimisation technique method, using Lagrange multipliers, to obtain details of those portfolios which sit on the "efficient frontier".

Exercise 8.7 *The reader may wish to develop their code to allow for the general case of n assets in the portfolio.*

Exercise 8.8 *Why may an investor choose a portfolio allocation which does not lie on the efficient frontier?*

Example 8.1 An investor owns a portfolio consisting of three assets with the expected returns and variance of returns as above. The portfolio is currently valued at £100 m. The investor wishes to allocate the assets such that there is less than a 0.1% probability that the portfolio value is less than £100 m in one year, whilst maximising the expected return. Assuming Normal distributions for each asset return determine an appropriate asset split given the investor's requirements.

```
VaR999<-mean_mpt_risky-qnorm(0.999,0,1)*sd_mpt2
combine_data1<-cbind(port_split,sd_mpt2,mean_mpt_risky,VaR999)
port_poss<-combine_data1[combine_data1[,6]>0,]
colnames(port_poss) <- c("allocA", "allocB", "allocC","st_dev", "exp_ret","99.9VaR")
(head(port_poss[order(-port_poss[,5]),],5))

      allocA allocB allocC   st_dev exp_ret    99.9VaR
[1,]    0.00   0.35   0.65 2.992407     9.3 0.05276704
[2,]    0.05   0.30   0.65 2.941088     9.2 0.11135412
[3,]    0.00   0.40   0.60 2.889983     9.2 0.26928210
[4,]    0.05   0.35   0.60 2.830459     9.1 0.35322315
[5,]    0.10   0.25   0.65 2.898362     9.1 0.14338933
```

These asset allocations may be considered appropriate given the investor's requirements set out above. Alternative portfolios may be preferable however if we include a risk-free asset (see Section 8.4).

Finally in this section we discuss the global minimum variance portfolio, which will often be of interest to the investor. The asset weightings of the global minimum variance portfolio are given by the following closed-form solution:

$$\frac{\Sigma^{-1}\mathbf{1}}{\mathbf{1}'\Sigma^{-1}\mathbf{1}} \tag{8.5}$$

The variance of this portfolio is given by:

$$\mathbf{1}'\Sigma^{-1}\mathbf{1} \tag{8.6}$$

where $\mathbf{1}' = [1, 1, 1]$

Example 8.2 Find the global minimum variance portfolio for the 3-asset portfolio discussed in this Section.

```
inv_cov<-solve(sigma3)    #the inverse of the covariance matrix
col_vec<-c(1,1,1)

numerator1<-inv_cov%*%col_vec
denominator1<-as.numeric(col_vec%*%inv_cov%*%col_vec)

(port_weight<-numerator1/denominator1) #closed form soln to weights of min var pfolio


          [,1]
[1,] 0.66964076
[2,] 0.24231040
[3,] 0.08804884

1/denominator1^0.5                        #min sd possible

[1] 1.749786

(t(port_weight)%*%sigma3%*%(port_weight))^0.5 #check

          [,1]
[1,] 1.749786

t(r_risky)%*%port_weight                  #expected return from min var portfolio

          [,1]
[1,] 6.836816
```

This point is highlighted in Figure 8.2. Thus the minimum standard deviation of returns which can be obtained is 1.75%, with a portfolio consisting of 67%, 24.2%, 8.8% in assets $A, B,$ and C respectively.

This minimum variance portfolio can also be determined using Lagrange Multipliers – see Appendix.

8.4 Introduction of a Risk-free Asset to the Portfolio

8.4.1 Adding a Risk-free Asset

Up to this point we have concerned ourselves with assets whose future investment return is uncertain i.e. a non-zero variance. What would our results look like if we could include a risk-free asset in the portfolio?

We must choose some further parameter values. The return from the risk-free asset should have zero variance (as defined), and have an expected return which is lower than that of the risky assets being considered (otherwise investors would have no incentive to invest in risky assets). We will use the same parameters from the 2-asset portfolio (A, B) described in Section 8.2 and add the risk-free asset (RF) to it:

```
(sigmalow2cash<-matrix(c(0,0,0,0,4,1,0,1,9),3,3)) # asset 1 is risk-free
     [,1] [,2] [,3]
[1,]   0    0    0
[2,]   0    4    1
[3,]   0    1    9
```

This is the new covariance matrix which now includes the risk-free asset. An appropriate expected returns vector (setting the return from the risk-free asset to be 2% p.a.) is:

```
r_cash2risky<-matrix(c(2,6,8),3,1)
```

Re-running the same code as that for the 3-asset portfolio above we obtain Figure 8.3; this is a particularly important graph.

Firstly we should note that the bold points represent those portfolios which do not have any risk-free asset allocation i.e. they are the points from Figure 8.1 (left); all the other points are in respect of portfolios which have some risk-free asset allocation. Before proceeding with further analysis of this plot we need to briefly cover a couple of related concepts.

Figure 8.3 3 Assets (including a risk-free asset); low correlation.

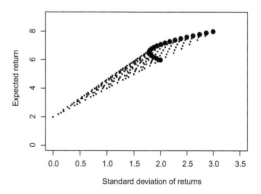

8.4.2 Capital Market Line and the Sharpe Ratio

The Capital Market Line identifies all the most "efficient" portfolios. These are the portfolios which, for a given expected return, there exists no alternative portfolios with a lower standard deviation (or "risk"). These portfolios can be identified by calculating the Sharpe ratio, which is defined as follows:

$$Sharpe\ Ratio = \frac{expected\ return - risk\ free\ return}{standard\ deviation\ of\ returns} \tag{8.7}$$

A higher Sharpe ratio reflects a more efficient portfolio. Calculating this ratio for each of the portfolios in Figure 8.3, and tabulating the ten highest Sharpe ratios (shown in Figure 8.4 as *):

```
sharperatiocash<-(mean_mpt_cash-2)/sd_mptlow2cash
output<-cbind(port_split,mean_mpt_cash,sd_mptlow2cash,sharperatiocash)
outputsort<-arrange(as.data.frame(output),desc(sharperatiocash))
colnames(outputsort)<-c("RF","A","B","mean_ret","st_devn","Sharpe_ratio")
(best10outputsort<-outputsort[1:10,])

      RF    A    B mean_ret   st_devn Sharpe_ratio
1   0.75 0.15 0.10     3.2 0.4582576     2.618615
2   0.50 0.30 0.20     4.4 0.9165151     2.618615
3   0.00 0.60 0.40     6.8 1.8330303     2.618615
4   0.25 0.45 0.30     5.6 1.3747727     2.618615
5   0.10 0.55 0.35     6.3 1.6424068     2.618109
6   0.15 0.50 0.35     6.1 1.5660460     2.618059
7   0.35 0.40 0.25     5.1 1.1842719     2.617642
8   0.40 0.35 0.25     4.9 1.1079260     2.617503
9   0.05 0.55 0.40     6.6 1.7578396     2.616849
10  0.60 0.25 0.15     3.9 0.7262920     2.616028
```

We can view these portfolios as potentially the most efficient under MPT i.e. they are expected to result in the greatest return in excess of the risk-free return, per unit of risk; whichever one we choose depends on how much risk we wish to take.

Do these asset allocations have anything in common? Calculating the ratio of the two risky assets' allocations:

```
best10outputsort[,2]/best10outputsort[,3]

[1] 1.500000 1.500000 1.500000 1.500000 1.571429 1.428571 1.600000 1.400000
[9] 1.375000 1.666667
```

The portfolios with the best Sharpe ratios have allocations of the two risky assets in the ratio 3 to 2 i.e. 1.5. (This is not quite true for rows below row 4 as we are constrained by the granularity of the portfolio allocations chosen i.e. every 5% – the reader may wish to rerun the calculations with 1% steps.) For example, when there is no risk-free asset in the portfolio, the "best" portfolio is split 60/40 (shown in row 3).

These portfolios form a straight line in expected return – standard deviation space, the Capital Market Line ("CML"):

$$Expected\ return = \frac{6.8 - 2.0}{1.833} \times sd + 2 = 2.6181818 \times sd + 2 \tag{8.8}$$

Figure 8.4 Capital market line, with borrowing.

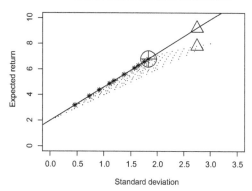

Risk–free asset and 2 risky assets low correlation

(The CML is shown in Figure 8.4.)

The "60/40 allocation" of risky assets with no risk-free assets (⊕ in Figure 8.4) is a special portfolio, and is known as the "market portfolio"; the CML is a straight line which joins the risk-free asset and the market portfolio in Figure 8.4. We can see that, according to MPT, we should *always* invest the risky assets in this proportion to get the "most efficient" portfolio, and invest the remainder of the portfolio in non-risky assets. The split of risky assets to non-risky assets depends on the investor's risk preference.

If we are particularly risk averse we could invest 90% in the risk-free asset, with 6% and 4% in the two risky assets. But if we are particularly "risk-seeking" we could invest only 10% in the risk-free asset, and 54% and 36% in the two risky assets.

However, it appears that if we want better expected returns than 6.8% p.a. we should solely invest in risky assets. For example we could invest 100% in Asset B; the expected return and standard deviation would be 8% and 3% respectively. However the Sharpe ratio for this allocation is somewhat lower than the efficient portfolios (Sharpe =2). Could we do better?

8.4.3 Borrowing to Obtain Higher Returns

A possible solution is to borrow money at the risk-free rate. If we can do this, we can invest our borrowings in A and B, provided the total allocation still adds up to 100%. Let's say we borrow an amount of cash (the risk-free asset) equal to 50% of the current asset value, and proceed to invest 90% in A and 60% in B. The relevant calculations give us the following:

Expected return $= -0.5 \times 2 + 0.9 \times 6 + 0.6 \times 8 = 9.2$
Variance $= 0.9^2 \times 4 + 0.6^2 \times 9 + 2 \times 0.9 \times 0.6 \times 1 = 7.56$
Standard deviation $= 2.7495454$

Going back to Figure 8.1 (left), the allocation with a standard deviation equal to 2.75% was a 10/90 allocation, but this only gave an expected return of 7.8%.

Exercise 8.9 *Calculate the Sharpe ratio of this portfolio and compare with those of the Capital Market Line and also the 10/90 portfolio.*

Using this as an example, MPT tells us that if we require expected returns of 9.2% and are willing to accept the higher uncertainty, then we should borrow 50% of the value of the fund and invest the total assets in a 60/40 proportion in A and B. This will result in a more efficient portfolio than if we were to invest 10% in A and 90% in B. We will have the same uncertainty (2.75%) but will expect a higher expected return of 9.2% compared to only 7.8%.

Figure 8.4 includes these two relevant points ($\triangle$). It is important to note, however, that in practice, it is unlikely that you would be able to borrow at the risk-free rate – banks will allow for default risk and the need to make profits.

Exercise 8.10 *List the assumptions included under mean-variance portfolio theory.*

8.5 Appendix: Lagrange Multiplier Method

This appendix outlines an optimisation method to determine the MPT efficient frontier. (A second example of this method is also included to determine the global minimum variance.) We will not cover the theory behind Lagrange multipliers (there are numerous textbooks covering this topic; see Brandimarte (2006) and de la Fuente (2000)). Using this optimisation technique we can obtain details of the portfolios (i.e. the asset allocations) which sit on the efficient frontier. By first setting our required expected portfolio return, γ, we can determine the portfolio which will result in the lowest variance of returns. We will consider a portfolio consisting of three assets.

Outline of method: Set up the Lagrangian with constraints. The two constraints here are:

1) The expected portfolio return is required to be γ where:

$$w_A \mu_A + w_B \mu_B + w_C \mu_C = \gamma$$

2) The proportions invested in the three assets must add to 1:

$$w_A + w_B + w_C = 1$$

where μ_X and w_X are the expected return on asset X and the proportion invested in asset X, respectively.

Our objective here is to minimise the variance of the returns from the portfolio subject to the above constraints. For example, the standard deviation of the portfolio returns may need to be limited to 3% due to regulations, or to it leading to an unacceptable probability of insolvency. We may be able to obtain a higher expected return – however this will require an unacceptable degree of risk to be taken. The variance is given by the 3-asset equivalent of Equation 8.3:

$$Var(P) = w_A^2 \sigma_A^2 + w_B^2 \sigma_B^2 + w_C^2 \sigma_C^2 + 2w_A w_B \sigma_{AB} + 2w_A w_C \sigma_{AC} + 2w_C w_B \sigma_{CB} \qquad (8.9)$$

where σ_X^2 and σ_{XY} are the variance of returns on asset X and the covariance of returns between assets X and Y, respectively.

Setting up the Lagrangian:

$$w_A^2\sigma_A^2 + w_B^2\sigma_B^2 + w_C^2\sigma_C^2 + 2w_Aw_B\sigma_{AB} + 2w_Aw_C\sigma_{AC} + 2w_Cw_B\sigma_{CB}$$
$$+ \lambda_1(w_A\mu_A + w_B\mu_B + w_C\mu_C - \gamma) + \lambda_2(w_A + w_B + w_C - 1) = 0 \qquad (8.10)$$

Taking derivatives of the Lagrangian with respect to w_A, w_B, w_C, λ_1, and λ_2 and writing the resulting 5 equations in matrix form gives us:

$$\begin{pmatrix} 2\sigma_A^2 & 2\sigma_A\sigma_B\rho_{AB} & 2\sigma_A\sigma_C\rho_{AC} & \mu_A & 1 \\ 2\sigma_A\sigma_B\rho_{AB} & 2\sigma_B^2 & 2\sigma_A\sigma_C\rho_{AC} & \mu_B & 1 \\ 2\sigma_A\sigma_C\rho_{AC} & 2\sigma_B\sigma_C\rho_{BC} & 2\sigma_C^2 & \mu_C & 1 \\ \mu_A & \mu_B & \mu_C & 0 & 0 \\ 1 & 1 & 1 & 0 & 0 \end{pmatrix} \begin{pmatrix} w_A \\ w_B \\ w_C \\ \lambda_1 \\ \lambda_2 \end{pmatrix} = \begin{pmatrix} 0 \\ 0 \\ 0 \\ \gamma \\ 1 \end{pmatrix} \qquad (8.11)$$

An important point to note is that γ is the required expected return; our task is to determine the portfolio allocations, w, which deliver this expected return for the lowest standard deviation.

At this point we should choose values for the parameters (the same values have been chosen as those in Section 8.3):

$\mu = [6, 8, 10]$ (vector of expected returns from A, B, and C)
$\sigma = [2, 3, 4]$ (vector of standard deviation of returns from A, B, and C)
$\rho_{AB} = 0.1666667$; $\rho_{AC} = 0.2$; $\rho_{BC} = 0.2$ (correlation coefficients)

The left-hand matrix in Equation 8.11 is therefore:

```
Lagrange_Markovitz

    V1    V2    V3 V4 V5
1 8.0   2.0   3.2  6  1
2 2.0  18.0   4.8  8  1
3 3.2   4.8  32.0 10  1
4 6.0   8.0  10.0  0  0
5 1.0   1.0   1.0  0  0

ind_asset_ret<-Lagrange_Markovitz[4,1:3] #to be used later
```

This resulting series of simultaneous equations from Equation 8.11 can now be solved, most easily by applying the inverse of the above matrix; we proceed to optimise the asset allocations for all expected returns from 6% p.a. to 10% p.a. (the extremes from the three assets). The results are plotted in Figure 8.5 (in bold):

```
inv_LM<-solve(Lagrange_Markovitz)   #obtain the inverse
solution_Leg<-matrix(0,41,5)    #set up a blank matrix for later

counter<-1
for(n in seq(min(ind_asset_ret),max(ind_asset_ret),.1)){
return_input<-n
RHS_matrix<-c(0,0,0,return_input,1)
answer<-inv_LM%*%RHS_matrix
weights<-answer[1:3]
cov_matrix<-Lagrange_Markovitz[1:3,1:3]/2
```

```
variance_port<-t(weights)%*%cov_matrix%*%(weights)
solution_Leg[counter,]<-c(return_input,variance_port^0.5,weights)
counter<-counter+1}

colnames(solution_Leg)<-c("return","sd","w1","w2","w3")
head(solution_Leg,2)

        return       sd       w1         w2          w3
[1,]    6.0 1.927638 0.9210526 0.1578947 -0.07894737
[2,]    6.1 1.889130 0.8910088 0.1679825 -0.05899123

tail(solution_Leg,2)

        return       sd       w1         w2          w3
[40,]    9.9 3.43882 -0.2506579 0.5513158 0.6993421
[41,]   10.0 3.52236 -0.2807018 0.5614035 0.7192982
```

Note that the bold points in Figure 8.5 are more "efficient" than those calculated in Section 8.3; these portfolios allow for short selling (**w** can be negative).

For example, if we require an expected return of (1) 8%, or (2) 10%, we should invest as follows:

```
solution_Leg[21,] #8%

   return        sd        w1         w2        w3
8.0000000 2.0797689 0.3201754 0.3596491 0.3201754

solution_Leg[41,] #10%

    return        sd        w1         w2        w3
10.0000000 3.5223597 -0.2807018 0.5614035 0.7192982
```

To achieve an expected return of 10% we should invest −28.1% (short), 56.1%, and 71.9% in A, B, C respectively, giving a standard deviation of returns of 3.52%. Without short selling an expected return of 10% could be obtained only by investing everything in Asset C, resulting in a standard deviation of returns of 4%.

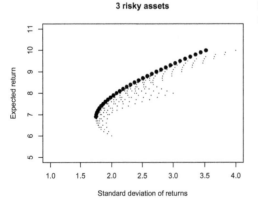

3 risky assets

Figure 8.5 Using a Lagrange Multiplier method.

Global minimum variance portfolio

Set out below is code to determine the global minimum variance portfolio, by applying the same Lagrange multiplier techniques as above. The calculation is similar to that above, except that the constraint on the expected return is not required and has been removed. This example is included to provide another example of how Lagrange multipliers can be used to solve such problems – the closed form solution was set out at the end of Section 8.3.

```
(Lag_Mark2<-rbind(cbind(as.matrix(Lagrange_Markovitz[1:3,1:3]),c(1,1,1)),c(1,1,1,0)))

     V1    V2    V3
1  8.0   2.0   3.2 1
2  2.0  18.0   4.8 1
3  3.2   4.8  32.0 1
   1.0   1.0   1.0 0

inv_LM2<-solve(Lag_Mark2)                    #obtain the inverse

RHS_matrix2<-c(0,0,0,1)
answer2<-inv_LM2%*%RHS_matrix2
(weights3<-as.vector(answer2[1:3]))          #minimum variance portfolio

[1] 0.66964076 0.24231040 0.08804884

variance_port2<-t(weights3)%*%sigma3%*%(weights3)
variance_port2^0.5                           #sd of min variance portfolio

          [,1]
[1,] 1.749786

Lagrange_Markovitz[4,1:3]%*%(weights3)       #expected return of min variance portfolio

          [,1]
[1,] 6.836816
```

8.6 Recommended Reading

- Benninga, S. (2014) *Financial Modeling* (4th edition). Cambridge, MA: MIT Press.
- Booth et al. (2005) *Modern Actuarial Theory and Practice* (2nd edition). Boca Raton, Florida, US: Chapman and Hall/CRC.
- P. Brandimarte (2006) *Numerical Methods in Finance and Economics* (2nd edition). Hoboken, New Jersey, US: Wiley
- Charpentier A. et al. (2016) *Computational Actuarial Science with R*. Boca Raton, Florida, US: Chapman and Hall/CRC.
- Cochrane, J. (2008) *Asset Pricing* (Revised Edition). New Jersey, US: Princeton University Press.
- Angel de la Fuente (2000) *Mathematical Methods and Models for Economists*. UK: Cambridge University Press.
- Hull, J. (2015) *Risk Management and Financial Institutions* (4th edition). Hoboken, New Jersey, US: Wiley.
- Ingersoll, Jr., J.E. (1987) *Theory of Financial Decision Making* Lanham, US: Rowman and Littlefield Publishers.

- Luenberger D.G. and Yinyu Ye (2016) *Linear and Nonlinear Programming* (4th edition). Switzerland: Springer International Publishing.
- Markovitz H. (1952) 'Portfolio Selection,' *The Journal of Finance* Vol 7. Hoboken, NJ: Wiley.
- Robert C. Merton (1972) 'An Analytic Derivation of the Efficient Portfolio Frontier,' *The Journal of Financial and Quantitative Analysis*, Vol. 7, No. 4 (Sep., 1972), UK: Cambridge University Press.
- Sweeting, P. (2017) *Financial Enterprise Risk Management* (2nd edition). Cambridge, UK: Cambridge University Press.
- Wilmott, P. (2007) *Introduces Quantitative Finance* (2nd edition). Chichester, West Sussex, UK: Wiley.

9

Duration – A Measure of Interest Rate Sensitivity

Peter McQuire

```
library(lifecontingencies)
```

9.1 Introduction

The concept of duration is important to actuaries in understanding the sensitivity of the present value of a series of cashflows to changes in interest rates. For example, an analyst should understand by how much the value of a bond portfolio may fall if bond yields increase by 1%. The cashflows could be income (e.g. from a bond portfolio) or outgo (e.g. pension scheme liabilities).

   This short chapter introduces the concept of duration, which will be developed in a number of later chapters in the book. As we will see, knowing the duration of our liabilities can go some way in helping us remove interest rate risk by ensuring the duration of our assets is similar to that of our liabilities. Indeed, it is usual for bond fund managers to publish the duration of their funds such that potential investors can choose appropriate funds to minimise interest rate risk.

9.2 Duration – Definitions and Interpretation

Before we proceed it is worth noting that there are a number of similar expressions related to duration which can lead to confusion. We will be discussing "Macaulay duration" (also known as "discounted mean term"), but similar expressions commonly used include "duration" itself (very confusingly), effective duration, modified duration, and volatility. We are also aware of some authors interchanging these terms, leading to further confusion. Care should therefore be taken when reading different texts to be clear exactly what is meant by "duration". We will mainly be discussing Macaulay duration in this chapter (and will refer to it as such); when the term "duration" is used we will be discussing it in its most general terms, which would apply to any variant definition.

R Programming for Actuarial Science, First Edition. Peter McQuire and Alfred Kume.
© 2024 John Wiley & Sons Ltd. Published 2024 by John Wiley & Sons Ltd.
Companion Website: www.wiley.com/go/rprogramming.com.

Macaulay duration, D, is defined as follows:

$$D = -\frac{1}{B}\frac{dB}{d\delta} \tag{9.1}$$

where B is the price of a bond (or more generally the present value of a series of future cashflows), and δ is the force of interest.

Remark 9.1 The expression for effective duration or volatility is usually defined as follows:

$$volatility = -\frac{1}{B}\frac{dB}{di}$$

where i is the effective rate of interest.

Loosely speaking, duration is given by the proportionate change in the price of a bond (or more generally any series of cashflows) divided by the small change in interest rates which caused the price change. We can see this by re-writing Equation 9.1 with a small interest rate change, $\Delta\delta$:

$$D \approx -\frac{\frac{\Delta B}{B}}{\Delta\delta} \tag{9.2}$$

Often, $\frac{\Delta B}{B}$ is the quantity we will be interested in, giving the proportionate change in the present value of our cashflows following a particular change in interest rates e.g. if the duration of a bond is 10 years, then an interest rate change of 1% will cause an approximate change in the bond price of $10 \times 0.01 = 10\%$. Of course the relationship is only exact as $\Delta\delta \to 0$.

It is trivial to show from Equation 9.1 that, given the formula for the price of a bond, $B = \Sigma c_t e^{-\delta t}$, the Macaulay duration, D, can also be written as:

$$D = \frac{\Sigma t c_t e^{-\delta t}}{\Sigma c_t e^{-\delta t}} = \frac{\Sigma t c_t e^{-\delta t}}{B} \tag{9.3}$$

where c_t is the cashflow amount at time t, and the sum is over all cashflows. Thus duration can also be thought of as the weighted average of the future times at which payments will be received (duration is measured in units of time). It is perhaps now clearer where the term "duration" comes from – it gives a measure of the average time when payments will be received. (Note that Macaulay duration is often referred to as "discounted mean term".)

Therefore, the duration of a zero coupon bond will simply equal the term of the bond i.e. a 6-year zero coupon bond will have a duration of exactly 6 years. In the situation where coupons are paid, as will usually be the case, the bond's duration will be less than the term. For example the Macaulay duration of a 2-year bond with 3% coupons payable annually in arrears, with $\delta = 8\%$ p.a., is 1.969 years. The calculation is shown below:

$$\frac{1\times3\times e^{-0.08}+2\times103\times e^{-0.08\times2}}{3\times e^{-0.08}+103\times e^{-0.08\times2}} = 1.969413 \text{ years}$$

Let's verify this using Equation 9.2 by calculating the change in price of this 2-year bond which results from a small change in interest rates. We will start with $\delta = 8\%$ p.a. and increase this by a very small amount:

```
term<-2;  coupon<-0.03;  delta_discount<-0.08;  discount_change<-0.000001
(delta_discount2<-delta_discount+discount_change)

[1] 0.080001

time<-seq(1,term);  payments<-c(rep(coupon,term-1),1+coupon)
(price<-presentValue(payments,time,exp(delta_discount)-1))

[1] 0.9054016

(price2<-presentValue(payments,time,exp(delta_discount2)-1))

[1] 0.9053998
```

The Macaulay duration follows from Equation 9.2:

```
(duration<-   -(price2-price)/price/discount_change)
[1] 1.969411
```

(The slight discrepancy compared with 1.969413 is due to the finite change in interest rates.) Therefore if interest rates increase by 1% (say from 8% to 9%) the price of this bond will fall (note the negative sign) by around 1.97%.

Exercise 9.1 *Calculate the actual change in bond price if the force of interest changed from 8% p.a. to 9% p.a. (instantly).*

Exercise 9.2 *Write your own Macauley duration function using Equation 9.3 and verify the duration of the above bond.*

As noted at the start of the chapter, the concept of duration can be applied to *any* series of cashflows e.g. bonds, pension scheme payments, life assurance liabilities, loan repayments, etc.

9.3   Duration Function in R

The reader may not be surprised to learn that a duration function exists in R, within the lifecontigencies package. Recalculating the Macaulay duration of the 2-year bond from above:

```
payments; time

[1] 0.03 1.03
[1] 1 2

duration(payments,time,exp(delta_discount)-1,k=1,macaulay = TRUE)

[1] 1.969413
```

The reader will note that the effective interest rate, i, is required for this function. Also note that macaulay = TRUE is the default so this argument is not required.

9.4 Practical Applications of Duration

Knowing the duration of a series of cashflows provides a good understanding of the sensitivity of their value to changes in interest rates. If we know the duration of our liabilities then we can perhaps buy assets of a similar duration; if we can achieve this then our asset and liability values will typically move by similar amounts following a change in interest rates, resulting in a stable financial position in the future.

The importance of duration, and how this process can be improved, will be further demonstrated in later chapters.

Exercise 9.3 *Consider a large defined benefit pension scheme with 50,000 members. Let's approximate our pension cashflows with an annuity certain for 30 years, with annual payments of £100 million (a fairly crude approximation).*

Given that $\delta = 5\%$ per annum, calculate the Macaulay duration using both Equation 9.2 and the duration function, and comment on the degree to which our liabilities are affected by a fall in interest rates.

Comment on solution: The Macaulay duration is 11.89 years. By knowing the duration we can quickly calculate that our liabilities will increase by *approximately* 1.2% if interest rates fall from 5% to 4.9% (i.e. about £18 million). The value of the liabilities actually changes from £1515.2 m to £1533.4 m.

Remark 9.2 Note that duration is a function of the level of interest rates. Thus when we calculate duration we should state the interest rate at which it has been calculated.

Exercise 9.4 *Recalculate the Macaulay duration of the above pension cashflows assuming $\delta = 10\%$ per annum.*

Example 9.1 The pension scheme advisors in Exercise 9.3 are considering investing an amount, equal to the present value of these pension payments (using $\delta = 5\%$ per annum), in a single government bond redeemable at par with a term of 10 years which pays 6% annual coupons in arrears. Comment on the amount of interest rate risk arising from such an investment strategy.

The duration of the asset is:

```
bond_receipts<-c(rep(6,9),106);   time2<-c(1:10);   delta_discount<-0.05
duration(bond_receipts,time2,exp(delta_discount)-1,k=1,macaulay = TRUE)
[1] 7.880761
```

Following a small change in interest rates the liability value will change by a greater proportion (as $D_L > D_A$). If interest rates fall then both our assets and liabilities will increase in value, but the liability value will increase by a greater amount resulting in $A < L$, leading to potential insolvency. (Note that if interest rates increase the scheme will be in surplus.) As noted above, investing in a bond fund with a duration of around 12 years may be considered to be a preferable strategy with regards to managing interest rate risk. This problem is related to that which is at least partially addressed by Redington Immunisation which is discussed in detail in later chapters.

9.5   Recommended Reading

- Booth et al. (2005) *Modern Actuarial Theory and Practice* (2nd edition). Boca Raton, Florida, US: Chapman and Hall/CRC.
- Hull, J. (2015) *Risk Management and Financial Institutions* (4th edition). Hoboken, New Jersey, US: Wiley.
- McCutcheon and Scott. (1986) *An Introduction to the Mathematics of Finance.* Oxford, UK: Butterworth Heinemnan.
- Wilmott, P. (2007) *Introduces Quantitative Finance.* (2nd edition). Chichester, West Sussex, UK: Wiley.

10

Asset-Liability Matching: An Introduction

Peter McQuire

10.1 Introduction

```
library(MASS)
```

The objective of this chapter is to illustrate, using as simple a model as possible, the effect various investments strategies may have on the solvency positions of annuity providers, insurance companies and pension schemes.

In particular, we will concern ourselves with analysing the stability, or predictability, of the entity's solvency position. To do this we will calculate the variance of the solvency position at a future point in time. To be clear, the solvency, or funding, position is defined here as follows:

$$Solvency = A - L \tag{10.1}$$

where A is the market value of our assets and L is the expected present discounted value of the future contractual payments, e.g. pensions (often referred to as "Technical Provisions" or simply the "liabilities"). Thus we will be interested in measuring, and minimising, the variance of $(A - L)$. Note that we will not concern ourselves with any technical discussions regarding different types of solvency measures, such as those required by regulators, shareholders or internal management; for our purposes we treat the term "solvency" in its most general sense.

The reader may first wish to briefly study the plots at the end of this chapter (Figures 10.1 and 10.2) to understand our objectives; these plots are typical of those presented by actuaries to describe the potential impact of various investment strategies on future solvency levels. The models in this chapter are, however, somewhat simplified to ensure the key points are made to the reader.

R Programming for Actuarial Science, First Edition. Peter McQuire and Alfred Kume.
© 2024 John Wiley & Sons Ltd. Published 2024 by John Wiley & Sons Ltd.
Companion Website: www.wiley.com/go/rprogramming.com.

10.2 What Interest Rates Do Institutions Use To Measure Their Liabilities?

Before proceeding it is important to understand the general methodology used by regulators, institutions and actuaries to measure the liabilities of defined benefit pension schemes, insurers, and annuity companies. The term "liabilities", in an actuarial context, refers to the discounted present value of the expected future payments which are due. As noted in earlier chapters, this is simply the estimated amount of money required, at the valuation date, which should be sufficient to meet these expected future payments, allowing for future investment returns and the probabilities of these payments. Thus we require an assumption for what this investment return may be.

The starting point in determining this assumption is usually the yield on a chosen sovereign bond, of a particular term. For example the UK Pensions Regulator requires reference to the gross redemption yield on UK gilts of a specified term. Thus if we have a payment to make 20 years from now it would be sensible to base our future return assumption on the 20-year spot rate (derived from gilt prices) as we are *almost* guaranteed to achieve this return. In this way we can therefore say that the money held now will be sufficient to meet our contractual payments in 20 years, given these *almost* guaranteed returns.

As a further example, the Pensions Protection Fund ("PPF") in the UK requires relevant company pension schemes to calculate their pension liabilities with reference to "the annualised yield on the FTSE Actuaries' Government 20-year Fixed Interest Index" (as of November 2018).

Thus institutions with long-term liabilities, such as annuity companies and pensions schemes, will tend to calculate their liabilities with respect to long-term interest rates. In the same way, institutions with shorter term liabilities will use short-term interest rates.

10.3 Variance of the Solvency Position

The variance of the solvency level, or funding level, is given by the following equation (the importance of the general form of this equation is noted several times throughout this book):

$$Var(A - L) = Var(A) + Var(L) - 2Cov(A, L)$$

or equivalently using more concise notation:

$$\sigma^2_{A-L} = \sigma^2_A + \sigma^2_L - 2\rho_{AL}\,\sigma_A\,\sigma_L \tag{10.2}$$

where the definitions are consistent with those in previous chapters. As noted above, we are aiming for stability in the solvency position, $A - L$; therefore a low value of σ^2_{A-L} is our objective. To achieve this, it is worth noting the following from Equation 10.2:

- A high positive value of ρ_{AL} is preferable.
  Thus holding assets, the prices of which are highly correlated with the liabilities, should result in a more stable solvency position.

- $\sigma^2_{A-L} = 0$ when both $\sigma_A = \sigma_L$ and $\rho_{AL} = 1$.
  We will prefer to hold assets whose returns have similar levels of uncertainty to our liabilities. Therefore, it is not necessarily the case that a low value of σ_A is desirable. An asset with a low σ_A will often be considered as a "low-risk asset" by many practitioners and laypersons – a low σ_A indicating more price stability and therefore, perhaps, less "risk" in some contexts. Critically here, however, we are concerned with the relative values of the assets and liabilities.

These points will be revisited later in the chapter. We will proceed to investigate how our solvency position, $A - L$, is affected by investing in various types of assets which have different degrees of uncertainty (i.e. standard deviation), and different degrees of correlation with our liabilities (ρ_{AL}).

10.4 Characteristics of Various Asset Classes and Liabilities

Our aim is to analyse the stability of solvency levels resulting from various investments strategies; these strategies will include investment in equity, bond and cash asset classes. We therefore need to understand some of the basic characteristics of these asset classes, such as the relative stability of asset prices, and the correlations of their price movements with the liabilities. Of course, we will also require an assumption in respect of the distribution of future values of the liabilities.

Bonds
The following general conclusions can be made from analysing the historical returns from bond markets:

- The standard deviation of returns from longer term bonds is higher than the standard deviation of returns from shorter term bonds, i.e. the prices of longer term bonds tend to be more uncertain than the prices of shorter term bonds.
    The reader is encouraged to research bond return data of different durations to verify these comments. (The reader may wish to visit the "FTSE Actuaries UK Gilts Index Series" website.)
- There is a high degree of correlation between yields of similar terms on the yield curve, and a lower degree of correlation between short-term yields and long-term yields.
- Following on directly from the above point, a similar pattern of correlation is seen in bond *prices* of different durations.

To illustrate these last two points, included below is a table of correlation coefficients between UK gilt yields of terms 1, 2, 5, 10, 15, and 20 years, in respect of the period from 1995:

	1	2	5	10	15	20
1	1.00	0.88	0.44	0.19	0.09	0.09
2	0.88	1.00	0.69	0.37	0.22	0.21
5	0.44	0.69	1.00	0.81	0.45	0.39
10	0.19	0.37	0.81	1.00	0.78	0.64
15	0.09	0.22	0.45	0.78	1.00	0.88
20	0.09	0.21	0.39	0.64	0.88	1.00

We will not carry out any detailed analysis of these correlations, nor will we concern ourselves with a detailed calculation of the uncertainty in the future price of bonds. We will simply use this information to analyse the effectiveness of particular investment strategies in stabilising solvency levels.

Equities

We will also make the assumptions that (1) equity market returns, in general, exhibit even greater uncertainty than returns from long-term bonds (the reader may wish to verify this by obtaining historical market data), and (2) the equity market is largely uncorrelated with the bond market.

Cash

For our purposes we will assume returns from cash are risk-free i.e. guaranteed. That is, the variance of the value of the cash fund at a future time is 0.

Liabilities

In the same way in which the duration of bonds affects the standard deviation of bond prices, so the duration of our liabilities affects the uncertainty in their discounted values; after all, both are just sets of cashflows, with assets being income and liabilities being outgo. To simplify matters we will assume that in the example which follows, our liabilities have the same duration as the long-term bond fund; the standard deviation of the value of the liabilities should therefore be similar to the standard deviation of the value of the long-term bond fund. (An exercise is included later in the chapter to analyse the effect of investment strategies where our liabilities are of much shorter duration.)

10.5 Our Scenarios

We will run four investment strategy scenarios:

1) long-term bonds (20yr)
2) short-term bonds (5 yr)
3) equities
4) cash

Our task is therefore to measure the uncertainty of our future solvency position at $t = 1$ year, and to analyse the effect of different investment strategies on the solvency position at that time. Based on our earlier discussion, we will use the following values for the standard deviations of assets and liabilities, and correlations between assets and liabilities (these are broadly consistent with historical market data):

```
covarx

             sd corrn with liabs
long          8            0.99
short         2            0.50
equity       20            0.00
cash          0            0.00
liabilities   8            1.00
```

It should be noted that the exact parameter values used are not particularly important; it is the relative values which the reader should take note. For example, the long-term bond fund has a high σ, similar to that of our liabilities. Equities exhibit the highest σ, with the short-term bond fund and cash exhibiting the lowest σ's.

To make our scenario more transparent we will define expected values of the assets and liabilities in one year's time; at $t = 1$ year, the liabilities have an expected value of £100, and the assets an expected value of £110.

The reader may be questioning that the same expected asset value at $t = 1$ is assumed irrespective of whether we are invested in equities, bonds or cash. For our purposes we are primarily interested in the *range* of likely solvency positions in one year. Introducing various different expected returns from our four strategies would simply move the distributions along the surplus axis, and is thus an unnecessary complication at this point.

10.6 Results

Calculating the theoretical standard deviation of the surplus in one year's time from each of the four investment strategies is straightforward, using Equation 10.2. The results from the four strategies are shown below:

```
tot_sd<- function(sd1,sd2,rho)(sd1^2+sd2^2-2*rho*sd1*sd2)^0.5
stand_dev_fund_level<-c(tot_sd(8,8,0.99),tot_sd(8,2,0.5),tot_sd(8,20,0),
                        tot_sd(8,0,0),0)

(solv_lev<-cbind(covarx,stand_dev_fund_level)[1:4,])

          sd corrn with liabs stand_dev_fund_level
long      8              0.99              1.131371
short     2              0.50              7.211103
equity   20              0.00             21.540659
cash      0              0.00              8.000000
```

From these results it is clear that the most predictable solvency position at $t = 1$ is obtained under the long-bond investment strategy, which has a solvency standard deviation of only 1.13. This is despite the long bond fund's returns exhibiting the 2nd highest standard deviation. As noted earlier in the chapter, the greatest stability in $A - L$ can be obtained where both $\sigma_A = \sigma_L$ and $\rho = 1$. The equity fund results in significant uncertainty. Adopting a 100% cash strategy will lead to a significant level of uncertainty in the solvency position, greater even than the short-term bond fund strategy, due to the lack of correlation with the liabilities.

Note that the mean solvency position for all strategies in this example is 10.

Exercise 10.1 *Assuming the solvency position at $t = 1$ can be represented by a Normal distribution, calculate the probability that $A < L$ at $t = 1$ for each of the four strategies.*

10.7 Simulations

It is a useful exercise to simulate possible results from our investment scenarios. In particular it is a stepping stone to more complex models (e.g. using copulas, GARCH models etc, which are discussed in later chapters) where the variance of $A - L$ may be of limited value to us due to the resulting skewed and/or leptokurtic distributions. In particular we may expect the future value of bond portfolios to have a skewed distribution, and for equity portfolios to exhibit excess kurtosis. Simulated results can also be extremely useful for presentational purposes, especially where our audience has a less mathematical background.

For now we will simply assume that our assets and liabilities follow a 2-dimensional multivariate Normal distribution; it should be noted that this assumption is made for convenience and is not particularly realistic. In particular, any analysis of tail-risk (e.g. 99% VaR) using such assumptions should be treated with care. We can simulate values from this distribution using the mvrnorm function, and defining a covariance matrix (called sigma below).

Scenario 1: long-term bonds
Description: assets with high σ, which are highly correlated with the liabilities. Simulated asset and liability values have been produced below:

```
no_sims<-100000
sdx<-8; sdy<-8; rhoxy<-.99;  mean<-c(110,100)

(sigma<-matrix(c(sdx^2,rhoxy*sdx*sdy,rhoxy*sdx*sdy,sdy^2),2,2)) #covariance matrix

      [,1]  [,2]
[1,] 64.00 63.36
[2,] 63.36 64.00

set.seed(500);  sim1<-mvrnorm(no_sims,mean,sigma) #simulating assets and liabilities
surplus1<-sim1[,1]-sim1[,2]
sim1<-cbind(sim1,surplus1)
colnames(sim1)<-c("long-term bond value","liability value","surplus")
head(sim1,3)      # sample simulations

     long-term bond value liability value  surplus
[1,]             118.1385        107.3186 10.81987
[2,]             126.0107        115.3565 10.65424
[3,]             116.9746        107.1711  9.80348

sd(surplus1)

[1] 1.126928
```

And for our three remaining investment strategies:

Scenario 2: short-term bonds
Description: assets with low σ which exhibit some degree of correlation with the liabilities.

```
sdx<-2; sdy<-8; rhoxy<-0.5
sd(surplus2)

[1] 7.177849
```

Scenario 3: equities
Description: assets with a very high σ which are uncorrelated with the liabilities.

```
sdx<-20; sdy<-8; rhoxy<-0
sd(surplus3)
```

```
[1] 21.49506
```

Scenario 4: cash
Description: assets with very low standard deviation which are uncorrelated with liabilities.

```
sdx<-0.1; sdy<-8; rhoxy<-0
sd(surplus4)
```

```
[1] 7.972188
```

The standard deviation values obtained from the simulations are consistent with those obtained earlier using Equation 10.2. Note that as we are using a multivariate Normal distribution, $A - L$ also follows a Normal distribution.

The plots of the simulated surpluses from each of the four strategies are shown in Figure 10.1. Given the relatively simple nature of these models the additional benefit of presenting these plots is perhaps somewhat limited – they are shown here to provide examples of how results can be presented to a non-financial audience. Indeed, Figure 10.1 is a fairly typical example of how a stochastic analysis of various pension scheme investment strategies may be presented.

Exercise 10.2 *Calculate the probability that the scheme will be in deficit (i.e. $A < L$) at $t = 1$ for each strategy, and comment on this methodology.*

Our conclusion from this exercise is that if we wish to minimise the uncertainty in the scheme's solvency position in one year's time we should invest in long-term bonds.

10.8 Exercise and Discussion – an Insurer With Predominately Short-Term Liabilities

Note that where our liabilities are deemed to be of a short-term nature, as may be the case in many non-life insurance lines of business, stability in our solvency position will be enhanced by investing in shorter term bonds. The reader should re-run the above calculations incorporating the following changes to three parameter values:

$$\sigma_L = 2, \qquad \rho_{longbond-liab} = 0.5, \qquad \rho_{shortbond-liab} = 0.99.$$

The results are shown in Figure 10.2. Thus the most stable solvency position is achieved, under this scenario, by investing in short-term bonds and cash.

Many insurers have short-term liabilities. For example, motor vehicle claims will predominantly be short-term in nature, with payments resulting from most accident claims paid within a few weeks, or even days. Even the larger, more contentious claims will

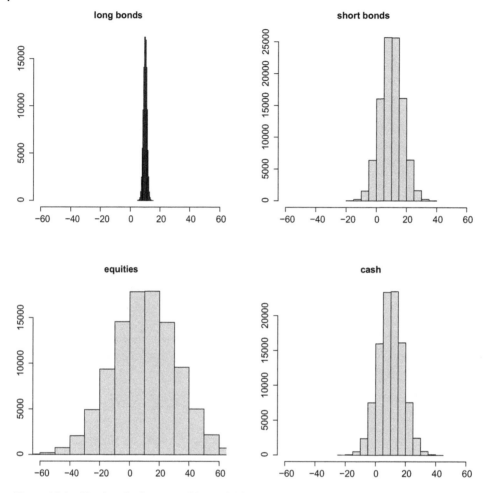

Figure 10.1 Simulated solvency positions with long-term liabilities under various strategies.

mainly be settled over the course of 2 to 3 years. These liabilities will therefore be measured using short-term yields, similar to those used to calculate the prices of short-term bonds. This is the fundamental reason why such insurers adopt investment strategies predominantly involving money market instruments and short-term bonds (or similar derivative instruments).

The interested reader who has little experience in non-life insurance companies is encouraged to access the report and accounts of such an insurer (plenty are freely available on the internet) and review their investment strategy.

10.9 Potential Exercise

Once familiar with the material included in this book, the reader is encouraged to develop, and simulate results from more complex asset–liability models. For example, a model could

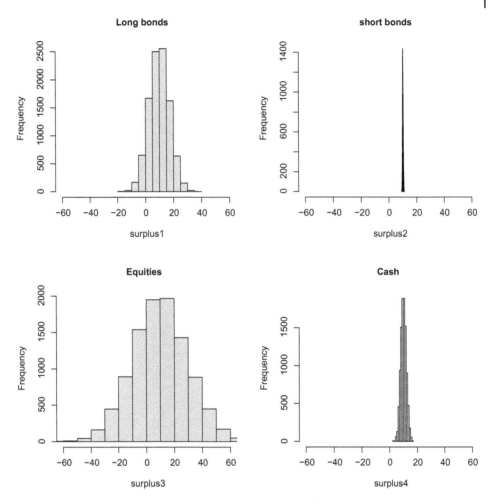

Figure 10.2 Simulated solvency position with short-term liabilities under various strategies.

include GARCH models, Extreme Value Theory, copulas, a bond portfolio model and a PCA interest rate model. An important part of this model development is understanding where the additional model complexity is warranted, and where a simpler model will suffice.

10.10 Conclusions

Our objective here was to find an investment strategy which produced the most stable future solvency position. We found that, for our institution with long-term liabilities, the long-term bond fund was likely to result in the most stable funding level. The reason for this is two-fold: the long bond prices are highly correlated with the liabilities, so when the liabilities increased it was highly likely that the assets would increase too. The second reason was that the long bonds had a similar variance to that of the liabilities. Hence when one increased the other was likely to increase by a similar amount.

10.11 Recommended Reading

- Booth et al. (2005) *Modern Actuarial Theory and Practice.* (2nd edition) Boca Raton, Florida, US: Chapman and Hall/CRC.
- Booth and Yakoubov (2000) *Investment Policy for Defined-Contribution Pension Scheme Members Close to Retirement: An Analysis of the "Lifestyle" Concept.* (April 2000) North American Actuarial Journal 4(2):1–19.
- S. Haberman, M. Z. Khorasanee, B. Ngwira, I. D. Wright (2003) *Risk measurement and management of defined benefit pension schemes: A stochastic approach.* IMA Journal of Management Mathematics, Volume 14, Issue 2 (April 2003).

11

Hedging: Protecting Against a Fall in Equity Markets

Peter McQuire

11.1 Introduction

Like diversification and matching, hedging is a key risk management strategy adopted by many institutions. Its aim is to minimise the net change in the future value of a position, by buying or selling an asset which behaves in the opposite direction to the existing position.

The concept of hedging is mathematically very similar to the concept of matching (see Chapter 10). In this chapter we focus on a specific example, that of how to hedge against a fall in the value of a portfolio of shares by taking a short position in equity index futures. In particular we will look at how we can minimise the uncertainty in our net position when exact hedging is not possible, which will usually be the case in practice, by making use of the "optimum hedge ratio".

A detailed understanding of futures pricing is not required to understand what follows – the key point is that the value of the "hedging" asset (in this case the futures contract) will tend to move in the opposite direction to that of the original asset or portfolio. For the reader who has not studied futures prior to reading this chapter, included below is a brief introduction. (Please see Hull (2012) chapter 3 for a more detailed explanation of financial futures contracts; the example below aims to explain the essence of futures contracts.)

11.2 Our Example

11.2.1 Futures Contracts – A Brief Explanation

In the fictitious country of Capland our client holds an equity portfolio consisting of shares in 20 of the largest listed companies in Capland. Our client is concerned about the potential of highly volatile equity markets over the next week resulting from a series of global political referendums and elections. Many analysts are predicting that markets could see unprecedented shocks (both positive and negative) depending on the results from those elections.

R Programming for Actuarial Science, First Edition. Peter McQuire and Alfred Kume.
© 2024 John Wiley & Sons Ltd. Published 2024 by John Wiley & Sons Ltd.
Companion Website: www.wiley.com/go/rprogramming.com.

Our client is risk averse and does not want to be exposed to the possibility of large losses from their equity portfolio which may result in insolvency (notwithstanding the potential for significant gains). Selling the entire equity portfolio is an option, but not a particularly attractive one as it will incur significant advisory fees and trading costs.

To hedge our position, one possibility is to "sell", or take a short position in, equity index futures contracts. (Note that the contract will be defined by the particular exchange which issues and regulates them. For example in the UK, a FTSE100 futures contract defines a point to equal £10, and a FTSE250 futures contract defines a point to equal £2. We use £1 for simplicity, but please see Section 11.4.) Let's say the equity index is currently at 7,500 points, the 3 month future price is 7,600, and each point on such a futures contract in Capland is defined to be worth £1. If we take a short position in one such contract and the index falls to, say 7,450 at expiry in 3 months, we will receive $(7,600 - 7,450) \times £1 = £150$ from the futures contract, that is we benefit from a fall in the index. If the index increases to 7,800 at expiry we would make a loss of $(7,800 - 7,600) \times £1 = £200$ on the futures contract, that is we make a loss from an increase in the index and would be required to make a payment of £200 to the exchange. The key point to note for our purposes is that the value of the short position in the futures contract will move, in general, in the opposite direction to the value of the client's equity portfolio.

Similarly the value of the contract following purchase will change continuously as the index level changes throughout the term of the contract. The simplest way to see this is to note that one can remove their position in the futures contract by taking the opposite position in an identical contract, in this case by buying the future. This is known as "closing out" your position. At the outset the futures contract should have nil value; however in the first example above the contract value just prior to expiry is close to +150, as the index has fallen. We could buy the same future at this point; its price will be close to 7,450 resulting in a net position which requires us to sell at 7,600 and buy at 7,450, thus receiving 150.

We can close out the position at any time. Let's assume that one day after entering into the above short position on the future contract (at 7,600), the index has increased by 10% from 7,500 to 8,250; your contract will have negative value to you – the price of the future may now be around 8,360 (the actual price will depend on market supply and demand). One can close out the position at this time by taking the opposite position on the identical future (i.e. a long position), the net position is that you have now agreed to buy at 8,360 and sell at 7,600, thus making a loss of 760. Thus the value now is around −760. Of course you will hopefully have gained from the underlying position in the equity portfolio, such that your net position may be relatively unchanged.

Provided that our portfolio of shares is to some extent representative of the shares included in Capland's equity index, then any portfolio losses may be at least partially offset by gains in the futures position. In the first example where the index fell to 7,450, the portfolio is likely to have fallen in value, but gains have been made from the short futures position. Similarly, if the equity market rises in value (as in the second example to 8,250) the gains from our portfolio would be offset from losses from the futures position. Thus any gains and losses from our portfolio are offset to some extent. This is hedging.

Critically, however, the degree to which this risk can be removed depends on the relative volatilities of our portfolio and the futures price, and also to what degree the movements in the portfolio and futures prices are correlated. This is demonstrated in our example.

11.2.2 Our Task

Our equity portfolio is currently valued at £100,000. How many futures contracts should we sell if we want to remove as much volatility as possible in the value of the portfolio at $t = 1$ week? (In our example all price changes are weekly.)

The answer is given by the optimal hedge ratio, h, which can be derived by minimising the variance of the change in value of the hedged portfolio. Given that the portfolio value and futures contract price at time t are P_t and F_t respectively, the change in the value of the hedged portfolio between times t and $t + 1$ is equal to

$$(P_{t+1} - P_t) - (hF_{t+1} - hF_t) = \Delta P - h\Delta F \tag{11.1}$$

and hence

$$Var(\Delta P - h\Delta F) = \sigma_{port}^2 + h^2 \sigma_{future}^2 - 2h\rho_{port,future}\, \sigma_{port}\, \sigma_{future} \tag{11.2}$$

where ΔP is the change in the value of the portfolio of assets, and ΔF is the change in the futures price. Also, σ_{future} is the standard deviation of the changes in the futures price, σ_{port} is the standard deviation of the changes in the portfolio value, and $\rho_{port,future}$ is the correlation coefficient between the price changes in the futures contract and the asset portfolio.

(Please also note that in this example, as will usually be the case, $E(\Delta P) = E(\Delta F)$. In this example both weekly expected returns are 0.)

Taking the derivative of Equation 11.2 with respect to h and setting this to 0, we arrive at the optimal hedge ratio, h:

$$h = \frac{\sigma_{port}}{\sigma_{future}}\, \rho_{port,future} \tag{11.3}$$

Thus selling h futures contracts minimises the uncertainty in the value of the net, hedged position.

Let's look at an example; our data, which shows historic weekly changes in the values of the portfolio and the futures price, can be found here:

```
data_hedge<-read.csv("~/Hedge_data.csv")
head(data_hedge,4)
```

```
    X       index   portfolio
1 1    8.529242   -285.8771
2 2  -70.141567 -2226.9231
3 3  -49.720619 -4427.2576
4 4   48.299563  1491.2088
```

For example, over the first week the index future price increased by 8.53 and the value of the portfolio fell by 285.88. To obtain the optimal hedge ratio we must calculate the standard deviations and correlations from this data:

```
(data_cor<-cor(data_hedge[,2],data_hedge[,3]))
[1] 0.5245584
(sd_index<-sd(data_hedge[,2]));   (sd_port<-sd(data_hedge[,3]))
[1] 73.32456
[1] 2013.815
(besthedgeratio<-sd_port/sd_index*data_cor)
[1] 14.40668
```

So our optimal hedge ratio, calculated from our data and using Equation 11.3, is:

$$h = \frac{2013.8150607}{73.3245633} \times 0.5245584 = 14.4066798$$

Remark 11.1 Note that the optimal hedge ratio is equal to the slope parameter from the appropriate linear regression model – we can determine our best hedging strategy by finding the slope of our best fitting line.

```
beta_obtain<-lm(data_hedge[,3]~data_hedge[,2])
beta_obtain$coeff[2] #our slope estimate

data_hedge[, 2]
        14.40668
```

Demonstrating how the standard deviation of the price of our hedged portfolio varies with the hedge ratio using Equation 11.2, and plotting the results (Figure 11.1):

```
hedge_var<-function(corrxy,sdx,sdy,h){sdy^2+(h*sdx)^2-2*h*corrxy*sdx*sdy}
h_sd<-hedge_var(data_cor,sd_index,sd_port,seq(-10,40,1))^0.5
```

We can see from Figure 11.1 that the variance (or standard deviation) is minimised at the optimal hedge ratio value calculated above. The standard deviation of changes of the net portfolio value of the best hedged position is

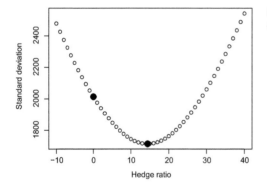

Figure 11.1 The effect of the hedge ratio on the uncertainty of our portfolio value.

```
hedge_var(data_cor,sd_index,sd_port,besthedgeratio)^0.5
[1] 1714.511
```

This compares to a standard deviation of equity portfolio returns of 2014, which is also shown in Figure 11.1 at $h = 0$.

Adopting this hedge ratio and running a number of simulations perhaps demonstrates the effect of hedging more clearly on our net position – the results are shown in Figure 11.2 (please note here the assumption of a multivariate Normal distribution):

```
#simulations
set.seed(100);   n<-100000
cov2<-data_cor*sd_index*sd_port
sigmax2 <- matrix(c(sd_index^2, cov2, cov2, sd_port^2), 2)   #covariance matrix
simsx <- mvrnorm(n, mu = c(0,0), sigmax2 )
net_portfolio_value<-simsx[,2]-besthedgeratio*simsx[,1]
sd(net_portfolio_value)      #sd of hedged portfolio (sims)

[1] 1709.816

sd(simsx[,2])                #sd of unhedged portfolio (sims)

[1] 2012.06
```

By hedging our position with the appropriate use of futures, we have managed to reduce the uncertainty of the value of our portfolio at $t = 1$ week. However, the reader may be somewhat underwhelmed by the reduction in the standard deviation of the change in portfolio values; the hedge has not significantly reduced the level of uncertainty. This is because the hedging instrument used is not highly correlated with our underlying asset ($\rho = 0.52$) and therefore the quality of the hedge is not particularly good. To achieve better results we need to use assets with a higher degree of correlation.

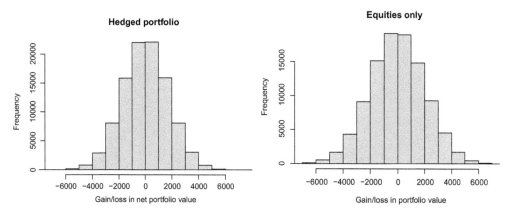

Figure 11.2   Simulations (assuming a multivariate Normal distribution).

Exercise 11.1 *Calculate the probability that our net position will fall by more than £4,000, under both the hedged and unhedged positions (using the same assumptions as above).*

11.3 Adopting a Better Hedge

In the example above the correlation coefficient between our portfolio returns and the futures contract was not particularly high ($\rho = 0.52$), resulting in a poor-quality hedge. This problem is commonly referred to as "cross-hedging"; here, this may have been due to a number of shares in the index not being included in the portfolio. If we can improve the quality of the hedge by using an instrument with a higher correlation we can reduce the variance of returns. Using $\rho = 0.95$, the hedge ratio is 26.09, and, from Equation 11.2, the standard deviation of net portfolios values is only 629.

Simulations from this "high-quality hedge", using the same assumptions as above, give the results shown in Figure 11.3. Comparing these plots with Figures 11.1 and 11.2 it is clear that the uncertainty in the net position at $t = 1$ has been significantly reduced by adopting this high-quality hedge.

Thus if we can find a "good hedge" the risk of incurring particularly bad losses is greatly reduced.

Exercise 11.2 *Compare the 95%VaR for the unhedged position with the "high-quality best-hedged" position, stating clearly any assumptions made.*

Exercise 11.3 *Calculate the probability that the portfolio falls by more than £2,000 over the next week if we are*

1. *not hedged*
2. *poorly hedged (where $\rho = 0.5$)*
3. *adopt a higher quality hedge (where $\rho = 0.95$)*

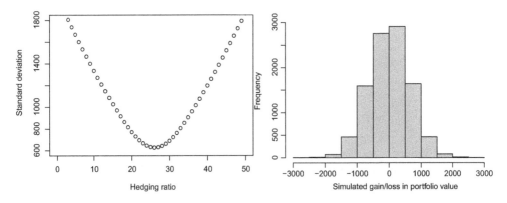

Figure 11.3 High-quality hedge.

11.4   Allowance for Contract and Portfolio Sizes

Due to the homogeneous characteristic of the standard deviation measure, (i.e. $Var(kX) = k^2 Var(X) \Rightarrow \sigma_{kX} = k\sigma_X$ where $k > 0$) it is straightforward to adjust Equation 11.3 to allow for the actual size of the underlying portfolio and futures contracts. For example, perhaps the future contract in Section 11.2.1 is based on £250 per index point, rather than just £1, and the underlying portfolio is actually five times bigger than stated above (i.e. £500,000), resulting in the following revised standard deviations:

$$\sigma_{future} = 250 \times 73.32 = 18331.14 \qquad \sigma_{port} = 5 \times 2013.82 = 10069.08$$

and hence the revised hedge ratio, h^*:

$$h^* = \frac{10069.08}{18331.14} \times 0.95 = 0.5218236$$

In this situation the analyst has a decision to make whether it is beneficial to adopt any hedging. Similarly, the futures exchange/regulator has to decide on an appropriate contract size, depending on market requirements.

11.5   Negative Hedge Ratio

Clearly our hedge ratio can be negative (see Figure 11.1). What does a negative hedge ratio mean? In our example above, instead of shorting the future we would be taking a long position in the futures contract such that its value to us is positively correlated with our underlying asset. This clearly makes our net position even more risky, and this is reflected in the higher variance when h is negative.

11.6   Parameter and Model Risk

A brief note about parameter risk is warranted at this point. The quality of the hedge depends significantly on the accuracy of the hedge ratio used, which in turn depends on the amount and relevance of the data used in determining the ratio, h. A key parameter which should concern us here is the standard error of our slope parameter estimator; any risk analysis should include the confidence we have in the estimated hedge ratio. Indeed it would be a useful exercise to repeat the above calculations but using less of the data to estimate the hedging ratio. Additionally, any risk measures such as Value at Risk need careful consideration – as noted on several occasions throughout the book care should be taking before assuming that a Normal distribution is appropriate.

11.7   A Final Reminder on Hedging

It is important to note that hedging is unlikely to result in entirely removing uncertainty (unless a perfect hedge exists). The above examples demonstrated that hedging can help to reduce the uncertainty in our net position. However the final result may turn out to be particularly bad even if we hold the optimal hedge ratio.

When communicating these ideas with a non-technical audience it is important to stress this point. In addition it should also be noted that, as with all types of hedging, the potential benefit from the underlying position will be reduced. This is a potentially embarrassing situation for the advisor if a competitor does not hedge their position and returns from the underlying position turn out to be positive. Communication is key!

11.8   Recommended Reading

- Hull, J. (2015) *Risk Management and Financial Institutions* (4th edition). Hoboken, New Jersey, US: Wiley.
- Hull, J. (2012) *Options, Futures and other Derivatives* (8th edition). Harlow, Essex, UK: Pearson.
- Sweeting, P. (2017) *Financial Enterprise Risk Management* (2nd edition). Cambridge, UK: Cambridge University Press.

12

Immunisation – Redington and Beyond

Peter McQuire

Remark 12.1 The reader may wish to ensure they are familiar with the material in Chapter 9 before studying this chapter.

```
library(matlib)
```

12.1   Introduction

In 1952 Frank Redington published an article which proposed a method for designing a bond investment strategy which could remove interest rate risk resulting from small parallel changes in the yield curve. In this chapter we discuss the key points of Redington's theory and look at various examples. Of equal importance, we also look at potential issues with the theory, and how to deal with these problems in practice.

First, we look at a number of simple scenarios, and construct asset portfolios consisting of one or two zero coupon bonds to demonstrate how Redington theory can be used to ensure that the value of the assets, A, always exceeds the value of the liabilities, L, (given that $A = L$ at $t = 0$) following a parallel change in interest rates.

We proceed by analysing a more realistic situation where the liabilities consist of several future payments and will aim to construct a larger portfolio of coupon-bearing bonds of various terms such that the solvency position is largely immune to any types of interest rate changes.

Finally we outline two problems with the theory – these relate to changes in the shape of the yield curve, and the relative timing of asset and liability cashflows which may result in liquidity issues.

R Programming for Actuarial Science, First Edition. Peter McQuire and Alfred Kume.
© 2024 John Wiley & Sons Ltd. Published 2024 by John Wiley & Sons Ltd.
Companion Website: www.wiley.com/go/rprogramming.com.

12.2 Outline of Redington Theory and Alternatives

The real purpose of Redington's theory is to construct a simple asset portfolio (e.g. two bonds) which, following a parallel change in a flat yield curve, always results in the value of these assets being at least as large as the liabilities.

It is useful at this stage to have a broad outlook on the various degrees of immunisation which are potentially available. We can view Redington immunisation as achieving partial immunisation against interest rate risk by holding a simple asset portfolio consisting of a small number of bonds, whose duration is equal to that of the liabilities whilst also ensuring that the convexity of the assets exceeds that of its liabilities. Simply by holding two zero-coupon bonds in particular proportions, it is possible to ensure an institution maintains a solvent position following parallel changes to the yield curve. However, as we shall see later in this chapter, if we hold too few bonds we may be at significant risk from non-parallel changes to the yield curve – the investment strategy resulting from applying Redington theory will not, in general, protect the solvency position from a change in the shape of the yield curve.

If we extend this idea to the extreme we arrive at complete cashflow matching; this involves purchasing a sufficient variety of bonds such that the investment income is exactly the same as the liability outgo in both size and timing (see Example 12.3). Clearly this is an impractical solution given the number of cashflow payments, the uncertainty of these payments, and the limited availability of specific types of bonds. In practice therefore, a compromise is usually made resulting in an investment strategy which consists of a sufficient range of bonds such that interest rate risk, including non-parallel yield curve changes, is managed to an acceptable degree.

Rather than deriving or stating Redington's conditions at this point, we will proceed by experimenting with various bond investment strategies to manage interest rate risk. Redington's conditions are stated at the end of this section for completeness. For a formal derivation of Redington's conditions please see Booth et al. (Chapter 5).

Example 12.1

The first example in this chapter involves an institution contracted to make the following two payments:

1) £10 million in 3 years, and
2) £25 million in 6 years.

Our aim is to adopt a bond investment strategy such that if we hold a portfolio of equal value to the present value of these two payments (i.e. such that we are 100% solvent), then we will continue to remain solvent following a parallel change in interest rates i.e. we will be immune to interest rate changes. An important caveat in these discussions is that we are only re-analysing the position before any actual cashflows are made, in this example before $t = 3$. We must be careful when analysing the position after $t = 3$; this practical issue is discussed later in the chapter.

First, we calculate the present value of these two payments (the liabilities). As noted above, a key assumption under Redington is that of a flat yield curve, such that only parallel movements are permitted. (We will see later in the chapter how non-parallel changes can result in significant problems.) Assuming an effective interest rate of 4% p.a. at all terms, the total present value of these two payments at $t = 0$ is:

```
interest<-0.04;        v<-1/(1+interest)
outgo<-c(10,25)   #payments due
outgo_time<-c(3,6)   #time of payments due
(liabs_pv<-presentValue(outgo,outgo_time,interest))

[1] 28.64783
```

The present value of the liabilities is £28.65 m. We will set the value of the assets at the start to be equal to this, such that we have exactly the required level of assets to meet the outgo.

Our task now is to invest £28.65 m in particular bonds such that, following a parallel change in the yield curve, the value of the assets does not fall below that of the liabilities. We will choose the simplest assets for the portfolio for now – zero coupon bonds (effectively a single redemption payment).

First we'll guess a simple strategy – a 20-year zero coupon bond. Therefore we will buy £28.65 m of this bond; the single payment that will be received from the bond in 20 years is:

```
(income<-liabs_pv/v^20)     # we'll receive this in 20 years

[1] 62.77092
```

The key questions is "what happens if interest rates change?" Recalculating the asset values, liability values, and funding levels over a range of interest rates at $t = 0$ (where the funding level is defined as the ratio of the asset value to the present value of the liabilities):

```
interestrange<-seq(0.01,0.08,0.01);      vrange<-1/(1+interestrange)
assets_pv<-income*vrange^20
liabs_pv<-outgo[1]*vrange^outgo_time[1]+outgo[2]*vrange^outgo_time[2]
assets_liabs<-assets_pv/liabs_pv
(al_data<-data.frame(interestrange,assets_pv,liabs_pv,assets_liabs))

  interestrange assets_pv liabs_pv assets_liabs
1          0.01  51.44356 33.25703    1.5468475
2          0.02  42.24303 31.62251    1.3358532
3          0.03  34.75473 30.08852    1.1550828
4          0.04  28.64783 28.64783    1.0000000
5          0.05  23.65770 27.29376    0.8667804
6          0.06  19.57227 26.02021    0.7521950
7          0.07  16.22120 24.82153    0.6535131
8          0.08  13.46739 23.69256    0.5684226
```

Thus when interest rates increase above 4% p.a. the present value of the liabilities and asset values decrease, but the asset value decreases by a greater amount; we would therefore be deemed to have insufficient assets i.e. we will be underfunded (or insolvent) if rates rise. Investing in 20-year zero coupon bonds would appear to be a poor strategy if we want to manage our interest rate risk. (See Figure 12.1.)

We encounter a similar problem if we invest in a shorter term bond.

Exercise 12.1 *Carry out similar calculations to the above but using a 4-year zero coupon bond (the key output is shown below).*

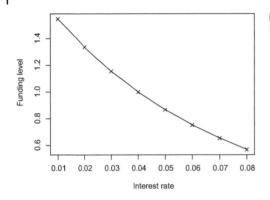

Figure 12.1 Funding level as a function of interest rates.

	interestrange	assets	liabs	assets_liabs
1	0.01	32.20620	33.25703	0.9684028
2	0.02	30.96167	31.62251	0.9791022
3	0.03	29.77667	30.08852	0.9896355
4	0.04	28.64783	28.64783	1.0000000

The problem here is when rates fall, the assets increase in value but the liabilities increase by a greater amount. Investing in a single zero-coupon bond appears to be a flawed strategy.

The above problems are good examples of term, or duration, mismatching. A common misconception is that financial institutions can manage interest rate risk simply by investing in bonds. Clearly, as demonstrated above, investing in bonds will not necessarily remove this risk.

Exercise 12.2 *The reader should try various alternative terms for bonds (preferably by first writing a suitable function). Of particular interest is to try a single zero-coupon bond with term equal to the duration of the liabilities in this section, measured at 4% p.a. (the reason for this will become apparent shortly).*

We have tried investing in 20-year bonds, and then in 4-year bonds, and, from Exercise 12.2, also with bonds with term equal to 5.07 years. Perhaps we should try investing in both the 4- and 20-year bonds?

12.3 Redington's Theory of Immunisation

Continuing with Example 12.1 we will now use Redington's conditions (which are shown at the end of this section) to calculate the proportions of the two bonds which will result in asset values which are greater than the liabilities, *irrespective of whether interest rates increase or decrease*. We will also look at some practical problems.

The central idea behind Redington's theory is to equate the change in asset value ("A") with the change in the value of the liabilities ("L"), following a small change in interest rates, *i*. Mathematically, this is simply:

$$\frac{dL}{di} = \frac{dA}{di} \qquad (12.1)$$

Note that, as $A = L$, this expression is equivalent to stating that we require the volatility or duration of the assets to equal that of the liabilities.

Exercise 12.3 *The reader should re-write all the equations in this section as a function of δ, the force of interest, to demonstrate that the use of i or δ results in the same solutions.*

The final requirement under Redington is to ensure that the second derivative of the assets with respect to interest rates is greater than the second derivative of the liabilities. The reason for this will become apparent shortly (see Figure 12.2), and was touched upon in Exercise 12.2. Redington himself described this intuitively as "the spread of the value of the asset-proceeds about the mean term should be greater than the spread of the value of the liability-outgo" [Redington (1952)].

The present values of the liabilities and assets, respectively, from Example 12.1 are as follows:

$$L = 10v^3 + 25v^6 \qquad A = Xv^4 + Yv^{20}$$

Our aim is to determine appropriate values of X and Y. Taking derivatives of both equations with respect to i:

$$\frac{dL}{di} = -30v^4 - 150v^7 \qquad \frac{dA}{di} = -4Xv^5 - 20Yv^{21}$$

As noted above we could alternatively equate the durations or volatilities of the assets and liabilities. We now have the following two equations, from the condition that the present values of the assets and liabilities are equal, and from Equation 12.1:

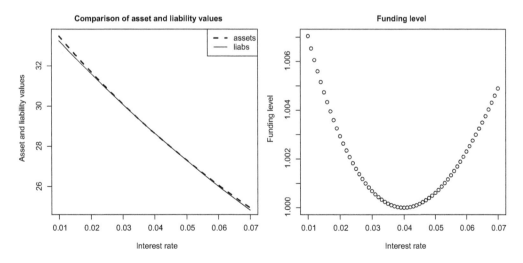

Figure 12.2 Comparison of assets and liabilities as a function of interest rates.

$$10v^3 + 25v^6 = Xv^4 + Yv^{20}$$

$$-30v^4 - 150v^7 = -4Xv^5 - 20Yv^{21}$$

With two equations and two unknowns it is straightforward to solve these simultaneous equations for X and Y at $i = 4\%$ p.a. (Here we use matrices to solve these.)

```
interest1<-0.04;  v<-1/(1+interest1)
a<- v^4;b<- v^20;c<- -4*v^5;d<- -20*v^21; e<- 10*v^3+25*v^6; f<- -30*v^4-150*v^7
matrix11<-matrix(c(a,b,c,d),nrow = 2,byrow = TRUE)   #matrix of coefficients
matrix22<-matrix(c(e,f),nrow = 2)
(bond_prop<-inv(matrix11)%*%(matrix22))

            [,1]
[1,]  31.27467
[2,]   4.19405
```

This gives us the solution:

$$X = 31.2746667, \qquad Y = 4.1940504$$

Hence we should buy £26.73 m of the 4-year bond and £1.91 m of the 20-year bond i.e. Xv^4 and Yv^{20} respectively.

Exercise 12.4 *Verify that the present values and first derivatives of the assets and liabilities are indeed equal under these conditions.*

We must also check the second derivatives or convexities (for reasons which we will address shortly). We can obtain these using the convexity function:

```
convexity(c(10,25),c(3,6),i=0.04)          #liabilities

[1] 30.22414

convexity(c(31.27,4.19),c(4,20),i=0.04)    #assets

[1] 43.18205
```

$$\text{Convexity of liabilities} = \frac{1}{L}\frac{d^2L}{di^2} = 30.2241415$$

$$\text{Convexity of assets} = \frac{1}{A}\frac{d^2A}{di^2} = 43.1820507$$

The convexity of the assets is greater than that of the liabilities, as required.

Exercise 12.5 *Verify both numerically, using a small Δi.*

Checking we are immunised against parallel changes in the yield curve:

```
interest<-seq(0.01,0.07,0.001);     v<-1/(1+interest)
#define our future cashflows
asset4payment<-bond_prop[1];  asset20payment<-bond_prop[2]
liab3payment<-10;             liab6payment<-25
```

```
asset4pv<-asset4payment*v^4;   asset20pv<-asset20payment*v^20
liab3pv<-liab3payment*v^3;     liab6pv<-liab6payment*v^6

totalassetpv<-asset4pv+asset20pv    #see plot
totalliabpv<-liab3pv+liab6pv        #see plot
fundlev<-totalassetpv/totalliabpv   #see plot
```

Producing a table of our results:

```
keystats<-cbind(interest,totalassetpv,totalliabpv,fundlev)
keystats[seq(1,61,10),]#select every 10th row

      interest totalassetpv totalliabpv  fundlev
[1,]     0.01      33.49155    33.25703 1.007052
[2,]     0.02      31.71543    31.62251 1.002939
[3,]     0.03      30.10928    30.08852 1.000690
[4,]     0.04      28.64783    28.64783 1.000000
[5,]     0.05      27.31044    27.29376 1.000611
[6,]     0.06      26.08019    26.02021 1.002305
[7,]     0.07      24.94312    24.82153 1.004898
```

Figure 12.2 (left) is fundamental to our understanding of immunisation – the plot shows there are no interest rates at which the $A < L$. This was our goal. Now we can see the importance of the second derivative – the curvature of the asset curve (i.e. the second derivative) at $i = 4\%$ is greater than that of the liabilities. Plotting the resulting funding level as a function of changes to i reinforces these ideas [see Figure 12.2 (right)].

The reader is reminded of the requirement under Redington immunisation that these interest rate changes must be equal at all terms; advising a client that "they are immunised against changes in interest rates" would be incorrect. We will consider this further in Sections 12.4 and 12.5.

The reader should contrast these results with those from earlier in the chapter, and particularly with Exercise 12.2 where we invested in a single bond of term 5.07 years. In that example, even though we had matched durations, we would always be underfunded following a parallel change in interest rates due to the convexity of the assets being less than that of the liabilities. It is worth noting however that the degree of underfunding may, depending on the circumstances, be considered immaterial.

For completeness, we set out below Redington's conditions for immunisation:

$$A = L \tag{12.2}$$

$$\frac{dA}{di} = \frac{dL}{di} \tag{12.3}$$

$$\frac{d^2A}{di^2} > \frac{d^2L}{di^2} \tag{12.4}$$

Exercise 12.6 *Consider what happens to the funding level between times 0 and 3 if the yield curve gradually falls from 4% to 1%.*

A common criticism of Redington theory is that the assets need continual adjustment to remain immunised, following parallel movements in the yield curve. However, provided

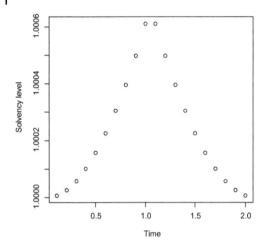

Figure 12.3 Solvency level evolution, with a moving flat yield curve.

there are no cashflows at these times (see below) we should not be overly concerned with not rebalancing our portfolio following a yield change – the solvency position improves following a yield change, and returns to the original level if the yield reverts to its original level.

Exercise 12.7 *Calculate the solvency position in the above example if the yield curve was to move gradually from 4% at t = 0 to 5% at t = 1, then gradually back to 4% at t = 2. (Note that there are no cashflows during this period.) See Figure 12.3.*

From Figure 12.3 we can see there is no particular problem; we remain solvent at all times, it's just that the solvency position deteriorates after $t = 1$.

However, some rebalancing would be required at $t = 3$ when the first payment is made, to remain strictly Redington immunised. For example, if interest rates had remained at 4% at $t = 3$, it turns out that we would need to sell at little more than £10 of the (now) 1-year bond to remain immunised – £10 to pay the benefit due, with the excess to buy more of the longer bond.

Exercise 12.8 *Determine the transactions required at t = 3 to remain immunised if interest rates are (a) 3 % or (b) 4%, or (c) 5% at t = 3.*

12.4   Changes in the Shape of the Yield Curve

We should reiterate exactly what is, and is not meant by "Redington immunisation": any asset / liability position is said to be Redington immunised if, following a parallel change in the flat yield curve, we have the position where $A > L$. But what is the effect on the funding level if the shape of the yield curve changes?

Example 12.2

Assuming the same cashflows as in the original example above, what would happen to the funding level after one day if the yield curve sloped upwards from *term* = 7 years?

In this situation only the bond with an initial term of 20 years is affected by this yield curve change (all other cashflows occur at times 6 years and earlier). The value of the 20-year bond will fall, resulting in the total value of assets falling below the value of the liabilities (which will be unchanged). Clearly we are not protected against particular types of yield curve changes, such as this. Although we are "Redington immunised," we are not immune to changes in the shape of the yield curve. Further examples will be discussed in more detail in Section 12.5.2.

However, we can generally manage interest rate risk *including changes in the shape of the yield curve* by adopting the principles of Redington, but, in addition, holding assets with cashflow timings which more closely align with those of the liability cashflows. This will leave us less exposed to changes in the shape of the yield curve – a major flaw of Redington's theory when applied in practice.

From our example above with liability payments due at times 3 and 6 years, a better investment strategy would include holding 2-year and 7-year zero coupon bonds in the correct proportions. Taking this to the extreme, we simply end up with "exact cashflow matching" such that the asset proceeds exactly match those of the liability payments (here holding 3-year and 6-year zero coupon bonds in the correct proportions). Of course the example above includes a particularly simple set of cashflows and is therefore straightforward to match precisely (assuming such bonds exist in the marketplace).

Exercise 12.9 *Demonstrate the effects on the funding level from the above change in yield curve (i.e. a change from year 7) if we were to hold the 4- and 20-year bonds (as compared to the 2- and 7-year bonds).*

12.5 A More Realistic Example

The majority of the remainder of this chapter considers an example with many cashflow payments over several years (there are 105 coupon receipts from assets and 19 benefit payments). Our objective remains the same – to maintain asset values which will not fall below that of the liabilities following changes to interest rates. However here we will also concern ourselves with practical issues.

12.5.1 Determining a Suitable Bond Allocation

In this example a financial institution has estimated it will pay the following cashflows (in £m) at future times $t = 1, 2,..., 19$ years (see Figure 12.4):

```
dataliabs<-c(57,76,81,113,123,135,146,167,189,188,176,203,184,156,149,124,101,70,30)
```

The institution's priority is to manage interest rate risk, and will do so by investing in bonds. However, in this example, there is a limited number of terms at which this institution can buy bonds, namely: 5, 10, 20, 30, and 40 years, with all bonds paying annual coupons of 4%. We will continue to use the concept of equating the duration of the assets and liabilities to help us determine an appropriate solution.

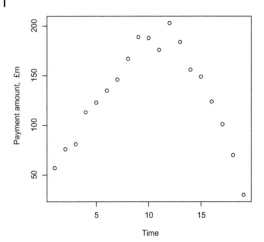

Figure 12.4 Liability cashflows.

We will initially assume that the yield curve is flat, with $\delta = 5\%$ p.a. Thus the present value of the liabilities is:

```
delta<-0.05;    (liab_pv<-presentValue(dataliabs,seq(1,19),exp(delta)-1))
[1] 1526.198
#the function requires effective rates to be used
```

The duration of the liabilities is (in years):

```
(liab_durn<-duration(dataliabs, seq(1,19), exp(delta)-1, k = 1, macaulay = TRUE))
[1] 9.102079
```

Our task is therefore to determine an appropriate bond portfolio allocation such that the duration of the portfolio is close to 9.1 years. The prices of the five bonds are (the calculations for only two bonds are shown):

```
bond1<-c(rep(4,4),104);    bond2<-c(rep(4,9),104)

bond1price<-presentValue(bond1,seq(1,5),exp(delta)-1)
bond2price<-presentValue(bond2,seq(1,10),exp(delta)-1)

(bondprice<-c(bond1price,bond2price,bond3price,bond4price,bond5price))
[1] 95.13730 91.35023 86.10388 82.92181 80.99179
```

And the durations of the bonds:

```
bond1durn<-duration(bond1,seq(1,5),exp(delta)-1, k = 1, macaulay = TRUE)
bond2durn<-duration(bond2,seq(1,10),exp(delta)-1, k = 1, macaulay = TRUE)
(bonddurn<-c(bond1durn,bond2durn,bond3durn,bond4durn,bond5durn))

[1]   4.619095  8.349792 13.622226 16.761418 18.547315
```

Now we need to choose the bond proportions not only to meet Redington's conditions, but also to ensure that any changes to the shape of the yield curve do not cause material problems. To do this we should aim to invest in such a way which more closely matches the timing of the liability payments, rather than simply investing in, say, 5-year bonds and 40-year bonds.

There are an infinite number of solutions given that we have five unknowns but only two equations; one method is to choose proportions for three bonds such that we are left with only two unknowns. For example:

```
choose_three<-2      #choose number of 10, 20, 30-year bonds

matrix_LHS<-matrix(c(bondprice[1],bondprice[5],bonddurn[1]*bondprice[1]/
liab_pv,bonddurn[5]*bondprice[5]/liab_pv),byrow=TRUE,nrow=2)

matrix_RHS<-matrix(c(liab_pv-choose_three*(bond2price+bond3price+bond4price),
liab_durn-choose_three*(bond2durn*bond2price+bond3durn*bond3price+
bond4durn*bond4price)/liab_pv),nrow = 2)

(bonds_1_5<-inv(matrix_LHS)%*%(matrix_RHS))

        [,1]
[1,] 8.609124
[2,] 2.301441

(no_of_bonds_x<-c(bonds_1_5[1],rep(choose_three,3),bonds_1_5[2]))

[1] 8.609124 2.000000 2.000000 2.000000 2.301441

(bond_prop<- ( no_of_bonds_x*bondprice/liab_pv))    #bond allocation

[1] 0.5366596 0.1197095 0.1128345 0.1086645 0.1221321
```

Checking that this gives the correct present value and duration:

```
(sum(no_of_bonds_x*bondprice))

[1] 1526.199

(sum(bonddurn*bond_prop))

[1] 9.102083
```

The Redington conditions for present values and duration are met.

(Alternatively we can combine the cashflows from all the bonds, and calculate the present value and duration directly:

```
x1<-no_of_bonds_x[1]*bond1 #cashflow from bond1
x2<-no_of_bonds_x[2]*bond2;    x3<-no_of_bonds_x[3]*bond3
x4<-no_of_bonds_x[4]*bond4;    x5<-no_of_bonds_x[5]*bond5
x1[(length(x1)+1):40]<-0;    x2[(length(x2)+1):40]<-0
x3[(length(x3)+1):40]<-0;    x4[(length(x4)+1):40]<-0
totalcashflow_in<-x1+x2+x3+x4+x5              #see plot
presentValue(totalcashflow_in,seq(1,40),exp(delta)-1)

[1] 1526.199

duration(totalcashflow_in,seq(1,40),exp(delta)-1, k = 1, macaulay = TRUE)

[1] 9.10208
```

The cashflows from the assets, calculated above, are shown in Figure 12.7, together with the liabilities.)

In addition, investing in bonds with these terms should meet the convexity requirements (Equation 12.4):

```
convexity(totalcashflow_in,seq(1,40),i=0.05)    # asset income
[1] 151.8218
convexity(dataliabs,seq(1,19),i=0.05)           # liability outgo
[1] 102.317
```

By investing in these five bonds in our chosen proportions we have met the Redington requirements and should be immune to parallel changes in interest rates. Also, as we have invested in several bonds over a range of durations we should also be reasonably immune to changes in yield curve shapes (see Section 12.5.2).

Exercise 12.10 *Demonstrate that we are immune to parallel changes in interest rates by calculating the ratio of assets to liabilities if interest rates change tomorrow, in the range 3% to 7%. A plot of the results is shown in Figure 12.5.*

By buying five bonds in the above proportions we have ensured that the funding level will not fall below 100% in the short term following parallel changes to the yield curve.

Exercise 12.11 *Demonstrate, by using similar calculations to those in Exercise 12.10, that if we buy too many shorter bonds, or too may longer bonds, we risk being underfunded in the future. Why may we consider such a strategy given these results?*

12.5.2 Change in Yield Curve Shape

As noted in Section 12.4 it is also important to assess the level of interest rate risk from changes in the *shape* of the yield curve. We should therefore investigate an appropriate range of such changes to assess our interest rate risk. For example, we could analyse the effect of the yield curve changes shown in Figure 12.6 (the code for these is shown below).

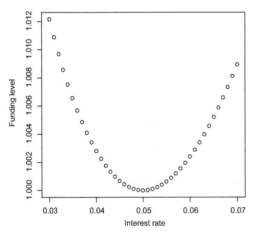

Figure 12.5 Funding level vs. interest rates.

Figure 12.6 Yield curve scenarios.

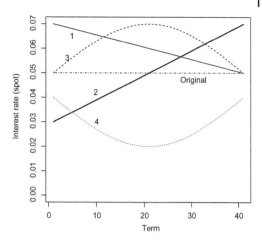

```
delta0<-rep(0.05,41)            #original flat yield curve
delta1<-0.07-seq(0,40,1)*0.02/40
delta2<-0.03+seq(0,40,1)*0.04/40
delta3<-sin(pi/40*seq(0,40))/50+0.05
delta4<- -sin(pi/40*seq(0,40))/50+0.04
```

Exercise 12.12 *Calculate the asset and liability values, and thus the funding level, using Scenarios 1 and 2.*

The funding level under Scenario 1 is 102.6% – not a concern. However, the funding level under Scenario 2 is 95.9%. Given that some asset cashflows in this example are received after 20 years, we are at risk from such a scenario occurring i.e. an upward sloping yield curve.

Exercise 12.13 *Calculate the resultant funding level from Scenarios 3 and 4.*

If we are not satisfied wth the level of interest rate risk, the bond allocation can be recalculated such that the level of interest rate risk is acceptable.

Exercise 12.14 *By investing in only the 5-year bonds and 40-year bonds above, calculate the proportion of bonds required such that we are Redington immunised, using an interest rate of 5% p.a. Discuss problems with this approach.*

Exercise 12.15 *The reader is encouraged to improve on the bond strategy suggested above.*

12.5.3 Liquidity Risk

Our job is not quite done. So far we have only been interested in maintaining the solvency position. Ultimately however, we must pay the correct benefits when they are due. It is important to compare the actual asset and liability cashflows to determine at what points we may need to sell assets in order to meet liability payments.

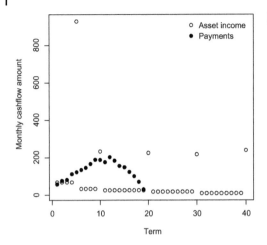

Figure 12.7   Asset and liability cashflows.

Figure 12.7 shows the asset and liability cashflows. (The code in respect of the asset cashflows was set out in Section 12.5.1.)

Although we may have largely solved our problem of changes in the yield curve, the actual cashflows do not look well matched. This is a typical problem when matching relatively stable cashflow payments, such as monthly pensions, with "lumpy" bonds; generally the annual payments exceed the annual income. We receive many small coupon payments plus the occasional very large redemption payment. Therefore in most years we will have negative cashflows which will require us to sell bonds, or hold some of the earlier redemption payments as cash (e.g. from receipts at times 5 and 10 years). For example, we are likely to be required to sell a number of bonds at times 2, 3, and 4 years to be able to make the pension payments at those times. We will also receive a significant redemption payment at five years which will require a decision on how to invest this income. Thus it is likely that we will regularly be changing the proportion of bonds held in our portfolio, incurring trading costs and in itself reducing the value of the assets held.

This issue can be partially resolved if there are bonds available with a greater range of redemption dates. The following exercise looks at the idealised position where 19 bonds exist with redemption dates from $t = 1$ to $t = 19$ years i.e. we can achieve exact cashflow matching if the liability payments are guaranteed.

Example 12.3

Write code to determine the bond allocation between these 19 bonds (with redemption dates from $t = 1$ to $t = 19$ years) which will result in exact cashflow matching of the payments set out in Section 12.5.1 i.e. `dataliabs`. Assume all bonds pay annual coupons of 3%.

First we match the final payment at $t = 19$ by purchasing the appropriate number of 19-year bonds, then calculate the number of 18-year bonds required, allowing for the coupons received from the 19-year bond. The process is repeated back to $t = 1$:

```
dataliabsx<-dataliabs*1000000
couponx<-3
lengthxx<-length(dataliabsx)
```

```
incomex<-rep(0,lengthxx)
no_of_bonds<-NULL

for (n in lengthxx:1) {
        no_of_bonds[n]<-dataliabsx[n]/(100+couponx) #no of bonds required
        incomex[n-1]<-no_of_bonds[n]*couponx+incomex[n]
        dataliabsx[n-1]<-dataliabsx[n-1]-incomex[n-1]
        }
no_of_bonds              #answer

 [1]    16155.7   206640.4   262839.6   590724.8   708446.5   849699.9   985190.9
 [8] 1224746.6 1481489.0 1515933.7 1441411.7 1754654.1 1617293.7 1385812.5
[15] 1357386.9 1148108.5   952551.7   671128.3   291262.1

check<-NULL

for (counter_x in 1:(lengthxx-1)) {
        check[counter_x]<-no_of_bonds[counter_x]*(100+couponx)+
        (sum(no_of_bonds[(counter_x+1):lengthxx]))*couponx
        }
check[lengthxx]<-no_of_bonds[lengthxx]*(100+couponx)
check/1000000   #agrees liab payments!!!

 [1]   57   76   81 113 123 135 146 167 189 188 176 203 184 156 149 124 101   70   30
```

Of course, such an approach would not be possible in practice; bonds with all the required terms would not be available, and there is likely to be considerable uncertainty regarding the actual amount and timing of the liability payments.

12.6 Conclusion

Approaches of varying complexity to manage interest rate risk have been demonstrated in this chapter, from investing in one bond to investing in potentially all bonds available in the market.

We can "Redington immunise" by investing in two bonds, resulting in an immunised position against parallel changes to the yield curve, but not against non-parallel yield curve changes. A better solution is to invest in assets which result in a closer cashflow match with the liabilities whilst continuing to meet the Redington requirements, reducing interest rate risk exposure to changes in the shape of the yield curve and easing liquidity issues.

Thus a life office or pension scheme should be aware of the duration of its liabilities; adopting a bond strategy with similar duration and approximate matching of cashflows will usually provide a practical strategy for managing interest rate risk at an acceptable level. The decision may also be influenced by the supply and demand of bonds; it may be considered reasonable to accept some interest rate risk if the potential returns from these bonds are sufficiently beneficial.

Any benefits gained from moving towards exact cashflow matching are likely to be more than offset by other factors e.g. (1) longevity risk is likely to be sufficient to make any accurate cashflow matching spurious, (2) availability of the required bonds, and (3) the extra administration costs involved in managing such a strategy.

Given the cashflow problems discussed above it will generally be the case, in practice, that the bond portfolio investment strategy is frequently adjusted to maintain an acceptable risk position.

12.7   Recommended Reading

- Booth et al. (2005) *Modern Actuarial Theory and Practice*. (2nd edition). Boca Raton, Florida, US: Chapman and Hall/CRC.
- Haynes A.J and Kirton R.J. (1952) *The financial structure of a life office*. Faculty of Actuaries (21) pp. 141–218.
- McCutcheon and Scott. (1986) *An Introduction to the Mathematics of Finance*. Oxford, UK: Butterworth Heinemnan.
- Redington F.M. (1952) *Review of the principles of life office valuations*. Journal of the Institute of Actuaries (78).
- Sweeting P. (2017) *Financial Enterprise Risk Management* (2nd edition). Cambridge, UK: Cambridge University Press.

13

Copulas

Peter McQuire

```
library(MASS)
library(copula)
```

13.1   Introduction

In many financial and actuarial settings, the modelling of multivariate distributions will form a crucial part in analysing the aggregated risks to which a financial institution is exposed. For example, an insurer will be interested in the incurred claims under each individual class or type of insurance it writes; however, of greater importance is the combined total claims amount from all its classes of business in a single time period. Crucially, the level of correlation between policies may vary substantially, often depending on varying economic conditions; insurer insolvency may occur if high claim levels from all insurance classes coincide at the same time. Similarly, a banking regulator will be interested in understanding the likelihood of a number of banks failing over a short period of time, and it may well be that losses in the banking sector are particularly highly correlated during periods of stressed financial conditions. To develop the best models we must aim to capture both the degree *and the pattern* of any dependencies which exist in our data. Copulas can help us do this.

The use of copulas in the financial sector has grown significantly since David Li published his paper in 2000, "On Default Correlation: A Copula Function Approach", in which he described how the Gaussian copula can be used to help in the pricing of collateralized debt obligations (CDOs). However their application was not entirely appropriate, prompting many warnings from academia. The resulting model risk is generally regarded to be one of the principal reasons behind the 2008 global financial crisis.

In this chapter we use copulas to model the dependent behaviour between two or more random variables. We will see how copulas can be used to model dependencies by separately analysing the dependency between the quantiles of each of the variables' distributions. In this way we can study the relationship between variables separately to analysing each of their underlying distributions. Thus the overall multivariate distribution with

R Programming for Actuarial Science, First Edition. Peter McQuire and Alfred Kume.
© 2024 John Wiley & Sons Ltd. Published 2024 by John Wiley & Sons Ltd.
Companion Website: www.wiley.com/go/rprogramming.com.

d dimensions (or variables) can be described using a *d*-dimensional copula together with *d* individual univariate distributions.

In this chapter, and the one that follows it, our focus is on the practical application of copulas. The fitting of copulas will be discussed in the following chapter.

13.2 Copula Theory – The Basics

Remark 13.1 The reader is strongly advised to review Appendix 1, following an initial reading of this chapter, which formally sets out the key properties of copulas. It is also expected that the reader will have a basic knowledge of copulas; an excellent text is McNeil, Frey, Embrechts (2005) chapter 5.

A *d*-dimensional copula, C, is a multivariate distribution in $[0, 1]^d$ space (which we refer to as "copula" space), with standard uniform marginal distributions. Thus copulas describe dependence purely on a quantile scale, which is often useful in risk management when estimating the probability of extreme events.

We will often use the term "marginals" or "marginal distributions" when discussing copulas; the "marginals" we refer to are the univariate distributions which are combined with a chosen copula to describe the multivariate distribution. It is important to note that there is no requirement for the various marginal distributions we decide to include in our final model to be similar in any way, nor is our choice of copula limited by the marginal distributions chosen. The reader should keep in mind that our ultimate objective will usually be to model and simulate data; thus it will normally be the case that we will map simulations using our chosen copula from $[0, 1]^d$ onto $d-$dimensional "data" space.

We will mainly be discussing 2-dimensional copulas in this chapter (i.e. $[0, 1]^2$ space), which can easily be represented graphically by a unit square (similarly a 3-dimensional copula can be represented within the unit cube i.e. in $[0, 1]^3$ space).

As noted above, the copula distribution function, C, is used in a multivariate environment. It is analogous to the distribution function for a univariate distribution, F, which is defined as follows:

$$F_X(x) = P(X \leq x) \tag{13.1}$$

Similarly for the copula:

$$C^{(\beta)}(F_{X_1}(x_1), F_{X_2}(x_2), ..., F_{X_n}(x_n)) = P(X_1 \leq x_1, X_2 \leq x_2, ..., X_n \leq x_n) \tag{13.2}$$

with a set of parameters, (β). The number of parameters required depends on the type of copula. The LHS of Equation 13.2 is often written in the more succinct notation $C(u_1, u_2, u_3, ...u_n)$, where $u_i = F_{X_i}(x_i)$, and $u_i \sim U(0, 1)$. We will use $F(x)$ and u interchangeably throughout the chapter. Given that C is a distribution function we can see that, as it is a probability, Equation 13.2 will return a number between 0 and 1.

Stating this more formally, we have Sklar's Theorem:

$$F(x_1, x_2, .., x_n) = C^{(\beta)}(F_{X_1}(x_1), F_{X_2}(x_2), ..., F_{X_n}(x_n)) \tag{13.3}$$

where $F(x_1, x_2, .., x_n)$ is the joint distribution function. It is perhaps easiest to develop an initial understanding of copulas by way of an example.

Example 13.1

An investment analyst has researched annual investment returns on UK and US equity markets and has concluded that UK annual returns may be modelled with a $N(5, 10^2)$ distribution, and US annual returns with a $N(4, 9^2)$ distribution.

A reasonable question to ask may be "What is the probability that both UK and US equity markets make losses next year?" This is given by the following expression:

$$C[F_{UK}(0), F_{US}(0)] \tag{13.4}$$

where $F_{UK}(0)$ is the probability that returns from the UK are less than zero next year, and $F_{US}(0)$ is the probability that returns from the US are less than zero next year. We can use the pnorm functions to find these values:

```
(ukloss<-pnorm(0,5,10));  (usloss<-pnorm(0,4,9))
[1] 0.3085375
[1] 0.3283606
```

The probability that UK equity returns are less than zero, $F_{UK}(0)$, is 0.3085375 , and the probability that US equity returns are less than zero , $F_{US}(0)$, is 0.3283606, giving our copula expression:

$$C(0.3085375, 0.3283606) \tag{13.5}$$

In words this means "what is the probability that UK returns will be in the bottom 30.9% of its distribution *and* US returns will be in the bottom 32.8% of its distribution." In general $C(u_1, u_2, u_3, ...u_d)$ gives the probability that all d random variables are less than the chosen set: $u_1, u_2, u_3, ...u_d$.

13.3 Commonly Used Copulas

The answer to Equation (13.5) depends on which particular copula, and parameter(s), are used. There are numerous copulas in the literature; amongst the most commonly used are:

- independent
- Gaussian
- Clayton
- Gumbel
- t
- Frank

The expression is then evaluated using whichever copula best models the dependency pattern in the data. We will discuss the first four copulas listed above in the following sections, which provide the reader with a good initial range of modelling possibilities.

13.3.1 The Independent Copula

Without any further discussion, we can arrive at an answer to Equation 13.5 if the independent copula is applied. (Indeed, no knowledge of copulas is required here.) The probability in this case would be:

$$0.301 \times 0.328 = 0.0987$$

This is often referred to as the product copula, $\prod$, for obvious reasons, and is defined below:

$$F(x_1, x_2, .., x_n) = \prod \left(F_{X_1}(x_1), F_{X_2}(x_2), ..., F_{X_n}(x_n) \right) = F_{X_1}(x_1) F_{X_2}(x_2)...F_{X_n}(x_n)$$

$$(13.6)$$

This copula may be appropriate if it was thought that UK equity returns and US equity returns were independent (which is certainly not the case).

The key point here is that we cannot know the probability of both events happening unless we know the pattern of the dependence behaviour. This is where copulas can be useful; in addition to allowing for varying degrees of dependence by varying the copula parameter, we can also allow for further flexibility in the *pattern* of dependency by choosing the most appropriate copula, ultimately improving our models.

In the following sections we shall consider three types of copulas – Gaussian (also known as "Normal"), Clayton, and Gumbel. Initially we shall consider only the 2-dimensional case for simplicity. Each of these 2-dimensional copulas has only one parameter which determines the degree of dependence. Our approach will be to look at simulations of these copulas to gain an initial appreciation of their distributions.

13.3.2 The Gaussian Copula

The Gaussian copula is defined as:

$$C^P_{Gauss} = \Phi^P(\Phi^{-1}(F_{X_1}(x_1)), ..., \Phi^{-1}(F_{X_d}(x_d)))$$

$$(13.7)$$

where Φ denotes the standard Normal distribution function, Φ denotes the joint Normal distribution function, and P is the correlation matrix of X. (A further discussion of this equation is included later in the chapter.)

To simulate copulas in R we can use the rCopula function in the copula package. 500 simulations from a (2-dimensional) Gaussian copula, with a single correlation coefficient, $\rho = 0.9$, can be simulated as follows (see Figure 13.1):

```
a <- rCopula(500,normalCopula( 0.9, dim = 2))
```

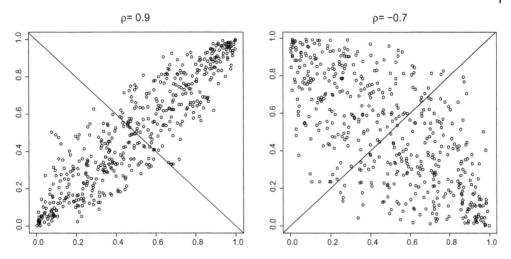

Figure 13.1   Two realisations of the Gaussian copula.

Also shown in Figure 13.1 is a simulated Gaussian copula with a negative correlation ($\rho = -0.7$). The following points should be noted:

- The Gaussian copula is symmetrical – the key line of symmetry is included in the plot. This contrasts with the Gumbel and Clayton copulas (see later). Indeed there have been several instances in the financial sector where the underlying data has not exhibited such symmetry but a Gaussian copula has nevertheless been included in the model, leading to potentially disastrous consequences.
- The range of values is between 0 and 1 (i.e. $[0, 1]^2$ copula space).

To aid understanding of the Gaussian copula, let's look at some values of the cumulative distribution function, C_{gauss}, which we can calculate using the pCopula function. Take the following expression:

$$C_{gauss}^{\rho=1}(0.5, 0.5) = 0.5$$

```
pCopula(c(0.5, 0.5), normalCopula(param=1.0))
[1] 0.5
```

The function calculates the probability that X_1 is in the bottom half of X_1's distribution and X_2 is in the bottom half of X_2's distribution; given that X_1 and X_2 are perfectly correlated with each other (i.e. $\rho = 1$) this is equal to 0.5.

Remark 13.2   This is an example of another type of standard copula, the *co-monotonicity* copula, M, where the random variables X_n are perfectly positive dependent:

$$M(F_{X_1}(x_1), F_{X_2}(x_2), ..., F_{X_n}(x_n)) = min(F_{X_1}(x_1), F_{X_2}(x_2), ..., F_{X_n}(x_n))$$

Perhaps confusingly, this is often referred to as the minimum copula (given the formula); however it gives the upper bound for the value of C for the set of $F_{X_i}(x_i)$.

Exercise 13.1 *Calculate the following probabilities in respect of the co-monotonicity copula:*

$P(F(x_1) \leq 0.3 \text{ and } F(x_2) \leq 0.2)$

$P(F(x_1) \leq 0.3 \text{ and } F(x_2) \leq 0.3)$

$P(F(x_1) \leq 0.3 \text{ and } F(x_2) \leq 0.8)$

$P(F(x_1) \leq 0.8 \text{ and } F(x_2) \leq 0.8)$

End of Remark 13.2.

And if $\rho = 0.5$ or $\rho = 0$?

```
pCopula(c(0.5, 0.5), normalCopula(0.5))  #with correlation coefficient = 0.5

[1] 0.3333333

pCopula(c(0.5, 0.5), normalCopula(0.0))  #same as the independent copula

[1] 0.25
```

We proceed to verify these probabilities by running simulations of a Gaussian copula with the above correlations and observe the proportion of simulations which are in the bottom left-hand quadrant.

Simulations of these three copulas are shown in Figure 13.2. We simply need to count the proportion of simulations where both $u_1 \leq 0.5$ and $u_2 \leq 0.5$:

```
set.seed(5000)
a<- rCopula(1000,normalCopula(1.0, dim = 2))
sum((a[,1]<0.5)*(a[,2]<0.5))

[1] 487

aa<- rCopula(1000,normalCopula(0.5, dim = 2))
sum((aa[,1]<0.5)*(aa[,2]<0.5))

[1] 333

aaa<- rCopula(1000,normalCopula(0, dim = 2))
sum((aaa[,1]<0.5)*(aaa[,2]<0.5))

[1] 254
```

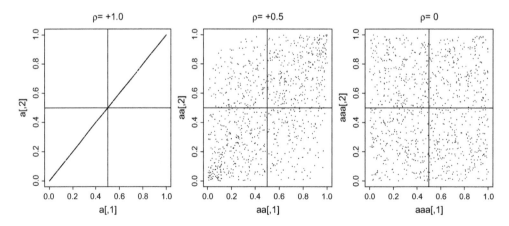

Figure 13.2 Gaussian copulas with varying degrees of dependence.

The copula plots are in agreement, allowing for stochastic uncertainty, with the calculations carried out above using the copula cumulative distribution functions. (The reader should re-run the above with more simulations.) The reader may find it instructive to analyse the ordered point values, using:

```
a_sorted <- a[order(a[,1]),]
```

The algorithms and formulae required to simulate the copulas discussed in this chapter are not included here. Please refer to McNeil et al. (2005) chapter 5 for these details.

13.3.3 Archimedian Copulas

An important group of copulas is the Archimedian copulas characterized by the generator function ϕ. These functions have the form (in 2 dimensions):

$$C(F_{X_1}(x_1), F_{X_2}(x_2)) = \phi^{-1}(\phi(F_{X_1}(x_1)) + \phi(F_{X_2}(x_2))) \tag{13.8}$$

where $\phi^{-1}(x)$ denotes the inverse function obtained by inverting the equation $x = \phi(y)$. We shall look at two examples of Archimedian copulas: Clayton and Gumbel.

13.3.4 Clayton Copula

In advance, two key points to note about the Clayton copula are:

- the Clayton parameter, θ, is a measure of the degree of dependence, with a high value indicating high dependence; note that it is not the correlation coefficient used for the Gaussian copula. For now we can note that:

$$\theta = \frac{2\tau}{1 - \tau}$$

where τ is Kendall's tau (see Appendix 2 for a brief discussion on Kendall's tau and rank correlations).
- crucially, the Clayton copula is not symmetrical – it has "lower tail dependency", meaning that there is greater correlation at the lower end of the distribution than at the upper end.

The Clayton generator function is $\phi = t^{-\theta} - 1$, giving the expression for 2-dimensional Clayton copula:

$$C_{Cl}^{\theta}(u, v) = (u^{-\theta} + v^{-\theta} - 1)^{-\frac{1}{\theta}} \tag{13.9}$$

Simulating and plotting two Clayton copulas, both in 2 dimensions, with (1) $\theta = 5$ and (2) $\theta = 20$ (Figure 13.3):

```
set.seed(600)
clay_5 <- rCopula(500,claytonCopula( 5, dim = 2)) #500 simulations
clay_20<- rCopula(500,claytonCopula(20, dim = 2))
```

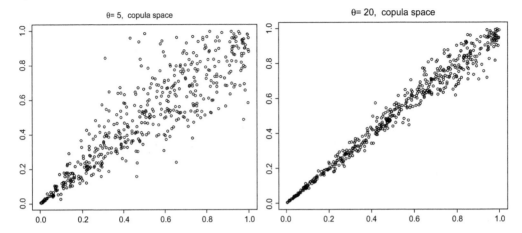

Figure 13.3 Clayton copula simulation.

Calculating the probability of both $F(x)$ and $F(y)$ being in the bottom left quadrant (i.e. $C(0.5, 0.5)$) from various Clayton copulas:

```
pCopula(c(0.5, 0.5), claytonCopula(5))

[1] 0.4366484

pCopula(c(0.5, 0.5), claytonCopula(20))#high correlation

[1] 0.4829682

pCopula(c(0.5, 0.5), claytonCopula(1))#lower correlation

[1] 0.3333333
```

Exercise 13.2 *Verify the above calculations by running appropriate simulations and calculating the number of points where $(F(x) < 0.5, F(y) < 0.5)$ (as performed earlier for the Gaussian copula).*

Exercise 13.3 *Also verify the above with Equation 13.9.*

As noted in the second bullet point above, the Clayton copula, in contrast with the Gaussian copula, is not symmetrical; there are far more points in the bottom-left square $[0, 0.1]^2$ than there are in the top-right square $[0.9, 1]^2$.

Exercise 13.4 *Verify visually, by re-running the above code with only 100 simulations and $\theta = 20$, that there are more points in $[0, 0.1]^2$, than there are in $[0.9, 1]^2$.*

This important characteristic of the Clayton copula is referred to as lower tail dependency. First look at $[0, 0.01]^2$ sub-space i.e. $C(0.01, 0.01)$, here using a Clayton (10) copula:

```
pCopula(c(0.01, 0.01), claytonCopula(10))

[1] 0.00933033
```

and then the extreme upper square $[0.99, 1]^2$ sub-space i.e. $1 - C(1, 0.99) - C(0.99, 1) + C(0.99, 0.99)$ (this is known as the "survival copula"):

```
1- pCopula(c(1, 0.99), claytonCopula(10))- pCopula(c(0.99, 1), claytonCopula(10))+
 pCopula(c(0.99, 0.99), claytonCopula(10))
```

```
[1] 0.00100062
```

This is a key concept with copulas. The lower tail of the Clayton copula behaves similarly to the co-monotonicity copula ($C(0.01, 0.01) \approx 0.01$) whilst the upper tail has approximately a tenth of this value i.e. 0.001. Thus the Clayton copula may be suitable to model situations where the degree of dependence exhibited in one tail is different to that in the other tail.

Exercise 13.5 *Run 1,000,000 simulations of a Clayton(10) copula. Calculate the number of points where X and Y are both less than 0.01, and also where both are greater than 0.99. Compare your findings with the theoretical values.*

Exercise 13.6 *Explain the reasoning behind the "survival copula" expression above.*

13.3.5 Gumbel Copula

The reader should carry out similar exercises in respect of the Gumbel copula to those set out earlier for the Gaussian and Clayton copulas. The Gumbel parameter, here denoted by α, is a measure of the degree of dependence in the same way that θ is a measure of dependence under the Clayton copula. Here we simply note that:

$$\alpha = \frac{1}{1 - \tau}$$

where, as per the Clayton copula, τ is Kendall's tau.

Simulating and plotting two Gumbel copulas, both in 2 dimensions, with (1) $\alpha = 3$ and (2) $\alpha = 12$ (Figure 13.4):

```
gum_3<- rCopula(500,gumbelCopula(3, dim = 2)) # 3 is the Gumbel parameter chosen

gum_12<- rCopula(500,gumbelCopula(12, dim = 2)) # 12 is the Gumbel parameter chosen
```

The Gumbel copula also exhibits asymmetry but, in contrast with the Clayton copula, has *upper tail* dependency. Again we can demonstrate the asymmetric nature using the survival copula:

```
pCopula(c(0.2, 0.2), gumbelCopula(3))
```

```
[1] 0.1316294
```

```
1-pCopula(c(1, 0.8), gumbelCopula(3))-pCopula(c(0.8, 1), gumbelCopula(3))+
pCopula(c(0.8, 0.8), gumbelCopula(3))                    #survival copula
```

```
[1] 0.1549202
```

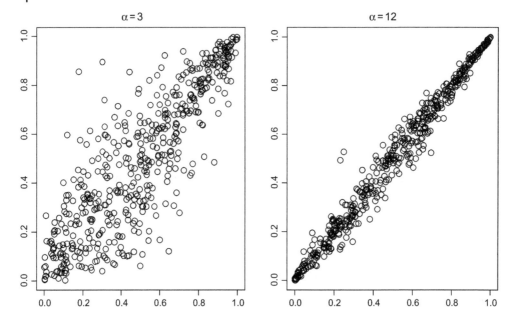

Figure 13.4 Realisations of Gumbel copulas.

In respect of the Gumbel(3) copula there are 18% more points in the top-right square than the bottom left square (as defined above).

The Gumbel generator function is $\phi(t) = (-lnt)^{\alpha}$, giving the expression for 2-dimensional Gumbel copula:

$$C_{Gm}^{\alpha}(u, v) = e^{-((-ln\,u)^{\alpha}+(-ln\,v)^{\alpha})^{\frac{1}{\alpha}}} \qquad (13.10)$$

Exercise 13.7 *Verify the above Gumbel copula probabilities.*

One should note that the copula parameter(s) determine the general degree of dependence. However the general shape of a particular copula in terms of asymmetry does not change.

13.4 Copula Density Functions

Before moving on, it should be noted that the density function, c, associated with the copula, C, is given by:

$$c(u, v) = \frac{\partial^2 C(u, v)}{\partial u \partial v} \qquad (13.11)$$

Included below are the copula density functions for the Clayton (Equation 13.12 and Figure 13.5 left) and Gumbel (Equation 13.13 and Figure 13.5 right)) copulas. These will be required when we are fitting copulas to data in the next chapter.

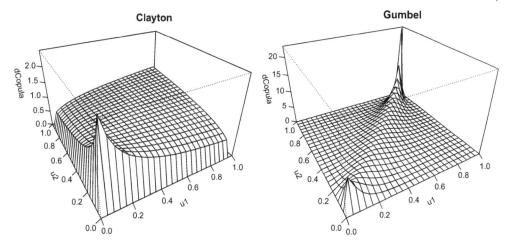

Figure 13.5 Copula density functions.

$$c_{Cl}(u,v) = (1+\delta)(uv)^{-(1+\delta)}(u^{-\delta}+v^{-\delta}-1)^{-\frac{1}{\delta}-2} \tag{13.12}$$

$$c_{Gbl}(u,v) = \frac{C(u,v)}{(uv)}((-\log u)^{\delta}+(-\log v)^{-\delta})^{\frac{1}{\delta}-2}\,(\log(u)\log(v))^{\delta-1} \\ \left(\delta-1+((-\log u)^{\delta}+(-\log v)^{\delta})^{\frac{1}{\delta}}\right) \tag{13.13}$$

Example 13.2

We can numerically estimate $C_{Cl}(u,v)$ using Monte Carlo integration and Equation 13.11. The following code estimates $C_{Cl}^{\theta=10}(0.2,0.2)$:

```
theta1 <-10
clay_dens<-function(u,v)(1+ theta1 )*(u*v)^(-(1+ theta1 ))*
(u ^(- theta1 )+v^(- theta1 )-1)^(-1/ theta1 -2)


nnn<-10000000
upx<-0.2; lowx<-0
upy<-0.2; lowy<-0
xa<-runif(nnn,lowx,upx); ya<-runif(nnn,lowy,upy)
den_vals<-clay_dens(xa,ya)


mean_val1<-mean(den_vals)
mean_val1*(upx-lowx)*(upy-lowy)                    #Monte Carlo simulation estimate

[1] 0.1865659

pCopula(c(upx, upy), claytonCopula( theta1 ))      #exact

[1] 0.1866066
```

Note that not all copulas have density functions; where the copula is not continuous the density function does not exist e.g. the co-monotonicity copula.

13.5 Mapping from Copula Space to Data Space

The above material has dealt with representations of copulas, not our actual data. When designing or running models which incorporate copulas we would usually include a further step in the modelling process – one which maps each simulated point in $[0,1]^d$ onto the marginal distribution for each dimension i.e. onto "data space". Ultimately we are interested in modelling our data, and thus simulating points in data space.

Example 13.3

Take the following three pairs of points in copula space, where, for simplicity, we will assume that both marginals are $N(0,1)$:

```
(z<-matrix(c(.01,.025,.5,.5,.95,.99),3,2,byrow=TRUE))

     [,1]  [,2]
[1,] 0.01 0.025
[2,] 0.50 0.500
[3,] 0.95 0.990
```

Mapping these copula points from $[0,1]^2$ onto $[-\infty, +\infty]^2$ data space using the qnorm function i.e. applying the appropriate inverse cumulative distribution function:

```
(zz<-qnorm(z))

          [,1]       [,2]
[1,] -2.326348 -1.959964
[2,]  0.000000  0.000000
[3,]  1.644854  2.326348
```

The points in copula and data space are plotted in Figure 13.6.

Thus when simulating such data it would be usual to first simulate copula data and then "map" it onto "data space" using the chosen marginal distributions.

In the above simplified example we have used standard Normal distributions for both X and Y, and a Gaussian copula. Any combination of copulas and marginal distributions can be used – we would choose the distributions which best fit our data.

Exercise 13.8 *Map the above copula points to data space using $X \sim Gamma(2,4)$ and $Y \sim Exp(0.02)$ marginal distributions. Note that our data space here is $[0, +\infty]^2$.*

Later in the chapter we will simulate data from chosen copulas and marginals using the simpler rMvdc function (this avoids the need to separately simulate copulas and then map those numbers onto our data space).

Before we move on it is worth reiterating the following general points:

- In copula space, $F(x)$ and $F(y)$ are uniformly distributed. Therefore, the number of simulated points (allowing for stochastic error) between $F(x_1)$ and $F(x_1) + a$, and $F(x_2)$ and $F(x_2) + a$ is the same (provided we stay within $[0,1]$ space). The same applies for Y values.

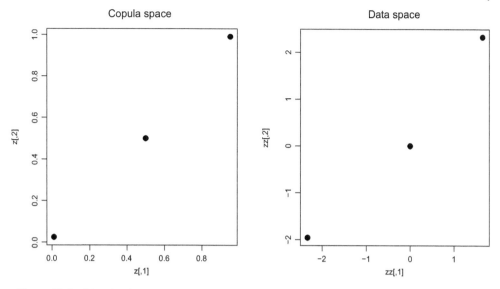

Figure 13.6 Mapping from copula to data space.

- Critically however the number of simulated points in each identically sized square within [0, 1] space will not, in general, be the same. This is perhaps the fundamental point of the chapter so far.

Exercise 13.9 *Demonstrate these two points by running suitable code.*

Exercise 13.10 *Using the* qnorm *function, map every point from the Clayton copula simulations shown in Figure 13.3 from* $[0, 1]^2$ *space onto* $[-\infty, \infty]^2$ *data space using* $X \sim N(5,1)$ *and* $Y \sim N(160, 40^2)$ *for the two marginal distributions. (A plot of the solution is shown in Figure 13.7.)*

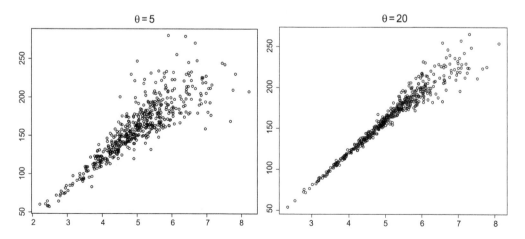

Figure 13.7 Clayton copula mapped to data space.

13.6   Multi-dimensional Data and Copulas

The examples discussed to this point have involved two variables (or dimensions). In this short section, we proceed to simulate copulas with more than two dimensions.

Note that different copulas require different numbers of parameters. Clayton and Gumbel copulas have exactly one each, irrespective of the number of dimensions, d. However, a Gaussian copula requires $d(d-1)/2$ terms in the correlation matrix. We define the number of dimensions we require using the dim argument.

Proceeding with 1,000 simulations of a 3-dimensional Gaussian copula with 3 different correlation coefficients:

```
d <- 3;        rho_a <- c(.9,.58,.3) #our 3 rho's
cc <- normalCopula(rho_a, dim = d,dispstr = "un")
U <- rCopula(1000, copula = cc)
```

Often we may simply use the same value for ρ between all our pairs in the Gaussian copula; rather than include a vector with repeated terms we simply set the dispstr argument to "ex" (where we do specify each value for ρ, as above, we use "un").

The splom function plots all available 2-dimensional plots, in this case for xy, xz, and yz. The plots reflect that the correlation between x and y is high (0.9), and relatively low between y and z (0.3).

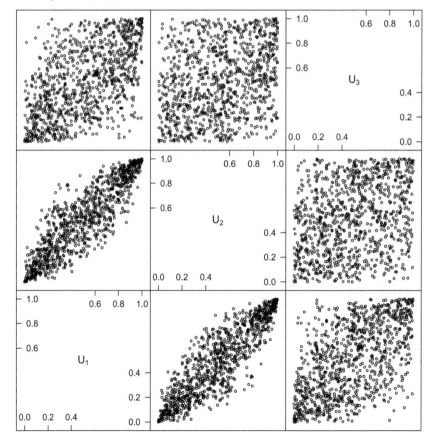

```
splom2(U, cex = 0.4, col.mat = "black") #produces 3 pairwise graphs
```

Simulating a 3-dimensional Clayton copula is straightforward – only one correlation parameter is permitted (this is a potential flaw of Clayton models and can be restrictive):

```
theta = 1.5;      d <- 3
cc <- claytonCopula(theta, dim = d)
U <- rCopula(1000, copula = cc)
```

The multiple-plot is redundant here and is not shown – all pairwise dependencies are the same as we have only one parameter.

Exercise 13.11 *Simulate and plot a 5-dimensional Gaussian copula with your own chosen correlations.*

13.7 Further Insight into the Gaussian Copula: A Non-rigorous View

We can think, informally, of a multivariate Normal distribution as the "product" of univariate Normal distributions and a Gaussian copula. In the same non-rigorous way, if we remove the Normal marginal distributions from the multivariate Normal distribution we are left with the Gaussian copula. We can remove the marginal by applying the cumulative distribution function; this maps the data onto [0,1] space.

Let's look at a simple 2-dimensional example; simulating three pairs from a 2-dimensional, multivariate Normal distribution using the mvrnorm function:

```
set.seed(1000)
(a<-mvrnorm(3,c(0,0),matrix(c(1,.9,.9,1),2,2)))
           [,1]         [,2]
[1,] -0.5774625 -0.29151931
[2,] -0.9994447 -1.35120250
[3,]  0.1262830 -0.04611306
```

Now mapping onto [0,1] space we arrive at three pairs simulated from a Gaussian copula:

```
(b<-pnorm(a,0,1))
          [,1]        [,2]
[1,] 0.2818135 0.38532709
[2,] 0.1587897 0.08831529
[3,] 0.5502460 0.48161007
```

More formally the Gaussian copula can be simulated by:

1) Simulating n realisations from a d-dimensional, zero mean, multivariate Normal distribution with a specified covariance matrix e.g. using Cholesky decomposition (see McNeil et al. (2015) chapter 3).
2) Apply the Normal cumulative distribution function to these $n \times d$ realisations, using the appropriate variance for each marginal.

Note that this is an alternative, and admittedly rather long-winded way of simulating a Gaussian copula – we have already done this using a simple R function, rCopula. However this does perhaps provide further insight into the Gaussian copula.

13.8 The Real Power of Copulas

Copulas allow us significant flexibility in modelling multivariate data. We can use whatever marginal distributions we require and incorporate the most appropriate copula to model the dependence pattern in our data. For example, below we simulate three claim amounts from a Gamma(6, 3) distribution and three claim amounts for an Exp(0.025) distribution, and allow for correlation between the two sets with a Gaussian copula (with $\rho = 0.5$):

```
set.seed(400);          (b<- rCopula(3,normalCopula(0.5, dim = 2)))
             [,1]          [,2]
[1,]  0.1998990 0.62780276
[2,]  0.8936272 0.39042855
[3,]  0.1758201 0.07183116

c<-qgamma(b[,1],6,3);    e<-qexp(b[,2],0.025)
(f<-cbind(c,e)) #our simulated data points

             c          e
[1,]  1.301000 39.533254
[2,]  3.052987 19.799965
[3,]  1.247021  2.981665
```

Thus by using copulas we have a flexible way to model multivariate distributions.

13.9 General Method of Fitting Distributions and Simulations – A Copula Approach

13.9.1 Fitting the Model

We set out below a general method to fit a copula model to data (we will see an application of this in the next chapter).

1) Obtain and verify your multi-variate data. For example we may be looking at claims from private car insurance (x) and commercial motor insurance (y).
2) Determine a best fitting distribution for each of the marginal distributions (usually obtained using Maximum Likelihood Estimation). For example, you may determine that Gamma(2.04, 0.34) and logN(1.007, 1.066) fit your x and y data best.
3) Map each data point onto $[0, 1]^d$ space using $F_X(x)$ for each marginal. This is your empirical copula.
4) Determine the copula which fits the empirical copula best, typically using Maximum Likelihood Estimation. Let's say the best fit we find is a Clayton(2.5).

5) Simulate an appropriate number of points, in $[0, 1]^d$, using the chosen copula.
6) Map these points into "data" space using the marginals found in step 2 by using the inverse cdf , $F_X^{-1}(F_X(x))$, in respect of each marginal.
7) This gives you simulated data points from a Clayton(2.5) with marginals Gamma(2.04, 0.34) and logN(1.007, 1.066).

   Note that points 5 and 6 are generally combined, as described in the following section.

13.9.2 Simulating Data Using the *mvdc* and *rMvdc* Functions

The simulations we carried out earlier in this chapter involved simulating the copula and marginals separately (steps 5 and 6 above); this method added transparency but is rather long-winded. A preferable method in practice is applying the rMvdc function, by first defining a multivariate distribution (i.e. the copula and marginals) using the mvdc function.

Below we simulate 1,000 2-dimensional data points with $N(6, 2^2)$ and Gamma(3, 0.2) marginals, and a Gumbel copula ($\alpha = 5$) (Figure 13.8):

```
model1 <- mvdc(gumbelCopula(5), margins = c("norm", "gamma"),
paramMargins = list(list(mean = 6, sd = 2),
list(3, 0.2)))

n <- 1000;     x <- rMvdc(n, mvdc = model1)
```

Typically we would proceed to evaluate various risk measures from such simulations; for example, calculating the 95% Value at Risk for aggregate claim amounts.

Exercise 13.12 *Simulate, using rMvdc, 2-dimensional data with N(5,2) and Exp(0.1) marginals, and a Clayton copula ($\alpha = 3$).*

13.10   How Non-Gaussian Copulas Can Improve Modelling

Gaussian copulas are extensively used in finance and insurance; they are well understood by practitioners and allow for significant flexibility as different degrees of correlations can be incorporated between different pairs.

As seen above, however, alternative copulas may be more appropriate in situations where the degree of correlation varies throughout the distribution. It may be the case that, under typical circumstances, there is little correlation between variables, but that this correlation increases during particular periods of time (e.g. due to war, recessions, extreme weather), affecting either the extreme right or left of the distributions.

In the following chapter we look at a simple example of total claim amounts from two departments within an insurer, where our objective is to model the insurer's total amount of potential claims next year, and assess the resulting tail risk i.e. the likelihood of insolvency and reserves required. Where the degree of correlation in the tails differs significantly to the central part of the distribution we may find that our estimates of tail risk are

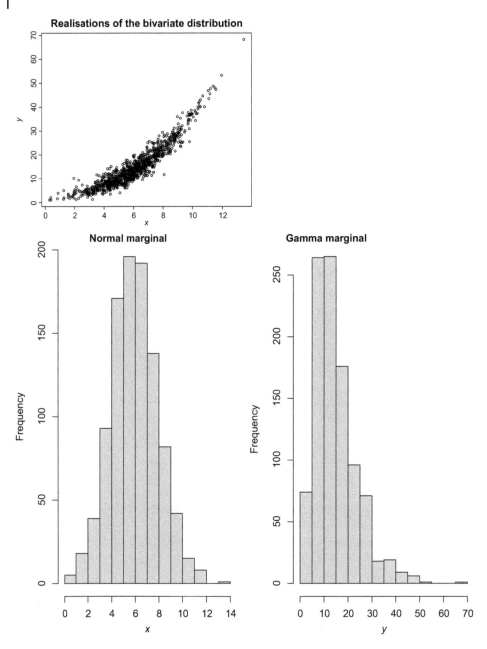

Figure 13.8 Simulated data using rMvdc and mvdc functions.

misleading if the default Gaussian copula is applied, resulting in inappropriate reserves being held and poor business decisions.

A similar example of model risk was a major contributing factor to the Global Financial Crisis (2007/2008) where the Gaussian copula was commonly used in modelling CDOs.

In this case the Gaussian copula underestimated the likelihood of extreme risks occurring together.

13.11   Tail Correlations

Finally, we will look at the tail correlation of copulas i.e. how the degree of correlation changes as we move further into the tail of the distribution. A commonly used measure is lower tail correlation, λ_L, which is defined as follows:

$$\lambda = P(F_Y(y) \leq u \mid F_X(x) \leq u) = \frac{C(u,u)}{u} \tag{13.14}$$

$$\lambda_L = \lim_{u \to 0} \lambda \tag{13.15}$$

Thus we can study tail correlation by investigating the above conditional probability, λ. Figure 13.9 shows $u = 10\%$ where the area of the rectangle is 0.1 and the area of the square is 0.01.

We will first look at a 2-dimensional Gaussian copula with $\rho = 0.5$. Applying Equation 13.14, using various extreme values of u:

```
lambda1<-NULL
for (ss in seq(1,10)) {
u<-c(10^-ss)
lambda1[ss]<-pCopula(c(u,u),normalCopula(.5))/u
}
round(lambda1,5)

 [1] 0.32402 0.12939 0.05426 0.02331 0.01016 0.00448 0.00199 0.00089 0.00040
[10] 0.00018
```

Figure 13.9   Representation of lower tail correlations.

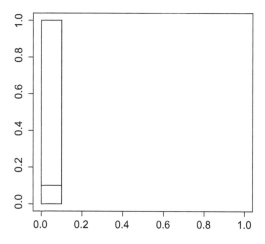

The reader may find it informative to run simulations in respect of these calculations. For example, looking at the lower 0.1% tail (corresponding to the third calculation in the output above i.e. 0.0542592):

```
set.seed(500);  rho_xx = 0.5;  d <- 2;  sims_tail<-10000000
U <- rCopula(sims_tail, copula = normalCopula(rho_xx, dim = d) )

xx<-which(U[,1]<0.001)
yy<-U[xx,]          #these are the points where x is less than 1%
zz<-sum(yy[,2]<0.001);    (zzz<-zz/length(xx))

[1] 0.04804146
```

(Even when running 10^7 simulations the stochastic error is material.)

Of all the pairs where $F(x) < 0.001$, 5.43% of them also have $F(y) < 0.001$. This gives a measure of tail correlation. Compare the situation where the two random variables are independent – one would expect just 0.1% to lie in the bottom 0.1%.

Risk actuaries are interested in this quantity. When one department, product, or geographical area experiences a critical event, what is the chance that something very bad will also happen to a different part of the company at the same time? These are the scenarios where corporates may experience insolvency i.e. lots of undesirable events occurring together.

λ falls as we move to the extreme tail of the Gaussian copula. It is often noted that the Gaussian copula does not exhibit tail correlation and indeed $\lambda \to 0$ as we move further into the tail (unless $\rho = 1$). Formally, for the Gaussian copula:

$$\lambda_L = 0$$

However it is misleading to say that the Gaussian copula does not exhibit correlation in the tails; it is potentially more illuminating to analyse the ratio of $\frac{\lambda}{u}$ (i.e. effectively comparing the copula with the independent copula).

```
rationew<-lambda1[1:5]/10^-seq(1,5);      round(rationew,1)

[1]    3.2   12.9    54.3  233.1 1016.4
```

Thus, comparing the Gaussian copula to the independent case (where $\lambda = u$), far more extreme combined events are expected in the tails of the Gaussian copula compared to the independent copula (where $\lambda = u$) e.g. there are approximately 54 times the events in the bottom 0.1% when compared to the independent case. The correlation is certainly not disappearing in the tail.

Repeating the analysis on the Clayton copula:

```
lambda2<-NULL
for (ss in seq(1,10)) {
q<-c(10^-ss)
lambda2[ss]<-pCopula(c(q,q),claytonCopula(2))/q
}
round(lambda2,5)

[1] 0.70888 0.70712 0.70711 0.70711 0.70711 0.70711 0.70711 0.70711 0.70711
[10] 0.70711
```

e.g. of all the pairs where $F(x) < 0.001$, 70.7% of them also have $F(y) < 0.001$.

For the Clayton copula λ_L does not approach zero; the theoretical coefficient of tail dependence for a Clayton copula is given by:

$$\lambda_L = 2^{-\frac{1}{\theta}}$$

In our examples above, $\theta = 2$, giving $\lambda_L = 0.70711$, in agreement with the above.

Thus, in contrast with the Gaussian copula, the Clayton copula does exhibit tail correlation. Indeed it can be said that the Clayton copula exhibits *significant* correlation in the tails – for an independent copula $\lambda = 0.001$ when $u = 0.001$; however for the Clayton ($\theta = 2$) copula $\lambda = 0.71$, and remains constant as we move further in the tails. We can say that the degree of correlation significantly increases as we move into the tails. Compare this to the case of the Gaussian copula where correlation does increase (as measured by λ/u) but not as significantly as it does for the Clayton copula.

Exercise 13.13 *Write a function using simulations to determine how tail correlation changes as we move further into the tail.*

As per the Clayton copula, the Gumbel copula also exhibits tail dependence. However here the correlation is in the upper tail. The theoretical coefficient of upper tail dependence for a Gumbel copula is given by:

$$\lambda_U = \lim_{u\to 1} P(F_Y(y) \geq u \mid F_X(x) \geq u) = 2 - 2^{\frac{1}{\alpha}} \tag{13.16}$$

Exercise 13.14 *Carry out a similar analysis (to that above on the Clayton copula) in respect of the Gumbel copula.*

13.12 Exercise (Challenging)

This exercise is intended to require more effort than typical exercises in this book.

A bank is considering issuing high-risk loans next month to various start-up ventures, mainly in the IT sector. It is anticipated that the size of each loan will be around £10 million. Subsequent to extensive research and discussion the bank has proposed the following time-to-default model for this loan business:

A 2-state solvent–insolvent Markov model with a time-dependent insolvency rate as follows:

$$\mu_t = 0.1 + 0.4e^{-3t} \tag{13.17}$$

where μ_t is the annual rate of insolvency, and t is the time since issuance of the loan. It is also proposed that the dependency of default times between company loans is modelled using a Gaussian copula, *where the same level of correlation is assumed between each pair of loans*.

Based on three 2-year loans, determine the distributions of the number of defaults of the three loans by times, $t = 0.5, 1.0, 1.5$, and 2 years. Three calculation scenarios should be carried out, based on the following correlation coefficients: $\rho = 0; \rho = 0.5; \rho = 1$.

13.13 Appendix 1 – Copula Properties

For a function to be a copula, it must exhibit the following three properties, set out below in Equations 13.18–13.20 (we adopt the simpler notation of u_i rather than $F_{X_i}(x_i)$):

$$C(1, u_i, 1, 1, ..., 1) = u_i \tag{13.18}$$

For example:

```
pCopula(c(1,1,1,1, 0.3), claytonCopula(5))
[1] 0.3
pCopula(c(1,0.65,1,1,1,1), gumbelCopula(5))
[1] 0.65
```

$$C(u_1, u_2, ..., u_n) \tag{13.19}$$

is an increasing function of each u_i. For example:

```
xxx<-NULL
for (i in 0:10) {xxx[i+1]<-pCopula(c(i/10, 0.5), claytonCopula(5))}
xxx
```

```
[1] 0.0000000 0.0999938 0.1996055 0.2956738 0.3785418 0.4366484 0.4694465
[8] 0.4858222 0.4938239 0.4978605 0.5000000
```

```
yyy<-NULL
for (i in 0:10) {yyy[i+1]<-pCopula(c(0.8,0.7,i/10), gumbelCopula(5))}
yyy
```

```
[1] 0.0000000 0.0999955 0.1999623 0.2998196 0.3992855 0.4973092 0.5896742
[8] 0.6612882 0.6911442 0.6952993 0.6954029
```

Lastly, where $a_i \leq b_i$, for all $(a_1, b_1), (a_2, b_2), ..., (a_d, b_d)$ in $[0, 1]^d$ space:

$$\sum_{i_1=1}^{2} \cdots \sum_{i_d=1}^{2} (-1)^{i_1 + \cdots + i_d} C(u_{1i_1}, u_{2i_2}, \ldots, u_{di_d}) \geq 0 \tag{13.20}$$

where $u_{j1} = a_j$ and $u_{j2} = b_j$. So for d-dimensions there will be 2^d terms in the summation.

Example 13.4

Applying a 2-dimensional Clayton copula with $\theta = 5$, calculate:

$$P(0.5 \leq u_1 \leq 0.8, 0.3 \leq u_2 \leq 0.7)$$

Figure 13.10   The rectangle inequality.

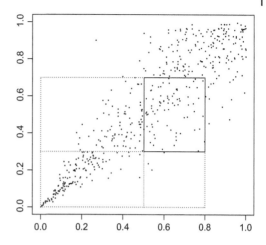

Thus $u_{11} = a_1 = 0.5$, $u_{12} = b_1 = 0.8$, $u_{21} = a_2 = 0.3$, $u_{22} = b_2 = 0.7$. Applying Equation 13.20 we obtain:

$$C(0.5, 0.3) - C(0.5, 0.7) - C(0.8, 0.3) + C(0.8, 0.7)$$

```
rec_prop<-(pCopula(c(0.5,0.3),claytonCopula(5))-pCopula(c(0.5,0.7),claytonCopula(5))
-pCopula(c(0.8,0.3), claytonCopula(5)) +pCopula(c(0.8,0.7), claytonCopula(5)))
rec_prop
```

```
[1] 0.1698765
```

The expression is shown graphically in Figure 13.10, where around 17% of the simulations will sit within the solid-sided rectangle:

Exercise 13.15 *Verify the above solution by running a suitably large number of Clayton(5) simulations, and counting the number of realisations obtained within the solid-sided rectangle above (this property is often referred to as the "rectangle inequality").*

13.14   Appendix 2 – Rank Correlation and Kendall's Tau, τ

The reader will be familiar with Pearson's coefficient, ρ, commonly referred to as the correlation coefficient. However, there are several measures of correlation. Pearson's coefficient gives a measure of the degree of linear correlation and is only valid if our marginal distributions are jointly elliptical.

Kendall's tau, τ, is a type of rank correlation, and can be a more appropriate measure of dependency given a particular situation being modelled e.g. where there is not a linear relationship between two variables. It is defined as follows:

$$\tau = \frac{2(p_c - p_d)}{T(T-1)} \tag{13.21}$$

where p_c is the number of concordant pairs in our data set, p_d is the number of discordant pairs in our data set, and T is the number of pairs. A pair of data points is defined to be concordant if the gradient of the line between the two points is positive, and discordant if the gradient is negative. It should be noted that τ only depends on the size ranking of our data points; the absolute size does not affect its value.

Example 13.5

The code below produces a small dataset ($T = 5$, so we can easily verify the coefficient calculations) shown in Figure 13.11. Counting the number of lines between all pairs of points, we have 9 positive gradients (p_c) and 1 negative gradient (p_d); this gives us $\tau = 0.8$, in agreement with the calculation below. This equates to an estimated Gumbel parameter of 5 (compared to $\alpha = 4$ used to simulate the data).

```
set.seed(49)
gum_norm <- mvdc(gumbelCopula(4), margins = c("norm", "norm"),
paramMargins = list(list(mean = 10, sd = 20),
list(mean = 0, sd = 1)))
sim_X <- rMvdc(5, mvdc = gum_norm)
(tau_sample <- cor(sim_X[,1], sim_X[,2], method = "kendall"))

[1] 0.8
```

The reader should vary the value of the Gumbel parameter and compare the sample τ's obtained. The serious student should read further on this topic.

Exercise 13.16 *Calculate Pearson's rho and Kendall's tau in respect of the correlation between ww1 and vv1 below (then ww2 and vv2), and comment on the results.*

```
ww1<-seq(0,2,.01); vv1<-exp(3*ww1+rnorm(length(ww1),0,.1))
ww2<-seq(0,2,.01); vv2<-exp(3*ww2+rnorm(length(ww2),0,.01))
```

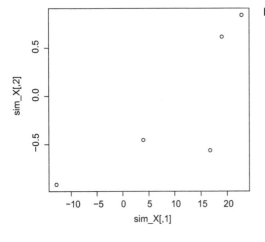

Figure 13.11 Calculation of Kendall's tau

13.15   Recommended Reading

- Charpentier A. et al. (2016). *Computational Actuarial Science with R.* Boca Raton, Florida, US: Chapman and Hall/CRC.
- Cherubini et al. (2004). *Copula methods in Finance.* Chichester, West Sussex, UK: Wiley.
- Danielsson J. (2011). *Financial Risk Forecasting.* Chichester, West Sussex, UK: Wiley.
- Frees E. and Valdez E. (1997). *Understanding relationships using copulas.* North American Actuarial Journal, January 1998.
- Hull, J. (2015). *Risk Management and Financial Institutions* (4th edition). Hoboken, New Jersey, US: Wiley.
- Li D.X. (2000). *On Default Correlation: A Copula Function Approach.* The Journal of Fixed Income Spring 2000, 9 (4) 43-54.
- McNeil, Frey, Embrechts. (2005). *Quantitative Risk Management.* Princeton, NJ: Princeton University Press.
- Nelsen R.B. (1999). *An Introduction to Copulas.* New York: Springer-Verlag.
- Patton A.J. (2009). *Copula-Based Models for Financial Time Series.* Handbook of Financial Time Series. Berlin, Heidelberg: Springer.
- Sweeting, P. (2017). *Financial Enterprise Risk Management.* (2nd edition). Cambridge, UK: Cambridge University Press.
- Wilmott, P. (2007). *Introduces Quantitative Finance.* (2nd edition). Chichester, West Sussex, UK: Wiley.

14

Copulas – A Modelling Exercise

Peter McQuire

```
library(MASS)
library(copula)
library(stats)
```

14.1 Introduction

In the previous chapter we looked at a number of fundamental concepts underlying copulas, together with various functions available in R which allow us to model multivariate data using copulas. In this chapter we develop those ideas to analyse a set of insurance claims data. The objective is to develop an appropriate copula model in respect of this data, estimate the level of future claims, and hence determine an appropriate amount of risk capital which the insurer should hold.

14.2 Modelling Future Claims

14.2.1 Data

The data we will use consists of total monthly claim amounts from the UK and US departments of an insurance company, spanning 200 months. We will use copulas to analyse any dependency pattern which may exist between UK and US claims. The process is similar to that set out in the previous chapter on copulas.

It is important not to lose sight of the company's key objective in this example – to analyse the *total, combined* claim amounts from the UK and US departments which may be incurred next month or year. This in turn helps us to determine, amongst other things, the likely distribution of the company profits and the likelihood of company insolvency. The central idea here is that the pattern of the dependency between UK and US claims will affect the total monthly claims distribution.

Let's download and plot the data (Figure 14.1):

```
data_2d<-read.csv("~/copuladataproject2.csv")
ukdata<-data_2d$UK;    usdata<-data_2d$US
```

R Programming for Actuarial Science, First Edition. Peter McQuire and Alfred Kume.
© 2024 John Wiley & Sons Ltd. Published 2024 by John Wiley & Sons Ltd.
Companion Website: www.wiley.com/go/rprogramming.com.

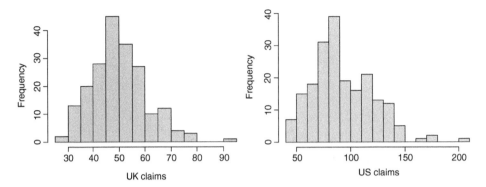

Figure 14.1 UK and US claims data

Before fitting a model it is important that the analyst/actuary reviews any apparent anomalies in the data. For example, it would be worth reviewing the extreme values in both data sets, in particular the 90 and 207 values in the UK and US data respectively.

14.2.2 Fitting Appropriate Marginal Distributions

Assuming we have verified that the data is accurate, we proceed to fit the marginal distributions for the UK and US claims. We will try fitting Normal, Gamma, and Lognormal distributions in this example, using the `fitdistr` function.

Analysing the UK data:

```
a=NULL
y1<-fitdistr(ukdata,"normal");     a[1]<-logLik(y1)#gets the logL
y2<-fitdistr(ukdata,"gamma");      a[2]<-logLik(y2)
y3<-fitdistr(ukdata,"lognormal");  a[3]<-logLik(y3)
a

[1] -756.2638 -752.6387 -753.1780
```

Based on these logL's, we would choose the Gamma distribution for the UK claims. Of course further analysis should be carried out (e.g. Shapiro-Kolmogorov, QQ plots, AIC, BIC test etc.) and the reader is encouraged to do so. Note that we can obtain the actual parameter values and associated standard errors by simply running the function:

```
fitdistr(ukdata,"gamma")

      shape          rate
 22.53427244    0.44857186
( 2.23675260) ( 0.04502361)
```

Exercise 14.1 *Repeat the method for the US data.*

The four parameters from our marginals can be obtained as follows (we will use these when running our simulations):

```
par1uk<-fitdistr(ukdata,"gamma")$estimate[1]
par2us<-fitdistr(ukdata,"gamma")$estimate[2]
par1us<-fitdistr(usdata,"lognormal")$estimate[1]
par2us<-fitdistr(usdata,"lognormal")$estimate[2]
c(par1uk,par2uk,par1us,par2us)

    shape        rate     meanlog        sdlog
22.5342724   0.4485719   4.4856798    0.2983503
```

We have our best fitting marginal distributions i.e. Gamma (22.53, 0.45) for the UK data, and logNormal (4.49, 0.3) for the US data. Of course, one should try fitting more distributions – our analysis would not end here in practice.

Exercise 14.2 *As a visual check, plot (1)QQ plots, and (2)pdf's of the selected distributions over the histograms of the data to check the goodness of fit.*

14.2.3 Fitting The Copula

To fit the copula we must first map the data onto $[0, 1]^2$ space using the cdfs of the selected marginal distributions. For the UK data:

```
copspaceuk<-pgamma(ukdata,  par1uk,  par2uk);      head(copspaceuk)

[1]  0.4053990 0.8628138 0.5137498 0.2959189 0.4122896 0.6858096
```

And for the US data:

```
copspaceus<-plnorm(usdata,par1us,par2us)
```

Plotting the resultant empirical copula can improve our understanding of any correlations which may exist (Figure 14.2, left):

```
empcop<-cbind(copspaceuk,copspaceus);    plot(empcop)
```

Indeed we shall be visually comparing the empirical copula with our chosen copula in Section 14.2.4 (Figure 14.2, right).

What copula does this look like? It certainly appears asymmetrical with lower tail dependence – it may therefore be reasonable to expect that a Clayton copula could be more appropriate than a Gaussian copula in modelling this set of claims data.

To find the MLE for each copula parameter we simply maximise:

$$\log L = \sum_{t=1}^{n} \log c_\theta(u_{1,t}, u_{2,t}) \tag{14.1}$$

where c_θ is the copula density function, and the sum is over all n pairs from the empirical copula data. Using the `fitCopula` function we proceed to fit the Clayton, Gaussian, and Gumbel copulas in our model (only the full results for the Clayton copula are included below).

```
tryClayton <- fitCopula(claytonCopula(), data = empcop, method = "ml")
summary(tryClayton)

Call: fitCopula(copula, data = data, method = "ml")
Fit based on "maximum likelihood" and 200 2-dimensional observations.
Clayton copula, dim. d = 2
      Estimate Std. Error
alpha    6.731      0.473
The maximized loglikelihood is 235.8
Optimization converged
Number of loglikelihood evaluations:
function gradient
     6        6

tryGaussian <- fitCopula(normalCopula(), data = empcop, method = "ml")
tryGaussian@loglik

[1] 174.0585

tryGumbel <- fitCopula(gumbelCopula(), data = empcop, method = "ml")
tryGumbel@loglik

[1] 139.0726
```

The copula with the highest logL is the Clayton copula (logL =235.8); this is consistent with our conclusion from the graphical inspection. The Clayton parameter is:

```
tryClayton@estimate

[1] 6.731294
```

We have our first attempt at a suitable claims model – a Clayton copula with Gamma (UK) and Lognormal (US) marginals.

Remark 14.1 Note that for an n-dimensional Gaussian copula (where $n \geq 3$), the estimated correlation matrix can be obtained using the `dispstr="un"` argument. For example, where we have four dimensions we obtain the six Pearson coefficients as follows:

```
d <- 4       #number of dimensions
nn<-1000;   set.seed(2000);
cc <- normalCopula(c(0.3,0.3,0.3,0.4,0.5,0.6), dim = d,dispstr="un")#6 correl coeffs
simGauss4 <- rCopula(nn, copula = cc)
fit_Gaus_cop4 <- fitCopula(normalCopula(dim=d, dispstr="un"), simGauss4, method="ml")
fit_Gaus_cop4@estimate    #6 Pearson coeffs from the 4-d data

[1] 0.3130921 0.2907870 0.2687685 0.3874786 0.5246406 0.6096799
```

If `dispstr="ex"` then only one "average" parameter is obtained:

```
fit_Gaus_cop1 <- fitCopula(normalCopula(dim=d, dispstr="ex"), simGauss4, method="ml")
fit_Gaus_cop1@estimate

[1] 0.4010603
```

End of remark.

14.2.4   Assessing Risk From the Analysis of Simulated Values

We now proceed to simulate copula points using our chosen model, and compare these simulations with our empirical copula:

```
d <- 2               #number of dimensions (uk and us)
nn<-1000             #number of simulations - very few here - for graph only!!!!
cc <- claytonCopula(tryClayton@estimate, dim = d)
set.seed(3000);   simclaytongraph <- rCopula(nn, copula = cc)
```

This looks quite reasonable (Figure 14.2). The simulations run above were only sufficient to obtain a reasonable graphical representation – we will run many more when carrying statistical comparisons.

Of course, we should at this point assess the goodness of fit of the chosen copula. This can be done using various methods, for example, using the gofCopula function. However a practical approach is often taken as the chosen copula will invariably be the one which simply performs the best, e.g. the lowest AIC score. The reader should bear in mind that often a Gaussian copula approach is taken without any further analysis of trying different copulas! An alternative approach could simply be to use the empirical copula in modelling the dependencies.

We now introduce the marginal distributions into our simulations, making use of the mvdc type functions (introduced in the previous chapter). Please note that Figure 14.3 (left) only contains 1000 simulated points, compared to one million simulations in the code below, for reasons of visual clarity.

```
#set up the multivariate distribution
mcc <- mvdc(claytonCopula(tryClayton@estimate), margins = c("gamma", "lnorm"),
paramMargins = list(list(par1uk,par2uk),
list(par1us,par2us)))


n <- 1000000;                      set.seed(3000)
sim_clay <- rMvdc(n, mvdc = mcc)
```

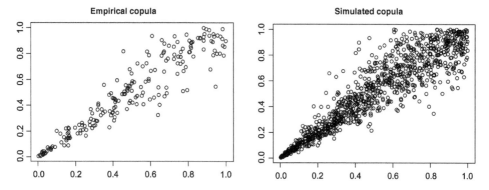

Figure 14.2   Comparing data with simulations from our chosen Clayton copula.

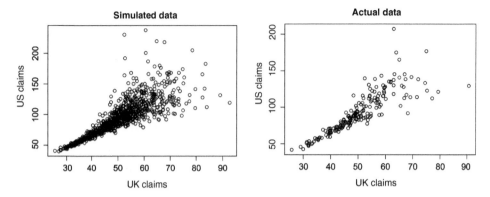

Figure 14.3 Comparison of simulated data and actual data.

From Figure 14.3 we can compare the actual data with our simulated data (note the values on the axes); visually this looks quite reasonable. Of course we should also make a quantitative comparison of our simulations with the data (we will do this later).

We now analyse the total simulated claim amounts in more detail. Adding the UK and US claims together (the first few rows are shown below), and calculating the Value at Risk at various percentiles from the simulations:

```
total_clay<-rowSums(sim_clay);       sim_clay_all<-cbind(sim_clay,total_clay)
head(sim_clay_all,3)

                        total_clay
[1,] 63.89573 108.54597    172.4417
[2,] 46.16112  83.70302    129.8641
[3,] 56.57058 102.05209    158.6227

(varclayton<-quantile(total_clay,c(.91,.93,.95,.97,.99,.995)))

     91%      93%      95%      97%      99%    99.5%
196.4432 202.6500 210.5730 222.0183 245.1123 259.1864
```

Based on our proposed model there is a 1% chance that our aggregate claims next month will exceed £245.1 million, and a 5% chance that they will exceed £210.6 million.

14.2.5 Comparison with the Gaussian Copula Model

The Gaussian copula is widely used and understood in the financial sector – it is quite possible that the company may simply use a Gaussian copula to model correlations without considering the possibility of asymmetric dependencies and other possible copulas. What would the reported risk position be if the Gaussian copula was included in the model in place of the Clayton copula?

Exercise 14.3 *Using the same process as above use the Gaussian copula to estimate various VaR metrics, and plot the equivalent of Figure 14.3 (left).*

The Gaussian results (1st row), together with the Clayton results as shown above, are as follows:

```
vargaussian

      90%      92.5%       95%     97.5%       99%     99.5%
193.3052 201.4263 212.3544 230.3486 253.2241 270.4711

varclayton

      90%      92.5%       95%     97.5%       99%     99.5%
193.7311 200.9940 210.5730 225.9168 245.1123 259.1864
```

Discussion of results

The worst losses that would be expected from using the best-fitting Gaussian copula model are consistently higher than from using the best-fitting Clayton copula model – this is a key point. As observed in the previous chapter, the Clayton copula exhibits particularly low levels of higher-tail dependency when compared with the Gaussian copula. Thus the Gaussian model results in simulations with higher total claims in our example. We can demonstrate this by looking at the highest value simulations from each model (obtained using the order function):

```
clay_order<-head(sim_clay_all[order(-sim_clay_all[,3]),],5)     #top 5
gaus_order<-head(sim_gaus_all[order(-sim_gaus_all[,3]),],5)     #top 5
(order_combine<-cbind(clay_order,gaus_order))

                       total_clay                           total_gaus
[1,] 76.32712 368.4847   444.8118 121.8932 397.3366          519.2299
[2,] 77.41553 341.0545   418.4700 100.8368 376.9312          477.7680
[3,] 58.66861 351.5329   410.2015 111.0507 356.0887          467.1394
[4,] 81.25570 327.8759   409.1316 108.6844 352.0182          460.7026
[5,] 86.41230 310.5687   396.9810 103.8059 332.6973          436.5033
```

Use of the Gaussian copula (which is commonplace) may lead the company to believe that this insurance business is more risky than it actually is, potentially resulting in the company allocating too much risk capital, taking out excessive reinsurance, or reviewing the type of business it writes.

Of course, the model we have arrived at is just the best of those which we have tried; we should investigate with further marginal distributions and copulas.

14.2.6 Comparison of the Models with the Data

Finally, we should compare the VaR results from our simulations with that of the data when deciding whether our model is reliable. A practical problem here is that we only have 200 data points, so the empirical VaR points are subject to considerable statistical noise. The VaR's in respect of the data can be calculated as follows:

```
datatotals<-rowSums(data_2d)
(vardata<-quantile(datatotals,c(.90,.925,.95,.975,.99,.995),type = 9))

      90%      92.5%       95%     97.5%       99%     99.5%
195.3575 199.5016 207.1199 217.1139 246.9591 263.4810
```

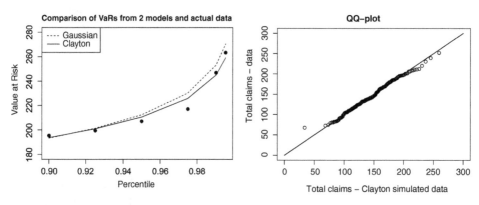

Figure 14.4 Comparison of simulations and data.

The VaRs calculated above from the data (•), the Clayton model, and the Gaussian model are shown in Figure 14.4 (left). Note that this is similar to a QQ-plot (Figure 14.4 (right)) except here we are only looking at the tails of the data.

The numerical analysis and the earlier graphical analysis suggest that the Clayton copula reflects the data better than the Gaussian copula.

Exercise 14.4 *Plot a QQ-plot comparing the data with the Gaussian model.*

Exercise 14.5 *Analyse the sensitivity of the model output to changes in the parameter values.*

14.3 Another Example: Banking Regulator

Included below is an exercise relating to the systemic risk which may exist within the banking sector. The risk here is when events occur which result in insolvency of one bank, this may result in some form of contagion, resulting in major consequences to the entire banking sector.

Exercise 14.6 *A banking regulator is analysing the probability of insolvency of the six largest banks, A-F, in the country. Banks A-C have calculated their individual probability of insolvency over the next year to be 1%, whilst D-F calculated this to be 2%. Further analysis by the regulator has estimated the dependency between the banks to be suitably modelled with a Clayton copula and a parameter value, $\lambda = 2.5$. (A previous analysis concluded that correlation in the banking sector could be reasonably modelled using a Gaussian copula ($\rho = 0.73$).)*

Calculate the probability distribution of the banking industry becoming insolvent over the next 12 months, by (1)using the pCopula function and (2)running simulations, and comment on the results. Compare these results if a Gaussian copula ($\rho = 0.73$) is used.

The table of insolvency probabilities obtained from running simulations, using the Clayton copula, is included below:

```
insolv_probs
       0        1        2        3        4        5        6
0.972348 0.007288 0.004960 0.003661 0.002782 0.002918 0.006043
```

The theoretical probability of all six banks defaulting given the above assumptions is 0.6%, in agreement with the simulations. The Gaussian copula model gives a probability of only 0.08%.

Even though individual probabilities of insolvency are 1% and 2%, the probability that all six banks default is estimated to be 0.6%, due to the high degree of lower-tail dependency from the Clayton copula. The best-fitting Gaussian copula model has predicted a significantly lower probability of banking catastrophe.

14.4 Conclusion

Clearly the insurance company claims model is not in complete agreement with the data at the extreme points. This is a common problem, particularly when fitting a model to a relatively small data set – the noise in the largest claims is often significant. A decision would be required on whether further analysis is required, for example, should we incorporate EVT into our model, or try alternative copulas?

However there would appear to be sufficient justification in this case to recommend inclusion of a Clayton copula in place of a Gaussian copula for this line of business.

15

Bond Portfolio Valuation: A Simple Credit Risk Model

Peter McQuire

15.1 Introduction

```
library(copula);library(lifecontingencies)
```

Several chapters in this book have discussed the importance of incorporating correlations in the models we have introduced. In Chapter 13 we introduced the concept of copulas which provides considerable flexibility in modelling multivariate distributions. In this chapter our objective is to model the distribution of the future value of bond portfolios; an important facet of this model is to incorporate correlations between the prices, and in particular the credit risk, of individual bonds. To do this we will make use of the Gaussian copula.

The model described in this chapter has some similarities with the *Credit Metrics* model which was introduced by JP Morgan in 1997, although the JP Morgan paper does not explicitly describe the Gaussian copula.

Our model uses bond ratings as a key variable in the valuation of bonds. Bonds are rated by many ratings agencies, such as Moody's and S&P, who analyse the likelihood that each bond will default on payments over the next year. For example a *Aaa* rating used by Moody's implies the bond is 'judged to be of the highest quality, and is subject to the lowest level of credit risk'. The issuing company (or country) is considered to be extremely strong, and unlikely to default in the next year. Compare a *Ca* rated bond, defined to be 'highly speculative and is likely in, or very near to, default, with some prospect of recovery of principal and interest'; such bonds are likely to have been issued by a much weaker company.

Different rating agencies use different ratings scales. For example, Moody's currently use 10 ratings, ranging from Aaa to C. (In addition they include numeric modifiers (1,2, and 3) to further improve the level of granularity.)

In our model each rating has its own yield curve which is then used to price the bond. In our simplified example we shall adopt only three rating levels, hence requiring three yield curves. We will adopt the following rating system:

R Programming for Actuarial Science, First Edition. Peter McQuire and Alfred Kume.
© 2024 John Wiley & Sons Ltd. Published 2024 by John Wiley & Sons Ltd.
Companion Website: www.wiley.com/go/rprogramming.com.

AAA – strong, very unlikely to default (low yield)

AA – reasonably strong, unlikely to default (medium yield)

A – speculative, with a reasonable chance of default (high yield)

Clearly this is a simplification of the actual credit risk of such bonds – bonds within the same rating group may have materially different risks of default and hence appropriate yields. The reader may wish to take this into account when developing their own model, and allow for the possibility of incorporating a greater number of rating levels and corresponding yield curves.

Choosing the parameter values for the *instantaneous* forward yield curves in our example (see Figure 15.1):

$$\delta_{AAA} = 0.04 - 0.01e^{-0.2t}$$

$$\delta_{AA} = 0.05 - 0.01e^{-0.2t}$$

$$\delta_A = 0.06 - 0.01e^{-0.2t}$$

```
delta1<-function(t){0.04-0.01*exp(-0.2*t)}
delta2<-function(t){0.05-0.01*exp(-0.2*t)}
delta3<-function(t){0.06-0.01*exp(-0.2*t)}
```

And hence the corresponding spot rates:

```
spot1<-function(t){(0.04*t+0.05*exp(-0.2*t)-0.05)/t}
spot2<-function(t){(0.05*t+0.05*exp(-0.2*t)-0.05)/t}
spot3<-function(t){(0.06*t+0.05*exp(-0.2*t)-0.05)/t}
```

And hence the forward rates at $t = 1$, which we will use to estimate bond prices at $t = 1$:

```
forward1<-function(t){(0.04*t+0.05*exp(-0.2*t)-0.04-0.05*exp(-0.2))/(t-1)}
forward2<-function(t){(0.05*t+0.05*exp(-0.2*t)-0.05-0.05*exp(-0.2))/(t-1)}
forward3<-function(t){(0.06*t+0.05*exp(-0.2*t)-0.06-0.05*exp(-0.2))/(t-1)}
```

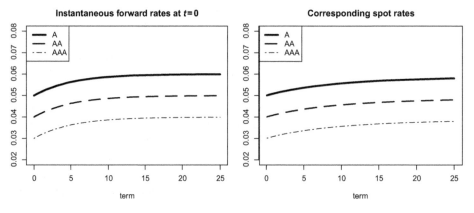

Figure 15.1 Yield curves.

Remark 15.1 With most of the calculations carried out in this chapter there are various alternative methods which can be used. The reader is encouraged to perform the calculations using more than one method.

15.2 Our Example Bond Portfolio

15.2.1 Description

Our simple portfolio consists of five corporate bonds of various terms and ratings. Four of these are rated 'AAA' (strong) and one is rated 'A' (weak). For simplicity all our bonds have coupons equal to 5% payable annually in arrears, with £100 redemption payment. The 5-, 10-, 15-, and 20-year bonds are all rated 'AAA', with the 25-year bond having an 'A' rating. We assume an equal 20% allocation of the total fund in each of the bonds.

We can now proceed to calculate the current prices of these bonds (only a sample of the code is shown below):

```
time1<-seq(1,5);      y1<-spot1(time1)
(bond1<-presentValue(c(rep(5,4),105),time1,exp(y1)-1))

[1] 107.1826

time2<-seq(1,10);     y2<-spot1(time2)
bond2<-presentValue(c(rep(5,9),105),time2,exp(y2)-1)
#etc.....
time5<-seq(1,25);     y5<-spot3(time5)
bond5<-presentValue(c(rep(5,24),105),time5,exp(y5)-1)

c(bond1,bond2,bond3,bond4,bond5)

[1] 107.18261 111.58637 114.73311 117.16969  88.61354
```

We now proceed to calculate the possible prices of our bonds at $t = 1$, immediately after the first coupon has been received, allowing for all potential ratings.

Remark 15.2 Note that we do not allow for any interest rate risk in this simplified model (although we do allow for this later using a simple method); the yield curves are assumed to be static in this model.

```
time1next<-time1[-length(time1)]
time2next<-time2[-length(time2)]
#etc
```

```
#yields for each bond and each possible rating at t=1
y1next1<-forward1(time1next+1)#bond1,rating 1
y1next2<-forward2(time1next+1)#bond1,rating 2
y1next3<-forward3(time1next+1)#bond1,rating 3

y2next1<-forward1(time2next+1)#bond2,rating 1
y2next2<-forward2(time2next+1)#bond2,rating 2
y2next3<-forward3(time2next+1)#bond2,rating 3

#etc
```

Now calculate next year's range of potential bond fund values; note that we must add £5 to the bond price in respect of the coupon received at the end of the first year:

```
bond1next1<-presentValue(c(rep(5,3),105),time1next,exp(y1next1)-1)+5
#next year fund value 1, rating 1
bond1next2<-presentValue(c(rep(5,3),105),time1next,exp(y1next2)-1)+5
#next year fund value 1, rating 2
bond1next3<-presentValue(c(rep(5,3),105),time1next,exp(y1next3)-1)+5
#next year fund value 1, rating 3

bond2next1<-presentValue(c(rep(5,8),105),time2next,exp(y2next1)-1)+5
#next year fund value 2, rating 1
bond2next2<-presentValue(c(rep(5,8),105),time2next,exp(y2next2)-1)+5
#next year fund value 2, rating 2
bond2next3<-presentValue(c(rep(5,8),105),time2next,exp(y2next3)-1)+5
#next year fund value 2, rating 3

#etc
```

For example, next year's values assuming they don't change ratings are:

```
c(bond1next1,bond2next1,bond3next1,bond4next1,bond5next3)

[1] 110.55030 115.09242 118.33803 120.85117  93.24414
```

Remark 15.3 Regarding the return if a bond was to default in the first year, we have taken a simplified approach: each bond would lose 20% of its value. The actual receipts on default are often unpredictable and can range from 0% to close to 100%.

Calculating the return from each potential ratings transition, and summarising them in a matrix:

```
return1_11<-(bond1next1/bond1)-1;          return2_11<-(bond2next1/bond2)-1
return1_12<-(bond1next2/bond1)-1;          return2_12<-(bond2next2/bond2)-1
return1_13<-(bond1next3/bond1)-1;          return2_13<-(bond2next3/bond2)-1

#etc...

return_def<-(-0.2)   #on default
```

Remark 15.4 Ultimately, we will be analysing portfolio returns later in the chapter; it will be simpler to do this if we work with annual effective rates (as we are only looking at values at $t = 1$).

```
return_matrix<-(matrix(c(return1_11,return1_12,return1_13,return_def,return2_11,
return2_12,return2_13,return_def,return3_11,return3_12,return3_13,return_def,
return4_11,return4_12,return4_13,return_def,return5_31,return5_32,return5_33,
return_def),5,4,byrow = TRUE))

round(return_matrix,4)    #annual effective rates for 5 bonds

        [,1]     [,2]     [,3] [,4]
[1,] 0.0314 -0.0046 -0.0393 -0.2
[2,] 0.0314 -0.0399 -0.1056 -0.2
[3,] 0.0314 -0.0671 -0.1542 -0.2
[4,] 0.0314 -0.0883 -0.1902 -0.2
[5,] 0.3865  0.2032  0.0523 -0.2
```

For example, the five-year bond will return 3.1% if its rating is unchanged over the year, and the 25-year bond (which starts as an 'A' rated bond) will return 20.3% if its rating improves to 'AA' over the year (both annual effective rates).

15.2.2 The Transition Matrix

We proceed to simulate bond ratings at $t = 1$ for each bond in our portfolio. To help us do this we use historic, empirical data of bond rating transitions over one year. An implicit assumption therefore is that rating transitions over the next year will follow a similar pattern to that in our historic data; care is required here during periods of economic volatility.

In our example, this results in obtaining a 3 × 4 'transition matrix', which sets out the probabilities of each possible ratings transition between the start and end of the year. (It is not a square 4 × 4 matrix as we do not analyse previously defaulted bonds.) We will use the following transition matrix:

```
trans_matrix

      1    2    3 default
1 0.85 0.12 0.02    0.01
2 0.10 0.73 0.15    0.02
3 0.05 0.30 0.45    0.20
```

Our calculations are easier if we create the following 'cumulative' matrix:

```
cum1<-cumsum(transition1);cum2<-cumsum(transition2);cum3<-cumsum(transition3)
(cum_transx<-matrix(c(cum1,cum2,cum3),3,4,byrow=TRUE))

       [,1] [,2] [,3] [,4]
[1,]  0.85 0.97 0.99    1
[2,]  0.10 0.83 0.98    1
[3,]  0.05 0.35 0.80    1
```

This is a key matrix; our simulated copulas will be applied to this matrix to determine each bond's new rating at $t = 1$. This is best explained by way of example.

From the above transition matrix, there is a 2% chance that a 'AAA' bond has a rating of 'A' at the end of the year, and a 73% chance that a 'AA' bond ends the year with the same rating. The fourth column relates to the chance of default; there is a 1% chance a 'AAA' rated bond defaults, there is a 2% chance a 'AA' rated bond defaults, and a 20% chance that an 'A' rated bond defaults. A feature of most bond transition matrices is that the highest probability usually corresponds to no transition occurring over the year.

To simulate a bond's rating at the end of the year we simulate numbers from a $U(0, 1)$ distribution and map it onto the appropriate column in the cumulative matrix above. Of course, as we need to allow for correlations between our bonds, we can simply use copulas to do this. We will use the Gaussian copula, but the reader is invited to experiment with various alternatives.

For example, for a 'AA' rated bond (row 2) at the start of the year,

- a simulated value of < 0.1 would correspond to a 'AAA' rating by the end of the year;
- a simulated value of between 0.1 and 0.83 would correspond to a 'AA' rating by the end of the year;
- a simulated value of between 0.83 and 0.98 would correspond to a 'A' rating by the end of the year; and
- a simulated value of > 0.98 would correspond to a 'default' rating by the end of the year.

It's worth checking the average returns (annual effective rates) from each bond we are expecting; this will be very useful later to verify our calculations.

```
exp_valuebond1<-t(return_matrix[1,])%*%(trans_matrix[1,])
exp_valuebond2<-t(return_matrix[2,])%*%(trans_matrix[1,])
exp_valuebond3<-t(return_matrix[3,])%*%(trans_matrix[1,])
exp_valuebond4<-t(return_matrix[4,])%*%(trans_matrix[1,])
exp_valuebond5<-t(return_matrix[5,])%*%(trans_matrix[3,])

(exp_value_combine<-c(exp_valuebond1,exp_valuebond2,exp_valuebond3,
exp_valuebond4,exp_valuebond5))

[1] 0.02336718 0.01780397 0.01356576 0.01030362 0.06380871
```

15.2.3 Correlation Matrix

The final piece of our model is the correlation matrix. We first look at a portfolio with little correlation between the bonds, and later, in Exercise 15.3, compare the results with a portfolio of highly correlated bonds. We will see that the level of correlation between our bonds will have a significant impact on the distribution of future portfolio values.

Remark 15.5 Please see 'CreditMetrics' Section 8.5 for a detailed discussion regarding how to develop a valid correlation matrix. A number of methods are applied in practice, perhaps the simplest of which is to calculate correlations of the equity returns of the bond issuers, and use this as a proxy for the correlations of bond ratings.

The correlations are set out below:

```
(correlations_low<-matrix(c(1,0.1,0.1,0.1,0.1,0.1,1,0.15,0.15,0.15,
0.1,0.15,1,0.2,0.2,0.1,0.15,0.20,1,0.25,0.1,0.15,0.2,0.25,1),5,5,byrow=TRUE))

     [,1] [,2] [,3] [,4] [,5]
[1,]  1.0 0.10 0.10 0.10 0.10
[2,]  0.1 1.00 0.15 0.15 0.15
[3,]  0.1 0.15 1.00 0.20 0.20
[4,]  0.1 0.15 0.20 1.00 0.25
[5,]  0.1 0.15 0.20 0.25 1.00

#not used directly
```

It is important to note that in simulating our Gaussian copula we only enter the non-diagonal terms in the matrix – these are set out below:

```
lowx<-c(0.1,0.1,0.1,0.1,0.15,0.15,0.15,0.2,0.2,0.25)   #very important
highx<-lowx+0.7                               #to be used in exercise
```

Setting up our copula with these parameters:

```
low_corrn_cop<-normalCopula(lowx,dim=5,dispstr = "un")
```

15.2.4   Simulations and Results

Now to run the simulations from the Gaussian copula (the first three sets are shown below):

```
set.seed(5555555);   sims<-1000000
simulatex<-rCopula(sims,low_corrn_cop)   #chosen low corrn here
head(simulatex,3)
           [,1]          [,2]        [,3]          [,4]          [,5]
[1,] 0.5694673 0.100675830 0.4349066 0.03519635 0.09452356
[2,] 0.6251861 0.586658212 0.8998339 0.53538227 0.63057032
[3,] 0.3424145 0.002442413 0.4590076 0.30619651 0.04213912
```

All that is left to do is to map these simulated copula values to our ratings. Our five bonds consist of four high-rated ('AAA'), and one low-rated ('A').

```
bondratings<-c(1,1,1,1,3)   #key parameter
num_bonds<-5
next_rating<-matrix(0,sims,num_bonds);   value_low<-matrix(0,sims,num_bonds)
```

Applying the simulated copula to the transition matrix:

```
#start loop
for(j in 1:num_bonds){
for(i in 1:sims){
z<-cum_transx[bondratings[j],] - simulatex[i,j]
z[z < 0] <- NA
next_rating[i,j]<-which.min(z)
}
value_low[,j]<-return_matrix[j,next_rating[,j]]
}
#end loop
```

A sample of simulated bond ratings at $t = 1$, and (effective) returns over the year, are set out below:

```
head(next_rating,3)#simulated new ratings

      [,1] [,2] [,3] [,4] [,5]
[1,]    1    1    1    1    2
[2,]    1    1    2    1    3
[3,]    1    1    1    1    1

head(value_low,3)#key
```

```
           [,1]       [,2]        [,3]       [,4]       [,5]
[1,] 0.03142005 0.03142005  0.03142005 0.03142005 0.20322351
[2,] 0.03142005 0.03142005 -0.06714559 0.03142005 0.05225611
[3,] 0.03142005 0.03142005  0.03142005 0.03142005 0.38652813
```

Exercise 15.1 *Verify that the simulated ratings at t = 1 are consistent with the transition matrix.*

Before calculating our results we should check the average return on each bond:

```
colMeans(value_low)    #from simulations

[1] 0.02334745 0.01776078 0.01354742 0.01033266 0.06372407

exp_value_combine      #excellent check

[1] 0.02336718 0.01780397 0.01356576 0.01030362 0.06380871
```

Finally, calculating the portfolio return for each simulation, and plotting them (Figure 15.2):

```
sim_portfolio_return<-rowMeans(value_low)
```

Key results from the simulations are as follows:

```
mean(sim_portfolio_return)

[1] 0.02574247

sd(sim_portfolio_return)

[1] 0.03952308

quantile(sim_portfolio_return,0.005)

      0.5%
-0.08854477

#using Normal approx.....
mean(sim_portfolio_return)+qnorm(0.005)*sd(sim_portfolio_return)

[1] -0.07606222
```

The best return is 10.2 %, where all four 'AAA' rated bonds remain 'AAA' rated, and the 'A' rated bond moves to 'AAA' rated. The worst returns are simulations which include several bond defaults:

```
max(sim_portfolio_return)    #best return

[1] 0.1024417

sim_portfolio_return[head(order(sim_portfolio_return))] #worst returns

[1] -0.2000000 -0.2000000 -0.1980491 -0.1734291 -0.1719509 -0.1714782
```

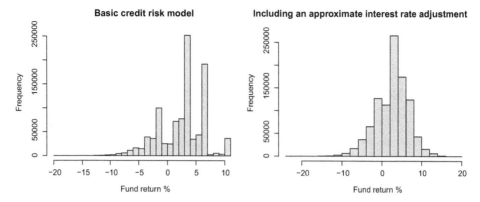

Figure 15.2 Simulated bond fund returns over one year.

It is clear from Figure 15.2 (left) that the simulations result in particular ranges with very few simulations; this is simply as a result of the limited number of rating categories used ('AAA', 'AA', and 'A') such that our bonds can only take particular values. In reality many more categories would be used.

Exercise 15.2 *Verify the mean and standard deviation of returns obtained from simulations with closed-form solutions.*

15.2.5 Incorporating Interest Rate Risk – A Simple Adjustment

As noted with Remark 15.2 we have assumed that each yield curve remains stable over the year. This is unrealistic. For the model to be considered valid, it is likely that an interest rate model should be incorporated to allow for the extra uncertainty resulting from changes in the yield curves. For our purposes, and not to detract from the prime objectives of this model, we have included a simple, approximate allowance for interest rate risk:

```
int_risk_adj<-rnorm(sims,1,0.2)#now allow for interest rate risk
sim_portfolio_return_int_risk<-sim_portfolio_return*int_risk_adj
```

Here we are adjusting the annual return by a random amount drawn from $N(1, 0.2^2)$. The corresponding results are as follows:

```
mean(sim_portfolio_return_int_risk)

[1] 0.02573582

sd(sim_portfolio_return_int_risk)

[1] 0.04063478

quantile(sim_portfolio_return_int_risk,0.005)

      0.5%
-0.09474318
```

The fund return has a 99.5% VaR of −9.5%. So if we invest £100 m in this bond fund we estimate that there is a 1 in 200 chance that the value of the fund in one year's time is less than £90.5 m. See Figure 15.2 (right).

15.2.6 Portfolio Consisting of Highly Correlated Bonds

Exercise 15.3 *Portfolio consisting of highly correlated bonds.*

This important exercise requires recalculating the above figures using the following correlation matrix:

```
correlations_high

     [,1] [,2] [,3] [,4] [,5]
[1,] 1.0 0.80 0.80 0.80 0.80
[2,] 0.8 1.00 0.85 0.85 0.85
[3,] 0.8 0.85 1.00 0.90 0.90
[4,] 0.8 0.85 0.90 1.00 0.95
[5,] 0.8 0.85 0.90 0.95 1.00
```

This may represent the situation where we have invested only in bonds issued from one company, industry, or economy. All the other assumptions remain the same.

The two key results, to be compared with earlier simulations, are as follows (the mean return is also shown):

```
mean(sim_portfolio_return_int_risk_high)

[1] 0.02568825

sd(sim_portfolio_return_int_risk_high)

[1] 0.05509955

quantile(sim_portfolio_return_int_risk_high,0.005)

      0.5%
-0.1785226
```

With the highly correlated bond portfolio the range of potential returns is much higher. For example, there is a 0.5% chance that our £100 m fund will have fallen to below £82.1 m in one year's time. Compare this with the diversified portfolio shown earlier – the 99.5% VaR fund value was only £90.5 m. As expected, the diversified portfolio exhibits less risk, with more predictable future fund values.

Exercise 15.4 *The code included in this chapter is specific for five bonds and three bond ratings. Write a function which allows for the analysis of a bond fund with n bonds and m different yield curves (and ratings). The reader is also encouraged to vary the composition of the transition matrix and compare results.*

15.3 Further Development of this Model

The model discussed in this chapter could be enhanced in a number of ways. In particular, development of the model could include:

- development of the methodology to determine appropriate levels of correlations;
- incorporate a more sophisticated default rate model;
- incorporate a more sophisticated interest rate model, incorporating possible changes to the shapes of yield curves, and also changes to yield curve spreads;
- currency risk model.

Note that during times of financial crises and/or global recession, it is often the case that money moves from high to low yield debt, widening the gaps between yield curves of various credit ratings. For example, in 1998 monies flowed from Russian bonds to lower yielding, more secure, US and German bonds, and in 2011–2012 a similar move was seen from Greek, Italian, and Spanish bonds. In a similar way, corporate bond yields tend to diverge during stressed financial periods, with monies flowing into more secure corporate entities.

15.4 Recommended Reading

- Gupton, G., Finger, C., and Bhatia, M. (1997) *CreditMetrics – Technical document.* 1997 JP Morgan Risk Metrics Group, J.P. Morgan and Co.
- Hull, J. (2015) *Risk Management and Financial Institutions* (4th edition). Hoboken, New Jersey, US: Wiley.
- McNeil, Frey, Embrechts. (2005) *Quantitative Risk Management* Princeton, NJ: Princeton University Press.
- Sweeting, P. (2017) *Financial Enterprise Risk Management* (2nd edition). Cambridge, UK: Cambridge University Press.

16

The Markov 2-State Mortality Model

Peter McQuire

16.1 Introduction

One of the fundamental tasks of the actuary who works in the life assurance or pension fields is to estimate the mortality characteristics of a policyholder or pension scheme member. For example, actuaries will be required to estimate the probability that a life which has taken out a pure endowment policy will survive to the end of the policy and be entitled to the payment. In the case of a pension scheme member, actuaries must estimate the probability that lives will survive successive future years when pension payments will be made.

(Please also see Chapter 22 which discusses more general multiple state Markov models; the model covered in this chapter is the simplest example of such a model.)

16.2 Markov 2-State Model

Many models have been developed to estimate these probabilities, the most fundamental of which is arguably the single decrement 2-state "alive-dead" Markov model ("2-state Markov model"), which is the subject of this chapter. In this model, lives can be in one of two states, alive (A) or dead (D), at time t, and transition at any time from state A to state D at the rate of μ per unit time. (Note that the term "transition rate" is also referred to as the "hazard rate", or "failure rate", and in this special case the "mortality rate".)

Figure 16.1 shows an example of the 2-state Markov model where the transition rate, μ, is equal to 0.01 per annum. Ultimately we will usually be interested in the probabilities of lives being in these states at a future time – we therefore need to define these probabilities.

Figure 16.1 2-state Markov model.

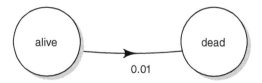

alive

dead

0.01

R Programming for Actuarial Science, First Edition. Peter McQuire and Alfred Kume.
© 2024 John Wiley & Sons Ltd. Published 2024 by John Wiley & Sons Ltd.
Companion Website: www.wiley.com/go/rprogramming.com.

To simplify the definitions we will use the following, standard actuarial notation for the 2-state model (we can do this here given the simple nature of the model):

$$_tp_x^{AD} = {}_tq_x \quad ; \quad _tp_x^{AA} = {}_tp_x$$

The first expression is the probability that a life, alive at age exactly x years, dies at some point in the next t years. The second expression is the probability that a life, alive at age exactly x years, is still alive t years later.

Critically, we can relate these probabilities and rates if we consider a very short time period. The mortality rate, μ_x, is defined under the Markov model as follows:

$$\mu_x = \lim_{dt \to 0} \frac{_{dt}q_x}{dt} \tag{16.1}$$

The probability that a life currently aged x is dead after a short time, dt, is therefore:

$$_{dt}q_x = \mu_x dt + o(t) \tag{16.2}$$

where μ_x is the mortality (or transition) rate at exact age x.

Remark 16.1 Formally, a function g(t) is said to be o(t) if $\lim_{t \to 0} \frac{g(t)}{t} = 0$.

Thus we can say:

$$_{\Delta t}q_x \approx \mu_x \Delta t \tag{16.3}$$

for small Δt. A further discussion of μ is included in the Appendix.

This equation is fundamental in allowing us to determine various probabilities over longer periods of time, which will, in general, be our main objective. By understanding what happens over very short periods of time we can build models to estimate what our system will look like over longer periods of time. (We also utilise Equation 16.2 in Chapter 22.)

By considering the probabilities that a life aged x in state A, remains in state A till age $x + t$, and then further survives a very small time dt, it is straightforward to show that, using Equation 16.2:

$$\frac{d\,_tp_x}{dt} = -_tp_x\mu_{x+t} \tag{16.4}$$

The solution to this differential equation is:

$$_tp_x = e^{-\int_0^t \mu_{x+s}ds} \tag{16.5}$$

(As defined, μ_x is required to be a continuous function of x so that the above integral is valid.) These are both fundamental equations of survival analysis – the reader should ensure that their derivations are understood.

So, for example, $_{14.5}p_{55.5}$ is the probability that a life aged exactly 55.5 years is alive in 14.5 years time, and is given by the following expression:

$$_{14.5}p_{55.5} = e^{-\int_0^{14.5} \mu_{55.5+s}ds}$$

Thus if we know the values of, or an expression for, μ at all ages, we can calculate the probability that a life aged x will survive for, or die within, a period of time. Clearly therefore it is important that we can estimate μ. We will look at the important task of estimating μ in Section 16.4, but first we will consider a few applications of Equation 16.5.

Remark 16.2 It is helpful at this point to formally define the positive continuous random variable T_x, which represents the future lifetime of a life aged exactly x. Then $F_x(t) = P[T_x \leq t] = {}_tq_x$ is the distribution function of T_x. The mortality rate, μ_{x+s}, may then be formally defined in terms of T_x as follows:

$$\mu_{x+s} = \lim_{ds \to 0} \frac{P[T_x \leq s + ds] \mid T_x > s}{ds} \qquad (16.6)$$

Equations 16.4 and 16.5 can be derived from this alternative formulation of the problem, using Equation 16.6.

16.3 Simple Applications of the 2-State Model

Example 16.1 Constant rate $\mu_x = \mu$

This is our simplest case, as $\int_0^t \mu_{x+s} ds = \mu t$. If $\mu = 0.05$ per annum for all ages, the probability that a life aged 55 years and 6 months survives till age 70 is as follows:

$$_{14.5}p_{55.5} = e^{-\int_0^{14.5} 0.05 ds} = e^{-0.05 \times 14.5} = e^{-0.725} = 0.4843$$

Exercise 16.1 *Plot a graph showing the probability of a life aged 55.5 years surviving to a range of ages up to age 110. (The solution is plotted in Figure 16.2.)*

By comparing the above plot with Equation 16.4, we can demonstrate that the points calculated are indeed the solutions to Equation 16.4. First, the LHS of Equation 16.4 – the gradient at age 65, say, is approximately:

```
(exp(-0.05*9.501)-exp(-0.05*9.499))/0.002
[1] -0.03109425
```

and the RHS, $_{9.5}p_{55.5}\,\mu_{65}$, equals:

```
-exp(-0.05*9.5)*0.05
[1] -0.03109425
```

However, the form of μ in Example 16.1 is clearly not appropriate for human mortality as mortality rates will, in general, increase with age.

Figure 16.2 Survival probability with constant mortality rate.

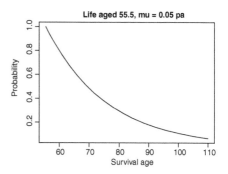

Exercise 16.2 *There are typically two exceptions to this observation. At what ages do human mortality rates tend to fall with increasing age?*

Example 16.2

Let's look at a better model for mortality rates – Gompertz Law. Here the mortality rate, μ_{age}, depends on age and is given by:

$$\mu_{age} = a \times b^{\,age}$$

Gompertz law

where a and b are parameters to be determined from a given data set. (We will see Gompertz Law again in the next chapter where we estimate values for the parameters a and b.) For now, let's say the parameter values which best fit a particular data set are $a = 0.0001$ and $b = 1.075$; these rates are plotted in Figure 16.3 (left).

Exercise 16.3 *In words, describe as simply as possible how these mortality rates change with age.*

Let's calculate the probability that a 60-year-old survives to their 65th birthday. This can be evaluated exactly (from Equation 16.5):

$$_5P_{60} = e^{-\int_0^5 0.0001 \times 1.075^{60+s} ds} = e^{-0.0001 \times 1.075^{60} \int_0^5 1.075^s ds} = 0.9548793$$

(since for $\alpha = \log(1.075)$, $\int_0^5 1.075^s ds = \int_0^5 e^{\alpha s} ds = \frac{1}{\alpha}(e^{5\alpha} - 1) = 6.0235805$.)
Alternatively, we can integrate this in R using the `integrate` function:

```
b<-0.0001;             c<-1.075
mu<-function(age) b*c^age
prob_surv<-function(lower,upper)
{x<-(integrate(mu,lower,upper))$value;
return(exp(-x))}
```

and calculate the probability of surviving from age 60 to age 65:

```
(surv60_65<-prob_surv(60,65))   #integrate between 60 and 65
[1] 0.9548793
```

Based on this Gompertz model the probability of surviving from age 60 to 65 is 95.5%.

Exercise 16.4 *Verify the answer by approximating μ as a linear function of age between 60 and 65 years.*

A plot showing all survival probabilities for a 60-year-old is shown in Figure 16.3 (right). Note the difference in shape between this plot and the plot using a constant mortality rate (Figure 16.2).

Therefore, once we have determined the mortality rates, μ_x, for a range of ages, we can calculate survival probabilities (which is our objective here). The reader may wish again to reconcile Figure 16.3 (right) with Equation 16.4.

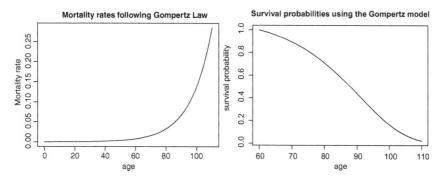

Figure 16.3 Mortality following Gompertz Law.

Exercise 16.5 *Calculate, using the Gompertz model above, the present value of a life annuity payable monthly in advance to a life aged 60 years using $\delta = 4\%$ per annum. Verify your answer approximately.*

Exercise 16.6 *A new electrical component is being developed for use in disposable technology. It is extremely cheap to manufacture, and is only required to last up to six days. An investigation has been carried out to estimate the rate at which the component fails over the first 10 days of its lifetime; the estimated hazard rate obtained from this investigation is shown in Figure 16.4 (left).*

Plot a graph of $_tp_0$, the probability that the component does not fail, against time, t, for $0 \le t \le 10$ (the solution is shown in Figure 16.4 (right)). Also comment on the likely validity of such a hazard rate function.

Exercise 16.7 *Calculate the probability that a 35 year old dies between the ages of 55 years and 65 years, based on the following mortality rates:*

$$\mu_{age} = 0.0005e^{0.09(age-20)}$$

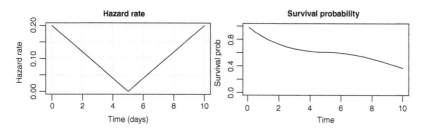

Figure 16.4 Mortality characteristics of component in first 10 days.

Exercise 16.8 *Without performing any calculations, try plotting a survival graph where the mortality rate takes the form of a sine function (shifted up such that rates at all times are positive).*

The burning question we must now address is how to estimate our mortality rates, μ_x.

16.4 Estimating Mortality Rates from Data

To estimate mortality rates we must analyse data. The two key data items we require to do this are:

- the number of deaths, and
- the length of time we observe all lives in the "alive" state during our investigation (this time is often referred to as the "exposed to risk" or "exposure time").

Data example:
An army was estimating the mortality rates of its soldiers during the period 1 Jan 2018 to 1 July 2018. This particular period may have been chosen for a number of reasons e.g. new practices may have been introduced on 1 Jan 2018, a war started on 1 Jan 2018 etc. An extract of three lives from the data is set out below:

Name	Start	End	Died?
Mark	1 April 2018	1 October 2018	No
Jane	28 August 2017	1 Nov 2018	No
Barry	16 July 2017	1 June 2018	Yes

The derived data we need for our analysis which follows is:

Name	Exposure time, v	d
Mark	3	0
Jane	6	0
Barry	5	1

where exposure time is in months (v will be defined shortly), and $d = 1$ defines a death.

Of course we may want to split our lives into more homogenous groups e.g. by age, gender, country of death. We will not concern ourselves with this important topic of heterogeneity in this chapter.

First we must set out the main theoretical ideas; the likelihood function of our data, assuming all lives are independent, is given by the most general expression:

$$L = \prod_i f_{x_i}(v_i)$$

where $f_{x_i}(v_i)$ is the probability density function of V_i, a random variable representing the time spent under observation for life i in our data (i.e. the exposed to risk). Our aim here is to write this in terms of μ's ; we can then find the value of μ which maximises L. Earlier in the chapter we noted the very important differential equation, Equation 16.4:

$$\frac{d_v p_x}{dv} = -_v p_x \mu_{x+v}$$

So we have the following for $f_x(v)$:

$$f_x(v) = \frac{d}{dv} F_x(v) = \frac{d}{dv}(1 -_v p_x) = -\frac{d}{dv}(_v p_x) = _v p_x \mu_{x+v} \tag{16.7}$$

and the likelihood function over all the lives in our data which are observed to die becomes:

$$L = \prod_i {_v p_x \mu_{x+v}} = \prod_i e^{-\int_0^v \mu_{x+s} ds} \mu_{x+v} \tag{16.8}$$

(Note that we have dropped some of the notation for clarity.) However, there is a potential problem here; what about the lives which are not observed to die during the investigation? (This is known as "censoring" and is discussed in more detail in Chapter 20.) That is, what expression should be included in the likelihood function for those lives which *don't* die in our investigation?

For these lives we simply include the probability they survived; the likelihood for each life, i, which survives the observed period of time v_i is therefore (from Equation 16.5):

$$L_i = e^{-\int_0^{v_i} \mu_{x_i+s} ds} \tag{16.9}$$

Thus all lives, whether they survive or die, contribute this exponential term to our likelihood, and only those lives which die also contribute the μ term (from Equation 16.8). The simplest way to express these terms is to use an indicator variable, d_i, which equals 1 if life i died, and 0 if life i survived; in this way the μ term only appears in the likelihood for deaths, as required. Thus we have:

$$L_i = e^{-\int_0^{v_i} \mu_{x_i+s} ds} \mu_{x_i+v_i}^{d_i} \tag{16.10}$$

This is a key equation which enables us to estimate the mortality rate over our chosen age range.

To move forward we must choose an expression for μ_x. For now we'll make the simplest assumption – that the mortality rate is constant over some chosen age range, and equals μ. (We will develop this in the next chapter.) Our likelihood function simplifies to:

$$L = \prod_i e^{-\mu v_i} \mu^{d_i} \tag{16.11}$$

By taking logs, we get the loglikelihood expression:

$$lnL = \sum_i -\mu v_i + d_i ln\mu \tag{16.12}$$

For example, the loglikelihood expression in respect of our earlier "army dataset" is:

$$lnL = (-\mu.3 + 0) + (-\mu.6 + 0) + (-\mu.5 + ln\mu) = -14\mu + ln\mu$$

Finally, differentiating Equation 16.12 with respect to μ, and setting the resulting expression to zero, we obtain the following expression for the maximum likelihood estimate of our mortality rate:

$$\hat{\mu} = \frac{d}{v} \tag{16.13}$$

where $d = \Sigma d_i$, the total number of deaths, and $v = \Sigma v_i$, the total time for which all lives are observed i.e. the total "exposed to risk". This is another important result and will be used several times in the book. The key assumption here is that the mortality rate is constant over the age period we have chosen.

From the army data above we have:

$$\hat{\mu} = \frac{1}{14} = 0.0714 \; per \; month$$

The reader may at this point be concerned with the assumption of constant mortality rates, given the comments in Section 16.3 above. Clearly a constant mortality rate is not particularly realistic. However, if we choose a sufficiently short age period the assumption may be considered to be reasonable e.g. it may be acceptable to assume that all lives between the ages of 30 years and 10 days, and 30 years and 20 days will have very similar mortality rates.

As previously noted we can choose any age range we wish. If the age range is too long however, the level of heterogeneity may be unacceptably high; adopting an age range of ten years would result in grouping together all lives between, say, 70 years and 80 years of age, and calculating one single mortality rate for all these lives, which would be unacceptable. However, in the army example, we may not be concerned about grouping together all lives between 20 and 25 years of age as we may consider all such lives to exhibit similar mortality characteristics.

In contrast, if we use too small an age range (e.g. one week) we may have very little data in each age group resulting in significant uncertainty in our parameter estimates (see Section 16.6).

Alternatively we could increase the size of our data sample by using a longer investigation period e.g. observe lives over five years. The problem here is that further heterogeneity is introduced from observing lives during a period which is likely to include changing underlying mortality rates due to socio-economic and medical factors.

Our choice should therefore be governed by the amount of data we have, and the level of heterogeneity in the data.

We proceed by assuming a constant mortality rate *over each year of age*, which is a common approach taken by such studies. For example, a life aged exactly 72.35 years of age will be assumed to have the same mortality rate as a life aged exactly 72.91 years, but different to a life aged 73.03 years.

Exercise 16.9 *Discuss the different choices that may be made by a government actuary in China with that of an actuary based in the UK when analysing human mortality rates.*

16.5 An Example: Calculating Mortality Rates for One Age Band

Consider observing a group of lives between the exact ages of 70 and 71 years. Our exposure and death data (the v's and d's) which we collect must relate to this age period; we only concern ourselves with deaths which occur between the exact ages of 70 years and 71 years.

To estimate μ for this age range we simply calculate the length of time each life was alive between these two exact ages, and note whether they died or not. We then sum all these times, calculate the total number of deaths, and apply Equation 16.13. Extracting our data (which consists of data on 1,600 lives collected over the 2-year period from 1 January 2018 to 31 December 2019):

```
data_for_mu<-read.csv("~/estimatemuatoneage.csv")
head(data_for_mu) #the first six lives only
```

(One can imagine that this data has been derived from original data similar to that of the army data shown earlier in this chapter.)

```
  deaths    exposure
1      0 0.67019169
2      0 0.53740100
3      1 0.40451532
4      0 0.08867825
5      0 0.86456997
6      0 0.87851133
```

Each row of data represents the data for one life; in respect of the first life in the data we observed the life for approximately eight months and the life did not die, and the third life died after approximately 0.4 years. Note that the exposure time in this dataset is in years.

Exercise 16.10 *It is important to understand the several possible types of underlying data from which the above exposure times were obtained. Given that the above data was obtained from an investigation covering the period 1 January 2018 to 31 December 2019 (i.e. 2 years) develop various possible dates and events for the first six lives above (for example, the first life may have had her 70th birthday on 1 May 2019 and did not die before 31 December 2019, or may have had her 70th birthday on 1 September 2017 and did not die before her 71st birthday).*

Applying Equation 16.13 to the dataset, the estimated mortality rate is:

```
(mu70<-sum(data_for_mu$deaths) / sum(data_for_mu$exposure))
```
```
[1] 0.02095332
```

Here $\hat{\mu}$ (our estimate of μ) has been calculated using data from lives between exact ages 70 and 71 years, so it should be reasonable to assume that this rate is an average of all the rates between 70 and 71. It is therefore usual to assume that the value of $\hat{\mu}$ calculated here is our best estimate of $\mu_{70.5}$. That is:

$$\hat{\mu}_{70.5} = 0.020953 \ pa$$

Similar calculations would be made in respect of all age ranges, say from age 20 to 110 years, such that we obtain $\hat{\mu}_{20.5}, \hat{\mu}_{21.5}, \hat{\mu}_{22.5}...\hat{\mu}_{109.5}$. This set of rates, which are the maximum likelihood estimates at each age, are commonly referred to as "crude rates". Given the data available for each age band we have chosen, these crude rates represent the best estimate of the mortality rate at each age. We will refer back to this important point in the next chapter.

16.6 Uncertainty in Our Estimates

In the previous section we demonstrated how we can estimate the true, underlying mortality rates. But how good are these estimates? This is a central question in the process of identifying a credible, reliable set of mortality rates.

It can be shown that the variance of the estimator for the mortality rate, $\tilde{\mu}$, under the Markov model is given by the following expression:

$$Var(\tilde{\mu}) = \frac{\mu}{E[V]} \tag{16.14}$$

where $E[V]$ is the expected exposure period from our investigation, and μ is the true, underlying mortality rate. One thing to note at the outset is that the more data we have, either from observing more lives, or by observing them for longer (or both of these), the larger the value of $E[V]$ will be, resulting in a smaller variance, and hence greater certainty regarding the value of μ.

But does Equation 16.14 help us? We cannot ever know μ – this is the entity we are trying to estimate – the best we can do is to use our best estimate for μ obtained from our data by applying Equation 16.13. Similarly we cannot know $E[V]$ for the same reason - we won't know the expected exposure period as we don't know μ. However we can use the actual exposure period from our data, v. So, given $\hat{\mu} = \frac{d}{v}$, we can *estimate* the variance of $\tilde{\mu}$ using d and v:

$$Est.Var(\tilde{\mu}) = \frac{\hat{\mu}}{v} = \frac{d}{v^2} \tag{16.15}$$

A fundamental property of MLE's is that its distribution is asymptotically Normally distributed. Thus we can write (approximately):

$$\tilde{\mu} \sim N(\hat{\mu}, \frac{\hat{\mu}}{v}) \tag{16.16}$$

Thus we can, approximately at least, make predictions regarding confidence intervals. As noted above, the longer we observe lives for, the smaller the variance is likely to be, thus resulting in more confidence in our estimate.

It should be noted that estimating the variance of $\tilde{\mu}$ is essential; without it we cannot understand how reliable our estimate of μ is, and ultimately, the reliability of the pricing of products and solvency estimates for life assurance companies and pension schemes.

Calculating the estimated variance and 95% confidence intervals for μ from our data:

```
(est_var70<-(mu70/sum(data_for_mu$exposure)))
```

```
[1] 2.582599e-05
```

```
mu70;mu70-1.96*est_var70^0.5;mu70+1.96*est_var70^0.5
```

```
[1] 0.02095332
[1] 0.01099275
[1] 0.0309139
```

Given that the MLE estimator is asymptotically Normally distributed, we may say, approximately, that we are 95% confident that $\mu_{70.5}$ lies between 0.010993 and 0.030914, and our best estimate is 0.020953.

Exercise 16.11 *Recalculate $\hat{\mu}_{70.5}$ and the 95% confidence levels based on increasing both the number of deaths and the exposure period in the above example by a factor of 100, and comment on the results.*

16.7 Next Steps?

In the next chapter we will find that these "best estimates" for the mortality rates at each defined age band (the "crude rates") can be improved if we assume that mortality rates should progress smoothly between ages. Each crude rate exhibits some inherent error, due to stochastic uncertainty. By estimating the confidence intervals of these rates we can make a better judgement regarding how acceptable any adjustments are that we make to these crude rates. Effectively we will try to obtain a smooth curve across all ages (e.g. from 20 to 100 years) which is acceptably close to each of these crude rates. We will continue to make use of the crude rates calculated in this chapter to ensure any curve we do ultimately fit is consistent with these best estimates.

16.8 Appendices

16.8.1 Informal Discussion of μ

The probability that a life aged x dies before age $x + dt$ is given by:

$$_{dt}q_x = -\frac{N_{x+dt} - N_x}{N_x} = -\frac{dN}{N_x}$$

where N_x is the number of lives alive aged x years, and dN is the change in the number of lives over time dt. (We require the negative sign as dN will be negative and the probability is positive.)

The definition of μ, the instantaneous, proportional rate at which lives die at exact age x is *defined* as:

$$\mu_x = -\frac{1}{N_x}\frac{dN}{dt}$$

Therefore,

$$_{dt}q_x = \mu_x dt \quad \text{in the limit as } dt \to 0$$

It may help to think of this in terms of a small period of time, Δt:

$$\mu_x \approx -\frac{1}{N}\frac{\Delta N}{\Delta t}$$

If we have 10,000 lives and two lives die in the next day, then

$$\mu \approx -\frac{-2}{10,000}\frac{1}{0.00274} \approx 0.0729927 \text{ per annum}$$

The reader should note that the above mathematics is analogous to the application of the force of interest, δ, to a fund value X. The fall in the population, dN, is given by $N\mu\, dt$; similarly, the increase in the fund value is $X\delta\, dt$.

16.8.2 Intuitive meaning of $f_x(t)$

The intuitive meanings of p, and q, are fairly clear. The mortality rate, μ_x, relates to the likelihood of a life currently aged x exact dying between the ages of x and $x + dt$ (Equation 16.6). However it is worth checking our understanding of $f_x(t)$ (Equation 16.7); in the context of mortality studies $f_x(t)$ gives a measure of the likelihood of a life currently aged x exact dying between the future ages $x + t$ and $x + t + dt$ ie $f_x(t) = {}_tp_x\mu_{x+t}$. We can see this from the following example for a life aged 60 years.

Using the same mortality rates as in Example 16.2, and writing a function for $f_x(t)$:

```
b<-0.0001;    c<-1.075;    mu<-function(age)b*c^age
surv_calcs<-function(age_now,age_end){
mu_y<-mu(age_end)
p_y<-exp(-integrate(mu,age_now,age_end)$value)
return(c(mu_y,p_y,mu_y*p_y))
}
#calculate 60 to 109
range<-65;   agenow<-60;     y<-matrix(0,range,3)
colnames(y)<-c("mu","p","f")
for(x in 1:range)y[x,]<-surv_calcs(agenow,agenow+x-1)
```

Our function, surv_calcs, returns three calculations: μ_x, ${}_tp_{60}$, and $f_{60}(t)$. A plot of $f_{60}(t)$, where $\mu_x = 0.0001 \times 1.075^x$, is shown in Figure 16.5, together with plots of the mortality rates and survival probabilities. The plot of $f_{60}(t)$ has a simple interpretation: a 60-year-old is most likely to die around the age of 91 years. In addition, the probability that a 60-year-old dies in a particular age range is given by calculating the area under the curve for the required age range.

Exercise 16.12 *Calculate the probability that a life, currently aged 60 years exact, dies between the exact ages of 65 and 70 years (assume the same Gompertz mortality rates as used above).*

Exercise 16.13 *Carry out similar calculations to the above but for a life now aged 64 years.*

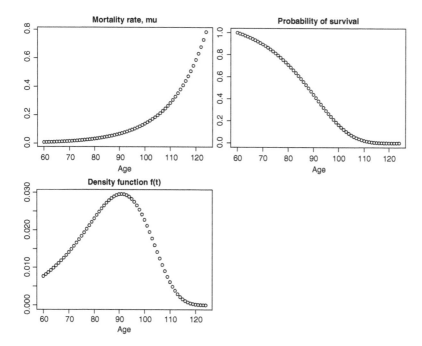

Figure 16.5 Mortality rates, survival probabilities, and density function: life aged 60 years.

Exercise 16.14 *Calculate the most likely age at which a 90-year-old will die given that:*

$$\mu_{age} = 0.005 \times age$$

16.9 Recommended Reading

- Dickson D., Hardy M., and Waters H. (2020) *Actuarial Mathematics for Life Contingent Risks.* 3rd edition. Cambridge, UK: Cambridge University Press.
- MacDonald A.S. et al. (2018) *Modelling Mortality with Actuarial Applications.* Cambridge, UK: Cambridge University Press.

17

Approaches to Fitting Mortality Models: The Markov 2-state Model and an Introduction to Splines

Peter McQuire

```
library(stats4)
library(splines)
```

17.1   Introduction

In the previous chapter we discussed the relationship between the mortality rate, μ_x, and the survival probability, $_tp_x$. This led to a general expression for the likelihood of our data. Furthermore, by assuming that the mortality rate was constant over a chosen age range, we obtained an expression for the "crude rate", $\hat{\mu}$, for that range. These crude rates represent the maximum likelihood estimates of the constant mortality rates over the chosen age range.

However, if we are to make the *a priori* assumption that mortality rates should increase smoothly between successive ages, there is a problem with these crude rates. The inherent uncertainty, or stochastic error, will result in crude rates which are higher than the true rates at some ages, and lower at other ages. Figure 17.1 is a plot of crude rates we shall analyse later in the chapter; we can see that there are many instances where we would want to improve on the crude rates, particularly when our population size is smaller e.g. at older ages. For example, at ages 27 and 105 years, the crude mortality rate falls a little (this is more easily seen from the plot of log rates, which are also plotted in Figure 17.1(right)). Therefore, we will aim to determine a smooth function for the mortality rates which is acceptably close to these crude rates.

The examples and exercises in this chapter aim to provide the reader with practice at fitting these simple mortality models to various types of mortality data, and how to apply the theoretical ideas to obtain real solutions. They also present the reader with an opportunity to practise writing more involved code, such as simulating lifetime data from particular mortality models (see Figure 17.3).

R Programming for Actuarial Science, First Edition. Peter McQuire and Alfred Kume.
© 2024 John Wiley & Sons Ltd. Published 2024 by John Wiley & Sons Ltd.
Companion Website: www.wiley.com/go/rprogramming.com.

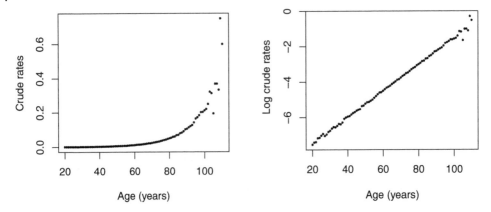

Figure 17.1 Crude rates calculated at each integer age.

17.2 Graduation of Mortality Rates

Our objective is therefore to fit better models to our mortality data such that we have more appropriate estimates of mortality rates over a larger age range. Various methods are set out in this chapter; the method one chooses will depend on the format of the data, and the degree of accuracy required. Broadly speaking, our data can be provided to us in two forms:

- individual member data, or
- summarised population data for each age band. For example:

```
  age exposure deaths
1  80    52953   2638
2  81    49468   2636
3  82    45276   2669
4  83    41368   2502
5  84    37534   2557
```

Here we have no information on any individual life, only the total exposure time and total number of deaths in each age band.

Remark 17.1 Terminology reminder: In what follows we shall assume that the age label used is "age last". This is simply a convenient way of defining each age band – we could alternatively have written "exact ages from 80 years to 81 years" etc., for each row in the age column. For example, there were 2638 deaths in our data which occurred between the exact ages of 80 and 81 years. [Note: the formal definition of age last is the life's exact age on their previous birthday.]

We now develop the key expressions discussed in the previous chapter, remembering our objective – to determine, from population data of exposure times and number of deaths, a suitable expression for μ_x, such that we can estimate survival probabilities, which can then be used to help estimate the solvency positions of life assurance companies and pension schemes.

In the previous chapter we identified the important expression relating mortality rates to survival probabilities under the 2-state Markov model:

$$_sp_x = e^{-\int_0^s \mu_{x+t}dt} \tag{17.1}$$

(Please refer to the previous chapter for definitions of the terms used in this section.) This enables us to calculate the probability that a life aged x exact will survive till age $x + s$ exact. This led to the general expression for the likelihood of the data which consists of n independent lives:

$$L = \prod_{i=1}^{n} e^{-\int_0^{s_i} \mu_{x_i+t}dt} \mu_{x_i+s_i}^{d_i} \tag{17.2}$$

and the log likelihood of the data:

$$\log L = \sum_{i=1}^{n} \log L_i = \sum_{i=1}^{n} \left(-\int_0^{s_i} \mu_{x_i+t}dt + d_i \log \mu_{x_i+s_i}\right) \tag{17.3}$$

where the sum is over all n lives, and s_i is the time for which life i is observed. We simply calculate the log likelihood for each life in our data and sum them using the above. (The data may take a form similar to that of the "army" data in the previous chapter.)

It may be the case that our expression for μ is easily integrable – we shall see an example shortly. However, it is useful to determine an approximate numerical method which will allow us to determine the parameters for our model under any proposed forms for μ; it also serves as an excellent check on our calculations and provides further insight into the calculations. We will use the equations which follow in this section in various examples and exercises later in the chapter.

Discretizing the integral in Equation (17.3) the log likelihood for life i becomes:

$$\log L_i = -\sum_{t=0,\Delta t,2\Delta t\ldots}^{s_i} \mu_{x_i+(t+0.5)\Delta t}\Delta t + d_i \log \mu_{x_i+s_i} \tag{17.4}$$

Remark 17.2 These equations, and the notation used, can appear a little clumsy. The reader may benefit from first studying and running the code which follows, then returning to the equations.

Effectively we are looking at each individual life and adding up the μ's over the period for which we observed that life; if the life is observed to die during our investigation ($d = 1$) we include the final term with the appropriate age at death. For example, if we observe a life, α, in our data from age 41.5 years who dies at age 42.7 years, using $\Delta t = 0.1$ years, our expression would include the addition of twelve μ terms and a single log $\mu_{42.7}$ term in respect of that life. Equation 17.4 becomes (noting the final term included below):

$$\log L_\alpha = -\mu_{41.55} \times 0.1 - \mu_{41.65} \times 0.1 - \mu_{41.75} \times 0.1 - \mu_{41.85} \times 0.1 - \mu_{41.95} \times 0.1 - \mu_{42.05} \times 0.1 - \mu_{42.15} \times 0.1 - \mu_{42.25} \times 0.1 - \mu_{42.35} \times 0.1 - \mu_{42.45} \times 0.1 - \mu_{42.55} \times 0.1 - \mu_{42.65} \times 0.1 + \log \mu_{42.7}$$

We must now sum the $\log L_i$ expressions for all n lives in our data (as per Equation 17.3):

$$\log L = - \sum_{lives,i}^{n} \sum_{t=0,\Delta t...}^{S_i} \mu_{x_i+(t+0.5)\Delta t}\Delta t + \sum_{lives,i}^{n} d_i \log \mu_{x_i+s_i} \tag{17.5}$$

(Note that some approximation is required here, as lives are assumed to start and leave the investigation in the middle of an age range.) We can swap the order of the summands in the first term above such that we first sum across all lives at each age range, then sum across all age ranges. We can simplify Equation 17.5 by introducing the term E_x^c:

$$\log L = - \sum_{x=0,\Delta t...}^{\infty} E_x^c \, \mu_{x+0.5\Delta t} + \sum_{lives,i}^{n} d_i \log \mu_{x_i+s_i} \tag{17.6}$$

where E_x^c is the sum of exposure times from a particular age range, x to $x + \Delta t$, across all our lives.

All the above formulae may be used where we have detailed individual member data. However, where we have *summarised data* we can use Equation 17.6 by incorporating the appropriate age range, Δt, consistent with the data provided. Often, as previously discussed, we will be given exposure and death data for each *integer* age band, such that $\Delta t = 1$, whereupon Equation 17.6 simplifies to:

$$\log L = \sum_{x=0,1,2..}^{\infty} (-E_x^c \mu_{x+0.5} + D_x \log \mu_{x+0.5}) \tag{17.7}$$

where D_x is the total number of deaths between exact ages x and $x + 1$. For example, given our dataset above, $E_{83}^c = 41368$ and $D_{84} = 2557$. This is the equation we will use in our first examples below.

To recap, if we are given detailed data on each life we can apply Equation 17.5, or 17.6, choosing an appropriate Δt which allows for run-time constraints. We will see this application later in Section 17.5. We can of course use Equation 17.3 (our most general equation) if our expression for the mortality rate is straightforward to integrate. Often however, the data we receive will be in this summarised form, leaving us no alternative but to use Equation 17.6 or 17.7. For example, a government mortality study may obtain data from hundreds of contributing insurance companies, each one providing summarised data of exposure time and number of deaths over each integer year of age. Provided the data set is large enough, the marginal benefits of receiving individual life data will be small. We will see an application of this in Section 17.3.2.

Brief note on the Poisson model

In many actuarial textbooks the Poisson model is used in this discussion. However, we shall continue with the Markov model. The Poisson process is a particular type of Markov process, and yields similar results to those derived in the previous chapter (which used the Markov 2 state model). However the Markov model more closely follows our actual process and is the more theoretically appropriate model.

17.3 Fitting Our Data

17.3.1 Objective

Our aim is to find an expression for μ_x which provides an appropriate representation of our mortality data. There are various expressions we could try. We will try the following two simple forms:

- $\mu_x = c$, a positive constant
- $\mu_x = e^{a+bx}$

As noted earlier, the choice of estimation method will depend on whether we have summarised data, or individual member data. We will look at both, starting with summarised data.

17.3.2 Summarised Data

Our summarised data consists of the total exposure time (in years) and total number of deaths for each integer age band. It is important to note that we have lost information here in so much that we do not know exactly when each life entered or left our investigation, nor which particular lives died.

```
getdata <- read.csv("~/graduation_Gompertz2.csv")
deaths<-getdata$deaths;   age<-getdata$age;   exposure<-getdata$exposure
head(getdata,3)
```

```
  age exposure deaths
1  20   100233     54
2  21   100218     61
3  22    99479     62
```

It's worth first explaining how this data may have been collated.

1. The analyst would determine an appropriate length of the investigation; it needs to be sufficiently long to ensure we have adequate data, but not too long such that the level of heterogeneity is unacceptable.
2. Collate complete data on each individual life.
3. Identify in which age band, if any, each life died.
4. Calculate the exposure time of each life in each age band.
5. All the death data and exposure data would then been summed for each age band, giving us data in a similar format to that provided in getdata.

Exercise 17.1 *Think of two extreme investigation periods which may have resulted in the above dataset (Hint: How would your method vary if you were conducting an investigation in China compared to New Zealand?).*

We will now look at two simple candidates for modelling the mortality rates, and find the parameter values which maximise the log likelihood (Equation 17.7) by using the `mle` function in R.

Example 17.1 – A constant rate of mortality

First let's try the simplest possible solution for μ and one which was covered in the previous chapter – that μ is constant and is not age dependent. The reader may already be questioning why we are trying such a model – however it is useful to try the simplest possible model rather than immediately proceeding to a more recognised model.

Note that we derived the expression for this in the previous chapter; allowing for practicalities, the MLE for the mortality rate under the 2-state Markov model when it is a constant is given by:

$$\hat{\mu} = \frac{d_x}{E_x^c}$$

This provides an estimate equal to:

```
(constantmu<-sum(deaths)/sum(exposure))

[1] 0.01422363
```

Writing out a function for the logL as per Equation 17.7 for integer years of age (remembering that `mle` locates the minimum value so we must multiply our function by −1):

```
lnlik1<-function(const_rate){(sum(-const_rate*exposure + deaths*log(const_rate))*-1)}
```

Applying the `mle` function in the `stats4` package to obtain the maximum likelihood estimate:

```
par_est1<-mle(lnlik1,list(const_rate=0.01),lower=0.001,method="L-BFGS-B") #min logL
summary(par_est1)

Maximum likelihood estimation

Call:
mle(minuslogl = lnlik1, start = list(const_rate = 0.01), method = "L-BFGS-B",
    lower = 0.001)

Coefficients:
            Estimate    Std. Error
const_rate  0.01424702  4.878569e-05

-2 log L: 887101.6
```

The value calculated above using the `mle` and `lnlik1` functions, $\mu = 0.014247$, is consistent with the theoretical value of the MLE calculated earlier.

Exercise 17.2 *Verify that this is indeed the maximum logL by looking at a range of values of μ, and plot a graph of logL against μ.*

If we were compelled to use just one rate for all ages we would perhaps use this value – this is our best estimate given the significant constraint that we can use only one number over the entire age range. (As discussed in the previous chapter, we can choose whatever age range we wish; the MLE expression, $\hat{\mu} = \frac{d_x}{E_x^c}$, is still valid.)

Similarly, if we choose single integer years of age we can determine the best estimate for the mortality rate for that single age range by using only data in respect of that age range. This is the same method as detailed in the previous chapter, but here we are calculating the crude rates for every integer age band (shown below are the rates for the four highest ages):

```
cruderates<-deaths/exposure   #at each age
tail(cruderates,4)

[1] 0.3684211 0.3333333 0.7500000 0.6000000
```

Remember that the crude rates are also MLEs; they are the best we can do using data solely from each single age band, from x to $x+1$, so we would hope that the rates produced from our final proposed model are reasonably consistent with these rates. The single, constant rate is smooth, but has poor adherence to our best estimates at single ages – see Figure 17.2, left. (Effectively the constant rate is a weighted average of the crude rates.) Clearly we can do a lot better!

Exercise 17.3 *Calculate MLE rates for each 10-year age group.*

Exercise 17.4 *Comment on the crude rates for (1)ages 78 and 79 years, and (2) all those > 100 years.*

Example 17.2 Gompertz model

Clearly our "constant rate" model and "crude rate" model are both inappropriate. We must try to find a more sensible model, one which not only exhibits smoothness in its age-dependent mortality rates, but also adheres sufficiently to the crude rates.

We will try the famous Gompertz mortality model. This model was proposed by Benjamin Gompertz in 1825 – it turns out that this is usually an exceptionally good fit, particularly for ages between 30 to 90 years. Gompertz Law is given by the following expression:

$$\mu_x = e^{a+bx} \tag{17.8}$$

where x is age exact and a and b are the parameter values to be determined. (Equation 17.8 is often written in the alternative forms $\mu_x = Bc^x$ or $\mu_x = Ae^{bx}$.)

Given that our data here is summarised within age bands we do not know, exactly, when each life died, nor over what exact ages we observed each life. Thus we need to make the reasonable assumption (if we believe our data is large enough and is not biased) that all lives die half way through the year of integer age, and that our lives were observed equally throughout each age range. Making a suitable adjustment and substituting Equation 17.8 into Equation 17.7:

$$\log L = \sum_{x=20,21..}^{110} -e^{a+b(x+0.5)}E_x^c + D_x(a + b(x + 0.5)) \tag{17.9}$$

It is important to understand that this sum is over all the age bands we have in our data; the sum over all lives has been done prior to receiving the summarised data.

Writing a function for this expression and calculating the MLE's for a and b as we did in Example 17.1:

```
lnlik2<-function(a,b){(sum(-(exp(a+b*(age+0.5)))*exposure+deaths
*(a+b*(age+0.5))))*-1}
par_gomp<-mle(lnlik2,list(a=-8,b=0.07))              #minimum log likelihood
gompertzrates<-exp(par_gomp@coef[1]+par_gomp@coef[2]*(age+0.5))
summary(par_gomp)

Maximum likelihood estimation

Call:
mle(minuslogl = lnlik2, start = list(a = -8, b = 0.07))

Coefficients:
      Estimate    Std. Error
a -8.98881690 0.0176585418
b  0.07439993 0.0002327059

-2 log L: 751726.1
```

Thus we have the following proposed expression, based on the Gompertz model, for the force of mortality:

$$\mu_{age} = e^{-8.9888169+0.0743999 \times age} \tag{17.10}$$

Calculating our Gompertz rates and plotting them against the crude rates (see Figure 17.2):

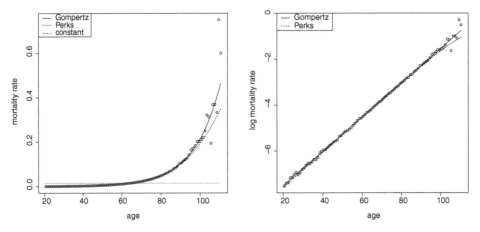

Figure 17.2 Comparing crude rates with Gompertz, Perks, constant.

As noted previously, the crude rates for each age band are effectively the best estimates of mortality rates using only data from that age band (x to x+1), so we should be comparing how closely our model agrees with the crude rates. From a visual inspection it appears to be a reasonable fit. A series of statistical tests should now be undertaken to test whether the fit, and hence the proposed model, is acceptable (see Chapter 18). The curve we finally agree on will exhibit a combination of both adherence to our crude rates, and smoothness. If our model "passes" all these tests, we may conclude that the proposed model is an accurate reflection of current mortality.

However we should also consider whether alternative, even better models exist, by comparing log-likelihood values and information criteria such as AIC and BIC. We should also ensure that our newly calculated rates are consistent with those calculated from previous years' analyses, or that any differences which do exist can be explained, e.g. changes due to COVID-19.

Exercise 17.5 *Fit the Perks hazard (Equation 17.11) to the above data (also plotted in Figure 17.2):*

$$\mu_x^{perks} = \frac{e^{a+bx}}{1 + e^{a+bx}} \tag{17.11}$$

Comparing the logL from these two models (Gompertz: -3.7586306×10^5 and Perks: -3.7589367×10^5), we may conclude that Gompertz is preferable in this case. As noted above, further statistical tests should be carried out to verify our choice.

Exercise 17.6 *We can also use a Generalized Linear Model to determine MLE's of the parameter values. (You may wish to skip this exercise if you have not yet covered the chapter on GLM's.) Calculate the Gompertz parameters using Poisson Regression.*

The parameter values obtained from the GLM exercise are -9.0180063 and 0.0747853, similar to the results obtained using previous methods.

17.4 Model Fitting with Least Squares

It is important to note that the Gompertz model fitted above is determined by using the exposure and death data *directly*, and estimating the parameters using maximum likelihood estimation given our chosen model; they are not obtained by calculating the crude rates first and fitting a curve to these crude rates.

In this section we fit a linear regression model using the log of the crude rates calculated at each age; this is perhaps more akin to what most practitioners would describe as "the graduation process", where we take a series of values (crude rates), from which we fit a model. Comparing the Gompertz expression, Equation 17.8, we have a least squares model:

$$\log \hat{\mu}_x = a + bx + e_x \tag{17.12}$$

where $e_x \sim N(0, \sigma^2)$. We can perform a linear regression of $\log \hat{\mu}_x$ against age, x, using the lm function (the crude rates for each integer year of age were calculated in Section 17.3.2 and are an estimate of $\mu_{x+\frac{1}{2}}$):

```
age_adj<-age+0.5;     log_rate<-log(cruderates)
data_for_lm<-as.data.frame(cbind(age_adj,log_rate))
model_linear<- lm(log_rate~age_adj,data_for_lm)
model_linear$coef

(Intercept)     age_adj
-9.02371927  0.07492688
```

Thus we have the regression coefficients, and our fitted Gompertz model using "least squares":

$$\mu_x = e^{-9.0237193+0.0749269\,x} \tag{17.13}$$

Often, we may be required to predict, or simulate, a future mortality rate at age x allowing for the uncertainty in this value. This should allow for the uncertainty in both the parameter estimates, and the stochastic uncertainty in the rate (reflected in σ). Fitted values of the log rates with confidence intervals, can be obtained as follows:

```
mort_conf<-predict(model_linear,interval="prediction",level=0.95)
mort_conf[1:5,]           #show ages 20.5 - 24.5

         fit       lwr       upr
1 -7.487718 -7.679732 -7.295705
2 -7.412791 -7.604675 -7.220908
3 -7.337865 -7.529620 -7.146109
4 -7.262938 -7.454568 -7.071307
5 -7.188011 -7.379520 -6.996502
```

Exercise 17.7 *Verify the above confidence intervals. From standard regression theory the variance of an individual response, or prediction, of the log rate at age x is given by (using standard regression notation, not repeated here):*

$$[1 + \frac{1}{n} + \frac{(x - \bar{x})^2}{S_{XX}}]\sigma^2 \tag{17.14}$$

and you are given that $t_{0.975,89} = 1.9869787$.

Exercise 17.8 *A one-year study observes 10,000 lives, aged 80 years exact at commencement. Estimate 95% confidence intervals for the number of deaths over the year, using the above results, and appropriate approximations.*

Remark 17.3 Alternatively we can carry out polynomial regression using the `poly` function; here using polynomials with a degree of one (as above), and degree of two:

```
poly_1<-lm(log_rate ~ poly(age_adj, 1, raw = TRUE));  poly_1$coefficients

            (Intercept) poly(age_adj, 1, raw = TRUE)
            -9.02371927                   0.07492688

poly_2<-lm(log_rate ~ poly(age_adj, 2, raw = TRUE));  poly_2$coefficients

            (Intercept) poly(age_adj, 2, raw = TRUE)1
           -8.966057e+00                   7.282876e-02
poly(age_adj, 2, raw = TRUE)2
           1.601613e-05
```

```
#alternative code
poly_alt<-lm(log_rate ~ age_adj+I(age_adj^2));          poly_alt$coefficients

   (Intercept)          age_adj  I(age_adj^2)
  -8.966057e+00   7.282876e-02   1.601613e-05
```

The Gompertz model parameters obtained from each of the methods above are indeed similar. See MacDonald et al. (2020) Chapters 5 and 10 for a further discussion regarding the comparison of these methods.

17.5 Individual Member Data

We now move onto analysing mortality data where detailed information on each life is available. Whereas previously in this chapter we only had summarised data for each age range, we now hold data on the length of time *each life* was observed for in our investigation and whether that life died.

Holding detailed individual member data is preferable to summarised data for two reasons: our calculations should be at least as accurate as fewer approximations are required e.g. no assumption is required regarding the date of death; secondly, we can ensure better scrutiny of the data, making it more likely that any errors in the data are detected.

Our objective remains the same – to determine an appropriate model for our mortality rates (here we will only adopt the Gompertz model). We will use two methods, by applying Equations 17.3 and 17.6.

For the example below we have simulated future lifetime data for 50-year-old lives (using the Gompertz model) – this is in the interest of simplicity. In reality we certainly would not follow 10,000 50-year-old lives until they died. (See Example 17.3.)

Exercise 17.9 *Why would this be an undesirable way to design a mortality investigation?*

The chosen form for our simulated mortality rates is as follows:

```
aaa<- -9.0;            bbb<-0.075
mux<-function(x)exp(aaa+bbb*x)
```

$$\mu_x = e^{-9+0.075x}$$

To simulate our data, we first need the distribution function, $F_{50}(t)$, which is calculated below using two different examples of code:

```
#method 1
survxx<-function(startx,delta_t){exp(-integrate(mux,startx,startx+delta_t)$value)}
agemax<-120;   agestart<-50;   dx<-0.01
no_of_timesteps<-(agemax-agestart)/dx

F_dist<-NULL                          #distribution function
for(i in 1:no_of_timesteps){
F_dist[i]<-1-survxx(agestart,i*dx)
}
```

```
#method 2
dist_exact<-1-exp((1/bbb)*(mux(50)-mux(51:120)))
```

Proceeding with the simulations (only using F_dist):

```
set.seed(500);  fut_time<-NULL;
no_lives<-100000 #simulated lives

for(step1 in 1:no_lives){
fut_time[step1]<-dx*which.min(abs(F_dist - runif(1)))}
head(fut_time)

[1] 43.76 39.57 53.16 30.71 42.87 19.42
```

The distribution functions calculated above, F_dist and dist_exact are plotted in Figure 17.3 (left), demonstrating their equivalence. A histogram of our simulated future lifetime data, fut_time, is shown in Figure 17.3 (right).

Remark 17.4 Simulations of future lifetimes using a life table rather than a Gompertz model are carried out in Chapter 23.

Applying Equation 17.3 (exactly)

Given the Gompertz expression is straightforward to integrate exactly, we can apply Equation 17.3 directly. The total log likelihood of our data, where X_i is the exact age of life i at death, is as follows:

$$\sum_i -[\frac{1}{b}e^{a+bx_i}]_{50}^{X_i} + a + bX_i \tag{17.15}$$

where the sum is over all i lives. Note that Equation 17.15 is a special case, reflecting our data where observation on all lives started at age 50 years exact, and all lives were observed until death (usually this would include a similar indicator term to that used in Equation 17.2 such that the $a + bX_i$ term was included only for each death – see Exercise 17.10). The following code includes this integrated Gompertz expression:

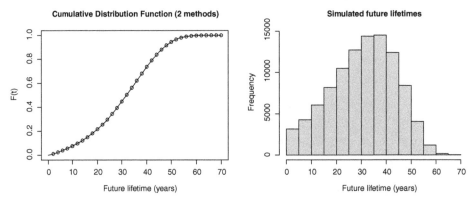

Figure 17.3 Lives aged 50 years at start of investigation.

```
lnlik4<- function(a,b) {
sum((-(1/b*exp(a+b*(agestart+fut_time))-1/b*exp(a+b*(agestart))) +
 (a+b*(agestart+fut_time))))*-1 }

par_est_ex<-mle(lnlik4,list(a=-9,b=0.075));      summary(par_est_ex)

Maximum likelihood estimation

Call:
mle(minuslogl = lnlik4, start = list(a = -9, b = 0.075))

Coefficients:
      Estimate     Std. Error
a -8.99999886 0.0199690144
b  0.07505742 0.0002440396

-2 log L: 788167.4
```

These are consistent with the actual parameter values used to simulate our data (-9 and 0.075).

Applying Equation 17.6 (approximation)

Calculating exposure times for each age band, using appropriate approximations often referred to as the census method (using $\Delta t = 0.01$ years):

```
dtt<-0.01
l_surv<-NULL
for(i in 1:((agemax-agestart)/dtt+1)){
l_surv[i]<-sum(fut_time> i*dtt)
}

lives_surv<-c(no_lives,l_surv[-length(l_surv)]) #remove last term and add first term
exp_time_censx<-(l_surv+lives_surv)/2*dtt #average between consecutive terms
exp_time_cens<-exp_time_censx[-length(exp_time_censx)] #remove last term
```

Death data is easily calculated:

```
deathx<-diff(-lives_surv)
```

Applying Equation 17.6 to this exposure data and death data, we obtain our parameter estimates:

```
agex<-seq(agestart,agemax-dtt,dtt)

lnlik5<- function(a,b) {(sum(-(exp(a+b*(agex+dtt*0.5)))*exp_time_cens +
deathx*(a+b*(agex+dtt*0.5))))*-1}

par_est2<-mle(lnlik5,list(a=-9.5,b=0.07));      summary(par_est2)

Maximum likelihood estimation

Call:
mle(minuslogl = lnlik5, start = list(a = -9.5, b = 0.07))
```

```
Coefficients:
      Estimate   Std. Error
a -8.96927060 0.0199435231
b  0.07467921 0.0002439233

-2 log L: 788166.8
```

The estimated parameter values are consistent with our inputed values.

Example 17.3

Earlier in this section we simulated lifetime data for lives aged 50 years, with no censoring. Here we simulate mortality data based on random starting ages, between 50 years and 80 years, but now for a 2-year investigation period (using the Gompertz model previously applied). Note that, in contrast with the previous data set, many lives in this data set are censored i.e. they do not die within the period of the investigation. The model is fitted in Exercise 17.10.

```
aaa<- -9.0;           bbb<-0.075;     sim_livesx<-100000
mux<-function(x)exp(aaa+bbb*x)
data_abcd<-matrix(0,sim_livesx,4)
ageminxx<-50;         agemaxxx<-100;  inv_len<-2

for (counter_pp in 1:sim_livesx) {
age_start_x<-runif(1,ageminxx,agemaxxx)
age_death_x<-(log(mux(age_start_x)-bbb*log(runif(1)))-aaa)/bbb
age_end<-min(age_start_x+inv_len,age_death_x)
death_abcd<-(age_end==age_death_x)
data_abcd[counter_pp,]<-c(age_start_x,age_end,age_end-age_start_x,death_abcd)
}
```

Exercise 17.10 *Fit a Gompertz model to the data generated in Example 17.3.*

17.6 Comparing Life Tables with a Parametric Formula

Perhaps we can represent an entire life table, such as *AM*92, with one simple formula including two parameter values? This would clearly be preferable as we could remove the need to hold an entire table of ages and l_x's.

Exercise 17.11 *Subject: the AM92 and AF92 life tables.*
The objective is therefore to fit the Gompertz model to the AM92 (am92.csv) and AF92 (af92.csv) life tables by applying linear regression as set out in Section 17.4. (Note: The reader will need to estimate the forces of mortality, μ_x, from the q_x's in the AM92 table, e.g. using Equation 17.1.)

Proposed solution to Exercise 17.11
A proposed model for the male table was found to be:

$$\mu_{age} = e^{-10.1971+0.0918 \times age}$$

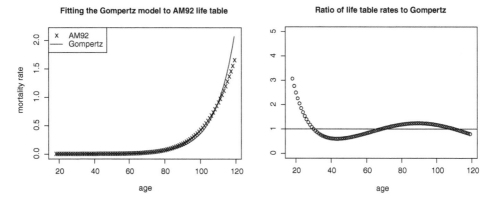

Figure 17.4   Comparing the Gompertz model to AM92 life table.

Figure 17.4 compares these "Gompertz" rates with the *AM*92 life table. From this analysis, the Gompertz model isn't a particularly good fit for the *AM*92 rates, although it may be considered a reasonable approximation over some age ranges.

Example 17.4

Calculate $\ddot{a}_{60}^{i=4\%}$ and $\ddot{a}_{40}^{i=4\%}$ using the Gompertz model fitted above for the *AM*92 mortality table, and Equation 17.1.

```
aaa1<- -10.197;          bbb1<-0.0918
mux1<-function(x)exp(aaa1+bbb1*x)
prob_gomp_92<-function(xx1,tt1){exp(1/bbb1*(mux1(xx1)-mux1(xx1+tt1)))}
deltayy<-log(1.04)
sum(prob_gomp_92(60,0:200)*exp(-deltayy*0:200))     #age 60 years

[1] 14.35531

sum(prob_gomp_92(40,0:200)*exp(-deltayy*0:200))     #age 40 years

[1] 19.83805
```

These compare with values of 14.134 (age 60) and 20.005 (age 40) using the *AM*92 life tables, l_x; the differences in these annuity values are consistent with Figure 17.4.

17.7   Splines: An Introduction

17.7.1   Overview

The final section in the chapter provides a brief introduction to splines. Various types of splines exist in the literature, such as cubic splines, B-splines, and P-splines – we shall restrict the discussion for this introduction to B-splines. (For a comprehensive introduction to splines, Micula (1999) is recommended.) Our objective is to demonstrate how we can fit a smooth curve to our crude rates using B-splines. The reader may wish to look ahead to Figure 17.7 to see an example of a fitted spline model.

As we have seen in earlier sections, the Gompertz model, and similar models, provide one function which is used to model our entire dataset. Gompertz, having only two parameters, is a particularly simple model. This idea was developed through the 20th century with several models incorporating further appropriate parameters for example, Makeham, Perks and Beard. However, by using a global function the rates at one age are determined, at least in part, by data far away from that age. An alternative approach is to fit several simple local functions, called splines, to several parts of the mortality curve. Thus splines may be preferred where the curve appears to consist of a number of idiosyncrasies which are best modelled using a number of distinct sections.

For example, we may wish to model the mortality rates of several different countries using the same underlying model. However some countries may exhibit particular characteristics, such as sudden changes in mortality rates at particular ages due to a particular social policies for e.g. children allowed to ride motorbikes at age 14 years, or an increase in healthcare at age 60 years. Splines will tend to cope better with such variations.

B-splines are piecewise polynomials; the basis functions are often referred to as "local" functions and take a value of zero outside defined domains. (See Carl de Boor, Chapter 10, for a detailed discussion on how these B-splines are determined; in particular, the recursion formulae.) Four examples are shown in Figure 17.5. When choosing the B-spline basis, the number and spacing of the "knots" must be chosen, depending on the data to which we are trying to fit our splines. Sufficient knots should be chosen to reflect the parts of the curve which appear idiosyncratic to some extent, and their spacing should be chosen accordingly. We must also decide on the degree of the polynomials to use i.e. linear (1), quadratic (2), or cubic (3). Cubic tend to be the most commonly used, and we will proceed with those here.

We will use the `splines` package in R, and in particular the `bs` function. For example:

```
bs(seq(1,100,1),knots = c(50,70),degree = 3)        #degree = 3 (default)
```

Here we have chosen two internal knots at 50 and 70. A variety of spline bases are plotted in Figure 17.5, using the code below; there is considerable flexibility with the `bs` function. The reader is encouraged to experiment with various alternatives:

```
xx<-seq(0,100,0.1)
matplot(xx,bs(xx),type='l',main="Default cub spline",ylab="",
xlab="Age")
matplot(xx,bs(xx,knots=c(40)),'l',main="Addnl knot at 40",ylab="",
xlab="Age")
matplot(xx,bs(xx,knots=seq(10,90,10),degree=3),type='l',
main="Nine additional knots",ylab="",xlab="Age")
points(seq(20,70,by=5),bs(xx,knots = seq(10,90,10),degree = 3)[
seq(201,701,by=50),5])
matplot(xx,bs(xx,degree=2),'l',main="Quad polynml",ylab="",xlab="Age")
```

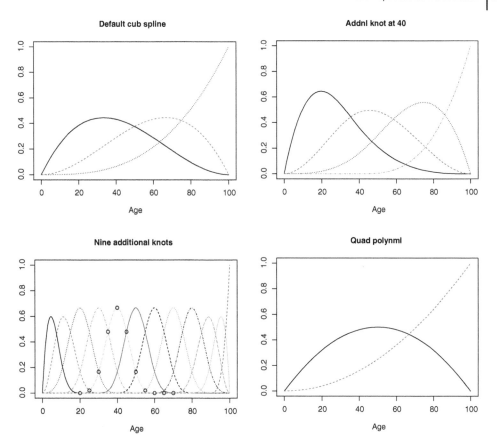

Figure 17.5 Chosen basis splines.

One can gain an understanding of the local nature of the basis functions by extracting the values of one of the functions in Figure 17.5 (denoted with ○):

```
bs(xx,knots = seq(10,90,10),degree = 3)[seq(201,601,by=50),5]  #(bottom-left)

[1] 0.00000000 0.02083333 0.16666667 0.47916667 0.66666667 0.47916667 0.16666667
[8] 0.02083333 0.00000000
```

The final model is determined by finding the best linear combination of basis splines which fits our data.

17.7.2 Data

The data for the example, which ranges from age 17 to 120 years, can be found here:

```
mydataexp<-read.csv("~/mort_table_splines1.csv")
log_ratesx<-log(mydataexp$mux);    agex<-mydataexp$age
```

17.7.3   Fitting the Model: Spline regression

For this example we shall use a basis spline with cubic polynomials and two additional knots at 40 and 60. The fitted regression, $\log \mu_t$, is obtained from:

$$\log \mu_t = \sum_{n=1}^{5} A_n B_{n,t} \tag{17.16}$$

where A_n is the vector of regression coefficients (`coeffs_A`), and $B_{n,t}$ are the basis splines (of which there are five in this case). The vector of regression coefficients (or "spline coefficients"), A_n, can be calculated using the following code:

```
(fit_bs<-lm(log_ratesx ~ bs(agex,knots = c(40,60))) )

Call:
lm(formula = log_ratesx ~ bs(agex, knots = c(40, 60)))

Coefficients:
                     (Intercept)   bs(agex, knots = c(40, 60))1
                        -7.28768                       -0.47322
bs(agex, knots = c(40, 60))2   bs(agex, knots = c(40, 60))3
                        -0.08095                        4.17849
bs(agex, knots = c(40, 60))4   bs(agex, knots = c(40, 60))5
                         6.66037                        7.13850
```

(This is similar to the method used in Section 17.4 for polynomial regression, where $\log \mu_t = \sum_n A_n^* x^n$.) For example, at age 60 years, $B_{1:5,60}$ is obtained from the following (see Figure 17.6):

```
bs(agex,knots = c(40,60))[44,1:5]
        1         2         3         4         5
0.0000000 0.4368932 0.5006068 0.0625000 0.0000000
```

The log of the mortality rate at age 60 years is given by $A_n B_{n,60}$, as follows (and highlighted in Figure 17.7):

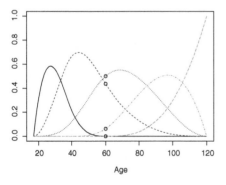

Figure 17.6   B(60) values.

Figure 17.7 Two knots at 40 and 60

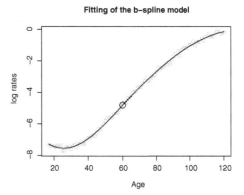

Fitting of the b-spline model

-7.287679 + -0.4732173 × 0 + -0.0809482 × 0.4368932 + 4.178494 × 0.5006068 + 6.6603695 × 0.0625 + 7.1385034 × 0 = -4.8149892

Applying Equation 17.16 across all ages, the fitted regression for $\log\mu$ can be obtained from the following code (rates for ages 60–65 years are shown):

```
t(bs(agex,knots = c(40,60))%*%fit_bs$coefficients[2:6])[44:49]+fit_bs$coefficients[1]
[1] -4.814989 -4.699067 -4.583945 -4.469650 -4.356205 -4.243637
```

Or more easily using the `predict` function (all log rates are plotted in Figure 17.7):

```
spline_rates<-predict(fit_bs,newdata = as.data.frame(agex));  spline_rates[44:49]
        44          45          46          47          48          49
-4.814989  -4.699067  -4.583945  -4.469650  -4.356205  -4.243637
```

Exercise 17.12 *The reader may find it instructive to re-plot the curve with adjusted coefficient values, for example, changing A_2 from -0.4732173 to 3, and A_6 from 7.1385034 to 2, to understand the effect of the coefficients.*

Exercise 17.13 *The reader should experiment with alternative sets of knots for example, at 25, 40, and 60.*

Exercise 17.14 *Calculate a_{60} with $\delta = 5\%$ pa and the spline model for mortality rates shown in Figure 17.7. Recalculate using a suitably fitted Gompertz model.*

Of course, we can calculate rates at smaller age intervals – here calculating log rates from ages 60 to 61 years at monthly intervals:

```
t(bs(seq(17,120,by=1/12),knots=c(40,60))%*%fit_bs$coefficients[2:6])[517:529]+
fit_bs$coefficients[1]
 [1] -4.814989 -4.805299 -4.795614 -4.785935 -4.776261 -4.766593 -4.756930
 [8] -4.747272 -4.737620 -4.727974 -4.718333 -4.708697 -4.699067
```

17.7.4   Adjusted Dataset

The reader is encouraged to fit splines to alternative data sets. For example, we have adjusted the above table of rates (`mort_table_splines1.csv`) to include a more pronounced hump between ages 10 and 20 years (Figure 17.8). Here we have used two approaches – one with three irregularly spaced knots, and another with 22 regularly spaced knots:

```
#first make up an alternative data set to fit
adj_fac<-c(seq(1,1.2,length=10),seq(1.2,1,length=10),rep(1,96))
rates_adj<-c(mydataexp$mux[12:1],mydataexp$mux)*adj_fac # our new adjusted rates
log_ratesx_adj<-log(rates_adj)
agex_adj<-c(seq(5,16),agex)

#fitting the B-spline
fit_adj<-lm(log_ratesx_adj ~ bs(agex_adj,knots = c(18,25,60)))    #3 knots
fit_adj2<-lm(log_ratesx_adj ~ bs(agex_adj,knots = seq(10,115,5)))   #22 knots
```

Clearly there is a balance to be struck between adherence to the data and adopting a reasonably smooth function. Splines with fewer knots will generally exhibit more smoothness, but increasing the number of knots usually increases the fit of the spline to the data.

Note how the spline method copes much better with this data set than the global Gompertz model, which would effectively fit a straight line to the log rates. However, the model in Figure 17.8 (right) is likely to be rejected – it has been over-fitted and produces a set of rates which do not progress smoothly between ages.

We have also included a fitted polynomial with a degree of 5 (see Remark 17.3) in Figure 17.8 (left), using the code set out below. The reader is encouraged to experiment with alternative models, for example, a polynomial with a degree of 6.

```
lm(log_ratesx_adj ~ poly(agex_adj, 5, raw = TRUE))    #poly with 5 degrees
```

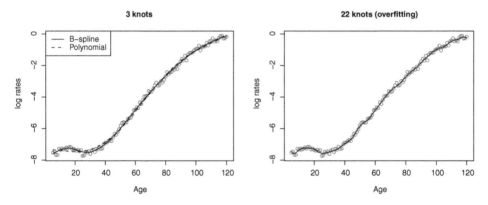

Figure 17.8   Pronounced accident hump.

Exercise 17.15 *Experiment with various B-splines in fitting simulations from a mixture distribution of* $N(100, 20^2)$ *and* $N(120, 10^2)$.

17.8   Summary

Our objective in this chapter was to fit mortality models given particular datasets. From these models we can estimate survival probabilities from Equation 17.1, and ultimately calculate, for example, insurer's premium rates and solvency positions, and pension schemes' required contribution rates and funding levels.

The method used will depend on whether we have detailed individual member data, or summarised membership data, and can assess the best fit by comparing logL's and information criteria. We must also assess whether the chosen model is appropriate by comparing the fitted rates with the crude rates, both visually and by undertaking a number of tests (see Chapter 18). If it is not we may decide to try alternative functions, or adopt alternative methods, such as splines. As noted throughout this chapter, some judgement is required when determining the most appropriate model.

17.9   Recommended Reading

- Carl de Boor. (1978). *A Practical Guide to Splines.* New York: Springer-Verlag.
- Forfar, D.O., McCutcheon, J.J., and Wilkie, A.D. (1988). *On graduation by mathematical formula. JIA* 115: 1–149.
- MacDonald, A.S. et al. (2018). *Modelling Mortality with Actuarial Applications.* Cambridge, UK: Cambridge University Press.
- Micula, Gheorghe and Sanda. (1999). *Handbook of Splines Mathematics and Its Applications.* (Kluwer Academic Publishers) Vol. 462. Dordrecht: Springer.
- Sweeting, P. (2017). *Financial Enterprise Risk Management*, 2e. Cambridge, UK: Cambridge University Press.
- *Continuous Mortality Investigations – working papers* (various). London: The Institute and Faculty of Actuaries.

18

Assessing the Suitability of Mortality Models: Statistical Tests

Peter McQuire

18.1 Introduction

```
library(stats4)
library(tseries)
```

In the previous chapter we undertook the process of determining a best-fitting, smooth function which represented our mortality rates given a particular set of mortality data. The objective of this chapter is to test these proposed rates to assess whether they are suitable, that is, do they represent the true mortality rates of the population from which the mortality data is obtained? We analyse each proposal using a series of statistical tests and thus decide on their suitability. We then consider which set of rates, out of all those which have "passed" our tests, is the most suitable.

Therefore, this chapter continues from the previous one where we fitted potentially suitable parametric models, such as the Gompertz model, to our mortality data. Here, we assess the suitability of the proposed model, and compare it with others (see Section 18.6).

We will be assessing proposed mortality rates produced from two graduation methods in this chapter:

- adjusting a standard mortality table which has previously been independently tested and verified;
- determine a suitable parametric model (this methodology was the subject of Chapter 17).

In reality, however, the actuary is first likely to adopt a different approach by testing the suitability of various standard mortality rate tables which currently exist in the literature, thus potentially saving us a great deal of work. If no suitable tables are found it is possible that a simple adjustment to one of these existing tables may still prove acceptable (the first method of graduation referred to above). The advantage of using these standard tables is

R Programming for Actuarial Science, First Edition. Peter McQuire and Alfred Kume.
© 2024 John Wiley & Sons Ltd. Published 2024 by John Wiley & Sons Ltd.
Companion Website: www.wiley.com/go/rprogramming.com.

that they will have been prepared from an extensive study involving a significant amount of data and analysis.

Please note that the objective of this chapter is to introduce a number of statistical tests which can be used to determine suitability – it is not to provide a comprehensive testing methodology. The interested reader should refer to the reading material at the end of the chapter, and in particular to Forfar et al. (1988) which provides a comprehensive introduction, including details of additional tests, for the serious student.

The objective of the final exercise in this chapter is to collate all the statistical tests and data under one function which outputs a summary of the results (see Exercise 18.11).

18.2 Theory

The central idea we are proposing, or hypothesising, is that the true underlying mortality rates of the population we are studying, μ_x, are equal to μ_x^*, the proposed mortality rates. The proposed rates we analyse in this chapter are one of three types:

- μ_x^S : A pre-existing table of mortality rates, e.g. S3PMA / S3PFA (Section 18.4);
- μ_x^{Sadj} : A set of rates obtained by applying a simple adjustment to a pre-existing table of mortality rates, e.g 99% of S3PMA / S3PFA (Section 18.5.2);
- μ_x^{par} : A set of rates defined by a parametric formula, e.g. Gompertz (Section 18.6).

An important part of our analysis is the calculation of the crude rates $\hat{\mu}_x$, the maximum likelihood estimate of the mortality rate at each individual age. From the previous chapter we have:

$$\hat{\mu}_x = \frac{d_x}{v_x}$$

where d_x and v_x represent the observed number of deaths and waiting time respectively (at age x). It is important to remember that $\hat{\mu}_x$ is an estimate obtained from the data, and is not a random variable.

Similarly the maximum likelihood estimator is given by:

$$\tilde{\mu}_x = \frac{D_x}{V_x}$$

where D_x and V_x are both random variables, representing the number of deaths and waiting time respectively (at age x).

As noted in the previous chapter, $\tilde{\mu}_x$ is asymptotically Normally distributed with mean μ_x and variance $\frac{\mu_x}{E(V_x)}$. That is, the asymptotic distribution is:

$$\tilde{\mu}_x \sim N\left(\mu_x, \frac{\mu_x}{E(V_x)}\right)$$

Our null hypothesis, H_0, is that the proposed rates, μ_x^*, are indeed the true underlying rates of our population, μ_x. Formally we write $H_0 : \mu_x = \mu_x^*$. Thus if our hypothesis is true, for practical purposes we can write (approximately):

$$\tilde{\mu}_x \sim N(\mu_x^*, \frac{\mu_x^*}{v_x})$$ (18.1)

and thus:

$$\frac{\tilde{\mu}_x - \mu_x^*}{\sqrt{\frac{\mu_x^*}{v_x}}} \sim N(0, 1)$$ (18.2)

We can test H_0 by calculating the key statistic, z_x, commonly referred to as the standardised deviation:

$$z_x = \frac{\hat{\mu}_x - \mu_x^*}{\sqrt{\frac{\mu_x^*}{v_x}}}$$ (18.3)

For example, at age 70 years, if $z_{70} > 1.96$ then we can conclude, with 95% confidence, that $\mu_{70}^* \neq \mu_{70}$. However given that we will have a large range of μ_x's to test, it is not particularity helpful to test each age – even if the table is appropriate our conclusion is likely to result in rejection of around 5% of the rates.

We will make use of Equations 18.2 and 18.3 in a series of statistical tests described in Section 18.4.

Remark 18.1 The standard approach involves modelling the number of deaths using a Poisson model, and formulating much of the material which follows in this chapter in those terms. That is $D_x \sim \text{Poisson}(\mu_x v_x)$ and hence, approximately : $D_x \sim N(\mu_x v_x, \mu_x v_x)$ giving the equivalent version of Equation 18.3:

$$z_x = \frac{d_x - \mu_x^* v_x}{\sqrt{\mu_x^* v_x}}$$ (18.4)

The reader may wish to re-formulate the material which follows in terms of $D_x \sim N(\mu_x v_x, \mu_x v_x)$ and Equation 18.4, comparing the two methodologies.

18.3 Our Mortality Data and Various Proposed Mortality Rates

Downloading our data and various sets of mortality rates, which we hope will be suitable:

```
stat_test_data<-read.csv("~/mortality_stats_tests.csv")
```

The first few rows of data are set out below:

```
head(round(stat_test_data,5),3)
```

```
  age  exposure  deaths  table1   table2   par_rates
1  20   62485     155    0.00113  0.00270  0.00243
2  21   62520     152    0.00121  0.00273  0.00246
3  22   62493     146    0.00128  0.00278  0.00250
```

The data extract includes three sets of proposed mortality rates: `table1`, `table2`, and `par_rates`. Note that the exposure period, or waiting period, is in years. Our objective is to determine which set of rates, if any, are suitable to use in modelling the underlying mortality rates of the population from which our data has been obtained, and whether we can adjust the standard tables in such a way to obtain alternative suitable rates.

We will assume that all data uses the "age nearest" definition. For example, from the top row of the data `stat_test_data`, there were 155 deaths aged between the exact ages of 19.5 and 20.5 years at date of death, i.e. at a life's date of death, they were (or would have been) exactly 20 years old on their nearest birthday. [Note: the formal definition of "age nearest" at any given time is the life's exact age on their nearest birthday.] The consequence of this definition is that the crude rate calculated at age x nearest is an estimate of the true rate at age x exact. The reader may wish to compare this with the "age last" definition employed in Chapter 17.

18.4 Testing the Standard Table Rates – Table 1, μ_x^{s1}

18.4.1 Data and initial plot

First we analyse the suitability of the mortality rates set out in the first pre-existing standard table, Table 1; we will be testing whether these standard table rates, μ_x^{s1}, adequately reflect the crude mortality rates calculated from our data. (Remember that these crude rates are, given the data, the best estimate of the actual mortality rates at each age.) The null hypothesis is therefore: $H_0 : \mu_x = \mu_x^{s1}$.

We'll define a number of vectors first, then calculate and plot the crude rates, comparing them with μ_x^{s1}:

```
agex<-stat_test_data$age_last;    exposure<-stat_test_data$exposure
deaths<-stat_test_data$deaths;    table1<-stat_test_data$table1

crude_rates<-deaths/exposure
```

Calculating the 95% confidence intervals:

```
mu_variance1<-table1/exposure
rateup<-table1+1.96*mu_variance1^0.5;    ratedown<-table1-1.96*mu_variance1^0.5
```

A plot of the crude rates and standard table rates is shown in Figure 18.1 – initial visual observations may conclude that the crude rates do broadly follow a similar pattern to that of the standard table. However:

- many points lie outside the 95% confidence limits, for example, between ages 55 years and 70 years,
- the crude rates appear to be of a different shape to the standard table; the crude rates tend to follow a more curved path.

The tests set out in the remainder of this chapter provide a more objective analysis.

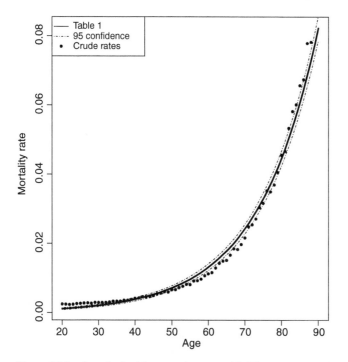

Figure 18.1 Standard table vs crude rates with 95 per cent confidence

18.4.2 χ^2 test

Our first statistical test is to analyse the complete set of rates such that we can form an initial judgement on the entire table. From Equation 18.2 (assuming that the data at each of the n ages are independent of each other):

$$\sum_{x=20}^{90} \left(\frac{\hat{\mu}_x - \mu_x^{s1}}{\sqrt{\frac{\mu_x^{s1}}{v_x}}} \right)^2 \sim \chi_n^2.$$

Carrying out this χ^2 test:

```
z_markov1<-(crude_rates-table1)/mu_variance1^0.5
z2_Markov1<-z_markov1^2;    sum(z2_Markov1)

[1] 1113.142

qchisq(0.95,length(z_markov1))    #71 d.o.f.

[1] 91.67024
```

Remark 18.2 When comparing crude rates with a standard table, the number of degrees of freedom to be used in the χ^2 test should be the number of age groups used; in this case there are 71 degrees of freedom i.e. $n = 71$.

As $1113.14 > 91.67$ we can clearly reject the hypothesis, with 95% confidence, that μ_x^{s1} represent the underlying mortality of our population; indeed we have much greater confidence than this. The crude rates calculated from our data are significantly different to the rates in this standard table, and we would certainly not recommend using such a table to model the mortality of our population. This appears consistent with Figure 18.1 – the standard table rates are too low at young and older ages, and too high between ages 45 to 80 years.

Problems with the χ^2 test

This χ^2 test is a useful, initial test in determining whether the proposed rates, in this case μ_x^{s1}, are potentially viable; however, it should certainly not be the sole test in determining whether they are appropriate. For example, all crude rates could be statistically very close to the standard table rates but all (or most) could be higher than the proposed rates, i.e. the rates may be biased. There may also be sub-ranges within our data which are biased, even though no overall bias exists.

The test could also, unfairly, result in rejection of H_0 due to just one crude rate being significantly different to the proposed standard table rate. These types of anomalies require further investigation into reasons for such differences. The following tests largely address these issues.

18.4.3 Signs Test – for Overall Bias

This simple test requires the calculation of the proportion of crude rates which are higher, and lower, than the proposed rates, and determines, using the Binomial distribution, how likely it is to obtain such a split. Given that $H_0 : \mu_x = \mu_x^{s1}$ we would typically expect half of the crude rates to be greater than, and half to be less than, the proposed rates μ_x^{s1}:

```
sum(z_markov1 > 0) # number of positive deviations
[1] 34
pbinom(sum(z_markov1 > 0),length(z_markov1),0.5)
[1] 0.4062943
```

As the p-value (40.6%) is greater than 2.5%, we have insufficient evidence to reject H_0 at the 95% level. Note that this is a 2-sided test – we are equally interested in rates being too high and too low.

Example 18.1

Given there are 71 ages, how many (or few) positive differences would we require such that we *would* reject H_0 at the 95% confidence level?

```
pbinom(seq(25,45),length(z_markov1),0.5)
```

```
 [1] 0.008497326 0.015963581 0.028407339 0.047961816 0.076956385 0.117548781
 [7] 0.171235499 0.238343897 0.317653821 0.406294325 0.500000000 0.593705675
[13] 0.682346179 0.761656103 0.828764501 0.882451219 0.923043615 0.952038184
[19] 0.971592661 0.984036419 0.991502674
```

Thus if we obtained ≤ 26 or ≥ 45 positive differences we would reject H_0.

18.4.4 Serial Correlations Test; Testing for Bias Over Age Ranges

Imagine the case where the χ^2 test and signs test have been used to analyse a set of mortality data, with both tests resulting in H_0 not being rejected; most of the crude rates could be acceptably close to the proposed rates, with close to half of them above (and below) the crude rates. However, perhaps all the crude rates at younger ages (under age 55 years) are slightly above the standard rates, and all the crude rates at older ages (above age 55 years) are below the standard rates? This would clearly be unsatisfactory. The test assesses this.

Exercise 18.1 *By observing Figure 18.1, anticipate the result of this test.*

As described by Forfar et al. (1988) p 46-47, approximately, for large m:

$$r \sim N(0, \frac{1}{m-1}), \text{ and therefore,} \quad r\sqrt{m-1} \sim N(0,1)$$

where r is the autocorrelation coefficient with lag $= 1$, and m is the number of age groups in the data; in this example $m = 71$. We have:

```
(r<-cor(z_markov1[1:70],z_markov1[2:71]))
```

```
[1] 0.9402485
```

```
(test_stat<-r*(length(agex)-1)^0.5);        round(qnorm(0.99),3)
```

```
[1] 7.866683
[1] 2.326
```

It's worth verifying the lag-1 sample autocorrelation, r:

```
z_markov1a<-z_markov1[1:70];    z_markov1b<-z_markov1[2:71]
sigma_a<-(z_markov1a-mean(z_markov1a));    sigma_b<-(z_markov1b-mean(z_markov1b))
sum((sigma_a)*(sigma_b))/(sum(sigma_a^2)*sum(sigma_b^2))^0.5
```

```
[1] 0.9402485
```

As $7.87 \gg 2.326$, we can reject this table of rates, μ_x^{s1}, with a high level of confidence; the z's have a high level of serial correlation.

Exercise 18.2 *Another common test applied to test for bias appearing over particular periods of ages is commonly referred to as "the runs test" or "groupings of signs test". Please see Bradley(1968), and Forfar et al. (1988) Section 9.4.*

18.4.5 Analysing the Distribution of Deviances

It is also worth comparing the distribution of standardized deviances, z_x, with that expected from a Normal distribution (see Equations 18.2 and 18.3). For example, one would expect around 68% of z's to be within one standard deviation of zero, and around 4.5% to be further than 2 standard deviations away from zero.

```
#Standard Normal distribution
(norm_vec<-c(1- pnorm(2), pnorm(2)-pnorm(1), pnorm(1)-pnorm(0),
pnorm(1)-pnorm(0),pnorm(2)-pnorm(1),1- pnorm(2)))

[1] 0.02275013 0.13590512 0.34134475 0.34134475 0.13590512 0.02275013

#Actual distribution of standardized deviances
c(sum(z_markov1< -2),sum((z_markov1< -1)*(z_markov1> -2)),
sum((z_markov1<0)*(z_markov1> -1)),sum((z_markov1< 1)*(z_markov1> 0)),
sum((z_markov1< 2)*(z_markov1> 1)),sum(z_markov1> 2))/length(z_markov1)

[1] 0.35211268 0.08450704 0.08450704 0.08450704 0.04225352 0.35211268
```

Only 16.9% of z's are within one standard deviation of zero, with 70.4% further than 2 standard deviations from zero.

Carrying out a Kolmogorov-Smirnoff test is also a useful test here:

```
ks.test(z_markov1, "pnorm")  #Kolmogorov-Smirnoff test


One-sample Kolmogorov-Smirnov test

data:  z_markov1
D = 0.3386, p-value = 9.136e-08
alternative hypothesis: two-sided
```

Clearly the crude rates do not appear to be distributed in line with Equation 18.2.

18.4.6 logL, AIC Calculations

The loglikelihood ("logL") of the data should be calculated and noted for later comparison with alternative candidate models. AIC can also be used to compare and rank models. The relevant discussion is included in Section 18.7.

We have calculated two logL's below (which should not be compared which each other) – one using the Markov model (from Chapter 16, $\log L = \sum_x (-v_x \mu_x + d_x \log \mu_x)$), and the other using the Normal approximation (Equation 18.1):

```
(logL_table1<-sum(-table1*exposure+deaths*log(table1)))    #using the Markov logL
[1] -259483.5
sum(log(dnorm(crude_rates,table1,(table1/exposure)^0.5)))    #using Eqn 1.1
[1] -71.70067
```

In Section 18.7 we compare the first of these values, and the AIC, with those obtained from alternative proposed rates analysed throughout this chapter to hopefully determine the most suitable set of mortality rates. The Appendix compares calculations of logL using the Normal distribution (the 2nd calculation of logL above, based on Equation 18.1).

Note that loglikelhood and AIC calculations tell us little about the absolute quality of a model, they only help in comparing models. Hence the importance of the statistical tests described earlier in this chapter.

18.4.7 Conclusions on μ_x^{s1}

The conclusion from the above battery of tests is that the mortality rates set out in Table 1, μ_x^{s1}, are not appropriate to represent the mortality of our population. We should at this point try various alternative standard mortality tables, using experience and judgement, with the hope of finding an appropriate table. However it may be the case that no such standard tables currently exist which fit our data, requiring us to proceed with the process of graduation. One option is to adjust the standard table rates in such a way that it more closely resembles the mortality rates of our population. This is the focus of the next section.

Exercise 18.3 *Create your own sets of mortality rates, with various characteristics, and perform the statistical tests described above.*

18.5 Graduation of Mortality Rates by Adjusting a Standard Table

18.5.1 Testing Table 2, μ_x^{s2}

As noted above, the initial standard table tested ("Table 1") was simply the wrong shape compared to our mortality data. We proceed to analyse rates from an alternative standard table ("Table 2"), μ_x^{s2}, shown in Column 5 in the dataset.

Exercise 18.4 *Plot a similar graph to that in Figure 18.1 using "Table 2", and perform the statistical tests set out in Section 18.4 on this alternative table, setting out your conclusions.*

The reader should conclude that these alternative rates, μ_x^{s2}, are also unsuitable; on this occasion the rates are in general too high, but do appear to be of a similar shape to the crude rates (see Figure 18.2). Perhaps this set of rates can be adjusted such that they are sufficiently close to the crude rates?

18.5.2 Adjusting Table 2

By using a simple proportionate adjustment, β, we can propose a set of rates $\mu_x^* = \mu_x^{s2adj} = \beta\mu_x^{s2}$, such that $H_0 : \mu_x = \beta\mu_x^{s2}$. We can determine β using maximum likelihood estimation:

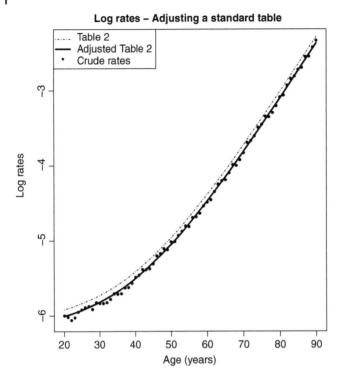

Figure 18.2 Graduation using a standard table

```
table2<-stat_test_data$table2

lnlik<-function(beta1){(sum(-(table2*beta1)*exposure+deaths*log(table2*beta1)))*-1}
beta_mle<-mle(lnlik,list(beta1=1.0),lower=0.8,upper = 1.1)
summary(beta_mle)@coef[1];    summary(beta_mle)@m2logL

[1] 0.9146005
[1] 518016.6
```

The logL can be confirmed as follows (note this is half of the above figure):

```
(logL_table2_adj<-(sum(-(table2*summary(beta_mle)@coef[1])*
exposure+deaths*log(table2*summary(beta_mle)@coef[1]))))

[1] -259008.3
```

Here we have determined, using maximum likelihood estimation, that applying an adjustment factor, $\beta = 0.9146005$ will result in the best fit. So we have $\mu_x^{s2_{adj}} = 0.9146005\mu_x^{s2}$. We have shown the log-rates, $\log(0.9146005\mu_x^{s2})$, in Figure 18.2, which tend to be easier to analyse visually, particularly at lower ages. This is the process of graduation – we have determined, or at least proposed at this stage, a set of smooth rates, which may be seen to be a good fit when compared to the crude rates.

We must now proceed to test the suitability of these adjusted rates as described in Section 18.4.

Remark 18.3 When performing a χ^2 test for graduated rates using a standard table, it is not a straightforward task to determine the number of degrees of freedom which should be used. When running χ^2 tests involving a standard table graduation in this chapter we will assume that the number of degrees of freedom should be reduced by 5, plus one for each parameter used. Thus in Exercise 18.5, 65 degrees of freedom should be used.

Exercise 18.5 *Carry out the above statistical tests on these graduated rates,* $\mu_x^{s2_{adj}}$, *setting out your conclusions.*

Conclusion to Exercise 18.5: all tests described in Section 18.4 result in H_0 not being rejected – we have insufficient evidence to reject the adjusted Table 2 rates, $\mu_x^{s2_{adj}}$.

Exercise 18.6 *Try a different adjustment to the alternative standard table rates of* $H_0 : \mu_x = \beta_{add} + \mu_x^{s2}$.

Exercise 18.7 *Obtain* $\mu_x^{s1_{adj}}$ *by adjusting the standard rates in Table 1 (analysed in Section 18.3) using similar methodology to that set out above. The reader should compare logL's, and re-run the tests.*

Example 18.2

As an alternative to determining the MLE parameters using the Markov model, we can alternatively use the Normal approximation to determine β (see Equation 18.1).

```
lnlik_abc<-function(adj1){
(sum(log(dnorm(crude_rates,table2*adj1,(table2*adj1/exposure)^.5))))*-1}   #Normal
zzzc_abc<-mle(lnlik_abc,list(adj1=1.0),lower=0.8,upper = 1.1)
summary(zzzc_abc)@coef[1];    summary(zzzc_abc)@m2logL

[1] 0.914468
[1] -909.7036
```

18.6 Testing Graduated Rates Obtained from a Parametric Formula, μ_x^{par}

As a further alternative, we could graduate our crude rates using a parametric formula, and test its suitability using the tests described above. The fitting of a parametric graduation was discussed in detail in Chapter 17 and is not repeated here (the reader is invited to fit an appropriate function). Such a method of graduation is usually reserved for very large datasets. For example, the production of standard mortality tables will be undertaken using this method, where the underlying data is obtained from aggregated life insurance or pension scheme data, or from otherwise available national statistics.

The 6th column in our data, "par rates", is a set of graduated rates (for our purposes we can assume that a Gompertz model was fitted) obtained using the method set out in Chapter 17. These graduated rates are denoted here by μ_x^{par}. So here, $H_0 : \mu_x = \mu_x^{par}$.

Exercise 18.8 *Plot a similar graph to that in Figure 18.1 using these Gompertz rates, and carry out the statistical tests described above on these rates.*

Remark 18.4 When performing a χ^2 test for graduated rates using a parametric formula, the number of degrees of freedom which should be used equals the number of age ranges less the number of parameters determined in the fitting process. In this example of the Gompertz model, two parameters are required, thus the number of degrees of freedom should be 69.

The reader should find that none of the tests resulted in the rejection of H_0; as such these proposed rates, μ_x^{par}, may also be a candidate for our final model.

18.7 Comparing Our Candidate Rates

We should, at this juncture, compare the various sets of rates analysed to this point. We will do this by comparing logL's and AIC's. (The logL from Table 1 rates was briefly discussed in Section 18.4.6.) Proceeding to compare logL's of the rates analysed up to this point:

```
#table 1
logL_table1   #calculated earlier

[1] -259483.5

#table 2 (unadjusted)
(sum(-(table2*1.0)*exposure+deaths*log(table2*1.0)))

[1] -259237.5

#table 2 (adjusted)
logL_table2_adj

[1] -259008.3

#parametric model
(logL_g<-(sum(-(gompertzrates)*exposure+deaths*log(gompertzrates))))

[1] -259007.2
```

Clear winners are the adjusted table 2 rates, $\mu_x^{s2_{adj}}$, and fitted Gompertz rates, μ_x^{par}, with comfortably the highest logL scores. Of course, there is little point comparing the logL's of those proposals which have been rejected by the statistical tests – these are only included here to demonstrate the consistency between logL values and quality of fit.

Which of these two models, $\mu_x^{s2_{adj}}$ or μ_x^{par}, is preferable? There is only one parameter included in the adjusted table, compared with two parameters in the Gompertz model – comparing AIC's is useful here:

```
(aic_adj<-2*(1-summary(beta_mle)@m2logL/(-2)))    #adjusted standard table 2

[1] 518018.6

(aic_gom<-2*(2-logL_g))                           #parametric model

[1] 518018.5
```

The AIC's suggests there is little to choose between the two models.

18.8 Over-fitting

An analysis of the distribution of standardized deviances z_x's which results in significantly more $|z_x| < 1$ than expected, and/or a low χ^2 score, may indicate over-fitting and a set of rates which do not progress smoothly across the age-range.

Exercise 18.9 *10-degree polynomial Fit the crude rates using a 10-degree polynomial, and analyse the distribution of the deviances z_x.*

Smoothness of the progression in the rates can also be quickly assessed by taking the 2nd differences of the proposed rates; imagine the case where proposed rates are the crude rates (an extreme example of over-fitting):

```
diff2_crude<-diff(diff(crude_rates))/crude_rates[1:69]
diff2_adj<-diff(diff
(table2*summary(beta_mle)@coef[1]))/table2[1:69]*summary(beta_mle)@coef[1]
```

(See Figure 18.3). This analysis can highlight any issues with over-fitting; clearly the crude rates do not represent a sufficiently smooth set of rates.

18.9 Other Thoughts

It may not be a straightforward decision as to which set of rates should be chosen following completion of the modelling and testing exercise; in our example it appears that either $\mu_x^{s2_{adj}}$, or the rates obtained from the parametric, "Gompertz" graduation, μ_x^{par}, would be acceptable.

Also where there is only a small amount of mortality data available, the range of suitable mortality rates may be substantial given the large variances due to the small amount of data. For example, if we have a pension scheme consisting only of a few hundred members, it is likely that a wide variety of standard tables with adjustments would be acceptable, and any detailed analysis may be considered spurious.

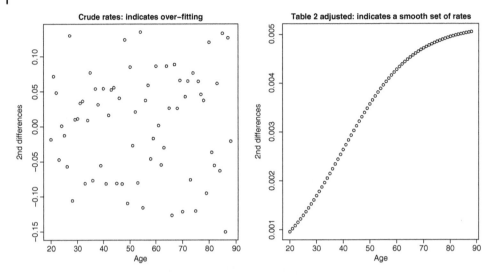

Figure 18.3 2nd differences: Assessing smoothness and over-fitting

Often, in practice, the mortality rates used will be set out in relevant regulations, either by government or by professional institutions or regulators.

The reader is encouraged to explore the various additional tests which may be applied to proposed mortality rates in deciding on a suitable set of rates. As noted several times throughout the chapter, the reader may benefit from a review of Forfar et al. (1988).

Exercise 18.10 *A useful exercise would be to reduce/increase the exposure and death data in the data file extract by a factor of 100, rounding the resultant data to nearest integer values, and repeating the tests in this Chapter.*

Exercise 18.11 *Write a function which applies a series of statistical tests to a proposed set of mortality rates and mortality data, and outputs a summary of key results, effectively incorporating much of the material in this chapter.*

18.10   Appendix – Alternative Calculations of LogL's

From Equation 18.1, $\tilde{\mu}_x \sim N(\mu_x^*, \frac{\mu_x^*}{v_x})$

```
#table1
sdqqq1<-(table1*1/exposure)^0.5
sum(log(dnorm(crude_rates,table1*1,sdqqq1)))

[1] -71.70067

#table2
sdqqq2<-(table2*1/exposure)^0.5
sum(log(dnorm(crude_rates,table2*1,sdqqq2)))
```

```
[1] 231.4121

#table2 adjusted
sdqqq2adj<-(table2*summary(beta_mle)@coef[1]/exposure)^0.5
(sum(log(dnorm(crude_rates,table2*summary(beta_mle)@coef[1],sdqqq2adj))) )

[1] 454.8512

(AIC1<-2-2*(sum(log(dnorm(crude_rates,table2*summary(beta_mle)@coef[1],sdqqq2adj))))))

[1] -907.7025

#Gompertz
sdqqqpar<-(gompertzrates/exposure)^0.5
(sum(log(dnorm(crude_rates,gompertzrates,sdqqqpar))) )

[1] 456.0761

(AIC2<-4-2*(sum(log(dnorm(crude_rates,gompertzrates,sdqqqpar))) ))

[1] -908.1521
```

And from Remark 18.1, $D_x \sim N(\mu_x \upsilon_x, \mu_x \upsilon_x)$, first we calculate the adjustment parameter, calculated previously in Section 18.5.2 (0.9146005) and Exercise 18.2 (0.914468) using this model:

```
lnlik_d<-function(adj2){(sum(log(dnorm(deaths,table2*adj2*exposure,
(table2*adj2*exposure)^.5))))*-1}
mle_d<-mle(lnlik_d,list(adj2=1.0),lower=0.8,upper = 1.1)
summary(mle_d)@coef[1]; summary(mle_d)@m2logL

[1] 0.914468
[1] 631.9718
```

And proceeding with the logL's:

```
#table1
sum(log(dnorm(deaths,table1*exposure,(table1*exposure)^0.5)))

[1] -842.5384

#table2
sum(log(dnorm(deaths,table2*exposure,(table2*exposure)^0.5)))

[1] -539.4256

#table2 adjusted
(sum(log(dnorm(deaths,table2*summary(mle_d)@coef[1]*exposure,
(table2*summary(mle_d)@coef[1]*exposure)^0.5))) )

[1] -315.9859

(AICx<-2-2*(sum(log(dnorm(deaths,table2*summary(mle_d)@coef[1]*exposure,
(table2*summary(mle_d)@coef[1]*exposure)^0.5)))))

[1] 633.9718

#Gompertz
(sum(log(dnorm(deaths,gompertzrates*exposure,(gompertzrates*exposure)^0.5))) )
```

```
[1] -314.7617

(AICx<-4-2*(sum(log(dnorm(deaths,gompertzrates*exposure,
(gompertzrates*exposure)^0.5)))))

[1] 633.5233
```

These results are consistent with those carried out throughout this chapter.

18.11   Recommended Reading

- Bradley, JV (1968). *Distribution-Free Statistical Tests*. Prentice-Hall.
- Forfar, D.O., McCutcheon, J.J., and Wilkie, A.D. (1988). *On graduation by mathematical formula*. JIA 115, pp. 1–149.
- MacDonald, A.S., Richards, S.J., and Currie, I.D. (2018). *Modelling Mortality with Actuarial Applications*. Cambridge, UK: Cambridge University Press.

19

The Lee-Carter Model

Peter McQuire

```
library(demography)
```

19.1 Introduction

In Chapters 16 and 17 we developed parametric models to analyse age-dependent mortality rates at a point in time. An example of the model is shown earlier in Figure 17.2. Such a model may be adequate if we are concerned with mortality rates in the immediate future, for example pricing one-year term assurance policies.

However, a major issue with these models is that they make no allowance for changes in the level of mortality rates with time. Mortality rates may change significantly in the future due to, for example, medical advances or increases in national obesity; by not projecting these rates we are implicitly assuming they will remain constant in the future.

Mortality rates in the developed world have generally been decreasing since population data records started to be collected (see Figure 19.1). In the 19th century UK life expectancy was around 40 years; currently UK life expectancy is in excess of 80 years. The improvements have been evident even in recent times, with UK infant mortality rates falling from 0.94% to 0.39% since the mid-80s (*Source: HMD*).

It is therefore vital to understand these trends and construct suitable models such that reasonable forecasts can be made about future mortality rates, allowing actuaries to calculate insurance premiums and pension scheme liabilities with more accuracy. The Lee-Carter model ("L-C model") was the first serious model to attempt to do this. There have since been several proposed models attempting to improve on the shortcomings of the L-C model, such as Cairns-Blake-Dowd, p-spline models, and CMI models which the reader is encouraged to study.

The objective of this chapter is to demonstrate how the Lee-Carter model can be used to analyse recent trends in mortality rates and thus estimate future rates, such that financial products can be valued more accurately. We will do this by:

R Programming for Actuarial Science, First Edition. Peter McQuire and Alfred Kume.
© 2024 John Wiley & Sons Ltd. Published 2024 by John Wiley & Sons Ltd.
Companion Website: www.wiley.com/go/rprogramming.com.

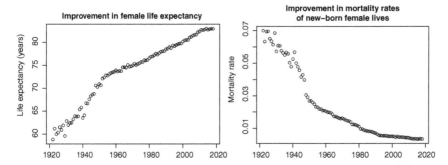

Figure 19.1 Changes in UK mortality from 1922 to 2018 : Source – the Human Mortality Database.

1. Creating fictitious mortality rates using the L-C model and then fitting the model, using Singular Value Decomposition ("SVD").
2. Fit the L-C model to actual mortality rates, using SVD and also the `lca` function from the `demography` package.
3. Estimate projected future mortality rates from the fitted L-C model.
4. Undertake a guided exercise to highlight some of the shortcomings of the L-C model.

19.2 Using the L-C Model to Create Data and Fit the Model

19.2.1 Introducing the Lee-Carter Model

We start by creating a set of fictitious mortality rates using the L-C model itself, for the period 1980 to 2012 covering ages 50 to 59 years – this will produce a 33×10 matrix of mortality rates. The mortality rates are functions of both age, x, and calendar year, t. The mortality rate, $\mu_{x+0.5,t+0.5}$, in the L-C model is given by:

$$\mu_{x+0.5,t+0.5} = e^{a_x + b_x k_t^T} \tag{19.1}$$

where, in this case:

- a_x is a column vector of length 10, for ages 50 to 59
- b_x is a column vector of length 10, for ages 50 to 59
- k_t is a column vector of length 33, for calendar years 1980 to 2012.

Remark 19.1 Note that the model is often presented using central mortality rates, $m_{x,t}$, where $m_{x,t}$ can be interpreted as the probability weighted average hazard rate over the chosen year of age. The crude rates (or original exposure and death data used to calculate these) on which the model is applied should be clearly defined when determining the age at which the crude rate is a best estimate. Any age labels used should be carefully noted. Throughout this chapter we assume data is grouped by "age last birthday" and by calendar year e.g. data collated in respect of "58-year olds" in 2017 relate to a best estimate of $\mu_{58.5,1\,July\,2017}$. This is revisited later in the chapter.

We will look at the interpretation of these vectors shortly. First we will create a set of rates; the values chosen for our three vectors are as follows:

```
a<-seq(-5.7,-4.8,length.out = 10)
b<-seq(-.3,-.5,length.out = 10)
k<-seq(-1,1,length.out = 33)
```

Including a degree of noise to produce realistic crude rates:

```
set.seed(100)
stoch_err<-matrix(rnorm(330,0,0.005),nrow=10,byrow=TRUE)

data_rates<-exp(a+b%*%t(k)+stoch_err) # these are the fictitious rates
```

A plot of these crude rates is shown in Figure 19.2; a selection of four calendar years are shown – 1980, 1990, 2000, and 2010. (Also shown in the plot are the fitted rates, obtained from the L-C model, which is the aim of this section.)

The objective is to use the Lee-Carter model to fit all 33 years of our data, by determining appropriate parameter values for a_x, b_x, and k_t. Ultimately we will want to forecast future mortality rates; we leave forecasting till later in the chapter when we analyse actual mortality rates.

19.2.2 Calculating the Parameter Values

There are several methods available to find values for a_x, b_x, and k_t. First we shall use Singular Value Decomposition (SVD) by applying the svd function. (For a detailed explanation of SVD please see Brunton, Kutz (Feb 2019).) Our aim here is to fit the model – we will simply proceed with the algorithm.

SVD allows us to equate any matrix, M, with the product of three matrices, U, D, and V:

$$M = UDV \qquad\qquad (19.2)$$

Taking logs of both sides, Equation 19.1 can be re-written:

$$\ln \mu_{x+0.5,t+0.5} - a_x = b_x k_t \qquad\qquad (19.3)$$

From Equation 19.2, $M = \ln \mu_{x+0.5,t+0.5} - a_x$, $U = b_x$ and $DV = k_t$. Note that without any constraints an infinite number of combinations of parameter values is possible. Many sets of constraints are possible, one of which is:

$$\sum b_x^2 = 1 \qquad \sum k_t = 0$$

This set of constraints is used in the svd function, which we apply in this section. Alternatively, the constraints used by Lee and Carter in their 1992 paper were as follows:

$$\sum b_x = 1 \qquad \sum k_t = 0$$

This second set of constraints is applied when using the lca function later in the chapter (Section 19.4). For completeness, code is also included later in this section which applies this second set of constraints using the svd function. The effect of which constraints are used is not a particular concern in the context of this chapter – the second constraints are usually preferred as they lead to a simpler interpretation of the output.

a_x can be obtained immediately; it is simply the average of the log rates over all T years, as shown below (here at age 58):

$$\ln \mu_{x+0.5,t+0.5} = a_x + b_x k_t$$

$$\sum_{x=58.5,t+0.5} \ln \mu_{58.5,t+0.5} = T\, a_{58} + b_{58} \sum_t k_t$$

$$a_{58} = \frac{\sum_{x=58.5,t+0.5} \ln \mu_{58.5,t+0.5}}{T}$$

Thus:

```
log_rates<-log(data_rates)#the first term above on LHS
ax<- rowMeans(log_rates)#the second term above on LHS
M<-log_rates-ax #LHS of the equation
```

M is the matrix to which we apply SVD to obtain U, D, and V. We will then have values for b_x and k_t. Applying SVD we have:

```
svd_lc<-svd(M)
```

U, D, or V are not shown here, given their size, but the reader may wish to view these now to verify their dimensions i.e. 10×10, 10×33, and 33×33, respectively.

```
svd_lc$u;  svd_lc$d;   svd_lc$v
```

Now we simply use the first vectors from U and V, and the first eigenvalue in D to provide a unique solution (this method is known as "truncated SVD"):

```
uu<-svd_lc$u[,1] #these are the b's

dd<-svd_lc$d[1];    vv<-svd_lc$v[,1]
dv<-dd%*%vv     #these are the k's
```

Alternatively we can incorporate this in the svd function, as follows:

```
svd_lc_alt<-svd(M,nu=1,nv=1)
```

Applying Equation 19.2 we obtain our L-C rates, plotted in Figure 19.2, together with the original rates:

```
udv<-uu%*%dv
log_lc_rates<-ax+udv
lcrates<-exp(log_lc_rates)    #the LC fitted rates
```

Figure 19.2   Fitted rates using L-C model.

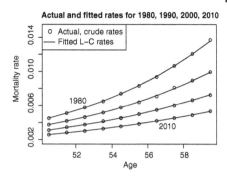

For example, for 2000:

```
lcrates[,21] #2000

 [1] 0.003109954 0.003418355 0.003754161 0.004125218 0.004538827 0.004991495
 [7] 0.005477680 0.006024471 0.006628555 0.007284581
```

We have verified the calculation for a 54-year-old in 2000 below:

$$\mu_{54.5,2000.5} = e^{-5.2997412+0.3038317\times(-0.3138101)} = 0.0045388$$

The crude rate was 0.0045287 (shown in bold in Figure 19.2).

Using the Lee-Carter methodology we can therefore model several years of historic mortality rates using just three vectors, and Equation 19.1.

(Alternative methods are commonly used to fit the Lee-Carter model, the most popular of which is using Poisson maximum likelihood estimation (see Brouhn et al. 2002).)

Included below is sample code which uses the alternative constraints referred to above i.e. $\sum b_x = 1$ (as noted earlier in this section).

```
svd_lc_u_new<-apply(svd_lc$u,2,function(x){x/sum(x)})
dd_all<-svd_lc$d
dd_full<-diag(10)*dd_all
dvx<-dd_full%*%t(svd_lc$v)

udv_new<-svd_lc_u_new[,1]%*%t(dvx[1,]*colSums(svd_lc$u)[1])
log_lc_ratesx<-ax+udv_new
lcratesx<-exp(log_lc_ratesx)

#check constraints
sum(svd_lc_u_new[,1])   #sum to one

[1] 1

sum(t(dvx[1,]*colSums(svd_lc$u)[1])) #sum to zero

[1] -2.54935e-14
```

19.2.3 Interpretation of $a_x, b_x,$ and k_t

It is important to understand the influence of each of the terms in the Lee-Carter model (Equation 19.1).

a_x is effectively the average log mortality rate at age x over the range of years in the data.

k_t adjusts the a_x for each calendar year, t. If we assume, for now, that rates have been gradually reducing over the period of our investigation, then e^{a_x} is representative of the mortality rates in some "middle year" in the data (as it is an average). The k_t values must therefore span a range in both negative and positive numbers.

However, with only a_x and k_t in the model all our mortality curves would be of a similar shape. b_x provides further flexibility to the model by allowing mortality rate improvements to vary by age. For example rates may not change for 100-year-old lives but reduce significantly for 60-year-old lives (or vice versa). So b_x allows our curve to change shape as we move forward in time.

Exercise 19.1 *The reader may wish to vary the entries in the vectors $a_x, b_x,$ and k_t to study the effect on the pattern of mortality rates, and re-fit the Lee-Carter model. For example:*

```
a<-seq(-4.5,-3.5,length.out = 10)
b<-seq(-0.5,+0.05,length.out = 10)
k<-seq(-2,+2,length.out = 33)
```

19.3 Using L-C to Model Actual Mortality Data from HMD

We now fit the Lee-Carter model to actual mortality data. We use the Human Mortality Database ("HMD") for the source of our data. The reader must also load the demography package.

First we download our data from the HMD website; the reader is required to register and submit an email address (username or login) and password:

```
uk1<-hmd.mx("GBR_NP", login, password, "UK")
```

This code extracts all the UK data in HMD:

```
Mortality data for UK
    Series: female male total
    Years: 1922 - 2018
    Ages:  0 - 110
```

Note that you may receive a "warning" message; this is not the same as an error. It means here that some parts of the data are missing – the user should take care when analysing results.

Here we have chosen to analyse UK data for the years 2000–2018 and ages 40–90 years.

```
uk1_sub<-extract.years(uk1,years = 2000:2018)
uk1extract<-extract.ages(uk1_sub,ages=40:90, combine.upper =FALSE)
uk1extract

Mortality data for UK
    Series: female male total
    Years: 2000 - 2018
    Ages:   40 - 90
```

It is important to note at this point that this data is held in a `demogdata` object (created from the above code):

```
class(uk1extract)

[1] "demogdata"
```

We shall briefly discuss this object type later in the chapter. The data can be viewed using, for example (the output is not shown here – it's huge):

```
head(uk1extract);   uk1extract$rate$male;   uk1extract$pop
```

Viewing a sample of the data is straightforward:

```
uk1_4090_18_leecarter<-uk1extract$rate$male
(uk1_4090_18_leecarter[1:4,1:7])

         2000      2001      2002      2003      2004      2005      2006
40 0.001682 0.001700 0.001692 0.001688 0.001605 0.001573 0.001569
41 0.001886 0.001821 0.001843 0.001694 0.001656 0.001748 0.001756
42 0.001843 0.001980 0.001891 0.001966 0.001770 0.001881 0.001862
43 0.002025 0.002037 0.002235 0.002248 0.002110 0.002033 0.002004

class(uk1_4090_18_leecarter)   #not a demogdata object now

[1] "matrix" "array"
```

It is important to note that the rates above are estimates of the mortality rate at age $x + \frac{1}{2}$ e.g. $\hat{\mu}_{42+\frac{1}{2}}^{2003+\frac{1}{2}} = 0.001966.$

(Details can be found in the HMD documentation, in particular in the "Methods Protocol" on the HMD website – the "age last birthday" definition is used for both the exposure date and death data). The reader should also note that when the data is manipulated as above it is no longer a `demogdata` object, and the code from earlier in the chapter can be applied to it.

To obtain the L-C vectors of parameters we can simply re-run the code from above to obtain the Lee-Carter fitted model.

Exercise 19.2 *Obtain the Lee-Carter parameters for this data set and plot a sample of graphs.*

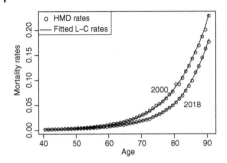

Figure 19.3 HMD and L-C fitted rates.

The HMD UK rates and L-C fitted rates from Exercise 19.2 in respect of 2000 and 2018 are shown in Figure 19.3. The model appears to be working very well. It's also clear that male UK mortality rates have improved significantly between the years 2000 and 2018.

Remark 19.2 It is worth noting the potential data issue at ages 80–82 in the year 2000, clearly visible in Figure 19.3:

```
(uk1_4090_18_leecarter[38:45,1])

     77       78       79       80       81       82       83       84
0.063338 0.071232 0.077169 0.091691 0.091798 0.104329 0.117361 0.128108
```

19.4   Using the *lca* Function in the *Demography* Package

As the reader may expect, there are several "Lee-Carter" functions within R which already exist, which make it much simpler to produce the "Lee-Carter" rates. For example, we can use the lca function from the demography package; to apply the lca function the data must be held in a demogdata object (see later for details on creating such an object from your own data). The above data, uk1extract, extracted from the HMD database, is automatically set up as such an object, and we can apply the fitted function to produce log rates for the required year (here for 2018 male rates):

```
class(uk1extract)

[1] "demogdata"

leecarter_uk1_male<-lca(uk1extract,series="male")    #find L-C parameters
fitted_all<-fitted(leecarter_uk1_male)        #fitted rates
fitted_rates_18<-exp(fitted_all$y[,19])       #fitted rates in 2018
```

These modelled rates, obtained using lca, and the actual rates for the year 2018 are plotted in Figure 19.4. The reader is invited to compare the L-C rates calculated using the lca function with those calculated using our code from earlier in the chapter.

Exercise 19.3 *Apply the Lee-Carter model for male, Italian lives for 1950 to 2017 over ages 20 to 100 years (from HMD). This particular exercise highlights common issues experienced at older ages.*

Figure 19.4 Fitted rates using the lca function.

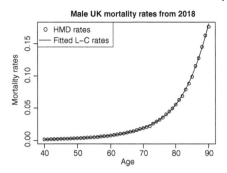

Note that we can obtain the Lee-Carter vectors a_x, b_x, and k_t using:

```
leecarter_uk1_male$kt ; leecarter_uk1_male$bx ; leecarter_uk1_male$ax
```

We leave it to the reader to fit further sets of mortality rates from HMD and elsewhere. The reader may wish to explore the relative rates of improvement in mortality rates in various countries.

19.5 Constructing Your Own *Demogdata* Object

Of course the reader may have their own mortality data to which they want to apply Lee-Carter's model. To use the functions in the demography package, data must be saved as a demogdata object, using the read.demogdata function, and use appropriate .*txt* files. The .*txt* files used below were downloaded from the *HMD* website by simply clicking on the required file. *The reader is advised to study these data files below to understand the required format.* The reader may wish to plot the k_t's to demonstrate the lca function does indeed work in this case (not shown here). This also demonstrates how straightforward it is to obtain output from a L-C analysis in R:

```
demog_own<-read.demogdata("~/uk_rates1.txt","~/uk_exposure1.txt",type="mortality",
label="uk")    #produces a demogdata object
class(demog_own)    #confirmation of demogdata object
plot(lca(demog_own,series="male")$kt)
```

19.6 Forecasting Mortality Rates

The main purpose of developing this model is to forecast future mortality rates. We proceed by analysing US mortality data from the HMD, for no other reason than to use alternative data to UK mortality data used earlier. Indeed we would recommend the reader to repeat the analysis in this chapter on data from various countries.

```
usa<-hmd.mx("USA", login, password, "USA")
```

```
Mortality data for USA
      Series: female male total
      Years: 1933 - 2019
      Ages:  0 - 110
```

We will analyse data only for male lives (ages 0 to 100 years) from the period from 1933 to 1990, and use Lee-Carter to forecast rates to 2018. This will allow us to assess the success of the model in this instance, by comparing these rates with the actual 2018 rates.

```
usa_1990<-extract.years(usa,years = 1933:1990)    #choose years
usa_1990_100<-extract.ages(usa_1990,ages=0:100)   #choose ages
usa_1990_100

Mortality data for USA
      Series: female male total
      Years: 1933 - 1990
      Ages:  0 - 100
```

Applying the lca function to obtain the L-C vectors as before:

```
leecarter_usa_male_1990<-lca(usa_1990_100,series="male")#find parameters
```

Exercise 19.4 *Compare the rates from the HMD database with the rates produced under the L-C model for various years between 1933 and 1990.*

Exercise 19.5 *Plot the values obtained for a_x, b_x, and k_t.*

The k_t values are plotted in Figure 19.5 using the following code; they imply that mortality rates, in general, fell over the periods from 1940 to 1955, and 1970 to 1990, with little overall improvement between 1955 and 1970.

```
kt<-leecarter_usa_male_1990$kt;    plot(kt)
```

We must now project these k_t values to future years i.e. beyond 1990. (Under the L-C model the a_x and b_x values are not functions of time.) In Lee and Carter's 1992 paper

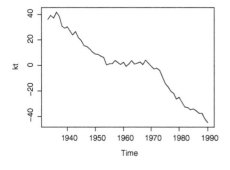

Figure 19.5 k parameter values.

the k_t's were modelled assuming a discrete random walk with drift i.e. an ARIMA(0,1,0) model:

$$k_{t+1} = k_t + \mu^d + \epsilon_{t+1} \tag{19.4}$$

where the drift, μ^d, is the average change in k_t over the period observed (in this case between 1933 and 1990), and ϵ_t are IID random variables from $N(0, \sigma^2)$. Note that μ^d should not be confused with the mortality rate μ.

Obtaining estimates of the parameters:

```
auto.arima(kt)

Series: kt
ARIMA(0,1,0) with drift

Coefficients:
         drift
       -1.4188
s.e.    0.3354

sigma^2 estimated as 6.525:  log likelihood=-133.83
AIC=271.66    AICc=271.89   BIC=275.75
```

It is straightforward to verify the drift, the standard error of the drift, and σ^2 (all of which are calculated above). The drift, μ^d, is simply the average change in k_t:

```
(drift<-mean(diff(kt)))

[1] -1.418833
```

An estimate of σ^2 (the variance of ϵ) can be verified from:

```
(sigma2<-sum((diff(kt)-drift)^2)/(length(kt)-2))

[1] 6.525227
```

and the standard error of μ^d:

```
(se<-sigma2^0.5/(length(kt))^0.5)

[1] 0.335416
```

(See MacDonald et al. (2018) for a further discussion of these estimates.)

Hence there are two key sources of uncertainty included the model – parameter error in μ^d and stochastic error (represented by σ).

Now to project this trend into the future. The forecast function calculates projected k_t values and mortality rates; here we have projected these over the next 28 years (from 1991 to 2018) by setting $h = 28$ (note that jumpchoice = fit is the default, which forecasts from the LC fitted rates in the last year; alternatively use jumpchoice = actual to forecast from the latest actual rates):

```
forecast_backtest<-forecast(leecarter_usa_male_1990,h=28,jumpchoice="fit")
plot(forecast_backtest$kt,ylab="k")
```

The plot of projected k_t's is shown in Figure 19.6.

Prediction intervals for each year (also included in Figure 19.6) can be estimated assuming a Normal distribution and incorporating the variances of both the drift estimator (0.1124695) and stochastic errors (6.525252), giving, for example, at $t = 1$, 6.6377215. Note that the forecast function produces prediction intervals with 80% as the default; the level argument should be included to choose an alternative prediction level e.g. level= 90.

For example, the 80% confidence range for k for the year 2018 ($t = 28$) is calculated as follows:

```
tt=28
drift*tt+qnorm(.9)*(tt^2*se^2+tt*sigma2)^0.5

[1] -18.63386

drift*tt-qnorm(.9)*(tt^2*se^2+tt*sigma2)^0.5

[1] -60.82081
```

A plot of the projected mortality (log) rates from 1991 to 2018 is shown in Figure 19.7. (Also shown in Figure 19.7 are the actual rates in 1990 for comparison – the black line.)

```
plot(forecast_backtest) #future log rates
```

Exercise 19.6 *By analysing Figure 19.7 what can we say about the values of* b_x? *Extract the values of this vector to verify your analysis.*

From Figure 19.7 we can see the model is predicting that, from 1990, the mortality rates at younger ages will reduce considerably, but that the rates at older ages will generally remain relatively constant. The model has detected a pattern of generally improving mortality rates between 1933 and 1990, but with the most improvement coming from the younger ages.

We can obtain the range of predicted rates, with prediction intervals, for 2018:

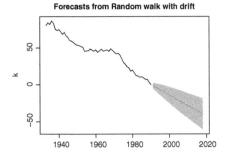

Forecasts from Random walk with drift

Figure 19.6 Projected k's from 1990.

Figure 19.7   Projection of rates to 1991–2018.

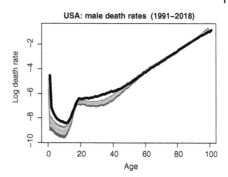

USA: male death rates (1991–2018)

```
proj_rates<-forecast_backtest$rate$male[,28]      #point estimate
est_rates_up<-forecast_backtest$rate$upper[,28]   #upper
est_rates_low<-forecast_backtest$rate$lower[,28]  #lower
```

These are plotted in Figure 19.8. It is interesting to compare the predicted and actual rates in 2018 (also included in Figure 19.8), remembering we are using data from before 1990. The *actual* mortality rates from year 2018 can be obtained from:

```
usa.9018<-extract.years(usa,years = 1990:2018)#choose years
usa.100<-extract.ages(usa.9018,ages=0:100, combine.upper =FALSE)#choose ages
```

The model would have underestimated the improvement in mortality rates at older ages, but overestimated the improvement of rates below the age of 50 years. Figure 19.9 compares the actual 1990 rates and the actual 2018 rates.

It is worth replicating the projected rates, here carried out for age 1 year for 1991:

```
forecast1991<-leecarter_usa_male_1990$ax+
leecarter_usa_male_1990$bx*(leecarter_usa_male_1990$kt[58]+drift)
(exp(forecast1991))[1]#our calculation

          0
0.008580078

forecast_backtest$rate$male[1,1]#agrees!

[1] 0.008580078
```

Figure 19.8   Comparison of 2018 actual and predicted rates.

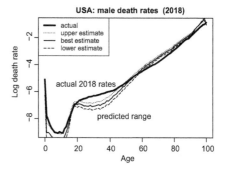

USA: male death rates (2018)

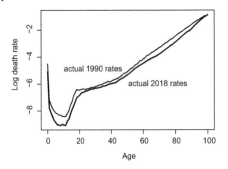

Figure 19.9 Comparison of actual rates from 1990 and 2018.

Exercise 19.7 *Obtain drift parameters for k in respect of Italy, France, Spain, and Japan data over similar periods.*

Exercise 19.8 *Project mortality rates for male and female lives using UK (HMD) life tables, and calculate a_{60} for a life born in 1960 (using an effective annual rate of interest of 2%pa).*

Thus we can produce estimates of future mortality rates which should help us analyse the solvency of pension funds and insurance companies more realistically.

Conclusions on the Model's Findings

The model predicted an overall reduction in mortality rates, with a greater improvement in younger ages. This can be seen from the b and k vectors. However the mortality rates at the older ages did actually reduce significantly between 1990 and 2018. From Figure 19.9 there appears to be a fairly consistent improvement in actual mortality rates across all ages, whereas the model predicted more significant improvements at the younger ages. The model effectively overestimated the improvements at younger ages and underestimated the improvements at older ages.

Several mortality projection models have been developed since Lee and Carter published their paper, such as the Cairns-Blake-Dowd model, and the Lee-Miller method. See Haberman and Renshaw (2011) for more details.

Exercise 19.9 *Under what circumstances may trends be ignored when modelling mortality rates?*

Exercise 19.10 *Calculate b_x values by analysing male and female mortality both pre and post 1990. What initial conclusions may be drawn from these results?*

19.7 Case Study: The Impact of the HIV Virus on Mortality Rates

The following example highlights how important patterns in mortality rates may be missed using the relatively simple L-C model. We will analyse how US male mortality rates between 1985 and 2005 changed at various ages; of particular interest are the rates for ages between 30 and 40 years.

First let's plot the mortality rates from 1985, 1995, and 2005 (the data was extracted earlier in the chapter).

```
plot(usa, year = c(1985,1995,2005), series = "male",col=c(1,1,1),
lty=c(1,2,3))
points(x = 35, y = -6, pch = 1, cex = 15)
legend(x = "topleft", legend = c("1985","1995","2005"), lty=c(1,2,3))
```

Figure 19.10 highlights an anomaly in the early 1990s around the ages of 30 to 40 where mortality rates appear to have worsened between 1985 and 1995 (but have improved at all other ages).

Exercise 19.11 *The reader should extract data for males rates between ages 30 and 40 years between 1985 and 2005, and apply the L-C model.*

Tabulating the rates in 1985 and 1995 from the original data shows the increase in mortality rates for this age group over this period:

```
rate_ratio<-cbind("1985"=usa_90_hiv3040$rate$total[,5],
"1995"=usa_90_hiv3040$rate$total[,15],
ratio=usa_90_hiv3040$rate$total[,15]/usa_90_hiv3040$rate$total[,5])
```

```
      1985      1995     ratio
30 0.001220 0.001394 1.142623
31 0.001291 0.001511 1.170411
32 0.001298 0.001588 1.223421
33 0.001382 0.001717 1.242402
34 0.001408 0.001748 1.241477
35 0.001460 0.001844 1.263014
36 0.001588 0.001930 1.215365
37 0.001632 0.002093 1.282475
38 0.001888 0.002242 1.187500
39 0.001898 0.002262 1.191781
40 0.002151 0.002414 1.122269
```

However, analysing the change in mortality rates over the entire age range (0 to 110 years of age) fails to pick up the effect of the HIV virus on mortality rates, although there is

Figure 19.10 Changes in US mortality rates from 1985 to 2005.

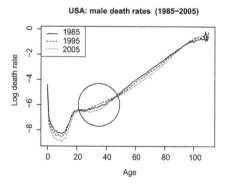

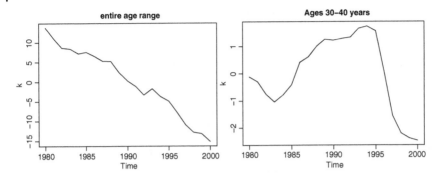

Figure 19.11 k vector analysis.

a blip in 1993 which may indicate an anomaly. The plots in Figure 19.11 show that when the entire age range is included in the analysis there appears to be a general improvement in mortality rates. However, the second plot clearly shows mortality rates in the 30 to 40-year-age group worsening between 1985 and 1995.

The higher rate of deaths in this age range is not highlighted from the results of the model when we look at the entire age range. This is due to the much greater number of deaths at older ages dominating the calculation; the mortality rates for males over 60 years are falling, and those rates dominate due to the larger number of deaths at those ages.

19.8 Recommended Reading

- Booth, H., and Tickle, L. (2008). *Mortality modelling and forecasting: a review of methods*, Australian Demographic and Social Research Institute Working Paper no. 3, Canberra, Australian National University
- Brouhns, N., Denuit, M., and Vermunt, J.K. (2002). *A Poisson Log-Bilinear Regression Approach to the Construction of Projected Life Tables.* Insurance: Mathematics and Economics, 31, 373-393.
- Brunton, Kutz (2019). *Data-Driven Science and Engineering.* Cambridge, UK: Cambridge University Press.
- Cairns, Blake, Dowd. (2008). *Modelling and Management of Mortality Risk:A Review*, Scandinavian Actuarial Journal, 2008, 2-3, 79-113.
- Charpentier A. et al. (2016). *Computational Actuarial Science with R.* Boca Raton, Florida, US: Chapman and Hall/CRC.
- *The CMI Mortality Projections Model.* CMI 2012. Working Paper 63 (Feb 2013).
- S. Haberman and A. Renshaw. *A comparative study of parametric mortality projection models.* Insurance: Mathematics and Economics, Volume 48, Issue 1, Jan 2011.
- R. Lee and L. Carter. *Modeling and Forecasting U.S. Mortality*, Journal of the American Statistical Association, Vol. 87, No. 419, Sep 1992.
- R. Lee. (2000). *The Lee-Carter Method for Forecasting Mortality, with Various Extensions and Applications.* North American Actuarial Journal, 4(1), 80-91, Jan 2000.

- MacDonald A.S. et al. (2018). *Modelling Mortality with Actuarial Applications.* Cambridge, UK: Cambridge University Press.
- Richards S.J. and Currie I.D. (2009). *Longevity risk and annuity pricing with the Lee-Carter Model.* British Actuarial Journal , Volume 15, Issue 2, July 2009.
- Sweeting, P. (2017) *Financial Enterprise Risk Management 2nd edition*, Cambridge, UK: Cambridge University Press.

20

The Kaplan-Meier Estimator

Peter McQuire

```
library(survival)
```

20.1 Introduction

In 1958 Edward Kaplan and Paul Meier published their paper "Nonparametric estimation from incomplete observations" which set out details of their famous non-parametric survival model. Their paper remains one of the most cited in medical survival research.

The objective of the model is to produce a non-parametric distribution of the survival function. An example of a typical survival function that may be obtained from a Kaplan-Meier ("K-M") analysis is shown in Figure 20.1 (left); the plot shows the survival distribution obtained from a study of lung cancer patients, together with 95% confidence limits.

Based on the results from a K-M analysis we can make estimates about the probability that a life will survive for a period of time. For example, from Figure 20.1 (left) we may estimate that a patient, with a particular set of characteristics, has a 41% probability of

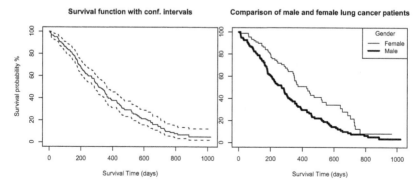

Figure 20.1 K-M survival function of lung cancer patients.

R Programming for Actuarial Science, First Edition. Peter McQuire and Alfred Kume.
© 2024 John Wiley & Sons Ltd. Published 2024 by John Wiley & Sons Ltd.
Companion Website: www.wiley.com/go/rprogramming.com.

surviving one year. We may also wish to compare the survival probabilities of two or more groups of lives, such as male/female, or where one group may have been administered with a recently developed medicine. Analysing two K-M curves can help us do this (see Figure 20.1 (right)). We will return to Figure 20.1 later in the chapter.

Remark 20.1 The simple `plot` function is used to obtain the K-M plots throughout this chapter.

The appeal of the K-M model is mainly two-fold. Firstly, it is a non-parametric model meaning that neither parameter estimation nor model fitting is required, thus potentially simplifying the process. Secondly, the model allows for the inclusion of censored data which may otherwise have been discarded, thus making the data set larger and the results more credible. Censored data is often highly significant in survival studies – these are the observations to which Kaplan and Meier referred to as "incomplete" in their paper heading.

20.2 What Is Censoring?

A major problem often encountered in survival analysis is that of how to deal with censored data. Censoring refers to incomplete or lost data, such that the time of a particular event in which we were interested (e.g. death) is unknown. For example when undertaking a medical study on lives with a life-threatening illness, patients may be monitored for a number of months; however it may not be possible to know when each patient died due to a number of reasons, such as:

- the patient may leave, or may not return to, the hospital for weekly checks such that the doctor is unaware if the patient has died or not (known as "lost to follow-up");
- the patient may die from another cause not relevant to the study e.g. car accident;
- the study may draw to a close before all patients have died e.g. financial funding was available such that doctors could study patients for only six months – if the patient died after seven months this would not be recorded under the investigation.

In all these cases where censoring occurs, the time at which the patient died, or may have died, from the illness would not be known.

With K-M methodology we use information on lives prior to the time at which the lives were censored and include this in our data and calculations. Although we do not know the time of the relevant event, we do know that it did not happen in a particular period of time and we can use this information to improve the quality of our analysis. Thus we do not discard all data in respect of our censored lives but can use it *up to the point of censoring* of the individual.

20.2.1 Non-Informative Censoring

The critical assumption made under Kaplan-Meier is that those lives which are censored exhibit the same characteristics as those lives which were not censored. This is known as "non-informative" censoring. The following is an example where informative censoring may be present.

Example 20.1 A number of patients are taking part in a medical study to investigate the effect of a new surgical procedure on mortality rates. Following surgery, a small number of these patients decide not to attend follow-up meetings with the doctor because they are feeling much better. It may be that these lives exhibit significantly lower levels of mortality than patients who do return for their weekly check-ups. Thus the healthy lives who do not return are censored and future data in respect of those lives is not recorded, leaving us with data consisting only of lives which are in relatively poor health i.e. our data is now biased and not representative of the original population. We are likely to conclude that the mortality rate is higher than the true underlying rate, and may, as a result, cease to use the new (possibly improved) surgical procedure. This is an example of informative censoring (which would make such an analysis invalid).

Example 20.2 To illustrate the usefulness of K-M with respect to censored lives, let us imagine a second study consisting of 1,000 lives spanning one year. The investigation is analysing the survival times of these lives, all of which have contracted a potentially deadly disease. 200 lives were observed to die from the disease within the first six months, such that we have 800 lives remaining in our study. At that point 550 of these lives subsequently leave the study of their own choice. The remaining 250 lives are observed for the final six months. The problem now arises of how to deal with the 550 lives which left the study; we do not know if they survived or died, and therefore we may need to remove them from our study, resulting in a study sample size now of only 450. This seems a pity as our results will be less credible given the smaller sample size (450 vs 1,000), particularly as we do know these 550 censored lives survived the first six months, information which must be of some use to us. The Kaplan-Meier method allows us to use this censored data in our analysis. (This assumes that the 550 lives which did leave are similar to the remaining lives i.e. "informative censoring" is not present.)

Exercise 20.1 *If those 550 lives which did leave at six months were all healthy (they left the hospital because they felt well) comment on how our conclusions may be incorrect if we assume non-informative censoring.*

20.3 Defining the Relevant Event

When conducting K-M calculations it is important to be clear exactly what our event of interest is under a particular study. The relevant event depends on the nature and objectives of the investigation. Examples of relevant events are:

- Death from a specific disease (if the life dies from another cause e.g. car accident, this would be censoring)
- Death from any cause
- Laboratory mice being cured from an illness
- Passing an exam
- Payment of an insurance claim

Once we have clearly defined our event of interest every other type of event or exit from the investigation is a censoring event.

20.4 K-M Theory

We provide below a brief review of the theory behind the K-M model. (For a full review of K-M methodology see Beyersmann et al. (2012), Chapters 2 and 4.) The data we require to carry out a K-M analysis are:

- the total number of lives observed under the investigation, both at $t = 0$ and at each subsequent monitoring of the position,
- whether each life which left the investigation did so by the defined event of interest or by a censoring event,
- the times when the position is monitored (in addition, the data must be time-ordered).

It is important to note the way in which data is collected may vary between investigations. For example we may start with 1,000 lives at $t = 0$ and monitor these lives each day for six months. Alternatively it may be that we accept new lives into the investigation throughout the six-month investigation period. The key measurement in both cases is the time spent by each object/life between entering the investigation and leaving the investigation. It is not necessary for the observation on all lives to start at the same time.

The Kaplan-Meier hazard rate estimate in respect of the jth event of interest at time t_j is given by the following expression:

$$\hat{\lambda}_{t_j} = \frac{d_{t_j}}{n_{t_j}} \tag{20.1}$$

where d_{t_j} is the number of non-censoring events (i.e. events of interest) which are observed at time t_j, and n_{t_j} is the number of lives present immediately before time t_j.

The maximum likelihood estimate of the probability of surviving to time t_i is given by the following:

$$\hat{S}(t_i) = \prod_{j=1}^{i}(1 - \hat{\lambda}_{t_j}) \tag{20.2}$$

(Note that $\hat{S}(t)$ is constant between events as λ_t is assumed to be equal to 0 at all times when there is no event occurring.)

Clearly the more frequently the position is monitored throughout the investigation the more accurate our analysis will be; the decision about frequency will be based on a cost-benefit analysis, together with the practicalities of increased monitoring.

We state below, without explanation, Greenwood's well-known estimator for the variance of $\tilde{S}_t$ at the time of the ith event:

$$Var(\tilde{S}_{t_i}) = (\hat{S}_{t_i})^2 \sum_{j=1}^{j=i} \frac{d_j}{n_j(n_j - d_j)} \tag{20.3}$$

The variance is constant between t_i and t_{i+1} as no further data is obtained during this time; thus we have an estimate of the variance at all times in the investigation.

20.5 Introductory Example: Monitoring Delays in Making Claim Payments

We first look at a simple example which uses a small data set. This is a slightly unusual example as the data are not living entities – we are analysing insurance claims. Our second example (see Section 20.6) uses actual data from a lung cancer study (two plots from this study were included in the Introduction).

An insurer is investigating how long it takes to settle its claims. The date of reporting each claim is recorded together with the date when the claim was paid by the insurer. Only claims reported between 31 Dec 2017 and 31 Dec 2018 are used in the data. Here the censored events arise because the analysis was over one calendar year and some reported claims had not been paid by 31 December 2018 (so we do not know when, or if, the claim was paid). The data collected in this introductory example are all month-end dates to simplify our calculations. The event of interest here is the payment of the claim.

Downloading our data:

```
datasimple<-read.csv("~/Kaplan_Meier_simple.csv")
```

```
  Claim Date.of.claim date.settled time status
1    1      12/31/2017     3/31/2018    3      2
2    2       1/31/2018     9/30/2018    8      2
3    3       1/31/2018    10/31/2018    9      2
4    4       2/28/2018    10/31/2018    8      2
5    5       6/30/2018     9/30/2018    3      2
6    6       6/30/2018    11/30/2018    5      2
7    7       6/30/2018     7/31/2018    1      2
8    8       6/30/2018           na    6      1
9    9       8/31/2018           na    4      1
```

Time is the period of time between the date a claim was reported and the date of payment. *Status* provides the censoring information, with "1" representing a censored event and "2" representing the event of interest i.e. in this case payment of a claim.

Exercise 20.2 *Carry out manual K-M calculations on this data using Equations 20.1 and 20.2.*

Functions in the Survival package allow for straightforward calculations of the Kaplan-Meier survival function – we will make use of those here. First we select the data required using the Surv function – under a Kaplan-Meier analysis the only data required is the time between the observation starting and the event, and whether it was a censoring event or not. We also must note how an event of interest is flagged (here with a "2"):

```
(survobjsimple <- with(datasimple, Surv(time,status==2)))
[1] 3  8  9  8  3  5  1  6+ 4+
```

A "+" sign indicates a censoring event (here at times 4 and 6). This is all the information that is required to carry out the K-M calculations. We now apply the survfit function to our newly created data object to obtain our K-M results:

```
calculations <- survfit(survobjsimple~1,conf.int = 0.95)
summary(calculations)    #produces the typical table from KM analysis

Call: survfit(formula = survobjsimple ~ 1, conf.int = 0.95)

 time n.risk n.event survival std.err lower 95% CI upper 95% CI
   1      9       1    0.889   0.105       0.7056        1.000
   3      8       2    0.667   0.157       0.4200        1.000
   5      5       1    0.533   0.173       0.2821        1.000
   8      3       2    0.178   0.156       0.0318        0.995
   9      1       1    0.000    NaN           NA           NA
```

Remark 20.2 Details relating to the calculation of the confidence intervals can be found in the "survival package" vignette, which can be found using:

```
browseVignettes(package = "survival")
```

(Note that the 95% confidence interval is the default value; this argument is not strictly required here.)

These are the Kaplan-Meier results; the column "survival" gives the estimated probabilities of survival. For example using Equations 20.1 and 20.2:

$$S(8) = \frac{8}{9} \times \frac{6}{8} \times \frac{4}{5} \times \frac{1}{3} = 0.178$$

So, for example, the estimated probability that a claim has not been paid within six months is 53.3%, and all claims are expected to have been settled by nine months. We can extract this information from summary:

```
x<-summary(survfit(survobjsimple~1), times = c(6,9))
x$surv
```

```
[1] 0.5333333 0.0000000
```

The K-M results are plotted in Figure 20.2 (using plot(calculations)). As one would expect from such a small dataset these confidence intervals are very large, making any conclusions from this dataset unreliable.

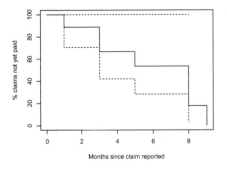

Figure 20.2 K-M plot of claims processing delays.

Exercise 20.3 *Starting with the original data set, increase the size of the data by a factor of 100 to obtain a bigger dataset. A Kaplan-Meier analysis on the new data set should show the same results as above but with much narrower confidence bands. As you have more data you can be more certain of the results.*

Now we've seen how to manipulate data and calculate the Kaplan-Meier survival function using a very simple data set, let's move on to something more realistic.

20.6   Lung Cancer Example

20.6.1   Basic Results

In this example we will analyse the survival probabilities of patients with lung cancer. The data is from a study of patients with advanced lung cancer undertaken by the North Central Cancer Treatment Group (Loprinzi et al. (1994)). A key part of the analysis involves the comparison of various K-M plots for different homogenous groups of lives e.g. comparing males and females, comparing patients with significant and mild symptoms as measured by their doctor.

The dataset is available within the Survival package. There is also a help document, help(lung), which is useful in understanding the data e.g. censored events defined as "1", females defined as "1", explanation of "ph.ecog" etc. The reader should first get a basic understanding of the data (output not shown here given the size of the data):

```
head(lung);  summary(lung)
```

Now create our K-M object using the Surv object, and carry out the same calculations as in our first example:

```
survobj_lung <- with(lung, Surv(time,status))
head(survobj_lung)

[1]   306   455  1010+  210   883  1022+

lungfit0 <- survfit(survobj_lung~1, data=lung)  #~1 - we are looking at all the data
```

The results are plotted in Figure 20.1 at the start of this chapter. I have not produced the full results output here given its size. Extracting survival probabilities at chosen times:

```
(survival_prob <- summary(lungfit0,time=c(100,200,300,400,500))$surv)

[1] 0.8639690 0.6802729 0.5306081 0.3768171 0.2932692
```

However, this analysis almost certainly contains too much heterogeneity. We may want to analyse, for example, the survival function for males and females separately, or the survival functions for patients with particular symptoms on diagnosis. It is not particularly helpful to inform a patient, who has a particular set of characteristics, of the prognosis of the "average outcome".

20.6.2 Comparison of Male and Female Rates

How do survival rates compare between males and females? We can analyse this by applying the following code:

```
lungfitgender <- survfit(survobj_lung~sex,data=lung)      #comparing genders
```

The plot is shown in Figure 20.1 (right) at the start of this chapter; it indicates that males with lung cancer tend to die earlier than women. We can test for any statistically significant difference between male and female mortality rates using the *logrank test*, with the surfdiff function:

```
survdiff(survobj_lung~sex, data=lung)

Call:
survdiff(formula = survobj_lung ~ sex, data = lung)

         N Observed Expected (O-E)^2/E (O-E)^2/V
sex=1 138      112     91.6      4.55      10.3
sex=2  90       53     73.4      5.68      10.3

 Chisq= 10.3  on 1 degrees of freedom, p= 0.001
```

Based on the above p-value we have sufficient evidence to reject the hypothesis that male and female lung cancer patients have the same mortality rates.

20.6.3 Doctor Assessment Scores – ph.ecog

It would certainly be useful to understand how good a doctor's initial assessment of the patient is in determining the patient's prospects. This is given by "ph.ecog" ; these are doctor's rating scores of 0, 1, 2, or 3 where, for example, 0 = "asymptomatic", and 3 = "in bed over 50% of the time". The K-M analysis is carried out below, together with the relevant plot (Figure 20.3):

```
table(lung$ph.ecog) #number of patients at t=0 with various assessments

  0   1   2   3
 63 113  50   1

fit1 <- survfit(survobj_lung~ph.ecog,data=lung)
```

A visual inspection of the plot suggests ph.ecog is a reliable covariate.

Exercise 20.4 *Test the credibility of ph.ecog scores using the logrank test and* surfdiff *function.*

However the plot does highlight some issues with this covariate e.g. a ph.ecog score of 3 initially has the best survival rate, then has the worst after 118 days; however there is only one patient with a score of 3, so should probably be removed from the analysis.

Exercise 20.5 *Carry out a K-M analysis on the GBSG2 dataset found in the* TH.data *package (note that "cens = 1" is an event), and determine whether hormone therapy may be regarded as a valid medical treatment for breast cancer.*

Figure 20.3 Survival Distributions using doctor score.

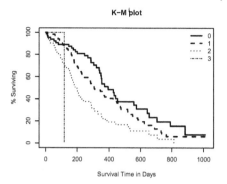

20.7 Issues with the Kaplan-Meier Model

K-M is an extremely useful methodology for survival studies, as can be demonstrated from its widespread and consistent use over recent decades (particularly in medical research fields). However it does have a number of limitations.

The original data will often contain significant heterogeneity which, ideally, we should address by stratifying our data i.e. split our original data into smaller, more homogenous sets of data. However, this could lead to results which carry less statistical credibility. For example, one can imagine that the insurance claims in our first example above come from one small, homogenous line of policies; the insurer could alternatively analyse all its claims together thus improving the credibility of the results, but at the risk of unacceptable heterogeneity.

As seen in the lung cancer example above, where we have a large number of covariates it can be difficult to make sensible comparisons; the issues are exacerbated when covariates can take a large number of values. Effectively each homogenous group has its own Kaplan-Meier curve. K-M methodology is therefore best applied when analysing one covariate which has perhaps 2 or 3 values, such as gender or ph.ecog score. Alternative regression models, such as the Cox model, are likely to be preferred if an analysis of several covariates is required (see Chapter 21).

As noted above, all censoring mechanisms which exist in an investigation need to be carefully studied and the impact of any non-informative censoring understood before reaching any conclusions.

20.8 Recommended reading

- Andersen P., Borgan Ø., Gill R., and Keiding N. (1993). *Statistical Models Based on Counting Processes*. Springer, New York.
- Jan Beyersmann, Arthur Allignol, Martin Schumacher. (2012). *Competing Risks and Multistate Models* with R. Springer, New York.
- Charpentier A et al. (2016). Computational Actuarial Science with R. Chapman and Hall/CRC.

- Crawley MJ. (2013). *The R book* (2nd edition). Wiley.
- Everitt BS and Hothorn T. (2006). *A Handbook of Statistical Analyses* using R. Chapman and Hall/CRC.
- Kaplan and Meier. (1958). Non-parametric estimation from incomplete observations. *JASA* Vol 53, 457-481
- Loprinzi CL. Laurie JA. Wieand HS. Krook JE. Novotny PJ. Kugler JW. Bartel J. Law M. Bateman M. Klatt NE. et al.. (1994). Prospective evaluation of prognostic variables from patient-completed questionnaires. North Central Cancer Treatment Group. *Journal of Clinical Oncology*. 12(3):601-7.
- MacDonald AS et al. (2018). *Modelling Mortality with Actuarial Applications*. Cambridge University Press.

21

Cox Proportionate Hazards Regression Model

Peter McQuire

```
library(survival);   library(dplyr)
```

21.1 Introduction

In this chapter we will look at one of the most cited statistical models used to analyse survival data – the Cox proportional hazards model (the "Cox model") developed by Sir David Cox in 1972. The principal objective of the Cox model is to determine the influence of various chosen covariates on the hazard rate, and hence understand the extent of heterogeneity present in a particular investigation.

In contrast with many statistical models we look at in this book, it is not the objective of the Cox model to determine the *absolute* hazard rate given a particular set of covariates. Instead, here we are interested in the *ratio* of hazard rates, or the relative effect which a covariate has on the hazard rate. For example we may use the Cox model to measure the efficacy of a trialled drug by comparing survival rates of patients diagnosed with a potentially fatal disease, where patients are given either the drug or a placebo (here the covariate is binary: drug or placebo). Our objective in this example is to compare the mortality rates of patients who were given the drug with those who were given the placebo.

The Cox model could also be used, for example, to address the following questions:

- What is the effect on human mortality rates from exercising, drinking alcohol or smoking cigarettes?
- Do prisoners' reoffending rates reduce if they receive certain types of education whilst in prison?
- How do bank loan default rates depend on various characteristics of the borrower e.g. salary, length of current employment, homeowner/renting etc.?

Remark 21.1 The term "hazard rate" tends to be used by mathematicians and scientists, whereas actuaries tend to use the terms failure rates, default rates, or mortality rates, depending on the context. All these terms are equivalent.

R Programming for Actuarial Science, First Edition. Peter McQuire and Alfred Kume.
© 2024 John Wiley & Sons Ltd. Published 2024 by John Wiley & Sons Ltd.
Companion Website: www.wiley.com/go/rprogramming.com.

Remark 21.2 Note that we will mostly refer to "lives" throughout this chapter when considering entities which "fail"; from the above examples clearly the hazard rates could apply to any entity which can be interpreted as "failing". Note also that "failing" could relate to a positive action e.g. passing an exam, acquiring a skill.

21.2 Cox Model Equation

Under the Cox model, the hazard rate, $\lambda_{i,t}$, for life, i, at time, t, is given by the following expression:

$$\lambda_{i,t} = \lambda_{0,t} e^{\beta_1 z_{1,i} + \beta_2 z_{2,i} + \dots + \beta_n z_{n,i}} = \lambda_{0,t} e^{\beta^T z_i} \tag{21.1}$$

where:

- $\beta = [\beta_1, \beta_2, \dots, \beta_n]^T$ is a vector of regression parameters which are to be determined from data,
- $z_i = [z_{1i}, z_{2i}, \dots, z_{ni}]^T$ is a vector of n covariates for life i which describes the life being studied,
- $\lambda_{0,t}$ is known as the "baseline hazard" at time t; we can see from Equation 21.1 that $\lambda_{i,t} = \lambda_{0,t}$ when all the z values are zero.

(Note: β^T is the transpose of the vector β.) Rearranging Equation (21.1):

$$\frac{\lambda_{i,t}}{\lambda_{0,t}} = e^{\beta_1 z_{1,i} + \beta_2 z_{2,i} + \dots + \beta_n z_{n,i}} \tag{21.2}$$

This is perhaps more informative, and gets closer to the purpose of the Cox model; $e^{\beta^T z_i}$ gives the ratio of two hazard rates i.e. the hazard rate of life i, and the hazard rate of the baseline. Thus by estimating the β values from data we can understand the effect on the hazard rate and the importance of each covariate being investigated.

For example if we were to analyse the effect that smoking and gender may have on mortality rates the z's may be defined as follows:

$z_1 = 0$: does not smoke $z_1 = 1$: does smoke
$z_2 = 0$: male $z_2 = 1$: female

Here a female, non-smoker would be identified by $(0, 1)$. If we want to understand the impact of smoking we could look at either of the following ratios:

$$\frac{\lambda_{1,1;t}}{\lambda_{0,1;t}} \text{ or } \frac{\lambda_{1,0;t}}{\lambda_{0,0;t}}$$

It is straightforward to show (from Equation (21.2)) that these are equal to e^{β_1}. As noted previously, our aim is therefore to determine the β values from our data; for example if we calculate that $\beta_1 = +0.7$, we may conclude that the effect of smoking is to increase the mortality rate by a factor of $e^{0.7}$ or 2.0137527.

Exercise 21.1 *The effect of gender could be represented by the value of β_2. What sign and value for β_2 might you expect in this case?*

A key feature of the Cox model

It should be noted that under this Cox model the ratio of hazard rates is not time dependent. This can be seen by noting that the expression $e^{\beta^T z}$ is not dependent on time (see Equation (21.2)). (A number of alternative time-dependent models do exist – see Martinussen and Scheike (2006)). It is instructive to think of a case where this assumption would not be appropriate.

Example 21.1 A new drug was given to patients prior to surgery which was thought to reduce mortality rates immediately following the operation. However it may be the case that after a period of time the effect of this drug would diminish such that there was little difference between mortality rates of those who were given the drug and those not given the drug. In this case the Cox model would be inappropriate (see Figure 21.1).

Figure 21.1 Non-constant ratio of hazard rates

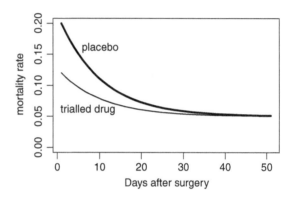

21.3 Applications

We now look at a number of applications of the Cox model. In our first two examples we apply the Cox model separately to two similar data sets of different sizes (one small and one large).

21.3.1 Smokers' Mortality: Small Data Set

The first data set consists of 25 lives, 23 of whom had died by the end of the investigation. For each life three pieces of information were recorded: (1) their date, or time, of death, (2) whether or not they were a smoker, and (3) whether or not the life died during the investigation ("status"). We have assigned a value of "1" to denote a smoker, and "0" to denote a non-smoker; note that this assignment is arbitrary.

In this introductory example we have included only one covariate i.e. smoker/non-smoker; our final example in this section is a little more involved, which allows for three covariates. A further, more complex example is included in Section 21.4, where we compare the Cox and Kaplan Meier models.

Extracting our data (a sample is shown below):

```
data<-read.csv("~/coxsmokersimple.csv");    head(data,5)
```

```
   time smoker status
1    1     1      1
2    2     0      1
3    5     0      1
4    7     1      1
5    8     0      0
```

It is instructive to note the form which our data takes for a Cox analysis. We only require: (1) the order in which each life left the investigation, (2) the covariate details of each life, and (3) the reason for leaving the investigation (either by death or by a censoring event – see later). Strictly speaking we do not require the actual times of the events, although these will usually be given so we can "order" our data.

"Status" tells us whether the relevant event of interest occurred or if there was censoring. In this example a death is denoted by "1" and a censoring event by "0". Note that the event of interest should be carefully defined in all survival studies; in this case it may be sensible to define the event as death excluding those caused by accidents, homicide or suicide. A more detailed discussion on censoring is included in Chapter 20.

It may be clear from a visual inspection of the data (e.g. boxplot) that smokers tended to die earlier on average, so a sensible estimate at this stage would be that smoking increases the mortality rate and that we may expect a positive value for β.

We will use the coxph and Surv functions from the survival package to obtain our results:

```
(smoke_small<-coxph(Surv(data$time,data$status)~data$smoker))
Call:
coxph(formula = Surv(data$time, data$status) ~ data$smoker)

                coef exp(coef) se(coef)     z    p
data$smoker 0.2252    1.2526   0.4330 0.52 0.603

Likelihood ratio test=0.27   on 1 df, p=0.6033
n= 25, number of events= 23
```

From the output, coef gives $\hat{\beta} = 0.2251966$. Additional statistical information can be extracted (not shown here) using summary(). It is also worth noting the content of the Surv object – note that the two censoring events, 5 and 7, at $t = 8, 14$ are flagged.

```
Surv(data$time,data$status)
 [1]  1    2    5    7    8+ 11   14+ 16   18   21   22   24   25   26   28   30   33   35   36
[20] 38   40   43   44   46   47
```

So, perhaps our model can be written:

$$\lambda_{i,t} = \lambda_{0,t} e^{0.2251966 \, z_i}$$

This result implies that smokers have a mortality rate which is $e^{0.2251966}$ or 1.2525689 times higher than non-smokers. (Note that the R output also gives this number under `exp(coef)`.)

Exercise 21.2 *The reader is encouraged to change the data to understand the effect on the estimated β. For example, re-order the lives so that all but one of the smokers die before any non-smokers.*

It is informative, although a little tiresome, to check our results here (this is partly why we chose a simple data set as our first example). The partial likelihood in respect of an event where life i dies, immediately prior to which n_i lives were observed, is given by:

$$L_i(\beta) = \frac{\lambda_i}{\sum_{j=1}^{n_i} \lambda_j} \tag{21.3}$$

and the likelihood of our entire dataset (here the product includes 23 terms):

$$L(\beta) = \prod_{events} L_i(\beta) \tag{21.4}$$

Using maximum likelihood estimation we can calculate MLE's for our parameters, in this case, β.

Exercise 21.3 *Write code for the partial loglikelihood of the above data and hence verify $\hat{\beta}$ calculated above.*

The solution is included below:

```
lnL<- function(b) (b-log(13*exp(b)+12)-log(12*exp(b)+12)-log(12*exp(b)+11)
+b-log(12*exp(b)+10)
+b-log(11*exp(b)+9)  +b-log(9*exp(b)+9)
+b-log(8*exp(b)+9) +b-log(7*exp(b)+9) -log(6*exp(b)+9) -log(6*exp(b)+8)
-log(6*exp(b)+7) -log(6*exp(b)+6) +b-log(6*exp(b)+5) +b-log(5*exp(b)+5)
+b-log(4*exp(b)+5)+b-log(3*exp(b)+5)
-log(2*exp(b)+5)-log(2*exp(b)+4)
+b-log(2*exp(b)+3)-log(1*exp(b)+3)
-log(1*exp(b)+2)-log(1*exp(b)+1)
+b-log(1*exp(b)+0))*-1

optimize(lnL,c(-2,2))$minimum

[1] 0.2251963
```

Thus the `coxph` function output (0.2251966) is consistent with our manual calculation of the estimate of β. We can further demonstrate this by calculating logL at various values of β (left as an exercise for the reader). Figure 21.2 shows a turning point at $\hat{\beta} = 0.2251963$.

However, we cannot reject the null hypothesis, H_0, that $\beta = 0$; given the p-value obtained (0.6029681) we have insufficient evidence to reject the hypothesis that smoking has no

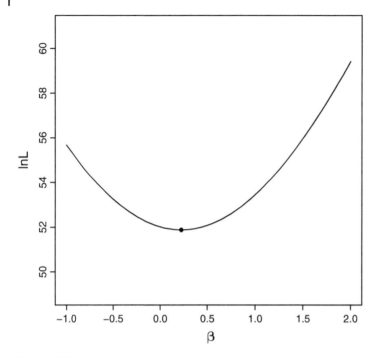

Figure 21.2   Loglikelihood function values

effect on mortality. The standard error of the estimator is large here perhaps due to the limited size of the dataset; we need to consider a larger dataset, which may provide a more credible estimator for β.

21.3.2   Smokers' Mortality: Larger Data Set

Remark 21.3   Note that this data set, in contrast with the initial, smaller dataset, contains events which occur at the same time, known as "tied" events e.g. events 4 and 5. Clearly this will, in general, be the case with real-world datasets we encounter. A method is required to deal with such "ties" as can be seen from Equation 21.3 – we will adopt Breslow's approximation for ties, which requires an extra argument to be included in our function (see below). Note that the default method in the survival package is the Efron approximation, which is the preferred method when there are a large number of ties. (See Ibrahim et al. (2001) for details regarding both methods.)

Extracting the data:

```
databig<-read.csv("~/coxsmokerbig.csv")
```

The new dataset contains data on 6,400 lives, rather than the 25 lives in our first example. Using the same methodology as above, but allowing for ties:

```
(smoke_large<-coxph(Surv(databig$time,databig$status)
~databig$smoker,databig,ties="breslow"))

Call:
coxph(formula = Surv(databig$time, databig$status) ~ databig$smoker,
    data = databig, ties = "breslow")

                  coef exp(coef) se(coef)     z       p
databig$smoker 0.28736   1.33291  0.02653 10.83 <2e-16

Likelihood ratio test=116.9  on 1 df, p=< 2.2e-16
n= 6400, number of events= 5791
```

From this dataset, $\hat{\beta}$ =0.2873626. Given the p-value = $2.4482845 \times 10^{-27}$ from this analysis, we can reject the hypothesis that $\beta = 0$, indicating a strong relationship between smoking and higher mortality. By using a larger dataset the confidence limits are much smaller, and we have a lot of confidence that the covariate under study is an important one.

Exercise 21.4 *Verify the 95% confidence limits for β by using* `summary()`.

Exercise 21.5 *Repeat the above calculations, swapping the z-covariate values in the dataset. What new value of β do you expect?*

21.3.3   Multiple covariates and interactions

Our third example involves the analysis of the rate at which bicycle tyres are punctured. Specifically we will analyse the effect which cycling on smooth or stony roads and/or buying cheap or expensive tyres has on tyre puncture rates. In particular, we will analyse if expensive tyres have the same effect on the puncture rate whether cycling on smooth or stony roads.

We will also analyse whether the cyclist's hair colour has any impact – we are expecting a parameter value close to zero for hair colour.

Our baseline covariate values ($z = 0$) are defined as "smooth road", "expensive tyres", and "blond". The $z = 1$ values are "stony road", "cheap tyres", and "dark-haired". Let's get our data:

```
data<-read.csv("~/coxtyres.csv")

time<-data$time; status<-data$status
haircolour<-data$haircolour;  road_surf<-data$road_surf;  tyres<-data$tyres
```

Here we have three covariates: type of road surface, quality of tyres and haircolour. Applying the coxph function to this dataset using the three covariates:

```
(bike_cox<-coxph(Surv(time,status)~road_surf+tyres+haircolour,ties="breslow"))

Call:
coxph(formula = Surv(time, status) ~ road_surf + tyres + haircolour,
    ties = "breslow")
```

```
            coef exp(coef) se(coef)      z        p
road_surf 0.53121   1.70099  0.03645 14.575  < 2e-16
tyres     0.19621   1.21678  0.03567  5.500 3.79e-08
haircolour 0.00429  1.00430  0.03539  0.121    0.904

Likelihood ratio test=229.8  on 3 df, p=< 2.2e-16
n= 3200, number of events= 3200
```

From these results it would appear that cycling on stony roads has a significant effect on puncture rates (70.1% higher rate compared to cycling on smooth roads). Buying cheap tyres also makes a large difference (21.7%).

However it may perhaps be the case that using expensive tyres is only really beneficial on stony roads. (This "interaction" effect was covered in more detail in Chapter 5.) We could stratify our data to investigate such effects; looking first at only the "stony" data, and then the "smooth" data (please note that I have not included the output details below given the substantial output produced):

```
stony_only<-filter(data,data$road_surf==1)
time1<-stony_only$time; status1<-stony_only$status;  tyres1<-stony_only$tyres
cox_stone<-coxph(Surv(time1,status1)~tyres1,stony_only,ties="breslow")
```

Here $\hat{\beta}_{tyres}$ =0.3272895, with p-value = 1.4846059 × 10^{-10}, suggesting that tyre type is significant for stony roads. And comparing the tyre coefficient on smooth roads:

```
smooth_only<-filter(data,data$road_surf==0)
time2<-smooth_only$time;  status2<-smooth_only$status;  tyres2<-smooth_only$tyres
cox_smooth<-coxph(Surv(time2,status2)~tyres2,smooth_only,ties="breslow")
```

Here $\hat{\beta}_{tyres}$ = 0.0639014. It's clear the original regression parameter obtained, $\hat{\beta}_{tyres}$ = 0.1962079, was some kind of average; the benefits of buying expensive tyres are much more pronounced on stony roads.

However, as noted in Chapter 20, stratifying data is not ideal. It's preferable to analyse our full original dataset without stratifying and simply include an interaction term, "roadsurf × tyres":

```
(bike_cox_int<-coxph(Surv(time,status)~haircolour+road_surf*tyres,data,
    ties="breslow"))

Call:
coxph(formula = Surv(time, status) ~ haircolour + road_surf *
    tyres, data = data, ties = "breslow")

                    coef exp(coef) se(coef)      z        p
haircolour        0.002739  1.002742 0.035392 0.077 0.938321
road_surf         0.399898  1.491673 0.050483 7.921 2.35e-15
tyres             0.063946  1.066035 0.050101 1.276 0.201830
road_surf:tyres   0.265935  1.304650 0.070896 3.751 0.000176

Likelihood ratio test=243.8  on 4 df, p=< 2.2e-16
n= 3200, number of events= 3200
```

```
bike_cox$loglik;        bike_cox_int$loglik

[1] -22631.85 -22516.96
[1] -22631.85 -22509.93
```

This reinforces our findings from the stratified data. We should first note the loglikelihood ratio test score when including the interaction term; the loglikelihood has increased from -22517 to -22509.9, an increase of 7.03, so we can comfortably reject the hypothesis that this model does not improve on our initial model.

The effect of buying the cheap tyres when cycling on smooth roads is only 6.6%. Indeed the p-value suggests that we do not have sufficient evidence to reject the hypothesis that the tyres alone have no impact on the puncture rate.

The type of road surface still has the biggest effect, with a puncture rate which is 49.2% higher on stony roads compared to smooth roads. But now, and this is the key point, the use of cheap tyres on stony roads has a huge additional effect (30.5%).

A reasonable conclusion is that if we are cycling on smooth roads the use of expensive tyres makes only a small reduction, if any, in puncture rates. However expensive tyres certainly appear worth the extra money if we are cycling on stony roads. Without the above analysis which included an interaction term our conclusion may have been very misleading e.g. buying expensive is always worth the extra cost.

Throughout this example the regression parameter for haircolour was close to 0, suggesting, as expected, that haircolour is unlikely to have an effect on puncture rates.

Our proposed model is therefore:

$$\lambda_{i,t} = \lambda_{0,t} \, e^{0.3998984 z_{road} + 0.2659351 z_{road*tyres}}$$

21.4 Comparison of Cox and Kaplan Meier Analyses of Lung Cancer Data

Please refer to Chapter 20 where an initial analysis of the lung dataset was carried out (using K-M methodology); here we use the Cox model to analyse the data.

As with the K-M analysis, we will analyse the relevance of *sex*, *ph.ecog* and the other covariates in modelling lung cancer survival rates. However great care should be taken using Cox analysis in this case (from a visual inspection of S the hazard rate ratios certainly do appear to depend on time). A detailed discussion of this important point is beyond the scope of this chapter (see Therneau and Grambsch (2000)).

There are numerous approaches which could be adopted to identify potential best candidates for our multivariate Cox model. For example we could apply univariate analysis or multivariate analysis first, or analyse stratified data. A detailed discussion of these is beyond the scope of this chapter, and refer the reader to the various texts on this subject.

First we apply a multivariate risk model, using all seven available covariates, to determine the likely candidates for our model:

```
library(survival)
surv_obj <- with(lung, Surv(time,status))
Cox_all_Model <- coxph(surv_obj~sex+age+ph.ecog+ph.karno+pat.karno+meal.cal+wt.loss,
data=lung)

Cox_all_Model

Call:
coxph(formula = surv_obj ~ sex + age + ph.ecog + ph.karno + pat.karno +
    meal.cal + wt.loss, data = lung)

                coef  exp(coef)   se(coef)      z       p
sex        -5.509e-01  5.765e-01  2.008e-01 -2.743 0.00609
age         1.065e-02  1.011e+00  1.161e-02  0.917 0.35906
ph.ecog     7.342e-01  2.084e+00  2.233e-01  3.288 0.00101
ph.karno    2.246e-02  1.023e+00  1.124e-02  1.998 0.04574
pat.karno  -1.242e-02  9.877e-01  8.054e-03 -1.542 0.12316
meal.cal    3.329e-05  1.000e+00  2.595e-04  0.128 0.89791
wt.loss    -1.433e-02  9.858e-01  7.771e-03 -1.844 0.06518

Likelihood ratio test=28.33  on 7 df, p=0.0001918
n= 168, number of events= 121
   (60 observations deleted due to missingness)
```

From the p-values obtained above it would appear *sex*, *ph.ecog*, and possibly *ph.karno* are the best candidates for our model. It may be useful to also apply a univariate risk model on each of these covariates – this is left as an exercise for the reader.

From these results we can build up a multivariate model, starting with *ph.ecog*, then *sex* (we have only included the full summary results for the first model run – the reader should carry these out for all models):

```
Cox_all_Model1 <- coxph(surv_obj~ph.ecog,data=lung)  #inc. only ph.ecog
coxph(surv_obj~ph.ecog,data=lung)$loglik

[1] -744.4805 -735.6967

summary(Cox_all_Model1)

Call:
coxph(formula = surv_obj ~ ph.ecog, data = lung)

  n= 227, number of events= 164
   (1 observation deleted due to missingness)

           coef exp(coef) se(coef)     z Pr(>|z|)
ph.ecog 0.4759    1.6095   0.1134 4.198 2.69e-05 ***
---
Signif. codes:  0 '***' 0.001 '**' 0.01 '*' 0.05 '.' 0.1 ' ' 1


        exp(coef) exp(-coef) lower .95 upper .95
ph.ecog      1.61     0.6213     1.289      2.01

Concordance= 0.604  (se = 0.024 )
Likelihood ratio test= 17.57  on 1 df,   p=3e-05
Wald test            = 17.62  on 1 df,   p=3e-05
Score (logrank) test = 17.89  on 1 df,   p=2e-05
```

Here the logL ratio test $(2(lnL_1 - lnL_0))$ shows *ph.ecog* is significant (as do the other tests above). Adding *sex* to our model:

```
Cox_all_Model2 <- coxph(surv_obj~ph.ecog+sex,data=lung)
coxph(surv_obj~ph.ecog+sex,data=lung)$loglik

[1] -744.4805 -729.9537

Cox_all_Model2$coefficients

  ph.ecog       sex
0.4874938 -0.5529698
```

The p-scores (not shown here) and logL values obtained suggest that both *sex* and *ph.ecog* are indeed important covariates for our model. The reader should continue adding covariates to the model. For example, adding *ph.karno* does not improve our model sufficiently – the increase in the logL is insufficient:

```
Cox_all_Model3 <- coxph(surv_obj~sex+ph.ecog+ph.karno,data=lung)
Cox_all_Model3$loglik

[1] -739.0605 -724.3774
```

And adding *pat.karno*:

```
Cox_all_Model4 <- coxph(surv_obj~sex+ph.ecog+pat.karno,data=lung)
Cox_all_Model4$loglik

[1] -728.3357 -713.2004
```

Again, the increase in the logL is insufficient to warrant the extra parameter for *pat.karno*. So, taking the parameter values for *sex* and *ph.ecog*, a reasonable model is perhaps:

$$\lambda_i = \lambda_0 e^{0.4874938\, z_{ph.ecog} - 0.5529698\, z_{sex}}$$

This is only an example of a simple analysis which may be carried out. Further analysis should be carried out before confirming the most appropriate model. For example, we should consider correlations and possible interactions between covariates.

21.5 Recommended Reading

- Boland PJ. (2007). *Statistical and Probabilistic Methods in Actuarial Science*. Boca Raton, Florida, US, Chapman and Hall/CRC.
- Crawley MJ. (2013). The R Book (2nd edition). Chichester, West Sussex, UK, Wiley.
- Charpentier A et al. (2016). Computational Actuarial Science with R. Boca Raton, Florida, US, Chapman and Hall/CRC.
- Everitt BS and Hothorn T. (2006). *A Handbook of Statistical Analyses using R*. Boca Raton, Florida, US, Chapman and Hall/CRC.

- Hosmer DW, Lemeshow S, and May S. (2008). *Applied Survival Analysis: Regression Modeling of Time-to-Event Data*. Hoboken, New Jersey, US, Wiley.
- Ibrahim JG, Ming-Hui Chen, and Sinha D. (2001). *Bayesian Survival Analysis*. New York, Springer.
- MacDonald AS et al. (2018). *Modelling Mortality with Actuarial Applications*. Cambridge, UK, Cambridge University Press.
- Martinussen T and Scheike TH. (2006). *Dynamic Regression Models for Survival Data*. New York, Springer.
- Therneau TM and Grambsch PM. (2000). *Modeling Survival Data: Extending the Cox Model*. New York, Springer.

22

Markov Multiple State Models: Applications to Life Contingencies

Peter McQuire

```
library(markovchain)
library(expm)
```

22.1 Introduction

Markov models are extensively used in actuarial work where we need to develop an understanding of how the state of a system may change in the future. The simplest Markov model actuaries will be familiar with is the "alive-dead" 2-state model discussed in Chapter 16, used to estimate probabilities that lives occupy these states at future points in time. In this chapter we develop this idea by looking at multiple-state systems.

Indeed, the focus of this chapter is the "Healthy-Sick-Dead" model, introduced in Section 22.5.2, and the related material which follows; such a model may be important to assess the length of time a policyholder or employee may be deemed as "sick" or in ill-health. Another example of a multiple state model in an actuarial context is that which models the state of a pension scheme member; they may be a contributing member, a member who has ceased to pay contributions but is not yet in receipt of a pension, or may be in receipt of this pension. They may also have transferred to an alternative arrangement or have died. By understanding the various probabilities of existing in particular states and of particular transitions occurring between states, we can better understand the expected future cashflows.

The chapter includes a section on Markov chain models, before developing these ideas by looking at Markov jump models. The focus of the chapter is to develop these ideas such that we can solve typical problems in an actuarial context. Some of the ideas have been discussed in Chapter 16; the reader may wish to familiarise themselves with that chapter before proceeding.

The reader who is familiar with the concept of Markov chain models may wish to briefly read Sections 22.2 and 22.4 before moving onto the Section 22.5 on Markov jump models (the focus of the chapter). A brief outline of the chapter is included below:

R Programming for Actuarial Science, First Edition. Peter McQuire and Alfred Kume.
© 2024 John Wiley & Sons Ltd. Published 2024 by John Wiley & Sons Ltd.
Companion Website: www.wiley.com/go/rprogramming.com.

- Markov chains and introducing the `markovchain` package (Sections 22.2 and 22.4)
- Markov jump models (Section 22.5 and 22.6)
- Premium calculations (Section 22.7)
- Multiple decrement models (Section 22.9)

The chapter includes little theory, with only the key ideas being highlighted as we proceed through the chapter. Rather than duplicate the theory available in numerous alternative texts, we will concentrate on applications and use several examples to illustrate the usefulness of Markov models in solving actuarial problems. The reader is strongly recommended to have access to one of the recommended texts. (It is assumed that the reader already has had some exposure to these key theoretical ideas.)

A note on verification

The calculations towards the end of the chapter can become quite involved; it is therefore important that the reader has an independent way to verify their calculations (a skill the trainee actuary must develop when addressing real-life situations). Often these alternative checks may be quite approximate. These are important skills which the student will do well to develop early in their career.

22.2 The Markov Property

The Markov property is a key idea which underpins much of what follows in this chapter. It is defined as follows:

$$P(X_n = x_n | X_{n-1} = x_{n-1}, X_{n-2} = x_{n-2}, ..., X_0 = x_0) = P(X_n = x_n | X_{n-1} = x_{n-1}) \qquad (22.1)$$

where X_n is a random variable representing the state occupied at time, n. Consider, for example, a system which changes state daily. The probability that the system, X, is in a particular state tomorrow can be determined solely by the state of the system today; knowing what has happened before today provides no additional useful information. This is often referred to as the memoryless quality of Markov models.

It is instructive to critique situations where this memoryless property may not be applicable, and would require a more advanced model. For example, Holywood actors who are currently married may be more likely to divorce if they have previously divorced. Thus, only having a "married-divorced" model may be inappropriate - the "memory" that the person has previously been married may be statistically significant. A better model may include states for "married", "divorced once", "divorced twice", "divorced more than twice".

22.3 Markov Chains and Jump Models

22.3.1 Examples

Figure 22.1 shows two examples of Markov models. The simplest Markov model is the 2-state alive-dead model (Figure 22.1 left); it has two states, alive and dead, with a single transition rate between these two states (here equal to 0.1 per annum). A more complex model may be required to model a life's marital status (Figure 22.1 right).

Markov jump process with mortality rate = 0.1   **Markov model with 5 states (rates shown)**

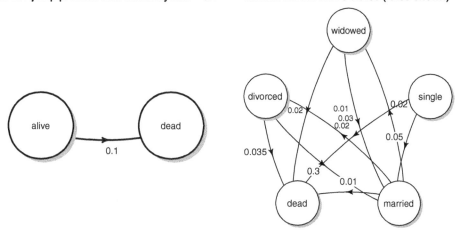

Figure 22.1   2 examples of Markov jump models.

Exercise 22.1 *Discuss how the second model in Figure 22.1 could be improved.*

Further examples of the application of Markov models to actuarial work are included below:

- From the model in Figure 22.1 (right) we may estimate how long a person, who is currently single, will be married for in the future.
- Sickness model - what is the probability a life which is currently healthy will be sick in two years time? What is the probability that a sick life will remain in the sick state for the next five years? For how many days may we expect a life to be ill in the next year given they are currently healthy?
- Given a life is currently employed, what is the total time we might expect that life to be unemployed for over the next 10 years?
- What level of premium should we charge for a combined sickness and life assurance policy?
- Is there a long-term stable position we can expect our system to exist in? We will see an example of this when we look at NCD systems later.
- What is the likelihood that a corporate bond with a given rating may default in the next five years.
- In Exercise 22.16, we discuss a similar model to the 2-state alive-dead model, where transitions in either direction are permitted.

Clearly the applications for actuaries are considerable. Many of the above problems will be discussed in examples and exercises in this chapter.

22.3.2   Differences between Markov Chain and Markov Jump Models

There appears to be no universal convention to distinguish Markov chains and jump models. Some texts only discuss Markov chains whereas others clearly define two different processes. Here we will define them as follows (which is consistent with their treatment in the majority of texts).

Markov chains are a discrete time Markov process with a discrete state space. Markov jump processes are similar but their processes run in continuous time; it is a continuous time Markov process with a discrete state space. As noted above, these definitions can vary between texts; often both the above models are referred to collectively as Markov chains.

Markov chain models tend to be used where a system is typically assessed at regular, discrete points in time, such that the most suitable way to represent these transitions is by using transition *probabilities*. Markov jump models are used where it is more convenient to estimate transition *rates* such that we can continuously monitor which states are occupied in our system. In particular, jump models lend themselves to scenarios where we may require to monitor the position at irregular points in time, or where transition rates tend to be non-constant and require a parametric function to be fitted.

For example, with non-life insurance policies such as house or car insurance, policies are usually reassessed annually (at renewal of the annual policy) - we will be interested in the probabilities of policyholders being in particular states at yearly intervals. Thus Markov chain models will tend to be used in these scenarios - we may not need to know what is happening between policy renewal dates.

In other situations it may be preferable to estimate the functional forms of rates at which particular decrements occur such that we can estimate the probabilities of existing in particular states at any point in the future, rather than only at discrete intervals. A chain model could be used, but would require a table of probabilities rather than one function. This is directly analogous to the choice between using life tables (i.e. probabilities) or a single function such as Gompertz Law (i.e. mortality rates) when modelling survival probabilities (see Chapters 16 and 23).

We will first briefly look at Markov chains, before turning our focus to Markov jump models. *The reader is reminded that the focus of the chapter is to apply Markov jump transition models to solve actuarial problems.* Ultimately we will analyse systems in which the transition rates between various states are not constant.

(The chapter includes a brief section on the Matrix Exponential Method which provides, in certain circumstances, a closed-form solution to Markov jump models with constant rates.)

22.4   Markov Chains (Discrete Time)

Markov chain models allow us to estimate the probability that an entity will be in a particular state at *discrete* points in time. Markov chains require us to set up a transition matrix which contains the probabilities, p_{xy}, that state y will be occupied at time $t + 1$ given that state x is occupied at time t. The models in this section include transition probabilities which are constant; later we will introduce models with time or age-dependent probabilities.

22.4.1   Applying Markov Chains to Estimate Future Probabilities

Example 22.1

Mary is hoping to play tennis every day for the next week, but can only play if it is not raining. She is interested in estimating the probability of playing tennis on each of the next seven days. First we set up a simple transition matrix which aims to estimate the type of weather there will be in future days:

```
transitions<-matrix(c(0.78, 0.14,0.08,0.22,0.6, 0.18,0.2,0.5,0.3),byrow=TRUE,
nrow=3, dimnames=list(c("sunny","overcast","rain"),c("sunny","overcast","rain")))
transitions

          sunny overcast rain
sunny      0.78    0.14 0.08
overcast   0.22    0.60 0.18
rain       0.20    0.50 0.30
```

This is the transition matrix, T. It contains all the possible transition probabilities for one time-step. (Note that each row is a probability distribution.) This particular matrix models how likely it is to be sunny, overcast or raining tomorrow. So, for example, the probability that tomorrow will be a rainy day given that it's sunny today is 8%.

Remark 22.1 When setting up large matrices, the reader may find it easier to type the rates into a spreadsheet and import them into R using "File->Import Dataset", and saving it as a matrix.

Alternatively, we can create a `markovchain` object using the above transition matrix. Defining our `markovchain` object "weather":

```
(weather <- new("markovchain", transitionMatrix=transitions))

Unnamed Markov chain
 A  3-dimensional discrete Markov Chain defined by the following states:
 sunny, overcast, rain
 The transition matrix  (by rows)  is defined as follows:
          sunny overcast rain
sunny      0.78    0.14 0.08
overcast   0.22    0.60 0.18
rain       0.20    0.50 0.30
```

In this example we proceed using the matrices and our own functions, then repeat the calculations using the `markovchain` package functions. Ultimately we would expect the reader to use the `markovchain` objects and functions.

Here we estimate the weather the day after tomorrow (t=2). We can calculate this using:

$$p_{ij}^{(2)} = \sum_k p_{ik} p_{kj} \tag{22.2}$$

where p_{ij} is the one-step probability - here that the weather is of type "j" tomorrow given that it is of type "i" today, and $p_{ij}^{(2)}$ is the probability that the weather is of type "j" the day after tomorrow given that it is of type "i" today. (Note that the "2" superscript does not refer to "p to the power of 2".) The probability of it raining at $t = 2$, given it is sunny today, is therefore:

$$p_{sr}^{(2)} = p_{ss} p_{sr} + p_{so} p_{or} + p_{sr} p_{rr}$$

For example, the probability it's rainy in two days given it's sunny today is 11.16%. This is calculated from:

$$0.78 \times 0.08 + 0.14 \times 0.18 + 0.08 \times 0.3 = 0.1116$$

Equation 22.2 is an example of a Chapman-Kolmogorov equation. For the general case:

$$p_{ij}^{(n)} = \sum_k p_{ik}^{(m)} p_{kj}^{(n-m)} \tag{22.3}$$

where $p_{ij}^{(n)}$ is the probability that the system is in state "j" after n time-steps, given it is in state "i" today , $n > m$ and both n and m are integer values. Equation 22.2 is the simple case where $n = 2$ and $m = 1$.

Exercise 22.2 *Manually calculate the remaining eight probabilities at $t = 2$ from the 3×3 matrix.*

The above calculation is more easily carried out applying matrix multiplication to the transition matrix, T, introduced earlier:

$$p^{(2)} = T \times T \tag{22.4}$$

where $p^{(2)}$ is a 3×3 matrix, consisting of elements $p_{ij}^{(2)}$ in the ith row and jth column. Applying Equation 22.4:

```
transitions%*%transitions

          sunny overcast    rain
sunny     0.6552   0.2332 0.1116
overcast 0.3396    0.4808 0.1796
rain      0.3260   0.4780 0.1960
```

Alternatively, we can apply our Markov chain object and functionality to get the same result:

```
weather^2

Unnamed Markov chain^2
 A  3 - dimensional discrete Markov Chain defined by the following states:
 sunny, overcast, rain
 The transition matrix  (by rows)  is defined as follows:
          sunny overcast    rain
sunny     0.6552   0.2332 0.1116
overcast 0.3396    0.4808 0.1796
rain      0.3260   0.4780 0.1960
```

Similarly, noting Equation 22.3:

$$p^{(3)} = p^{(2)} \times T = (T \times T) \times T$$
$$p^{(4)} = p^{(3)} \times T = ((T \times T) \times T) \times T$$

and so on. Thus we can calculate all future probabilities of existing in particular states by applying this transition matrix to the initial occupancy probabilities. In general we can determine probabilities at any future discrete time, and incorporate an initial state row vector, p^0, the probability distribution at $t = 0$, by applying:

$$p^{(n)} = p^0 T^n \qquad\qquad (22.5)$$

T^n is the matrix multiplication of the transition matrix, n times. Often with actuarial problems we will have information about the starting state. For example, when a new pol-icyholder takes out a sickness policy there will usually be a requirement for the life to be "healthy" at the start of the policy. In our example above, if we know the starting state is "sunny":

$$p^0 = [1, 0, 0]$$

Note that we can have as many scenarios of the starting probabilities as we wish (by incorporating extra rows/vectors). Given that we start in a particular state, we can calcu-late the probability of being in any state at a future time. Let's say we start in the "rainy" state:

```
future_weather_rainy<-function(t){c(0,0,1) * (weather^t)}
future_weather_rainy(2)

      sunny overcast  rain
[1,] 0.326    0.478 0.196
```

Or we could just run all starting scenarios together:

```
future_weather_general<-function(t){diag(3) * (weather^t)}
future_weather_general(2)

       sunny overcast    rain
[1,] 0.6552    0.2332 0.1116
[2,] 0.3396    0.4808 0.1796
[3,] 0.3260    0.4780 0.1960
```

Exercise 22.3 *Determine the probabilities that the future weather over the next 10 days, given that today is (a)rainy (b)sunny, and plot the results (see Figure 22.2).*

The following code may be used to determine a particular transition probability:

```
transitionProbability(weather^2,"rain","overcast")

[1] 0.478
```

The reader is encouraged to explore the functions within the markovchain package.

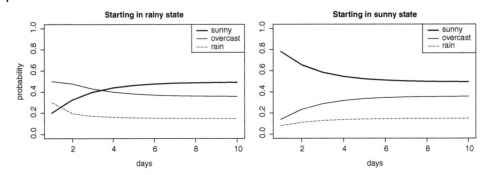

Figure 22.2 Weather forecast using Markov chains.

We can see from Figure 22.2 that, in the long term, the system has a stationary, or steady state distribution. This distribution is found at a sufficiently large value of t, where $p^{(n+1)} = p^{(n)}$. Thus we need to find $p^{(n)}$ such that:

$$p^{(n+1)} = p^{(n)}T = p^{(n)} \tag{22.6}$$

From this we can understand the probability of the system existing in each state in the long term. The stationary state distribution can be found using the steadyStates function:

```
steadyStates(weather)
          sunny   overcast       rain
[1,] 0.4932503 0.3582555 0.1484943
```

Or simply by brute force:

```
weather^100

Unnamed Markov chain^100
 A  3-dimensional discrete Markov Chain defined by the following states:
 sunny, overcast, rain
 The transition matrix  (by rows)  is defined as follows:
            sunny overcast      rain
sunny    0.4932503 0.3582555 0.1484943
overcast 0.4932503 0.3582555 0.1484943
rain     0.4932503 0.3582555 0.1484943
```

Remark 22.2 Noting that we require $p^{(n)}T = p^{(n)}$, and this is in the form $vA = v$ (where $\lambda = 1$), we can solve Eqn 22.6 by performing an eigendecomposition of T.

```
aa<-(eigen(transitions));   bb<-ginv(aa$vectors)

bb[1,]/sum(bb[1,]) #normalising

[1] 0.4932503 0.3582555 0.1484943
```

End of remark.

Irrespective of the starting state, the long term distribution is always (0.4932503, 0.3582555, 0.1484943). Note that a Markov chain which has a finite number of states will always have at least one stationary state distribution.

Exercise 22.4 *What are the major flaws with this particular model of the weather?*

22.4.2 Markov Chain Model - NCD

Before moving on to Markov jump chains (which is our principal focus in this chapter) let's briefly look at an actuarial example of a Markov chain model.

Insurance companies use a range of ratings factors to estimate risk and calculate insurance premiums. Underwriting for motor accident policies includes obtaining information on the model of car, miles driven, where it is kept overnight, the age of the driver and how long they have been qualified to drive.

However there is substantial evidence to suggest that one of the best rating factors is the number of claims recently made by the driver. Thus insurance companies may charge lower premiums to those drivers who have made no claims in their recent history. These "no claims discount" (NCD) systems are also commonly referred to as "Bonus malis" systems. There are a wide variety of "Bonus malis" systems adopted by insurance companies - we will look at a particularly simple example.

Example 22.2

If a driver made a claim last year they will move to (or stay in) level 1. If they don't make a claim they will move to or stay in level 2. Drivers in level 2 are entitled to a 25% discount to the full premium of £100.

The transition matrix which reflects the probabilities set out in Figure 22.3 is as follows:

```
(transition<-matrix(c(0.2,0.1,0.8,0.9),2,2,
dimnames=list(c("level 1","level 2"),c("level 1","level 2"))))

         level 1 level 2
level 1    0.2     0.8
level 2    0.1     0.9
```

For example, there is an 80% probability that a driver in level 1 will not make a claim and therefore move to level 2 next year. Assuming a driver starts in level 1 what is the pattern of average premiums paid over the next six years? And what is the steady state?

Figure 22.3 NCD system.

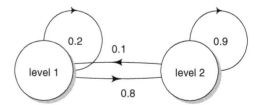

```
premium<-c(100,75)
position<-c(1,0)#starting in class 1
number_of_years<-6
income<-NULL
for(j in 1:number_of_years){
income[j]<-sum(position*premium)
position<-position%*%transition
}
income#expected annual income

[1] 100.000  80.000  78.000  77.800  77.780  77.778

sum(income)#expected total income

[1] 491.358
```

And to get the steady state:

```
NCD1 <- new("markovchain", transitionMatrix=transition)
steadyStates(NCD1)

        level 1    level 2
[1,] 0.1111111 0.8888889
```

22.4.3   Coding Exercise for Markov Chains

Before moving on to the main section of this chapter, try the following exercise:

Exercise 22.5 *By using a Markov chain, write code to model the following random walk in discrete time:*

$$X_t = X_{t-1} + Z_t$$

where Z_t is a random variable taking the values 1 or -1 with equal probability, and $t = 1, 2....$ Given that $X_0 = 0$, determine the probability distribution of X_{10}, and calculate $P(X_{100} = 0)$.

22.5   Markov Jump Models

In contrast with Markov chains, Markov jump models allow us to monitor a multi-state system at any point in time, i.e. continuously. As noted earlier, these models are used where it is more natural to study the continual state of a system, rather than, for example, at the end of each year. Such models are particularly relevant where the rates are time-dependent.

As such, Markov jump models first require the estimation of transition *rates* between the states defined in a particular model, rather than probabilities as we do with Markov chains. The matrix of transition rates is known as a "generator matrix", from which the required probabilities can be derived. It is worth noting at this point that numerical methods used to estimate occupancy probabilities using Markov jump models are effectively Markov chains with small time intervals.

Remark 22.3 The estimation of these rates is covered in Chapter 16, and also briefly in Section 22.8.

Markov jump models are associated with a series of differential equations known as "Kolmogorov forward equations"; these are discussed in more detail in Example 22.3. The method set out in this chapter allows us to solve these Kolmogorov differential equations numerically (see also Macdonald, Chapter 14).

We will proceed by looking at a number of multiple state models. We start with a 3-state model where all transitions between states are possible.

22.5.1 Example - Simple 3-State Model (All Transitions Possible)

In this example we model whether a baby is likely to be sleeping, eating, or playing at a future time, t (Figure 22.4). Choosing parameter values for the hourly rates:

```
mu12=0.25;  mu13=0.1;  mu21=0.35;  mu23=0.40;  mu31=0.3;  mu32=0.15
```

Noting that the sum of rates in each row must sum to zero (this will be justified shortly), the diagonal terms can be calculated from:

```
mu11=-mu12-mu13;   mu22=-mu21-mu23;   mu33=-mu32-mu31
```

The generator matrix which contains all the possible rates is equal to:

```
t_ratesx=c(mu11,mu12,mu13,mu21,mu22,mu23,mu31,mu32,mu33)
(gen_matx=matrix(t_ratesx,3,3,byrow=TRUE,
dimnames=list(c("sleep","eat","play"),c("sleep","eat","play"))))

        sleep    eat   play
sleep  -0.35   0.25   0.10
eat     0.35  -0.75   0.40
play    0.30   0.15  -0.45
```

Figure 22.4   3-state model : all transitions are possible.

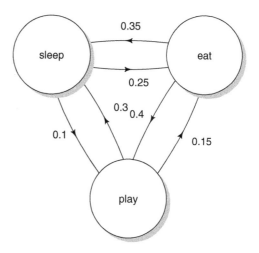

This matrix is composed of transition *rates*, not probabilities and, therefore, the entries can exceed 1.0. The units of these rates in this example are "per hour" (probabilities, of course, do not have units).

Ultimately our aim is to calculate the probabilities of existing in particular states at future points in time - we can determine these by applying μ^{ab}, the transition rate of moving from state a to state b. The probability of being in state a at age x and in state b a short time, Δt, later is given by the following approximate expression:

$$_{\Delta t}p_x^{ab} \approx \mu_x^{ab}\Delta t \tag{22.7}$$

for $a \neq b$, where the expression is exact as $\Delta t \to 0$.

Remark 22.4 This is the same equation discussed in Chapter 16. Please refer to the Appendix in that chapter for a further discussion of this expression.

By using a sufficiently small time interval, Δt, we can calculate a transition matrix which contains the (approximate) probabilities of transitioning over Δt using Equation 22.7 and the above generator matrix. Here we will use $\Delta t = \dfrac{1}{60}$ hour:

```
dt=1/60   #in hours
(P_dt <- diag(3) + gen_matx*dt)    #this is the transition matrix

              sleep        eat        play
sleep  0.994166667 0.004166667 0.001666667
eat    0.005833333 0.987500000 0.006666667
play   0.005000000 0.002500000 0.992500000
```

These are the approximate probabilities of being in one state at t_1 and a second state at $t_2 = t_1 + \Delta t$. The probability that a sleeping baby is playing one minute later is thus, approximately, 0.0016667, and the probability that a playing baby is sleeping one minute later is 0.005. Note that the sum of probabilities for each row is one, as required (this provides a justification for our calculation of the diagonal terms in the generator matrix).

Now we have our transition matrix, we can continue as we did for Markov chains by repeatedly multiplying the matrix with itself to get probabilities at time intervals of Δt. The calculated probabilities of a sleeping baby being in the various states in the future are shown in Figure 22.5. The various probabilities at $t = 240dt = 4$ hours for each of the three possible starting conditions are calculated as follows:

```
current=diag(3) #allows for every initial scenario
for(i in 1:240){current =current%*%P_dt
}
current
           sleep       eat       play
[1,] 0.5157123 0.2234943 0.2607934
[2,] 0.4497322 0.2327614 0.3175065
[3,] 0.4402800 0.2045900 0.3551300
```

Exercise 22.6 *Repeat the above calculations using a* markovchain *object and using a smaller* Δt.

Exercise 22.7 *Calculate the probability that a sleeping baby wakes up in the first hour.*

Figure 22.5 Probabilities of sleeping, playing, and eating (initially sleeping at t=0).

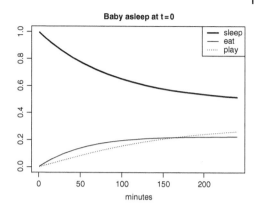

Exercise 22.8 *What is the probability that a sleeping baby will eat something in the next hour?*

Exercise 22.9 *If the baby eats 1ml of babyfood every minute whilst "eating" how much babyfood should you keep in the house to meet demand for one day?*

Exercise 22.10 *Vary the transition rates used to study the effects on the population of the three states.*

Note that this model allows for all six possible transitions between the three states to be permitted. In general, not all possible transitions will be permitted, as is the case with the "Healthy-Sick-Dead" model which we discuss in the next section (and much of the remainder of the chapter).

22.5.2 Example – H-S-D Model

We now look at the well-known *"Healthy-Sick-Dead"* model (Figure 22.6, left), and variations thereof. Under this model lives can repeatedly transition between the healthy and sick states, and die from both of these states. Note the differences compared to the example in Section 22.5.1; there are no transitions here from the dead state, whereas in the previous example all transitions were possible.

This model allows us to analyse, for example, the time spent in the sick state – this may be useful to employers who want to understand how many back-up staff are required to cover employees on sick-leave, or to insurance companies when pricing sickness policies under which benefits are payable during periods of sickness.

If we initially ignore the sick state, we have the even simpler 2-state alive-dead Markov model discussed in Chapter 16 (see Figure 22.6, right); it is a useful preliminary exercise to compare the results in Chapter 16 with those using the Markov jump model, here using a mortality rate equal to 0.12 per annum. Defining our rates (only one is required here):

```
mu13<-0.12      # healthy-dead
mu23<- mu21<- mu12<- mu31<-mu32<-0.00
mu11=-mu12-mu13;    mu22=-mu21-mu23;    mu33=-mu32-mu31
```

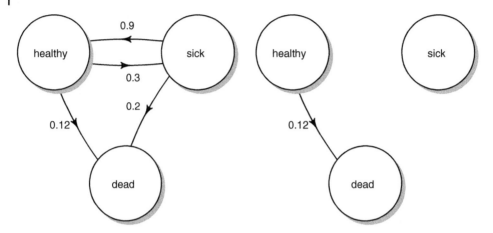

Figure 22.6 Healthy - sick - dead model.

Our generator matrix, and probability matrix over a small time interval, Δt, are as follows (here we have chosen $\Delta t = \dfrac{1}{365}$ years; smaller Δt's will result in more accuracy):

```
t_rates=c(mu11,mu12,mu13,mu21,mu22,mu23,mu31,mu32,mu33)
(gen_matrix=matrix(t_rates,3,3,byrow=TRUE,
dimnames=list(c("healthy","sick","dead"),c("healthy","sick","dead"))))

        healthy sick dead
healthy   -0.12    0 0.12
sick       0.00    0 0.00
dead       0.00    0 0.00

dt=1/365;    (P_dt =diag(3) + gen_matrix*dt)

          healthy sick         dead
healthy 0.9996712    0 0.0003287671
sick    0.0000000    1 0.0000000000
dead    0.0000000    0 1.0000000000
```

Given that all lives are alive at $t = 0$ we can calculate the probabilities at (almost) continuous times of being alive or dead over, say, the next 20 years.

```
years<-20;  current=c(1,0,0) #lives initially alive in 2-state model
fut_prob2<-matrix(0,years/dt,3)
for(i in 1:(years/dt)){
current =current%*%P_dt
fut_prob2[i,]<-current}
```

The solution is plotted in Figure 22.7 (left). From the earlier example, the probability of a life being alive at $t = 10$ years is obtained from the following:

```
fut_prob2[10/dt,1]

[1] 0.3011348
```

Comparing the solution at $t = 10$ years with the exact solution obtained from the methodology developed in Chapter 16:

$$_{10}p_x = e^{-\int_0^{10} 0.12dt} = e^{-0.12 \times 10} = 0.3011942$$

Exercise 22.11 *Using appropriate code, calculate the theoretical solution (from $t = 0$) using the 2-state model discussed in Chapter 16 and plot the solutions on top of the numerical solution shown in Figure 22.7 (left).*

From this exercise we can see that our Markov jump model provides us with the solution to the differential equation for the alive-dead model (see Chapter 16). Of course we don't need to employ this numerical method here as an exact theoretical solution is available due to the simple nature of the differential equation. We will, however, use this to solve the slightly more complex 3-state H-S-D model below.

Example 22.3 Including the sick state (Figure 22.6, left).

Assigning values to the required rates under this model:

```
mu12=0.3;    mu13=0.12;    mu21=0.9;    mu23=0.2;    mu31<-mu32<-0.0
mu11=-mu12-mu13;           mu22=-mu21-mu23;          mu33=-mu32-mu31
```

Thus the generator matrix, and probability matrix over a small time interval, $\Delta t = \frac{1}{365}$ years, are now:

```
gen_matrix    #generator matrix

          healthy sick dead
healthy   -0.42  0.3 0.12
sick       0.90 -1.1 0.20
dead       0.00  0.0 0.00

P_dt            #prob matrix

          healthy          sick           dead
healthy 0.998849315 0.0008219178 0.0003287671
sick    0.002465753 0.9969863014 0.0005479452
dead    0.000000000 0.0000000000 1.0000000000
```

Proceeding to calculate the probabilities that the life is in the healthy, sick, and dead states, given the life was healthy at $t = 0$, and applying alternative code using the markovchain object:

```
years<-20; dt<-1/365; fut_probs<-matrix(0,(years/dt),3)
colnames(fut_probs)<-c("healthy","sick","dead")
hsd_probs<-new("markovchain",states=c("healthy","sick","dead"),transitionMatrix=P_dt)
for(i in 1:(years/dt)){
fut_probs[i,]<-(hsd_probs^i)[1,]}   #initial state = healthy
```

A plot of the results is shown in Figure 22.7(right); these are the approximate solutions using the previous simple, numerical method to the following set of three differential

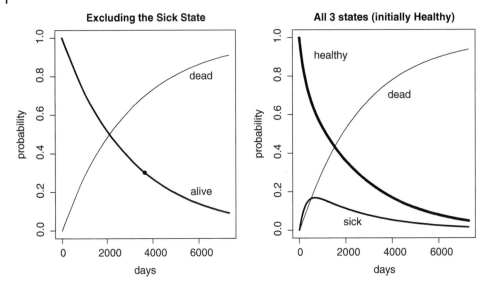

Figure 22.7 Solutions to H-S-D model (constant rates).

equations, which describe the H-S-D model, given the life was healthy at $t = 0$ (the initial condition):

$$\frac{d_t p_x^{HH}}{dt} = {}_t p_x^{HS} \mu_{x+t}^{SH} - {}_t p_x^{HH}(\mu_{x+t}^{HS} + \mu_{x+t}^{HD})$$

$$\frac{d_t p_x^{HS}}{dt} = {}_t p_x^{HH} \mu_{x+t}^{HS} - {}_t p_x^{HS}(\mu_{x+t}^{SH} + \mu_{x+t}^{SD})$$

$$\frac{d_t p_x^{HD}}{dt} = {}_t p_x^{HH} \mu_{x+t}^{HD} + {}_t p_x^{HS} \mu_{x+t}^{SD}$$

(See *MacDonald Section 14.6* for a derivation of these equations.) The relevant differential equations in respect of a life which is in the sick state at $t = 0$ are similar:

$$\frac{d_t p_x^{SH}}{dt} = {}_t p_x^{SS} \mu_{x+t}^{SH} - {}_t p_x^{SH}(\mu_{x+t}^{HS} + \mu_{x+t}^{HD})$$

$$\frac{d_t p_x^{SS}}{dt} = {}_t p_x^{SH} \mu_{x+t}^{HS} - {}_t p_x^{SS}(\mu_{x+t}^{SH} + \mu_{x+t}^{SD})$$

$$\frac{d_t p_x^{SD}}{dt} = {}_t p_x^{SS} \mu_{x+t}^{SD} + {}_t p_x^{SH} \mu_{x+t}^{HD}$$

Exercise 22.12 *Calculate the probabilities of existing in the three states over the next 20 years for lives who are sick at $t = 0$.*

For completeness, the relevant differential equations for lives in the dead state:

$$\frac{d_t p_x^{DH}}{dt} = \frac{d_t p_x^{DS}}{dt} = \frac{d_t p_x^{DD}}{dt} = 0$$

These last equations simply state that once a life is dead nothing changes.

Exercise 22.13 *Derive the above nine differential equations from first principles.*

In this case, all rates μ_{x+t}^{XY}, between states X and Y, are constant i.e. $\mu_{x+t}^{XY} = \mu^{XY}$; often referred to as a time-homogenous model. In Section 22.6 we solve for the more realistic case where the rates are dependent on age.

Markov jump models can therefore easily provide solutions to such models. The reader who has studied numerical solutions to differential equations will recognise this as Euler's method, and is encouraged to satisfy themselves that the two methods are exactly equivalent.

Below we verify the Markov chain calculations by using the discretized versions of the differential equations set out earlier. Calculating the probabilities of existing in each of the three states using the same parameter values (given the life was healthy at $t = 0$):

$$\mu^{HS} = 0.3 \quad \mu^{SH} = 0.9 \quad \mu^{HD} = 0.12 \quad \mu^{SD} = 0.2 \quad \Delta t = \frac{1}{365}$$

At $t = 1$ day:

$$_1 p_0^{HH} = 1 - (\mu^{HS} + \mu^{HD})\Delta t = 0.9988493$$

$$_1 p_0^{HS} = \mu^{HS}\Delta t = 0.0008219$$

$$_1 p_0^{HD} = \mu^{HD}\Delta t = 0.0003288$$

And at the next time, $t = 2$ days:

$$_2 p_0^{HH} = (1 - \mu^{HS}\Delta t - \mu^{HD}\Delta t)^2 + \mu^{HS}\Delta t.\mu^{SH}\Delta t = 0.997702$$

$$_2 p_0^{HS} = (1 - \mu^{HS}\Delta t - \mu^{HD}\Delta t)\mu^{HS}\Delta t + \mu^{HS}\Delta t.(1 - \mu^{SH}\Delta t - \mu^{SD}\Delta t) = 0.0016404$$

$$_2 p_0^{HD} = (1 - \mu^{HS}\Delta t - \mu^{HD}\Delta t)\mu^{HD}\Delta t + \mu^{HS}\Delta t.\mu^{SD}\Delta t + \mu^{HD}\Delta t = 0.0006576$$

This is simply a recursive application of Euler's method. These figures can be verified from the results calculated earlier using Markov chains:

```
fut_probs[1:2,]

        healthy        sick         dead
[1,] 0.9988493 0.0008219178 0.0003287671
[2,] 0.9977020 0.0016404128 0.0006576063
```

Alternative numerical solutions

The Euler scheme described above is one of the simpler numerical methods available to solve differential equations such as those shown earlier. Various alternatives exist that can provide more accurate solutions to systems of non-linear equations, such as Runge-Kutta methods, which are beyond the scope of this book.

Remark 22.5 Matrix exponential method

There exists a closed form solution to the above model with constant transition rates – the "matrix exponential method".

$$\frac{d}{dt} \boldsymbol{p}_t = \boldsymbol{p}_t \boldsymbol{\mu} \tag{22.8}$$

where $\boldsymbol{p}_t$ is the matrix containing occupancy probabilities at time t, and $\boldsymbol{\mu}$ is the generator matrix; both are square matrices of equal dimension. This has a solution:

$$\boldsymbol{p}_t = e^{\mu t} \boldsymbol{p}_0 \tag{22.9}$$

where $\boldsymbol{p}_0$ include the initial occupancy probabilities. It can be shown that:

$$e^{\mu} = \sum_{k=0}^{\infty} \frac{\mu^k}{k!} \tag{22.10}$$

with direct analogy to Taylor series expansion. Thus given a generator matrix, $\boldsymbol{\mu}$, we can determine e^{μ}, and the matrix of occupancy probabilities $\boldsymbol{p}_t$, at time, t.

Next, we have compared the solutions using the expm function in the package of the same name, with the numerical solution obtained above using Markov chains:

```
expm(gen_matrix)      # t=1
            healthy      sick      dead
healthy 0.7301916 0.1494920 0.1203164
sick    0.4484759 0.3913431 0.1601810
dead    0.0000000 0.0000000 1.0000000

fut_probs[365,]                # numerical method

  healthy      sick      dead
0.7300250 0.1496452 0.1203298

expm(gen_matrix*10)   # t=10
            healthy       sick      dead
healthy 0.1926628 0.06014621 0.747191
sick    0.1804386 0.05633134 0.763230
dead    0.0000000 0.00000000 1.000000

fut_probs[365*10,]             # numerical method

   healthy       sick       dead
0.19261173 0.06013029 0.74725798
```

Also shown is our own code aiming to duplicate the expm function output:

```
gen_matrix2=diag(3);  gen_matrix3=diag(3);  gen_matrix4=diag(3)

for (rr in 1:20) {gen_matrix2<-(gen_matrix2%*%gen_matrix)
gen_matrix3<-gen_matrix2/factorial(rr)
gen_matrix4<-gen_matrix4+gen_matrix3#for unit time
}

gen_matrix4  #answer at t=1

         healthy      sick      dead
[1,] 0.7301916 0.1494920 0.1203164
[2,] 0.4484759 0.3913431 0.1601810
[3,] 0.0000000 0.0000000 1.0000000

mat_time5<-gen_matrix4%*%gen_matrix4%*%gen_matrix4%*%gen_matrix4%*%gen_matrix4 #t=5
```

```
(mat_time10<-mat_time5%*%mat_time5)    #t=10

         healthy       sick       dead
[1,] 0.1926628 0.06014621 0.747191
[2,] 0.1804386 0.05633134 0.763230
[3,] 0.0000000 0.00000000 1.000000
```

However, no closed form solution exists where the parameter values are time-dependent; thus other numerical methods such as those used throughout this chapter will prove invaluable.

End of Remark 22.5

It may be useful for the reader to note that the comparison of discrete time and continuous time models is analogous to our discussion in Chapter 3 regarding the comparison of the force of interest with effective rates of interest; there it was noted that the force of interest allows us to determine values at any point in time, whereas effective rates only allow us to determine values at those discrete times consistent with the time interval used in the effective rate (unless simplifying assumptions are made).

Exercise 22.14 *The above transition rate matrix includes rates consistent with a country with poor levels of healthcare resulting in relatively high morbidity and mortality rates. Calculate solutions for the H-S-D model where healthcare standards and availability of medicine is at a higher level, and comment on the different appearance of the plot with that given previously.*

Exercise 22.15 *The H-S-D model discussed above is often referred to as the "H-S-D with recovery" model, as lives which become sick can become healthy again. Re-run the model with no probability of recovery from sickness (often referred to as the "H-S-D without recovery" model) and compare the solutions.*

Before moving onto the next section in which we discuss the calculation of insurance premiums, the reader may wish to try the following two exercises.

Exercise 22.16 *2-state, forward and backward transitions model*

This is another 2-state model, similar to the "alive-dead" model (Figure 22.6, right), but here transitions can occur in both directions and the initial state can be either of the two states.

Let's model internet activity over a period of 24 hours. A user can either be "on" the internet ("n"), or "off" ("f") the internet. Calculate the probabilities of being on or off the internet given either initial state, using the following constant, hourly rates: $\mu_{fn} = 0.2$; $\mu_{nf} = 0.05$.

This Exercise, like our previous example of the 2-state model, has a closed-form solution. The reader should derive the following exact solutions, and compare with the solutions obtained numerically using Markov chains (shown in Figure 22.8):

$$p_{nn} = 0.8(1 + 0.25e^{-0.25t})$$

$$p_{ff} = 0.8(0.25 + e^{-0.25t})$$

$$p_{fn} = 0.8(1 - e^{-0.25t})$$

$$p_{nf} = 0.2(1 - e^{-0.25t})$$

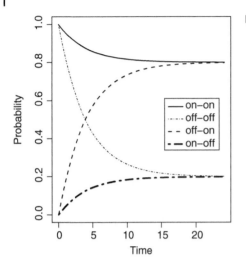

Figure 22.8 Solution to 2-state 2-way model.

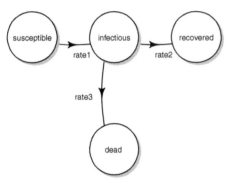

Figure 22.9 SIRD model.

Exercise 22.17 *The SIRD model (Figure 22.9)*

The SIRD model is one of the fundamental models used to model populations subject to disease, such as COVID-19. Under the model, susceptible lives (S) have not previously contracted COVID, and can catch COVID from infectious lives (I). Infectious lives can either recover (R) or die (D). The following rates are required under this model:

$$\text{rate1} = bp_t^i \; ; \quad \text{rate2} = k \; ; \quad \text{rate3} = \mu$$

where b is the transmissivity rate, p_t^i is the probability that a life is infectious at time t, k is the rate of recovery, and μ is the rate of death from the infectious state. Note that the rate at which susceptible lives become infectious, bp_t^i, is proportional to the proportion of the population which is infectious, p_t^i; this is a key assumption of this model. The value of parameter b will depend on the degree of social distancing, face coverings etc. Using the following parameter values and a Markov jump model, calculate the proportions in each of the four states from $t = 0$ to $t = 100$, assuming that 99% of the population are initially susceptible and 1% are infectious.

$$b = 0.3\text{pa}; \quad k = 0.13\text{pa}; \quad \mu = 0.04\text{pa}$$

Figure 22.10   Solutions to SIRD exercise.

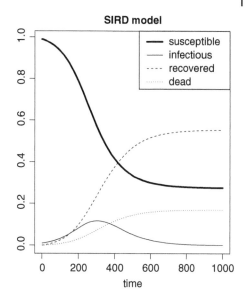

The solution to Exercise 22.17 is shown in Figure 22.10.

Remark 22.6 Exercise 22.17 involves a non-linear system, so would produce different results under Euler and Runge-Kutta schemes. The reader may wish to explore such differences.

22.6   Non-Constant Rates

We noted earlier that most of the models studied in this chapter will include transition rates which are constant and not dependent on time, that is, time-homogenous. This may be considered reasonable where the actual rates are not expected to vary materially (e.g. for short term products such as one-year term assurance), or for verification purposes.

For other scenarios however, this approach will not do, and we should allow for non-constant rates. We now develop the above ideas to allow for non-constant rates with a simple example. (Note that the alternative closed-form method – the matrix exponential method – cannot be applied to such models.)

Example 22.4 Consider the simple 2-state alive-dead model discussed at the start of Section 22.5.2; clearly an age-dependent rate should be used here. Using the following expression for mortality rates for a life aged exactly x years, calculate the probability that a new-born ($x = 0$ years) will survive to age 80 years:

$$\mu_x = 0.0001 \times 1.075^x$$

The calculations are very similar to those shown earlier; the key difference is that the generator matrix is recalculated at each point in our loop, at the mid-point of each discrete age band (solutions plotted in Figure 22.11 (left)):

```
dt1=0.01;          lastage<-100
P_dt1=diag(2);     current1=diag(2);        mat_soln<-NULL
#start loop
for(j in 1:(lastage/dt1)){
age_step<-j*dt1
mu12=0.0001*1.075^(age_step-dt1/2)
mu11= -mu12
t_rates1=c(mu11,mu12,0,0)
gen_matrix=matrix(t_rates1,2,2,byrow=TRUE)
P_dt1 =diag(2) + gen_matrix*dt1
current1<-current1%*%P_dt1
mat_soln[j]<-current1[1,1]}
mat_soln[80/dt1] #surv prob at age 80

[1] 0.6383532
```

Remark 22.7 A similar example is also discussed in Chapter 16, where the exact result was calculated.

The following exercise develops the simple 2-state model above, using the H-S-D 3-state model to calculate occupancy probabilities.

Exercise 22.18 *Calculate healthy, sick and dead state occupancy probabilities for a 60-year-old healthy life up to age 100 years, using an H-S-D model with the following age-dependent rates. The solution is plotted in Figure 22.11 (right).*

$$\mu_x^{HS} = 0.0004 + 0.0000001e^{0.18x}$$

$$\mu_x^{HD} = 0.0005 + 0.000055e^{0.09x}$$

$$\mu_x^{SH} = 0.5\mu_x^{HS}$$

$$\mu_x^{SD} = 2\mu_x^{HD}$$

A visual comparison of Figure 22.7 (right) and Figure 22.11 (right) is useful here.

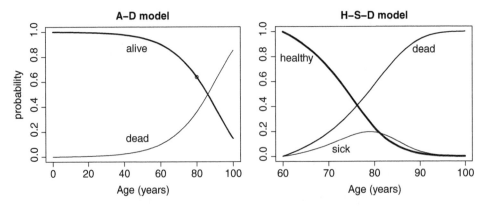

Figure 22.11 Probabilities from models with non-constant parameters (using Markov chains).

22.7 Premium Calculations

We now develop the above calculations to carry out premium calculations for sickness policies. We shall use the same H-S-D model introduced in Section 22.5.2.

Example 22.5

Under an annual sickness contract the policyholder receives daily payments at the rate of £1,000 per annum while sick. No death benefits are payable under the policy. Calculate, using the (constant) transition rates set out in Example 22.3, the single premium which is due at the start of the contract. Assume the policyholder is healthy at the start of the contract, and ignore the effect of discounting, and any loadings such as expenses and profit.

Here, we assume the single premium payment is simply the EPV of the sickness benefit; we need to calculate the probability that a life which is healthy at the start of the contract is subsequently sick on each day. Setting the calculations to run over one year (some of the code here is repeated from Example 22.3):

```
mu12=0.3; mu13=0.12; mu21=0.9; mu23=0.2; mu31<-mu32<-0.0
sick_ben1<-1000;  dt=1/365;    years<-1;   a<-matrix(0,(years/dt),3)
hsd_probs[1:3,1:3]   # hsd_probs was calculated in earlier example

              healthy         sick         dead
healthy 0.998849315 0.0008219178 0.0003287671
sick    0.002465753 0.9969863014 0.0005479452
dead    0.000000000 0.0000000000 1.0000000000

for(i in 1:(years/dt)){
a[i,]<-(hsd_probs^i)[1,]}

(ann_prem<-sum(sick_ben1*dt*a[,2]))    #answer

[1] 94.89696
```

Exercise 22.19 *A competitor believes that it has much higher quality data, and estimates sickness and recovery rates which are both half of those set out in Example 22.5. Calculate the competitor's premium rate.*

The following is similar to Example 22.5, but requires weekly premiums to be paid only when the policyholder is healthy.

Example 22.6

A new annual sickness policy is being designed which pays out £100 each week if the policyholder is deemed to be sick. No death benefits are payable under the policy. Premiums are paid weekly in advance only whilst the policyholder is healthy, with benefits payable at the end of each week. Using the same, constant transition rates as in Examples 22.3 and 22.5, calculate the weekly premiums due assuming an annual force of interest of 5%.

In this example we use functions from the `markovchain` package (the reader may wish to verify their answer using the same method as earlier).

```
mu12=0.3;    mu13=0.12;    mu21=0.9;    mu23=0.2;    mu31<-mu32<-0.0
mu11=-mu12-mu13;        mu22=-mu21-mu23;        mu33=-mu32-mu31
t_rates=c(mu11,mu12,mu13,mu21,mu22,mu23,mu31,mu32,mu33)

dt_week=1/52
gen_matrix=matrix(t_rates,3,3,byrow=TRUE)
P_dt =diag(3) + gen_matrix*dt_week
(MC_week <- new("markovchain", states=c("H","S","D"), transitionMatrix=P_dt))

Unnamed Markov chain
 A  3 - dimensional discrete Markov Chain defined by the following states:
 H, S, D
 The transition matrix  (by rows)  is defined as follows:
        H            S            D
H 0.99192308 0.005769231 0.002307692
S 0.01730769 0.978846154 0.003846154
D 0.00000000 0.000000000 1.000000000
```

This gives us the required matrix of the "weekly probabilities" (this is a further approximation which is likely to be acceptable). Thus we can calculate the probability that a life, which is healthy at the start of the policy, is healthy at the start of each week:

```
delta<-0.05/52       #weekly force of interest
prob_hh<-NULL
for(j in 1:51){prob_hh[j]<-transitionProbability(MC_week^j,"H","H")}
head(prob_hh)#prob of being healthy each week

[1]  0.9919231 0.9840112 0.9762602 0.9686659 0.9612243 0.9539314
```

And hence calculate the expected present value of a premium of £1 payable at the start of each week if the life is healthy:

```
discount1<-exp(-delta*seq(1,51))
pv_prem<-NULL
pv_prem<-c(1,prob_hh*discount1) #premiums payable in advance
sum(pv_prem)

[1] 43.02299
```

Calculating the expected present value of the benefits, and the required premium:

```
prob_hs<-NULL
for(j in 1:52){prob_hs[j]<-transitionProbability(MC_week^j,"H","S")}
discount2<-exp(-delta*seq(1,52))
pv_ben<-100*prob_hs*discount2
sum(pv_ben)

[1]  488.3709

(Sickness_premium<-sum(pv_ben)/sum(pv_prem))

[1] 11.3514
```

Hence a weekly premium of £11.35 is payable at the start of each week whilst the policyholder is healthy under this policy.

Exercise 22.20 *(Challenging)*

It has been to decided to offer a policy similar to the above but which also pays a death benefit of £10,000, where death can be from any state. Calculate the value of this death benefit.

Note that for the Examples and Exercises included earlier in this section, constant transition rates may be deemed acceptable given the short nature of the contract. If the term is longer, it becomes more likely that age-dependent rates are required. Next we look at a longer term product with non-constant rates.

Example 22.7 *Premium calculation with non-constant rates*

Using the same age-dependent rates as set out in Exercise 22.18, calculate the required premium, payable weekly in advance, in respect of the following policy: Weekly sickness benefits (paid in arrears) equal to £10,000 pa while sick, plus a lump sum payment of £20,000 payable immediately on death from any state. The policy is for a life aged 50 years, and a term of 20 years. Assume $\delta = 5\%$ pa. Ignore expenses and profit.

```
sick_benx<-10000;         death_benx<-20000
agestart<-50;             lastage<-70

a1=0.0004;    b1=0.0000001;    c1=0.18
a3=0.0005;    b3=0.000055;     c3=0.09

mu21_12_ratio<-0.5;    mu23_13_ratio<-2

delta<-0.05;    dt=1/52    #weekly
P_dt=diag(3);    current=diag(3)
jj<-matrix(0,(lastage-agestart+1)/dt,3)
jj[1,]<-c(1,0,0)
pv_sick<-NULL;  pv_death<-NULL;  pv_prem<-NULL

for(j in 1:((lastage-agestart)/dt)) {
step_count<-(j-1)*dt
mu12=a1+b1*exp(c1*(agestart+step_count))
mu21=mu21_12_ratio*mu12
mu13=a3+b3*exp(c3*(agestart+step_count))
mu23=mu23_13_ratio*mu13
mu11=-mu12-mu13;  mu22=-mu21-mu23;  mu33=-mu32-mu31

t_rates=c(mu11,mu12,mu13,mu21,mu22,mu23,mu31,mu32,mu33)
gen_matrix=matrix(t_rates,3,3,byrow=TRUE)
P_dt =diag(3) + gen_matrix*dt
current<-current%*%P_dt
jj[j+1,]<-current[1,]
discount<-exp(-delta*j*dt)
pv_prem[j]<-dt*(jj[j+1,1])*discount #further adjustment below
pv_sick[j]<-dt*(jj[j+1,2])*discount
pv_death[j]<-(jj[j,1]*mu13+jj[j,2]*mu23)*discount*dt*exp(delta*dt/2)
}

pv_prem_adv<-c(dt,pv_prem[-length(pv_prem)])     #want in advance
(premHSD_week<-(sick_benx*sum(pv_sick)+death_benx*sum(pv_death))/sum(pv_prem_adv))*dt

[1] 10.24735

#weekly premium due in advance
```

Exercise 22.21 *State the assumptions used in the above code and make adjustments which may be warranted on grounds of accuracy.*

22.8 Transition Rate Estimation

Before completing our discussions with the closely related subject of multiple decrement models, we should comment on the fundamental requirement of estimating the transition rates in the jump models; without these estimates we cannot obtain solutions to our models.

This topic was covered in Chapter 16 when we discussed the 2-state Markov model, and the same principles apply here (hence the relatively brief discussion). In the 2-state model we noted the following important expression for the maximum likelihood estimate of the constant mortality rate:

$$\hat{\mu} = \frac{d}{v} \tag{22.11}$$

where, from a particular investigation or dataset, d was the total number of observed deaths and v was the total time for which all lives were observed alive. Similarly it is straightforward to show in the multiple state model the corresponding general expressions for each transition rate $\hat{\lambda}_{ij}$ between states i and j:

$$\hat{\lambda}_{ij} = \frac{t_{ij}}{v_i} \tag{22.12}$$

where t_{ij} is the total number of transitions between states i and j, and v_i is the total time for which all lives are observed in state i. In the same way as we obtained a parametric form for the mortality rate in Chapter 17 we can obtain parametric functions for each of the transition rates on our model (as included in Exercise 22.18).

22.9 Multiple Decrement Models

Remark 22.8 As with most examples and exercises in this chapter, the reader is reminded of the benefit of tackling problems using a number of approaches.

22.9.1 Introduction

Multiple decrement models represent a particular subset of multiple state models, where the initial, or starting, state is defined and the only transitions permitted are decrements from that starting state. Hence decrement models are typically simpler than most multiple state models. Often we will not be concerned with events after the initial transition, making such decrement models particularly relevant. For example, we may wish to model how an employee may leave a company's employment e.g. by retirement, resignation, dismissal, illness, or death - it may not be of concern to the employer what happens to the life after employment ceases.

Figure 22.12   Multiple decrement model.

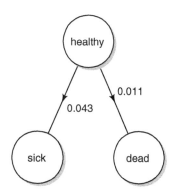

Figure 22.12 is an example of a multiple decrement model, and is used in Example 22.8; such a model may be useful where a healthy life can purchase a sickness policy which includes a single benefit payment due on the first occurrence of sickness or death (the policy ceases once a payment has been made). Note that the H-S-D model discussed in Section 22.5.2 is not a multiple decrement model.

We first solve these models, with constant rates, numerically, and proceed to compare with an exact method. We finish the chapter with an example which incorporates age-dependent rates.

22.9.2   Using a Numerical Approach for the above Fixed Rate Problems

It is straightforward to calculate occupancy probabilities in situations where the transition rates are fixed by using similar Markov jump models and code introduced in Section 22.5.

Example 22.8

Calculate the value of benefits payable under a 10-year policy where £75,000 is payable on the onset of sickness and £150,000 is payable on death, whichever occurs first. Assume an annual force of interest of 5%pa, and that transition rates are equal to those set out in Figure 22.12. (A maximum of one payment can be made under the policy.)

```
mu12<- 0.043;        mu13<- 0.011 #hs and hd resp
mu21=0.0; mu23=0.0; mu31=0.0; mu32=0.0
mu11=-mu12-mu13; mu22=-mu21-mu23; mu33=-mu32-mu31
```

Note that there are now only two non-zero rates in the model. The generator matrix is given by:

```
t_rates=c(mu11,mu12,mu13,mu21,mu22,mu23,mu31,mu32,mu33)
(gen_matrix=matrix(t_rates,3,3,byrow=TRUE))

         [,1]   [,2]   [,3]
[1,] -0.054 0.043 0.011
[2,]  0.000 0.000 0.000
[3,]  0.000 0.000 0.000
```

Proceeding to calculate the expected present value of these benefits:

```
sick_ben<-75000;        death_ben<-150000
deltax<-0.05;      term<-10;      dt=1/365
P_dt =diag(3) + gen_matrix*dt

current=diag(3)
value_sickness<-matrix(0,term/dt,3);    value_death1<-matrix(0,term/dt,3)
a<-matrix(0,(term/dt+1),3);             a[1,]<-c(1,0,0)

for(i in 1:(term/dt)){
value_sickness[i]<-a[i,1]*mu12*dt*exp(-deltax*dt*i)
value_death1[i]<-a[i,1]*mu13*dt*exp(-deltax*dt*i)
current =current%*%P_dt
a[i+1,]<-current[1,]
}

sum(value_sickness)*sick_ben;    sum(value_death1)*death_ben

[1] 20048.9
[1] 10257.58

(totalcost<-sum(value_sickness)*sick_ben+sum(value_death1)*death_ben)

[1] 30306.48

a[3651,] #probs at end - we shall compare these with exact calculations later

[1] 0.58272497 0.33227456 0.08500047
```

The total value of this contract at inception is therefore £30306.

Regarding the occupancy probability of the healthy state, the reader should note the difference between this calculation and those performed earlier under the H-S-D model where more transitions were possible.

Exercise 22.22 *The policy described in Example 22.8 is now adjusted to include an extra death benefit of £75,000, payable on death from the sick state (this would seem sensible as it may be considered unfair if the life died shortly after being declared ill and received only £75,000). Calculate the additional premium required assuming a death rate from the sick state of 0.08 per annum. (Here we need to revert to the multiple state model.)*

22.9.3 An Exact Approach

Remark 22.9 Note that this methodology is typically included in a Contingencies-type chapter in textbooks covering the topics of survival analysis and life-contingent risks.

This section is closely linked to Chapter 16. Recalling the important equation from the 2-state Markov model, where μ is the mortality rate:

$$_nP_x = e^{-\int_0^n \mu_{x+t}dt} \tag{22.13}$$

This 2-state model is a single decrement model; it is straightforward to derive the equivalent "survival" equation for the multiple decrement model, and is left as an exercise for the reader. Here we show the probability of a life, aged exactly x, remaining healthy for n

years under the 2-decrement H-S-D model (Figure 22.12):

$$_nP_x^{HH} = e^{-\int_0^n (\sigma_{x+t} + \mu_{x+t})dt} \tag{22.14}$$

where σ is the transition rate from healthy to sick and μ is the transition rate from healthy to dead. The probability of a healthy life, aged x, becoming sick before age $x + n$ is given by:

$$_nP_x^{HS} = \int_0^n {}_tP_x^{HH} \sigma_{x+t} dt \tag{22.15}$$

and thus:

$$_nP_x^{HS} = \int_0^n \sigma_{x+t} e^{-\int_0^t (\sigma_{x+t'} + \mu_{x+t'})dt'} dt \tag{22.16}$$

If the transition rates are constant the above simplifies to:

$$_nP_x^{HS} = \sigma \int_0^n e^{-(\sigma+\mu)t} dt \tag{22.17}$$

and finally:

$$_nP_x^{HS} = \frac{\sigma}{(\sigma + \mu)} (1 - e^{-(\sigma+\mu)n}) \tag{22.18}$$

And similarly for $_nP_x^{HD}$.

Exercise 22.23 *Derive equations 22.14 to 22.18, stating all assumptions.*

To calculate the expected present value of the benefits payable at the time of a transition, we need to include the force of interest, δ, in Equation 22.16. For example, the expected present value of a payment of 1 ("EPV") made on first becoming sick is given by:

$$EPV^S = \int_0^n \sigma_{x+t} e^{-\int_0^t (\sigma_{x+t'} + \mu_{x+t'} + \delta_{x+t'})dt'} dt \tag{22.19}$$

If the transition and interest rates are constant, this becomes:

$$EPV^S = \frac{\sigma}{(\sigma + \mu + \delta)} (1 - e^{-(\mu+\sigma+\delta)n}) \tag{22.20}$$

We shall use Equations 22.17 to 22.20 to solve the examples and exercises which follow in this chapter, where our objective is to calculate the probabilities of a healthy life either becoming sick, or dying, using the 2-decrement model. Subsequently, we will calculate the expected present value of the potential cashflows payable under such an agreement using both fixed transition rates and more realistic age-dependent rates.

First, looking at examples with constant rates:

Example 22.9

Verifying the calculations of $_{10}P_x^{HS}$ and $_{10}P_x^{HD}$ from Example 22.8, using Equation 22.18

```
sigma<-0.043;      mu<-0.011;      term<-10
(pHS<-sigma/(sigma+mu)*(1-exp((-sigma-mu)*term)))
```

[1] 0.332256

```
(pHD<-mu/(sigma+mu)*(1-exp((-sigma-mu)*term)))
```

[1] 0.08499573

This is consistent with the solution obtained from Example 22.8.

Example 22.10

Solving Example 22.8 exactly using (1) Equation 22.19, and (2) Equation 22.20.

```
deltax<-0.05;     sigma<-0.043;      mu<-0.011;      term<-10
sick_int<-function(t){sigma*exp(-(mu+sigma+deltax)*t)}
sick_ben*integrate(sick_int,0,10)$value
```

[1] 20049.12

```
#OR
sick_ben*sigma/(mu+sigma+deltax)*(1-exp(-(mu+sigma+deltax)*10))
```

[1] 20049.12

```
death_int<-function(t){mu*exp(-(mu+sigma+deltax)*t)}
death_ben*integrate(death_int,0,10)$value
```

[1] 10257.69

```
#OR
death_ben*mu/(mu+sigma+deltax)*(1-exp(-(mu+sigma+deltax)*10))
```

[1] 10257.69

These are consistent with the values calculated previously from Example 22.8 (a total cost of £30306).

22.9.4 Age-Dependent Rates

Now we reset our question using more realistic age-dependent rates, and use Equation 22.16 given that the rates are not constant. Note that this is perhaps the real power of jump models where transition rates are used – it is far more convenient to calculate the transition probabilities for various scenarios using parametric functions for the time-dependent rates than it would be to interpolate using tables of probabilities, e.g. life tables.

Example 22.11

Using the 2-decrement H-S-D model with the transition rates for a life aged x years exact set out below, calculate the probabilities of a healthy life aged 60 years remaining healthy, becoming sick, or dying, by age 70 years. Also calculate the expected present value of the sickness/death 10-year policy detailed in Example 22.8 for such a life, using a constant annual force of interest equal to 5%pa.

$$\sigma_x = 0.004 \times 1.035^x \qquad \textit{sickness rate}$$

$$\mu_x = 0.00015 \times 1.075^x \qquad \textit{death rate}$$

```
sick<-function(t){0.004*1.035^t};    death<-function(t){0.00015*1.075^t}
sick_ben<-75000;                     death_ben<-150000
age_start<-60;    term<-10;    dt<-0.01;    steps<-term/dt
combine_interest<-function(t){death(t)+sick(t)+delta}
```

The probabilities of remaining healthy, becoming sick, and dying (under the multiple decrement model), calculated by applying Equations 22.14 and 22.16, are:

```
delta<-0.00    #set to zero - we only need probs for now
exp(-integrate(combine_interest,age_start,age_start+term)$value)   #prob survival

[1] 0.5799596

#probability of becoming sick
HSx<-NULL
for(t in seq(1,steps)){HSx[t]<-sick(age_start+(t-0.5)*dt)*dt*
exp(-integrate(combine_interest,age_start,age_start+t*dt)$value)
}
sum(HSx)

[1] 0.2914345

#probability of death
HDx<-NULL
for(t in seq(1,steps)){HDx[t]<-death(age_start+(t-0.5)*dt)*dt*
exp(-integrate(combine_interest,age_start,age_start+t*dt)$value)
}
sum(HDx)

[1] 0.1284919
```

Calculating the expected present values of the benefits:

```
delta<-0.05                 #now include discounting factor

#sickness benefit
HSpv<-NULL
for(t in seq(1,steps)){
HSpv[t]<-sick(age_start+(t-0.5)*dt)*dt*
exp(-integrate(combine_interest,age_start,age_start+(t-1)*dt)$value)
}
sum(HSpv)*sick_ben

[1] 17353.43

#death benefit
HDpv<-NULL
for(t in seq(1,steps)){
HDpv[t]<-death(age_start+(t-0.5)*dt)*dt*
exp(-integrate(combine_interest,age_start,age_start+(t-1)*dt)$value)
}
sum(HDpv)*death_ben

[1] 15065.44

sum(HDpv)*death_ben+sum(HSpv)*sick_ben

[1] 32418.87
```

Remark 22.10 Note that under the multiple decrement model, the probability of death calculated above does not include a healthy life first becoming sick and subsequently dying.

Exercise 22.24 *Verify the above calculations using numerical estimation, as described in Section 22.6.*

Exercise 22.25 *Recalculate the value of the policy in Example 22.11 using a non-constant force of interest equal to $\delta_t = 0.03 - 0.01e^{-0.5t}$.*

Exercise 22.26 *Repeat these calculations in Example 22.11 allowing for assumed policy lapses at the rate of 5% each year, following which no benefit is payable.*

22.10 Recommended Reading

- J Beyersmann, A Allignol, M Schumacher. (2012). *Competing Risks and Multistate Models with R.* New York: Springer.
- Boland PJ. (2007). *Statistical and Probabilistic Methods in Actuarial Science.* Boca Raton, Florida, US: Chapman and Hall/CRC.
- Z Brźniak and T Zastawniak. (1999). *Basic Stochastic Processes.* London: Springer-Verlag.
- Charpentier A et al. (2016). *Computational Actuarial Science with R.* Boca Raton, Florida, US: Chapman and Hall/CRC.
- Shailaja Deshmukh. (2012). *Multiple Decrement Models in Insurance.* New Delhi: Springer.
- Dickson DCM, Hardy MR, Waters HR. (2013). *Actuarial Mathematics for Life Contingent Risks.* Cambridge, UK: Cambridge University Press.
- Haberman, S., Pitacco, E. (1998). *Actuarial Models for Disability Insurance: A multiple state approach (1st ed.).* New York: Routledge.
- MacDonald AS et al. (2018). *Modelling Mortality with Actuarial Applications.* Cambridge, UK. Cambridge University Press.

23

Contingencies I

Peter McQuire

```
library(lifecontingencies)
library(moments)
```

Remark 23.1 This chapter is one of those recommended at the start of this book which can be used by readers who are already familiar with the mathematical material included in this chapter, but are new to the R language, and who wish to practise a basic level of coding. Much of this chapter takes the form of a set of examples and exercises, with the emphasis on writing code and tackling typical problems.

23.1 Introduction

The objective of this chapter is to develop an understanding of actuarial contingency functions, such as term assurance and permanent annuities, with the use and application of R code. Such functions will lead to the pricing of, and reserving for, typical annuity and life assurance policies, and also the calculation of actuarial liabilities of defined benefit pension schemes. This chapter is followed immediately by the Contingencies II chapter, which develops ideas covered in this chapter.

We will make use of the *AM92* and *AF92* life tables throughout this chapter, copies of which are included on the website. These tables were chosen as many readers will previously have used these, and already have access to them, allowing convenient verification of many of the calculations in this chapter. (The reader should of course apply the coding in this chapter to a variety of life tables available to them.) We will also introduce the simulation of future lifetimes from life tables which should reinforce the central ideas.

There is also some emphasis on checking our results. In particular, the reader may wish to study the later sections of Chapter 22, which provides alternative methods of pricing life assurance type products.

R Programming for Actuarial Science, First Edition. Peter McQuire and Alfred Kume.
© 2024 John Wiley & Sons Ltd. Published 2024 by John Wiley & Sons Ltd.
Companion Website: www.wiley.com/go/rprogramming.com.

Indeed, there is inevitably some overlap with material covered in other chapters of the book. It is important for the reader to understand alternative methods which can be applied to solve actuarial problems; being able to solve problems using a variety of methods should lead to a greater understanding of the underlying mathematics.

The layout of this chapter is as follows:

- The life table;
- Summary of common contingency expressions;
- Writing code for annuities and whole life assurance;
- `lifecontingencies` package applications: annuities and life assurance;
- Simulations of future lifetimes and discounted values.

However, the topics typically covered by life contingency courses are too vast to be comprehensively covered in this and the following chapter, and we have therefore limited the coverage to a smaller variety of topics.

In this chapter we introduce a number of functions from the `lifecontingencies` package which are convenient to use – however it is important that the reader fully understands the calculations behind the functions. As such, we first set a number of coding exercises which should enhance the reader's understanding of various types of annuities and assurance contracts. The reader's coding can be checked using the pre-existing functions in the `lifecontingencies` package.

If the reader is familiar with these actuarial concepts they may wish to start with Section 23.6 (although this means missing out on coding practice).

23.2   What is Meant by "Contingencies" in an Actuarial Context?

The subject of Life Contingencies is concerned with analysing payments which are due to a life (or lives) at some point in the future. Such payments will be contingent, or dependent on, the life (or lives) being alive, or being in some other state e.g. sick or dead. Typically, these lives will hold policies sold by a life assurance company or annuity company, or be members of a pension scheme.

Valuing such cashflows can therefore be considered to be an extension of valuing the guaranteed cashflows discussed in earlier chapters (e.g. an annuity certain); in addition to applying appropriate discounting factors to the cashflows we must also allow for the probability of each payment being made.

23.3   The Life Table

The life table is a simple construction which allows its user to calculate probabilities of survival and death over various age intervals. An example is the *AM92* life table (which we will use extensively in this chapter), excerpts of which are set out below:

	Age_x	lx
1	17	10000.000
2	18	9994.000
3	19	9988.064
	Age_x	lx
63	79	5619.758
64	80	5266.460
65	81	4901.479
	Age_x	lx
102	118	1.08e-05
103	119	2.19e-06
104	120	4.00e-07

Here the life table starts at age 17, with $l_{17} = 10000$ known as the radix. The radix can be interpreted as the number of identical lives in our population which are aged 17 years exact. We can define the probability of surviving between two exact ages, x and $x + t$ as follows:

$$S_x(t) = \frac{l_{x+t}}{l_x} \tag{23.1}$$

For example, based on *AM92*, the probability of a 17-year-old surviving to age 80 years, $S_{17}(63)$, is 0.526646, and the probability of an 80-year-old surviving to age 81, $S_{80}(1)$, is 0.930697. It is assumed under this table that no lives will survive past the age of 120 years.

Similarly, we can introduce another quantity, d_x, which can be interpreted as the number of deaths between exact ages x and $x + 1$.

Of course the radix value is arbitrary; it is the relative values of the l_x's which matter. The starting age of the table also can vary considerably e.g. life tables used by pension schemes will typically start around the age of 18 years as the vast majority of lives will be of working age or older. For departments concerned with retirement annuities, the life table may commence at a later age for example, 45 years.

The life table is a relatively intuitive construction and is simple in its application; it also offers flexibility to allow for particular features of populations' mortality rates for example, the "accident hump", the inclusion of which may be more problematic with a parametric model. It is also practical and appropriate for use still to this day by actuaries who will often be concerned with analysing annual payments to policyholders and pension scheme members.

Many methods have been used by actuaries to develop life tables; the texts by McCutcheon, Benjamin, and Tetley are recommended. The reader may also benefit from studying many of the CMI reports available on the IFoA's website. The reader may wish to refer to Chapters 16 and 17 which discuss fitting splines and parametric models such as the Gompertz model. The reader who is familiar with parametric models of mortality should consider the relative benefits and drawbacks of the life table.

23.4 Expected Present Values of the Key Contingency Functions

Set out below are the formulae for the key contingency functions together with a brief explanation for each one. The discussion here is rather brief – for a more detailed discussion of these functions please see Dickson et al. Chapters 4 and 5.

Each formula is based on the following general equation for the expected present value of a series of n future cashflows ("EPV"):

$$EPV = \sum_{m=1}^{n} v^{t_m} \, p_m \, C_m$$

where C_m is the amount of cashflow m, p_m is the probability of cashflow m occurring, and v^{t_m} is the discounting factor applicable for the payment at time t_m. (As per standard actuarial notation, $v = \frac{1}{1+i}$ where i is the annual effective rate of interest.) Note that here we are assuming a flat yield curve, such that v is constant – we shall see an example later in the chapter where the interest rate is not constant. In the following equations all cashflows, C, are for the amount of 1.

1. An annuity payable to a life aged x with payments made at regular periods (e.g. annually) in arrears until death, commonly referred to as an "annuity-immediate":

$$a_x = \sum_{t=1}^{\infty} v^t \, {}_t p_x$$

Example usage: to value all pensions currently in payment to annuity policyholders and to existing pensioners who are members of defined benefit schemes.

2. As above, but with payments made at the start of the period (known as an "annuity-due" or "in advance"):

$$\ddot{a}_x = \sum_{t=0}^{\infty} v^t \, {}_t p_x$$

3. Temporary annuity: An annuity payable to a life aged x with payments annually in arrears until death, but with a maximum of n payments (this is easily adjusted to allow for "in advance" as seen earlier):

$$a_{x:\overline{n}|} = \sum_{t=1}^{n} v^t \, {}_t p_x$$

Example usage: Used in the calculation of premiums in respect of term assurance policies, or to value a child's pension (paid following the death of a parent) which will cease when the life reaches a particular age, say 21 years.

4. Deferred annuity: An annuity payable to a life currently aged $x - z$ commencing from age x with payments annually in advance until death:

$$_{z|}\ddot{a}_x = {}_z p_{x-z} \, v^z \sum_{t=0}^{\infty} v^t \, {}_t p_x$$

Example usage: used to value "deferred pension members" of pension schemes; for example, a former employee with retained pension benefits accrued under their former employer's company pension scheme, currently aged 50 years, is entitled to receive a pension payable from the pension scheme from age 65 $(_{15|}\ddot{a}_{65})$.

And the common assurance functions:

5. Term assurance: the sum assured is paid only if the life dies within the specified term, n, of the policy; here the payment would be made at the end of the policy year:

$$A^1_{x:\overline{n|}} = \sum_{t=0}^{n-1} v^{t+1} \, {}_tp_x \, q_{x+t}$$

Example usage: This type of policy is commonly purchased, for example, by a homeowner who has recently taken out a mortgage in respect of a house purchase – if the homeowner dies during the mortgage repayment period a payment is made to help in meeting the outstanding loan amount.

6. Pure endowment: this is effectively a savings policy, with a payment being received by the policyholder if they survive the defined period, n:

$$A_{x:\overline{n|}}^{1} = v^n \, {}_np_x$$

7. Endowment assurance is simply a combination of the above two policy types:

$$A_{x:\overline{n|}} = \left(\sum_{t=0}^{n-1} v^{t+1} \, {}_tp_x \, q_{x+t}\right) + v^n \, {}_np_x$$

8. Finally, whole life assurance (where a payment is made at the end of the year of death):

$$A_x = \sum_{t=0}^{\infty} v^{t+1} \, {}_tp_x \, q_{x+t}$$

All the above expressions are in respect of the *expected* present value ("EPV") of future cashflows. Also of concern to us should be the uncertainty in these present values; this discussion is deferred to the following chapter.

Several variations commonly exist, in particular where payments increase, or decrease in the future, and, in the case of life assurance benefits where payments may be made immediately rather than at the end of a defined period (see later).

Exercise 23.1 *For a payment of £1,000 and a term (where applicable) of 10 years, rank the above four assurance policies for a 30-year-old life in order of value.*

We will shortly make use of the lifecontingencies package to calculate these annuity and insurance benefit expressions, and variations thereof. However, it's beneficial first to write our own code in respect of these.

23.5 Writing Our Own Code – Some Introductory Exercises

This section is mainly a series of exercises. Here we develop our own code for annuities and whole life assurance benefits, leaving the other types of policies as exercises for the reader. This develops understanding and also helps with our coding skills. The reader will also then be able to value specific policies which exhibit unique features. As mentioned earlier, the reader who is familiar with these concepts and does not wish to practise coding in this context may wish to move to Section 23.6.

We will be able to easily verify our calculations by using the lifecontingencies functions (covered later in the chapter).

(Tip: the cumprod function will be useful for a number of calculations which follow.)

Example 23.1 Whole life assurance

Write a function to calculate the EPV of a whole life assurance policy, A_x. We will initially write basic, fairly cumbersome code to calculate the whole life assurance for a 17-year-old male life using the *AM92* table and $i = 4\%$ *pa*. Downloading the life table:

```
am92_entire<-read.csv("~/am92_durn2.csv")
```

We proceed by first calculating the q_x's and p_x's and then applying the relevant equation:

```
am92<-am92_entire[1:length(am92_entire[,1]),c(1,2)]
q<-1-am92[seq(2,nrow(am92)),2]/am92[seq(1,nrow(am92)-1),2]
am92[,3]<-c(q,1)             #add column of q's
am92[,4]<-c(1-am92[,3])      #add column of p's
colnames(am92)[3:4]<-c("qx","px")
head(am92,3)

  Age_x      lx       qx        px
1    17 10000.000 0.000600 0.999400
2    18  9994.000 0.000594 0.999406
3    19  9988.064 0.000587 0.999413
```

and then apply the above equation for A_{17}:

```
interest_rate<-0.04;    v<-1/(1+interest_rate)
cum_prob<-c(1,cumprod(am92[,4])) #we need to shift this column down one row
am92[,5]<-cum_prob[-length(cum_prob)] #...and remove the last term
am92[,6]<-am92[,5]*am92[,3]
am92[,7]<-v^seq(1,length(am92[,1]))*am92[,6]
sum(am92[,7])

[1] 0.1012692
```

(We will shortly verify this figure using the Axn function from the lifecontingencies package.)

Exercise 23.2 *The previous code is cumbersome, and limited. Write a function to calculate A_x for any age x, as concisely as possible, and run your function to calculate the assurance function for all ages, 17 to 70 years, with any interest rate.*

Example 23.2 Annuity due

Set out below is example code for $\ddot{a}_{17}$, an annuity payable annually in advance to a 17-year-old, with $i = 4\%$ *pa* (duplicated code from the previous example is not repeated here):

```
sum(v^seq(0,length(am92[,1])-1)*am92[,5])

[1] 23.367
```

Exercise 23.3 *Write your own annuity function and calculate an annuity for all ages, 17 to 70 years, with any interest rate.*

Exercise 23.4 *By developing these functions from Exercises 23.2 and 23.3, calculate the following seven terms:*

$$\ddot{a}_{60} \quad a_{60:\overline{10|}} \quad {}_{15|}a_{75} \quad A^{1}_{30:\overline{10|}} \quad A_{30:\overline{20|}}^{1} \quad A_{25} \quad A_{30:\overline{10|}}$$

Exercise 23.5 *Under a 5-year term assurance policy a payment will be made at the end of the year in which death occurs; £100 will be paid if death occurs in the first year, increasing by £10 each year. If the policyholder survives 5 years no payment is made. Calculate the EPV of the benefits under this policy. (First estimate your answer.)*

Assumptions: The probability of the life surviving each year is 98%; constant force of interest, δ = 4% pa.

Exercise 23.6 *Calculate the EPV of a 5-year increasing temporary annuity, with payments annually in advance starting at £100 and increasing each year by £10.*

Assumptions: The probability of the life surviving each year is 98%; constant force of interest δ = 4% pa.

Exercise 23.7 *Calculate $a_{30:\overline{10|}}$, the EPV of a 10-year temporary annuity for a 30-year-old, payable in arrears, with an effective rate of interest i = 3% pa, and using the following life table:*

	age	lx	age	lx	age	lx
1	30	1000	34	960	38	920
2	31	990	35	950	39	910
3	32	980	36	940	40	900
4	33	970	37	930		

Exercise 23.8 *Mary is starting university in 12 months' time, taking a 4-year Masters course in Actuarial Science. Her parents want to ensure that the course fees can be paid in the event of her father's death. Annual university fees and living expenses are £15,000. The benefits required on death are therefore £60,000 at the end of year 1, reducing by £15,000 each year. Design and price a suitable assurance policy (use appropriate assumptions).*

23.6 The *Lifecontingencies* Package

In this section we will use various functions which will help us value annuities and life assurance products – a central role of the actuary. When applying these functions it is good practice for the reader to occasionally check the black-box nature of the functions using your code developed in the preceding part of this chapter.

We will make use of the functions in the lifecontingencies package and demonstrate the use of the lifetable and actuarialtable objects created under this package.

23.6.1 The Lifetable and Actuarialtable Objects

Before we use actual mortality tables we'll first construct our own simple `lifetable` and `actuarialtable` objects. All we require are two vectors; one for *age* and one for l_x:

```
age<-seq(0,90)
lx<-seq(9000,0,length.out=91)
```

We create the `lifetable` object as follows:

```
mort_table<-new("lifetable",age,lx)
tail(mort_table)

    x  lx
85 84 600
86 85 500
87 86 400
88 87 300
89 88 200
90 89 100
```

We then create the `actuarialtable` object from our `lifetable` object.

```
act_table<-new("actuarialtable",mort_table@x,mort_table@lx,
interest=0.03,"exampleactuarialtable")
```

Note that we are required to include an effective interest rate at this point; we have used 3% pa here. However, this can be easily changed when running calculations by including a different interest rate *at that point* (we do not need to recreate a new object). Note that it may be considered best practice to include the rate of interest in the code for reasons of clarity; this is usually the approach taken in this chapter.

We now proceed to calculate the expected present value of annuities and insurance benefits using these functions from the `lifecontingencies` package. The values can be checked using the functions developed earlier in the chapter.

Example 23.3

Calculate $ä_{88}$ using an effective interest rate of 3% pa – this annuity only has 2 payments, at ages 88 and 89 years as our simple life table only extends to age 89 (not very realistic). The answer below is easily verifiable.

Using the `axn` function and checking the answer:

```
axn(act_table,88)
```
```
[1] 1.485437
```
```
1+100/200/1.03   #check
```
```
[1] 1.485437
```

Note that the annuity is calculated with an interest rate of 3% pa (this was incorporated when writing the `actuarialtable` above). We can change this without rerunning our table by entering an extra argument in the function e.g. for 4% :

```
axn(act_table,88,i=0.04)    #extra argument included -- 4% overides the 3%

[1] 1.480769

1+100/200/1.04              #check

[1] 1.480769
```

Example 23.4

Calculate A_{87} with an interest rate of 3% pa:

```
Axn(act_table, 87)

[1] 0.9428705

(100/300)*(1/1.03+1/1.03^2+1/1.03^3)

[1] 0.9428705
```

23.6.2 Application to Actual Mortality Tables: AM92 and AF92

We now apply the above methods to actual mortality tables – the *AM92* and *AF92* tables used earlier in this chapter. (We strongly recommend that the reader carries out the calculations in this chapter using various alternative tables, particularly those included on the excellent *HMD* website (*https://www.mortality.org/*).)

Constructing the `lifetable` and `actuarialtable` objects, and setting our chosen interest rate to $i = 4\%$ pa:

```
am92<-read.csv("~/am92_durn2.csv") #previously downloaded
am92_lt<-new("lifetable",  x=am92[,1],lx=am92[,2],name=" am92 males")
am92_at<-new("actuarialtable",x=am92_lt@x,lx=am92_lt@lx,interest=0.04,name="am92")
```

Here we simply verify the annuity and assurance function calculated in Examples 23.1 and 23.2:

```
Axn(am92_at,17, i = 0.04);     axn(am92_at,17,i=0.04)

[1] 0.1012692
[1] 23.367
```

Remark 23.2 The `actuarialtable` object is, however, neither particularly easy to access nor verify, both of which the reader should do at this point; we will, therefore, also save these objects as a dataframe object to access individual numbers and thus validate our calculations. (This step is purely for verification purposes – we would not usually expect to carry out these steps.)

```
am92lt_dframe<-as(am92_lt,"data.frame")#now we have the lifetable as a dataframe
am92at_dframe<-as(am92_at,"data.frame")#now we have the actuarialtable as a dataframe
```

End of remark.

And creating corresponding tables for female lives using the af92_durn2.csv file:

```
af92<-read.csv("~/af92_durn2.csv")
af92_lt<-new("lifetable",  x<-af92[,1], lx=af92[,2], name = " af92 males")
af92_at<-new("actuarialtable",x=af92_lt@x, lx=af92_lt@lx,  interest=0.04,
name="af92")
```

Much of the remainder of this chapter uses the material covered earlier and applies it to the calculation of annuities and assurance benefits; first we look at annuities.

23.6.3 Annuities

Calculating a variety of types of annuities for example, deferred annuities (m), temporary annuities (n), payable monthly (k) (discussed later):

```
axn(am92_at,50,i=0.04)        #50year old - annuity in advance (default)

[1] 17.44418

axn(am92_at,50,i=0.04,m=1)    #50year old - annuity in arrears

[1] 16.44418

axn(am92_at,50,i=0.04,payment="arrears")   #alternative code

[1] 16.44418

axn(am92_at,50,i=0.04,m=20)  #deferred annuity, starting at age 70

[1] 3.926605

axn(am92_at,50,n=5,i=0.04)   #temporary annuity, payable for max 5 years

[1] 4.604848

axn(am92_at,50,i=0.04,k=12)  #payable monthly in advance (discussed later)

[1] 16.98151
```

Exercise 23.9 *Verify the previous calculations using the dataframe object, or otherwise.*

Exercise 23.10 *Often we will require a deferred annuity calculation which uses different mortality tables for the deferral period to that for the payment period. At the time of writing, this option was unavailable from the package. Write a function for this option.*

Example 23.5

Write a table of annuity rates (annually in advance) for ages 17 to 90 years, using annual effective interest rates of 3%, 4%, 5%, 6% (effectively duplicating parts of the Formula and

Tables produced by the Institute and Faculty of Actuaries). Shown here is the calculation of one set of rates at 4% pa for ages 17 to 22 years:

```
rangeannuity<-axn(am92_at,17:90,i=0.04)
head(rangeannuity)

[1] 23.36700 23.27565 23.18044 23.08121 22.97783 22.87014
```

23.6.4 Annuities Paid more Frequently than Annually

All calculations shown previously have allowed only for yearly payments. However, it is common for annuities to be paid monthly, not yearly. Given that the life tables only have annual entries (integer ages), this presents a problem in calculating the probabilities of a life being alive at non-integer ages. An assumption is therefore required to deal with payments made at more frequent intervals. There are various assumptions that can be adopted and we refer the reader to Dickson et al. Section 3.3 for a detailed discussion.

Remark 23.3 Alternative assumptions are available under the lifecontingencies package i.e. "linear", "hyperbolic", "constantforce" ; see the lifecontingencies vignette (Recommended Reading).

Here, we use the assumption of a "uniform distribution of deaths" throughout the year; the main advantage of this method is that of simplicity. Thus:

$$_sq_x = s.q_x \text{ where } s < 1 \text{ and } x \text{ is an integer.} \tag{23.2}$$

Example 23.6

To demonstrate this we calculate $a^{(12)}_{50:\overline{1}|}$, a 1-year temporary annuity, payable monthly in arrears, to a life aged 50 years exact, with $i = 4\%$ pa, using both code within the lifecontingencies package (using the k argument), and our own code:

```
disc_x<-0.04;     pthly<-12    #discount rate and frequency (monthly)
qxt(am92_at,50,1)              #from the package

[1] 0.002508

q[34]  #alternatively, taken from earlier in the chapter

[1] 0.002508

q_list<-1:pthly*q[34]/pthly    #as per above equation
sum((1-q_list)/(1+disc_x)^(1:pthly/pthly)/pthly)    #term assurance value

[1] 0.9777198

#alternative calculation from the package, agrees
axn(am92_at,50,n=1,i=0.04,k=12,m=1/12)

[1] 0.9777198

axn(am92_at,50,n=1,i=0.04,k=12,payment="arrears")    #alternative code

[1] 0.9777198
```

Exercise 23.11 Calculate $a_{60}^{(12)}$ using the same methodology and assumptions as in Example 23.6.

23.6.5 Increasing Annuities

It is also commonplace for annuities to increase while in payment. This may be in line with a defined index (e.g. some measure of inflation), or at a fixed rate. This is a straightforward application of the above.

Example 23.7

Calculate $\ddot{a}_{50:\overline{3}|}$, increasing at 2.5% pa, payable annually in advance (with $i = 4\%$ pa).

```
net_i_star<-1.04/1.025-1
axn(am92_at,50,n=3,i=net_i_star)
```

```
[1] 2.949309
```

Exercise 23.12 Verify this using the annuity function developed in Section 23.5.

Exercise 23.13 Calculate a_{50}, increasing at 2.5% pa (payable annually in arrears).

In the following exercise we bring the earlier ideas together:

Exercise 23.14 Calculate the expected present value of an annuity, payable monthly in arrears to a life aged 50 years, with annual increases equal to 2.5% pa. Use $i = 4\%$ pa. (Note that this type of contract is particularly common in company pension schemes.)

Exercise 23.15 The reader may wish to write their own function to incorporate all the annuity options discussed in this chapter to this point.

23.6.6 Reversionary Annuities

It is common for annuity policies to include payments which continue to the policyholder's surviving spouse following the policyholder's death. To value these policies we must calculate the EPV of a reversionary annuity, $a_{x|y}$; this is the present value of 1 per annum payable to life aged y years which begins on the death of a life aged x years.

$$a_{x|y} = a_y - a_{xy} \tag{23.3}$$

where a_{xy} is the EPV of a joint life annuity of 1 per annum payable while both lives, x and y, are alive. The axyzn function calculates the value of joint-life annuities; we can value an annuity paid while both lives are alive (status = "joint"), or until the last survivor dies (status = "last").

Example 23.8

Using the male and female tables AM92 and AF92, calculate $a_{65|60}$, a reversionary annuity of 1 per annum in respect of a male life aged 65 years and a female life aged 60 years, commencing on the death of the male life and ceasing on the death of the female life:

```
(joint_6560<-axyzn(list(am92_at,af92_at),c(65,60),
i=0.04,status="joint",m=0))        #jointlife
```

```
[1] 11.22468
```

```
(ann_fem<- axn(af92_at,60,i=0.04))
```

```
[1] 16.00578
```

```
(aa_fem_rev<-ann_fem-joint_6560)
```

```
[1] 4.781103
```

Thus, $a_{65|60} = 4.781103$. Note we have the same result if both terms are "in arrears";

```
axn(af92_at,60,i=0.04,m=1)-axyzn(list(am92_at,af92_at),c(65,60),i=0.04,
status="joint",m=1)
```

```
[1] 4.781103
```

(See Section 23.7 where the previous calculation is carried out by running simulations of future lifetimes in respect of the male and female lives.)

Exercise 23.16 *The male reversionary annuity should be lower in value. Why? Verify this.*

Example 23.9

Annuity companies and pension schemes often provide annuities with an attaching spouse's pension, typically equal to around two-thirds of the policyholder's annuity amount; such an annuity is valued using the above reversionary annuities. Here we calculate the EPV of a pension, payable annually in advance, to a male life aged 65 years exact with a 50% attaching female spouse's pension who is 5 years younger than her husband:

```
(ann_male<- axn(am92_at,65,i=0.04))
```

```
[1] 12.27561
```

```
(pen_with50_f_spouse<-ann_male+0.5*aa_fem_rev)
```

```
[1] 14.66617
```

Functions similar to these will be used by pension funds and annuity companies to calculate their liabilities.

Remark 23.4 In the above calculations, we have assumed the two lives are independent; that is, the mortality rate of each life does not depend on whether their spouse is alive. This will generally not be the case in reality. To deal with the case of dependent lives it is most straightforward to set up a 4-state Markov model with different transition rates depending on whether each spouse is alive. The reader familiar with Chapter 22 is encouraged to design such a model with suitable transition rates, and compare with the results obtained previously.

23.6.7 Example: Annuity Company Valuation

In this example we undertake a more realistic practical application by performing a valuation exercise on a book of annuities. We've simulated below a dataset, "annlives", using random ages between 50 and 90 years and random pension amounts between £2,000 and £50,000 per annum to produce 1,000 annuity policies:

```
no_simulations<-1000;    set.seed(100)
ann_lives<-matrix(c(signif(runif(no_simulations,50,90),2),
+signif(runif(no_simulations,2000,50000),2)),no_simulations,2)
colnames(ann_lives) <- c("age", "annuity amount")
head(ann_lives,2)   #sample

     age annuity amount
[1,]  62           5600
[2,]  60           7400
```

Calculating the present value of our annuity book using the *axn* function, a discount rate of 3% pa and the same mortality table as used previously:

```
nlives<-nrow(ann_lives)
liabs<-0
liabs<-axn(am92_at, ann_lives[,1], i = 0.03, type = "EV")*ann_lives[,2]
head(liabs)

[1]   82138.41 115101.20 328220.59 643877.73 231028.89 127065.89

sum(liabs)  #this is the total reserves

[1] 288596954
```

Thus the total liability for this book of policies is £288.6 m. Of course we should initially perform some broad checks to ensure our calculations are reasonable for example, what is the range of liabilities, is the total value reasonable etc.:

```
summary(liabs)

   Min. 1st Qu.  Median    Mean 3rd Qu.    Max.
  11382  131865  226776  288597  401117  966718

data_with_values<-cbind(ann_lives,liabs);    head(data_with_values,2)

     age annuity amount     liabs
[1,]  62           5600  82138.41
[2,]  60           7400 115101.20

data_with_values[which.min(liabs),]#good

          age annuity amount          liabs
        90.00        2700.00       11381.75

axn(am92_at, ann_lives[which.min(liabs),1],
i=0.03, type = "EV")

[1] 4.215463
```

This is consistent with the minimum individual liability figure shown earlier. It is also reasonable that the minimum value would be obtained for a 90 year-old with a small annuity.

Exercise 23.17 *Carry out a similar check to the top five liability values.*

We could also check the total liability value is broadly consistent with the total annuities payable and the average age of an annuitant:

```
(av_age<-mean(data_with_values[,1]))   #average age

[1] 70.725

(total_pen<-sum(data_with_values[,2]))  #total annual annuity payments

[1] 26125500
```

Now calculate an annuity for our "average" life and estimate the total liabilities:

```
(estimatedtotal<-total_pen*axn(am92_at, av_age, i = 0.03,
type = "EV", power = 1))

[1] 282264033
```

Comparing this with our calculated value:

```
estimatedtotal/sum(liabs)

[1] 0.9780562
```

Given the approximate nature of the check, this is reasonably close. (The reason for the discrepancy is due to the slightly non-linear nature of the annuity values with age.) These would all be standard preliminary checks before more detailed checking is performed.

Exercise 23.18 *Carry out a similar valuation to that above, but for both existing annuitants and deferred annuities, with a range of annual pension increase rates, and with attaching spouse's pensions.*

Exercise 23.19 *A pension scheme has proposed an increase to the benefits it provides by guaranteeing the first 10 years of payments made to a pensioner – if the pensioner dies within 10 years of retiring the pension continues in payment to the legal dependants (e.g. spouse) until 10 payments have been made. Calculate the percentage increase in expected costs from this proposal. All pensions are paid monthly, and increase each year at the rate of 3%. Assume all pensioners retire at age 60, there are an equal number of male and female pensioners, and the effective rate of interest is 4% pa.*

That concludes our discussions on annuities for now. We proceed with a discussion on assurance functions.

23.6.8   Life Assurance functions

Set out here are various types of assurance benefits which were discussed briefly in Section 23.4, together with corresponding code. The reader should verify each output; we have included checks for the first example only.

$$A_{x:\overline{n}|} = (\sum_{t=0}^{n-1} v^{t+1} \, _tp_x \, _1q_{x+t}) + v^n \, _np_x$$

```
AExn(am92_at, 50, n=10, i = 0.04, type = "EV", power = 1)  # endowment insurance
[1] 0.6802421
with(am92at_dframe,(Mx[34]-Mx[44]+Dx[44])/Dx[34]) # check 1
[1] 0.6802421
1-0.04/1.04*axn(am92_at,50,10,i=.04) # check 2
[1] 0.6802421
```

$$A^1_{x:\overline{n}|} = \sum_{t=0}^{n-1} v^{t+1} \, _tp_x \, q_{x+t}$$

```
Axn(am92_at, 50, n=10, i = 0.04, type = "EV", power = 1) #term assurance
[1] 0.03423063
```

$$A_x = \sum_{t=0}^{\infty} v^{t+1} \, _tp_x \, q_{x+t}$$

For whole life assurance we use the same function but exclude any term for *n*:

```
Axn(am92_at, 50, i = 0.04, type = "EV", power = 1) #whole life (no term required)
[1] 0.3290702
```

23.6.9   Assurance Policies with *immediate* Payment on Death: $\overline{A}_x$

It is worth noting at this point that we have made two key assumptions throughout this chapter when valuing life assurance benefits:

1. Payments are made at the end of the year in which the death occurred;
2. Interest rates are constant (this assumption was also made in respect of the annuities section).

Both these assumptions are often made in exam-type questions, and also when verifying real-world calculations. However both of these are unrealistic. It is unlikely that the death benefit will be paid at the end of the year of death – a key reason for the provision of term assurance benefits is that financial dependents should not be placed in a position of financial hardship immediately after the death of a spouse.

Often a simple approximation is made in respect of (1) above – we can assume all deaths occur halfway through the year by dividing our value by $v^{0.5}$.

$$\overline{A}_x \approx \sum_{t=0}^{\infty} v^{t+0.5} \, {}_tp_x \, q_{x+t} \tag{23.4}$$

More accurately we should use the following to value a whole life assurance for a life aged exactly x (for a constant interest rate):

$$\overline{A}_x = \int_0^{\infty} {}_tp_x \mu_{x+t} v^t dt \tag{23.5}$$

where μ_x is the force of mortality at age x. The mathematics is similar to that in Chapters 2 and 3 where we were valuing a varying payment with a varying interest rate.

Exercise 23.20 *(part 1)*

1. *Given $\mu_x = 0.0001 \times 1.07^x$ pa and using a constant force of interest equal to 3% pa, calculate $\overline{A}_{60}$ using Equation 23.5, where the death payment is payable immediately on death.*
2. *Repeat the calculation, first setting up an approximate daily life table based on μ_x and using the appropriate expression for whole life assurance from Section 23.6.8.*
3. *Compare your answer with that obtained by applying Equation 23.4 (approximate method).*

As noted above it is unrealistic to assume a flat yield curve.

Exercise 23.20 *(part 2)*
Repeat *(part 1)*, but instead of a constant force of interest, assume the following annual force of interest (ie instantaneous forward rate):

$$\delta_t = 0.03 - 0.01e^{-0.5t}$$

Again, check your answer using an approximation.

It is important for the reader to appreciate the effect of making such approximations; there is often a balance to be struck between model accuracy, costs incurred, and delays from adjusting, and checking, models.

23.7 Simulation of Future Lifetimes

The reader may find it a useful exercise to simulate future lifetimes given a particular life table. Simulations of future lifetimes, using a Gompertz model rather than a life table, are carried out in Chapter 17. Here we simulate future lifetimes for a male life aged 45 years, using a life table (*AM92*) and the rLife function (see Figure 23.1):

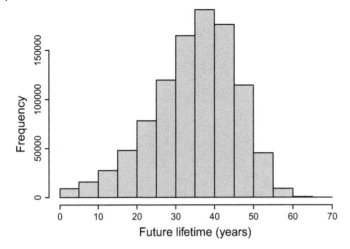

Figure 23.1 Simulated Future Lifetimes for a 45-year-old using AM92.

```
sims_y<-1000000     #no of simulations used throughout this section
set.seed(100)
rand_life<-rLife(n = sims_y, object = am92_at, x = 45, type = "Tx") #100000 sims
head(rand_life,10)

[1] 42.5 41.5 45.5 39.5 44.5 31.5 22.5 33.5 45.5 36.5
```

(type="Tx", the default, approximates the use of a continuous distribution for future lifetimes; type="Kx" simulates whole years.) The expectation of life, $\mathring{e}_{45}$, can be estimated from these simulations:

```
mean(rand_life)

[1] 34.77586
```

It is a useful exercise to write a function which duplicates the rLife function. Here is suggested code in respect of a 45-year-old (note the use of runif to generate lifetimes across all ages):

```
set.seed(100);    sim_lt<-NULL
am92_45<-am92[29:104,]  #from age 45 years
am92_45_cdf<-NULL
am92_45_cdf<-1-am92_45[seq(1,nrow(am92_45)-1),2]/am92_45[1,2]   #cdf
sim_unif<-runif(sims_y)
for (nnn in 1:sims_y) {
sim_lt[nnn]<- max(which(am92_45_cdf<=sim_unif[nnn]))-runif(1)
}
#sim_lt are the simulated lifetimes
head(sim_lt);      mean(sim_lt)

[1] 30.09755 28.10727 37.66299 15.63805 35.46827 35.08107
[1] 34.78699
```

Exercise 23.21 *Calculate the theoretical value for $\mathring{e}_{45}$ and compare with the above values obtained from simulations.*

Such simulations of future lifetimes may be used to simulate present values of the various contingency functions we are interested in. This is covered in more detail in the following chapter where we will be particularly interested in the range of present values; however an example is also included here:

Example 23.10 Calculate A_{45} and $\ddot{a}_{45}$ using both simulations, and the Axn and axn functions from the lifeContignencies package (use *AM92* and $i = 4\%$ pa).

```
int_rate<-0.04
mean((1+int_rate)^-ceiling(sim_lt))              #life assurance, simulations

[1] 0.2758105

(whole_life_45m<-Axn(am92_at, 45, i = int_rate))    #compare

[1] 0.2760495

ann_sim<-NULL
for (tt in 1:sims_y) {
       ann_sim[tt]<-annuity(int_rate,ceiling(sim_lt[tt]),type = "advance")}

mean(ann_sim)                          #annuity, simulations

[1] 18.82893

(annuity45<-axn(am92_at,45,i=int_rate))   #compare

[1] 18.82271
```

Example 23.11

In Section 23.6.6 we calculated a reversionary annuity, $a_{65|60}$; here is a check on the calculation of the same annuity, using simulated future lifetimes:

```
disc_rev<-0.04
set.seed(1000)
rand_lifem<-rLife(n = sims_y, object = am92_at, x = 65, type = "Kx")
rand_lifef<-rLife(n = sims_y, object = af92_at, x = 60, type = "Kx")
diff_xy<-pmax(rand_lifef-rand_lifem,0)
rev_ann_sim<-NULL

for (hhh in 1:sims_y) {
rev_ann_sim[hhh]<-annuity(disc_rev,n=diff_xy[hhh],m=rand_lifem[hhh])
}
mean(rev_ann_sim)       #from simulations

[1] 4.779543

aa_fem_rev              #from earlier in the section

[1] 4.781103
```

Exercise 23.22 *Using simulations of future lifetimes, verify the value calculated from Example 23.9, and also carry out calculations for the following:*

$$a_{60} \quad a_{60:\overline{10}|} \quad {}_{15|}a_{75} \quad A^{1}_{30:\overline{10}|} \quad A_{30:\overline{20}|}^{1} \quad A_{30:\overline{10}|}$$

23.8 Recommended Reading

- Benjamin, B. (1968). *Health and Vital Statistics* (1st ed.). London, UK. Routledge.
- Charpentier A et al. (2016). *Computational Actuarial Science with R*. Boca Raton, Florida, US: Chapman and Hall/CRC.
- Dickson DCM, Hardy MR, Waters HR (2013). *Actuarial Mathematics for Life Contingent Risks*. Cambridge, UK: Cambridge University Press
- McCutcheon, J. (1969). *A Method for Constructing an Abridged Life Table*. Transactions of the Faculty of Actuaries, 32, 404-411.
- MacDonald AS et al. (2018). *Modelling Mortality with Actuarial Applications*. Cambridge, UK. Cambridge University Press.
- Tetley, H. (1953) *Actuarial Statistics: Volume I (Statistics and Graduation)* (2nd ed). Cambridge University Press, Journal of the Institute of Actuaries.
- Readers may also find the following vignette useful:
  https://cran.microsoft.com/snapshot/2022-04-13/web/packages/lifecontingencies/vignettes/anintroductiontolifecontingenciespackage.pdf
     by GA Spedicato
- List of Life Tables online at: https://www.actuaries.org.uk/

24

Contingencies II

Peter McQuire

```
library(lifecontingencies);library(moments);library(tseries);library(readxl)
```

24.1 Introduction

This chapter follows on directly from Chapter 23; it is assumed that the reader has some familiarity with this subject matter (Life Contingencies), and with most of the terms used. Before proceeding with the material below the reader should familiarise themselves with the material in that chapter.

We start the chapter with a discussion on the uncertainty of present values of life contingent benefits, and the resulting probability distributions of both individual policies and portfolios; in the previous chapter we concerned ourselves principally with the expected present value of such benefits. By calculating appropriate premiums we can then determine the distribution of profits from selling such policies. We proceed to analyse how profits typically emerge throughout the term of policies, and how, by calculating policy values and setting up reserves, we can minimise the likelihood of having insufficient assets to pay benefits under these policies. We close the chapter by looking at unit-linked assurance policies.

Whilst a brief discussion is included on key subject matter, the emphasis in this chapter is to run code for key tasks, analyse results, and suggest additional tasks in the form of exercises which often require developing previous code. A comprehensive theoretical study of all topics relating to Life Contingencies cannot be undertaken in this relativity short chapter, and the reader should not use this chapter as their sole source of learning material on this topic, instead using this chapter to augment and enhance their studies. The reader is referred to the Recommended Reading, and in particular, to Dickson et al. (2013), and Booth et al. (2005) for further details.

R Programming for Actuarial Science, First Edition. Peter McQuire and Alfred Kume.
© 2024 John Wiley & Sons Ltd. Published 2024 by John Wiley & Sons Ltd.
Companion Website: www.wiley.com/go/rprogramming.com.

The reader may find it useful to approach some problems in this chapter using a variety of techniques and models to further their understanding of the material, in particular, from the perspective of applying Markov models (see Chapters 16 and 22).

Remark 24.1 Note that throughout the chapter we will often use shorthand to set parameter values and information for examples and exercises. For example, a question including a term assurance policy with a term of 20 years issued to a male life aged 60 years exact, with annual premiums payable in advance and a benefit of £10,000 payable at the end of the year of death, and calculations should use an interest rate of 4% pa, will simply be described as:

"term, age 60, 20 year term, benefit £10,000"

Unless otherwise stated it is assumed that premiums will be paid annually in advance, deaths benefits will be payable at the end of the year of death, *AM*92 mortality tables will be used, the life in question is male, with an effective annual rate of interest of 4% pa. Additional terms may be required depending on the particular scenario for e.g. non-constant values, expense details.

Remark 24.2 As noted in the Introduction chapter, there are a number of topics in this book for which Microsoft Excel can be a viable alternative to R when approaching tasks. The reader is encouraged to tackle a variety of the examples and exercises in this chapter using Microsoft Excel.

24.2 Mortality Tables: AM92

The reader should download the mortality table, *AM*92, which will be used extensively throughout this chapter:

```
am92_entire<-read.csv("~/am92_durn2.csv")
am92<-am92_entire[,c(1,2)]
```

and set up an `actuarialtable` object using a default interest rate of 4% pa (as outlined in Chapter 23):

```
am92_lt<-new("lifetable",  x=am92[,1],lx=am92[,2],name=" am92 males")
am92_at<-new("actuarialtable",x=am92_lt@x,lx=am92_lt@lx,interest=0.04,name="am92")
```

We have chosen to use only one mortality table to simplify calculations performed in this chapter. The reader is encouraged to apply the code in this chapter to a variety of mortality tables. The reader may also find it useful to have available to them a copy of "Formulae and Tables" (published by the Institute and Faculty of Actuaries, and available at *www.actuaries.org.uk* at the time of publication) whilst undertaking calculations included in the chapter.

24.3   Uncertainty in Present Values: Variance

In the previous chapter we concerned ourselves with calculating expected present values of annuities and various life assurance benefits. However, we must also analyse the degree of uncertainty in our results. Policyholders' actual mortality will turn out to be different to expectations; it is important to understand to what extent our actual experience may differ compared to expectations and the impact these differences may have on the finances of the insurance company or pension scheme.

Remark 24.3 Uncertainty in these values will also arise due to uncertainty in interest rates. The emphasis in this chapter is to analyse the level of uncertainty in mortality rates due to general volatility. (Note that we address trend mortality risk in Chapter 19.) The reader should therefore assume that interest rates are constant throughout the chapter unless otherwise stated; we do allow for variability in interest rates in a number of examples/exercises.

To help us do this we first calculate variances (and standard deviations) of the present value of futures cashflows, and later compare these results with simulations, which in general will provide a more accurate picture of the tails of the distributions (provided sufficient simulations are run).

We set out below a brief derivation in respect of whole life assurance benefits only, leaving derivations in respect of other insurance benefits and annuities as exercises for the reader. Such derivations are included in many textbooks on this topic (see Dickson et al. (2013)).

The variance of a random variable, X, is given by:

$$Var(X) = E((X - E(X))^2) = E(X^2) - E(X)^2 \tag{24.1}$$

We define the random variable, K_x, as the future lifetime of a life, currently aged x, measured in complete years. Thus K_x is any non-negative integer. The present value of a payment of 1, payable at the end of the year of death in respect of a life currently aged x, is also a random variable, given by v^{K_x+1}, where $v = \frac{1}{1+i}$, and i is the constant effective rate of interest. To simplify our notation, let $Y = v^{K_x+1}$.

The variance of the present value of the payment due under a whole life assurance policy, where a payment of 1 is made at the end of the year of death, and p_t is the probability that the payment is made at time, t, is therefore (from Equation 24.1):

$$Var(Y) = \sum_{t=1}^{\infty} v^{2t} p_t - \left(\sum_{t=1}^{\infty} v^t p_t\right)^2 \tag{24.2}$$

Using standard actuarial notation this can be written:

$$Var(Y) = {}^2A_x - A_x^2 \tag{24.3}$$

where A_x is the expected present value of payments with an effective rate of interest, i, under the whole life policy, and ${}^2A_x$ is calculated using an effective rate of interest equal to $(1 + i)^2 - 1$. Similar expressions exist for term and endowment assurance.

The variance calculation of various insurance benefits and annuity products, is straightforward in R; to calculate ^2A_x we can use the power argument from the lifecontingencies package, or alternatively use $(1 + i)^2 - 1$, as follows:

```
Axn(am92_at, 60, i = 0.04, type = "EV", power = 2)

[1] 0.2372307

Axn(am92_at, 60, i = 1.04^2-1, power = 1)

[1] 0.2372307
```

For example, the variance of "whole, age 60, $i = 4\%$" (using the agreed shorthand description from Remark 24.1) can be calculated by applying Equation 24.3:

```
(VarAxn<-(Axn(am92_at, 60, i = 0.04, type = "EV", power = 2)
-(Axn(am92_at, 60, i = 0.04, type = "EV", power = 1))^2))

[1] 0.02892991
```

where $^2A_x = 0.2372307$, $A_x = 0.4563998$, and $A_x^2 = 0.2083008$. To simplify coding throughout the chapter the following functions have been written for whole life, term, and endowment assurance:

```
#variance function for whole life
var_whole<-function(tablex,age,interestx)
{Axn(tablex, age, i = interestx, type = "EV", power = 2)-
(Axn(tablex, age, i = interestx, type = "EV", power = 1))^2}
var_whole(am92_at,60,0.04)        #example

[1] 0.02892991

#variance function for term assurance
var_term<-function(tablex,agex,termx,interestx){
Axn(tablex, x = agex,n=termx, i = interestx, type = "EV", power = 2)-
(Axn(tablex, x = agex,n=termx, i = interestx, type = "EV", power = 1))^2}
var_term(am92_at,60,10,0.04)        #example

[1] 0.07208955

#variance function for endowment
var_endw<-function(tablex,age,termx,interestx){
AExn(tablex, age, i = interestx, n=termx, type = "EV", power = 2)-
(AExn(tablex, age, i = interestx, n=termx, type = "EV", power = 1))^2}
var_endw(am92_at,60,10,0.04)        #example

[1] 0.002413478
```

Hence the standard deviation of this whole life assurance is $\sqrt{0.0289299} = 0.1700879$. We will see shortly that knowing the standard deviation of such life assurance functions can be useful in estimating the uncertainty of the future financial position of life assurance companies.

Exercise 24.1 *Comment on the relative values of the variances of these 3 types of assurance benefits.*

Example 24.1

It's useful to verify the above, calculating the variance directly from the life table, using Equation 24.2.

```
vb<-60    #length of vector
age_stx<-60
prob_death_each_year<-(am92_at@lx[(age_stx-16):((age_stx-16)+vb-1)]-
am92_at@lx[(age_stx-15):((age_stx-15)+vb-1)])/am92_at@lx[(age_stx-16)]
disc_y<-1/(1.04^(1:vb))
EX<-sum(prob_death_each_year*disc_y)
sum(prob_death_each_year*(disc_y-EX)^2)    #agrees

[1] 0.02892991
```

Exercise 24.2 *Using various methods, calculate the standard deviation of whole life assurance for male lives, aged from 20 years to 90 years, using AM92 mortality rates and a range of interest rates from 2% pa to 8% pa.*

As a reminder, the expected present values of whole life (A_{60}) and term assurance ($A^1_{60:\overline{10}|}$), which were seen in Chapter 23, are calculated as follows:

```
(meanAxn<-Axn(am92_at, 60, i = 0.04, type = "EV", power = 1))

[1] 0.4563998

(meanAxn_term<-Axn(am92_at, 60, 10, i = 0.04, type = "EV", power = 1))

[1] 0.1043154
```

However, as we do not know the underlying distribution, the standard deviation is not necessarily an appropriate risk measure. For example, we may want to know the probability that the present value exceeds a particular level. Often the simplifying assumption of a Normal distribution is applied inappropriately, leading to potentially dangerous risk assessments.

Remark 24.4 We will calculate $_{t|1}q_x$ (the probability that a life aged x survives t years then dies the following year) frequently during the chapter; we write a function for this below, using the *AM*92 table (functions for q_x and $_tp_x$ are also written).

```
#  prob a life aged x dies in a future year:
fx_fn1<-function(x0,range_x){(am92_at@lx[(x0-16):((x0-16)+range_x-1)]-
am92_at@lx[(x0-15):((x0-15)+range_x-1)])/am92_at@lx[(x0-16)]}

#  table of q rates:
qx_rates1<-function(x0,range_x)(am92_at@lx[(x0-16):((x0-16)+range_x-1)]-
am92_at@lx[(x0-15):((x0-15)+range_x-1)])/am92_at@lx[(x0-16):((x0-16)+range_x-1)]

#  surviving from age x to various future ages:
surv_fn1<-function(x0,range_x){am92_at@lx[(x0-15):((x0-15)+range_x-1)]/
am92_at@lx[(x0-16)]  }

fx_fn1(60,5)

[1] 0.008022000 0.008936730 0.009940513 0.011038855 0.012233579
```

```
qx_rates1(60,5)
```

[1] 0.008022 0.009009 0.010112 0.011344 0.012716

```
surv_fn1(60,5)
```

[1] 0.9919780 0.9830413 0.9731008 0.9620619 0.9498283

```
# e.g. checking the whole life and term assurance figures above:
sum(fx_fn1(60,60)/1.04^(1:60));    sum(fx_fn1(60,10)/1.04^(1:10))
```

[1] 0.4563998
[1] 0.1043154

Figure 24.1, left (•), shows the probability distribution of present values of the benefits from this whole life assurance, produced from the following code:

```
start_age<-60;        rangez<-60
pv_whole<-1/(1.04^(1:rangez))    #pv of benefit in each year, at start of contract
prob_death_fut_yr<-fx_fn1(start_age,rangez)
```

Clearly, adopting a Normal distribution to model the probability distribution is inappropriate here. Here, the "Normal" assumption leads us to conclude that there is a 1% chance that the value exceeds 0.85, whereas the true value is approximately 0.92. Adopting a Normal distribution to assess the probability of large losses (i.e. from early deaths) would significantly underestimate this risk.

Exercise 24.3 *Verify the above 99% VaRs.*

Exercise 24.4 *Write similar code for a 10-year term assurance policy (for the same life), and plot the probability distribution of present values.*

Figure 24.1, right (o), shows the probability distribution of present values from term assurance (from Exercise 24.4).

Exercise 24.5 *The probability distribution for an endowment policy shows some similarity to that of a term assurance policy. Plot the relevant probability distribution for a 10-year endowment policy, using the same parameter values as used above for the term assurance policy.*

In Section 24.4 we run simulations using mortality tables to obtain a better understanding of the shape of the distribution of benefits from both single policies, and portfolios of policies.

Variance of annuities

The variance of an annuity, for a life aged x, is easily shown to be:

$$\frac{{}^2A_x - A_x^2}{d^2} \tag{24.4}$$

where $d = \frac{i}{1+i}$.

Writing a function for the variance of an annuity, and calculating the standard deviation for a male life, age 60 years, at $i = 4\%$ pa:

```
var_ann<-function(tablex,age,intx)
{(Axn(tablex, age, i = intx, type = "EV", power = 2)-
Axn(tablex, age, i = intx, type = "EV", power = 1)^2)/(intx/(1+intx))^2}

var_ann(am92_at,60,0.04)^0.5

[1] 4.422286
```

Exercise 24.6 *Write alternative code to verify the variance of an annuity as we did in Example 24.1 (for the whole life assurance policy).*

24.4 Simulations

In this section we simulate present values of various types of policies, comparing the output with the theoretical calculations above. Again, we will look at whole life assurance, term assurance, and annuities; the reader should also develop code for other types of contracts. The objective of this section is to demonstrate the consistency between results obtained from running simulations with those from theory, and also to compare results obtained from simulations in respect of portfolios of policies with results using common approximations.

Remark 24.5 As noted elsewhere in this book it is important when running simulations to ensure the results obtained have sufficient stability such that we can make reliable conclusions. It may be the case that more simulations are required than shown in this chapter, simply as a result of memory constraints when compiling the chapter.

24.4.1 Single Policy

Whole life assurance

Here we simulate 10,000 present values in respect of a single whole life assurance policy (£1 assured), issued to a male life, aged 60 years, using the rLifeContingencies function together with the Axn argument, calculating the mean, variance, and standard deviation of these samples:

```
set.seed(1000);     sims<-10000
randomwholelife<-rLifeContingencies(sims,"Axn",am92_at,60,999,parallel=FALSE)
mean(randomwholelife);     var(randomwholelife);     var(randomwholelife)^0.5

[1] 0.4576702
[1] 0.0289407
[1] 0.1701197
```

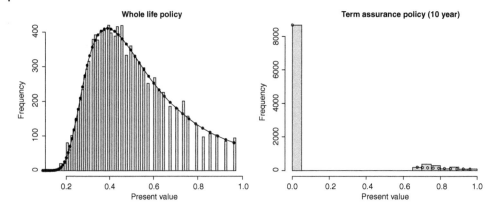

Figure 24.1 Probability distributions of a single policy: Whole life and 10-year-term – theory and simulations (histogram).

(Alternatively we can use `replicate(10000,Axn(am92_at,60,type="ST"))`, but the above code is faster.) These statistics are consistent with the theoretical values calculated earlier; both the simulated values (histogram) and theoretical values (•) are plotted in Figure 24.1, left.

The plot of the simulated values (Figure 24.1, left) clearly shows a skewed distribution. As noted earlier, simply knowing the standard deviation and assuming a Normal distribution can be very misleading; for example, comparing the 99% percentile assuming the best-fitting Normal distribution and the more appropriate 99% percentile from simulations (which is also consistent with the value from Exercise 24.3):

```
meanAxn+qnorm(.99,0,1)*VarAxn^0.5  #Normal approx

[1] 0.8520835

quantile(randomwholelife,0.99)     #simulations

     99%
0.9245562
```

To be clear, the highest simulated values (those close to 1.0) represent lives which die shortly after taking out a policy; they have been discounted by a small amount and therefore have the largest values.

Term assurance

Similarly, running simulations for a single term assurance policy give statistics consistent with theory (ideally more simulations should be run here):

```
sims<-10000;   set.seed(1000);   term_ins<-10;
random_1policies_t<- rLifeContingencies(sims,"Axn",am92_at,i=0.04,60,t=term_ins)
mean(random_1policies_t);   var(random_1policies_t);   var(random_1policies_t)^0.5

[1] 0.1051102
[1] 0.07257741
[1] 0.269402
```

The results from theory are 0.1043 and 0.2685 (mean and standard deviation respectively).

Example 24.2 As an alternative to using the function in the `lifecontingencies` package, the reader may wish to write their own function to simulate lifetimes and related values. For example, the following code can be used to simulate present values in respect of a whole life policy:

```
set.seed(100);        sim_no<-10000
abc<-runif(sim_no)
pv_wholeyy<-cdf_whole<-NULL
cdf_whole<-c(0,1-surv_fn1(60,60))    #this is F(x)

for (nnn in 1:sim_no) {
pv_wholeyy[nnn]<- pv_whole[max(which(cdf_whole<=abc[nnn]))]}
mean(pv_wholeyy);   var(pv_wholeyy)^0.5

[1] 0.4553101
[1] 0.1684222
```

We use this alternative code in various parts of this chapter.

Next we compare the uncertainty from writing one policy (as above) to that from writing a portfolio of independent policies.

24.4.2 Portfolios with 100 Policies – Portfolio Claim Distribution from Simulations

In Section 24.4.1, by running simulations, we looked at the distribution of claims arising from writing a single policy (Figure 24.1). In this section we look at the claims distribution from a portfolio of 100 policies (Figure 24.2).

Before we proceed to run simulations on our portfolio of 100 whole life assurance policies and term assurance policies, an important assumption is required regarding the degree of dependence between our policies. We will assume that all our policies are independent; this is a simplifying, but perhaps not unreasonable, assumption. Most insured lives will tend to be largely independent, although some degree of dependence may exist, for example employees working in the same building are exposed to a similar risk from fire, citizens in one geographical area to a natural disaster, globally to a deadly virus etc. (Note that we can incorporate a degree of correlation by incorporating copulas into our model – see Chapters 13 and 14.) However, if the degree of dependence is significant, we risk underestimating the probability of large losses, and hence insolvency. (This was first highlighted in Chapter 6.)

For simplicity we will also assume that our policies are identical, e.g. all policyholders exhibit the same rates of mortality. Thus the policies are independent and identically distributed, i.e. "iid".

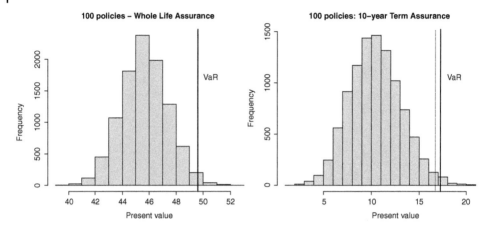

Figure 24.2 Portfolios (100 policies): Whole life and Term assurance policies: Simulated distribution of present value of benefit.

Whole Life Assurance

From a portfolio of 100 iid policies (each with £1 assured), we would expect that both the mean and variance will both have values 100 times those for one policy. Running 10,000 simulations of the 100 policies produces results consistent with theory:

```
random_100policies<-totalyyyy<-NULL;   set.seed(600)
sims<-10000;    policies<-100

rand_wh<-matrix(rLifeContingencies(policies*sims,"Axn",am92_at,x=60,t=80),sims,
policies)
random_100policies<-apply(rand_wh, 1, sum)
mean(random_100policies)

[1] 45.6646

var(random_100policies);   sd(random_100policies)

[1] 2.906841
[1] 1.704946

(sdAxn_port<-(var_whole(am92_at,60,0.04)*policies)^0.5) #s.d. of 100 policies -
theory

[1] 1.700879
```

Comparing these values with those for the single policy, whilst allowing for stochastic error, the standard deviation only increases by a factor of 10 – this is the $\sqrt{n}$ rule for independent events we see many times throughout the book. This is a key point; although we're writing 100 times as many policies such that our expected payments and liabilities are 100 times larger, the uncertainty, measured here by standard deviation, has only increased ten-fold. The simulations in respect of this 100-policy whole life portfolio are shown in Figure 24.2 (left).

We noted above that the Normal distribution is a poor model for the present value of a single whole life assurance policy. In accordance with the Central Limit Theorem, it should perform better when modelling the present value of the sum of 100 whole life policies.

Exercise 24.7 *Calculate the 99% VaR from 1)the above simulations and 2)the best-fitting Normal distribution to the claim amounts (as we did above for the case of just one policy). Comment on the appropriateness of using a Normal distribution approximation.*

The 99% percentile from the simulations is 49.6, and from the best-fitting Normal distribution is 49.6. (Of course care is required here when estimating VaRs using simulations – the reader should test the stability of this result.) Adoption of the Normal distribution model for this portfolio clearly appears appropriate in this scenario.

Note that the expected present value per policy is the same irrespective of the number of policies sold (assuming there are no expenses); however we are more certain of the results by writing more independent policies. Of course, this emphasises one of the central themes of the book, that the more independent policies you can write the greater certainty you will have about the result. The same principle can, of course, be applied to assets or debt.

Exercise 24.8 *Using the same parameters as in the above examples calculate the single premium required, payable at the start of the policy, such that we would be 95% confident of making a profit if we sell (1) 100 policies, (2) 1,000 policies, or (3) 10,000 policies. Ignore expenses.*

Term assurance policies (10-year term)

We now compare the relative variances of term assurance with that of whole life assurance, and again consider the appropriateness of the Normal distribution approximation.

Exercise 24.9 *Run simulations for a portfolio of 100 10-year term assurance policies, similar to those simulations for the whole life portfolio above, and demonstrate the mean and standard deviation agree with theory. Also estimate the 99% VaR.*

This term assurance has a significantly greater standard deviation, as a proportion of the expected present value, than whole life assurance (issued to lives of the same age); from the 100 policy portfolios, the expected present value and standard deviation of the term assurance portfolio are 10.43 and 2.68, and 45.64 and 1.7 from the whole life portfolio. Of course, similarities between the term assurance distribution and whole life distribution increase as the term, and age at issue, increase.

Using 10,000 simulations, we obtained the following results for the 99%VaR for term assurance (from Exercise 24.9, simulations in `term100_sims`), also comparing with the Normal approximation:

```
VaRyy<-0.99
quantile(term100_sims,VaRyy)                          #from simulations

     99%                                                 |
17.22184

(sdnorm_port<-round(meanterm10*policies+qnorm(VaRyy)*(varterm10*policies)^0.5,2))

[1] 16.68

#Normal approx
```

From these simulations, the 100-policy term assurance portfolio has a 99% VaR of 17.22, compared to 16.68 from the Normal distribution approximation (compare the whole life portfolio where the Normal approximation VaR is significantly closer to the VaR from simulations). Given the skewness of the distribution from the term assurance policies, greater care is required in adopting a Normal distribution approximation than for whole life assurance.

The skewness of the various distributions is calculated below (the reader may also wish to calculate the kurtosis of the data and undertake a Jarques-Bera test, using `jarque.bera.test`):

```
skewness(randomwholelife);        skewness(random_100policies)    #whole life

[1] 0.8893958
[1] 0.06246109

skewness(random_1policies_t);     skewness(term100_sims)          #term

[1] 2.216828
[1] 0.2169788
```

As noted several times in this book, an appropriate number of simulations should be run such that the output has acceptable stability; the required number here is likely to be relatively large given the skewness and kurtosis of the distributions.

Exercise 24.10 *An analyst has calculated, using the Normal approximation, that if we charge a single premium at the start of the contract of £16,680 for a 10-year term assurance policy to a male life aged 60 years which pays £100,000 on death, there is a 99% probability of making a profit. Comment on the statement.*

Exercise 24.11 *Carry out the same calculations for a portfolio of 1-year term assurance policies.*

Exercise 24.12 *Run simulations in respect of 10-year term assurance policies, for books of 1,000 policies, and examine appropriate VaRs.*

Exercise 24.13 *Carry out the same simulations and calculations included in this section but for a range of endowment policies. Before calculating, how do you think the variance of endowment insurance compares with term and whole life?*

24.5   Simulation of Annuities

It is straightforward to simulate annuity values using the `rLifeContingencies` function. For a single policy, aged 60 years, payable in advance:

```
set.seed(5000);    age_annz<-60;    sims_ann<-10000
sampleaxn <- rLifeContingencies(sims_ann, "axn", am92_at,x=age_annz)   # simulations
```

`sampleaxn` is plotted in Figure 24.3, left (together with the theoretical values). Comparing sample statistics from the above simulations with theoretical values:

```
mean(sampleaxn); sd(sampleaxn)                              # simulations

[1] 14.18785
[1] 4.384531

axn(am92_at,age_annz); var_ann(am92_at,age_annz,0.04)^0.5   # theory

[1] 14.1336
[1] 4.422286
```

It is straightforward to simulate annuities payable in arrears, and non-annually, for example monthly, as follows: `rLifeContingencies(sims_ann,"axn",am92_at, x=age_annz,m=1)` and
`rLifeContingencies(sims_ann,"axn",am92_at,x=age_annz,k=12)` respectively.

Example 24.3

Running simulations similar to those carried out in Section 24.4.2, but for a portfolio of 100 annuity policies:

```
rand_100_ann_policies<-NULL;    set.seed(10000)
sims<-1000;     policies<-100

rand_annx<-matrix(rLifeContingencies(policies*sims,"axn",am92_at,x=60),sims,policies)
rand_100_ann_policies<-apply(rand_annx,1,sum)
mean(rand_100_ann_policies);    sd(rand_100_ann_policies)

[1] 1413.452
[1] 43.53999
```

See Figure 24.3, right. From these simulations, the mean present value obtained was 1,413, with standard deviation of 43.54, consistent with the single policy values calculated above (the exact values will depend on the method and `set.seed` used).

Exercise 24.14 *Compare 99% VaRs obtained from simulations with the Normal approximation for the above portfolio of 100 annuity policies.*

Exercise 24.15 *An annuity company wants to ensure there is less than 5% chance of making losses on its annuities it sells. Calculate the premium which should be charged assuming it sells 10,000 policies. Adopt "annuity, age 60, benefit £10,000 pa, $i = 4\%$". Ignore loadings for expenses and state your assumptions.*

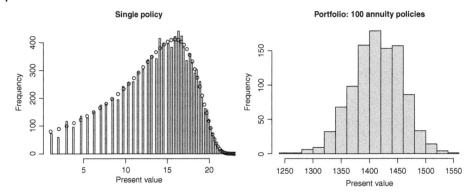

Figure 24.3 Probability distribution of annuity present values.

24.6 Premium Calculations

In the previous section we analysed the distribution of the present values of benefit payments of various policy types. In this short section we discuss how we may assess the price, or premium, which should be paid for such policies. This will allow us to analyse profits (which, in its most general sense, equals premiums less benefits less expenses), which may be considered to be the primary focus of this chapter.

The fundamental method used by actuaries to determine premiums is to apply the equivalence principle, which equates the expected present value of benefits and expenses payable under the contract with the expected present value of premiums receivable under the contract. We also briefly look at an example of the percentile method of calculating premiums, which becomes particularly relevant when assessing portfolios of policies (see Exercise 24.18).

Alternatively, a more modern approach assesses particular profit criteria (such as the net present value of profits, or internal rate of return), analysing various levels of premiums and reserves to determine an appropriate premium to charge such that there is a particular probability of achieving required profits.

There are two types of premium calculations traditionally covered on this topic: the gross premium, and the net premium. The gross premium, P, here payable annually in advance, can be considered the more general calculation. Applying the equivalence principle, we have:

$$P\ddot{a}_x = S A_x + F A_x + f S A_x + I + e\ddot{a}_x \tag{24.5}$$

where I is initial expenses, f is claims handling expenses, F is fixed claims handling expenses, e is renewal expenses, and S is the benefit payable at the end of the year of death. The precise form of the equation can vary depending on the scenario, and may include additional terms.

Example 24.4

Calculate the gross premium, payable annually in advance, for a whole life policy with sum insured of £100,000, for a male life aged 40 years, using *AM92* mortality tables and $i = 4\%$, allowing for £200 initial expense plus 5% of each premium payable in respect of renewal expenses and claim expenses of £500.

```
(premium_whole_gross<-(100500*Axn(am92_at, 40)+200)/(0.95*axn(am92_at, 40)))
[1] 1229.731
```

The net premium can be considered a special type of gross premium, and differs in two respects. Its calculation equates the expected present value of benefits payable under the contract with the expected present value of premiums – we ignore any cashflows in respect of expenses. In addition, the net premium is calculated using the reserving basis in place at the valuation date which will, in general, be different to the premium that is actually paid. The net premium, P, payable annually in advance for a whole life assurance benefit of £S payable at the end of the year of death is given by:

$$P = S \frac{A_x}{\ddot{a}_x} \tag{24.6}$$

Example 24.5

Adopting "whole, age 40, benefit £100,000, $i = 4\%$", the annual net premium is:

```
(premium_whole<-100000*Axn(am92_at, 40)/axn(am92_at, 40))
[1] 1152.485
```

Exercise 24.16 *Calculate the annual net premium due for term assurance, endowment assurance, pure endowment where premiums are payable annually in advance throughout the policy whilst alive. Calculate for a range on interest rates (e.g. $i = 3, 4, 5, 6\%$) and ages (e.g. 40 to 60 years) at purchase.*

As our focus is primarily on applying coding to solve problems we will tend to simplify calculations and ignore most, if not all, expenses.

24.7 Profits – Probability Distributions of Single Policies and Portfolios

In this section we look at the profit distribution arising from various types of insurance policies, looking at both single policies and portfolios. This will help in understanding the characteristics of these various policy types, and the advantages of writing larger portfolios. We will also see that we can determine an appropriate premium rate to charge by analysing the range of possible profits that may arise from selling groups of policies, and ensuring there is a required probability that a profit will be made by charging a sufficient premium.

In this section we will view profit arising from a policy simply as the present value of the premiums received less benefits and expenses paid. For example, under a term assurance policy, if the life survives the term of the policy the insurer will have received a series of premiums and paid no benefits, thus making a profit. Alternatively, if the life dies then the contractual benefit will be paid and a loss incurred (this is discussed in more detail later – see Figure 24.6, right). We will see shortly that the expected profits are in general positive in the early years of a contract, and negative in later years of contract (losses).

We will discuss profit in some detail in later sections, where policy values, or reserves, are set aside affecting the discounted value of these profits.

Profit distribution: Single whole life policy

Here we analyse the probability distribution of the present value of profits, or losses, under "whole, age 60, benefit £10,000, $i = 4\%$".

The profit is calculated as the present value (at the issue of the policy) of premiums received less the benefits paid. Note that this does not require any simulations.

Remark 24.6 Note that these are similar plots to the plots shown earlier of the present value of benefits; here we are showing profits, that is we are allowing for premiums received.

```
start_age<-60;    range_z<-120-start_age;    ben_wh1<-10000
(premium_wh1<-ben_wh1*Axn(am92_at, start_age)/axn(am92_at, start_age)) #prem payable
```
```
[1] 322.9182
```
```
pv_ben_whole<-ben_wh1/(1.04^(1:range_z))      #pv benefit in each year, at t=0
prem_pv_wh<-premium_wh1*(1-1.04^-(1:range_z))/0.04*1.04   #pv of prems
profit_whole<-prem_pv_wh-pv_ben_whole        #resulting profit from death each year
prob_death_fut_yr<-fx_fn1(60,60)             # prob of life dying in a future year
```
```
year_of_deathyt<-1:range_z
head(cbind(year_of_deathyt,prob_death_fut_yr,pv_ben_whole,prem_pv_wh),3)
```
```
      year_of_deathyt prob_death_fut_yr pv_ben_whole prem_pv_wh
[1,]               1       0.008022000     9615.385    322.9182
[2,]               2       0.008936730     9245.562    633.4165
[3,]               3       0.009940513     8889.964    931.9725
```
```
(whole_sd_single_profit<-sum(profit_whole^2*prob_death_fut_yr)^.5) #sd of profit
```
```
[1] 3128.916
```
```
cor(pv_ben_whole,prem_pv_wh)
```
```
[1] -1
```

See Figure 24.4. The potential profits/losses (present value at $t = 0$) range from £-9,292 to £6,647, in respect of a policyholder death between ages 60 to 61 years (i.e. in the first year of the policy), and 119 to 120 years respectively; death in the first year will result in losses of £10,000 less the initial premium (plus interest). Profits will be made only from policies where death occurred after age 80 years (denoted with • in Figure 24.4, right).

The variance of the present values of the profits is given here by:

$$Var(\text{Profit}_{\text{NPV}}) = P^2 \frac{^2A_x - A_x^2}{d^2} + B^2(^2A_x - A_x^2) + P * B \frac{^2A_x - A_x^2}{d} \tag{24.7}$$

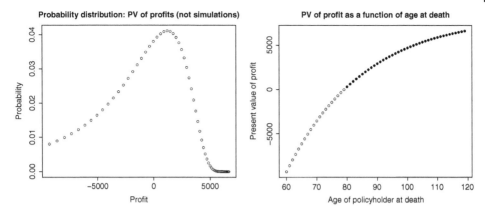

Figure 24.4 Single whole life assurance policy: Present value of profits.

where P is the premium, and B is the benefit payable. The standard deviation is calculated below (and is consistent with the earlier calculation):

```
(sd_theory<-(var_whole(am92_at,60,0.04)*(premium_whl*1.04/0.04+10000)^2)^0.5)
[1] 3128.916
```

Exercise 24.17 *Derive Equation 24.7, and simplify further.*

Example 24.6 Single whole life policy: It is a useful exercise to duplicate the above results using simulations. Below we calculate the standard deviation and VaR from simulations. (Note: the following code is similar to that included in Example 24.2.)

```
set.seed(100);          sim_no<-100000
ddd<-runif(sim_no)
sim_profit_whole<-termx<-cdf_whole<-NULL
cdf_whole<-c(0,1-surv_fn1(60,60))

for (nnn in 1:sim_no) {
sim_profit_whole[nnn]<- profit_whole[max(which(cdf_whole<=ddd[nnn]))]
}
var(sim_profit_whole)^0.5          #sd from sims, consistent with theory

[1] 3130.669

quantile(sim_profit_whole,0.01); qnorm(0.01)*sd_theory #99%VaR sims vs Normal approx

      1%
-8612.146
[1] -7278.947
```

Profit distribution: Portfolio of 100 whole life policies

See Figure 24.5 (left). Re-running the calculations, but with 100 policies in a portfolio instead of just one, and comparing the standard deviation and VaR calculations.

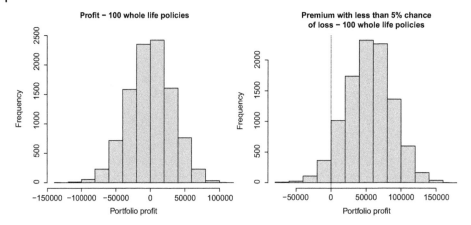

Figure 24.5 Present value of profits: Portfolio of 100 whole life assurance policies.

```
set.seed(10000)
no_polxz<-100              #no of policies
sim_toplevel<-10000        #no of simulations

ddd<-sim_profit_whole100<-book_profit_whole100<-NULL
for (vvv in 1:sim_toplevel) {
ddd<-runif(no_polxz)
for (nnn in 1:no_polxz) {
sim_profit_whole100[nnn]<- profit_whole[max(which(cdf_whole<=ddd[nnn]))]
}
book_profit_whole100[vvv]<-sum(sim_profit_whole100)}
var(book_profit_whole100)^0.5   # sd of profits - simulations (100 policies)

[1] 31274.84

100^0.5*sd_theory             # sd of profits - theoretical (100 policies)

[1] 31289.16

quantile(book_profit_whole100,0.01); qnorm(0.01)*10*sd_theory #VaR:sims vs Norm appx

        1%
-73932.29
[1] -72789.47
```

As demonstrated many times in this book, the importance of using simulations to estimate VaR results is clear from the above results relating to the single policy, and the portfolio.

Remark 24.7 The reader may prefer to use the rLife function from the life contingencies package to simulate future lifetime data, here running 100,000 simulations of complete years (type=''Kx'') survived for a male, 60-year-old life.

```
rand_life<- rLife(n=100000, object=am92_at, x=start_age, type="Kx")
```

Exercise 24.18 *Using trial and error, estimate, by running an appropriate number of simulations, the required premium such that there is less than a 5% probability that a loss is made from the 100 whole life policy portfolio. Compare this with only one policy in force.*

Discussion from Exercise 24.18 – see Figure 24.5, right. By charging a premium of £361, from the simulations run we estimate that there is a 5% probability of incurring a loss; this compares with charging a premium of £323 which would yield an expected profit of zero. One can, therefore, in theory at least, determine a suitable premium which should be charged such that the insurer has a particular probability of making a profit from the portfolio of 100 policies. This is similar to the method discussed in more detail in Chapter 25.

Remark 24.8 This method of calculating the life assurance premium is often referred to as the percentile method.

Exercise 24.19 *Verify the reasonableness of this premium calculation using a Normal approximation, and comment on the difference.*

Note that a large number of simulations is required here to obtain results that may be considered stable. As is often required in this chapter, the reader may wish to experiment with the number of simulations.

Exercise 24.20 *Repeat Example 24.18 but with a portfolio of 1,000 policies. Review the accuracy of the Normal approximation and compare with the scenario where we only have 100 policies.*

Single Term assurance policy: 10-year term

Turning now to look at term assurance, we repeat the above calculations for a 10-year policy.

Exercise 24.21 *Carry out a similar analysis to that above for a single whole life assurance policy, this time for "term 10 years, age 60, benefit £10,000, $i = 4\%$", and produce corresponding plots to those in Figure 24.4.*

See Figure 24.6 – note the highly skewed distribution of profits from the term assurance policy, compared to that of the whole life policy. Profit is highly probable from this single policy (87%), but only if the life survives to the end of the term.
The standard deviation of profits from this single term assurance policy is £2,824 – this will be used in the calculations which follow.

Profit distribution from a portfolio of 100 term assurance policies

Exercise 24.22 *By running an appropriate number of simulations, estimate the standard deviation and 99%VaR of profits/losses from a portfolio of 100 term assurance policies. Compare the results with those obtained from theory.*

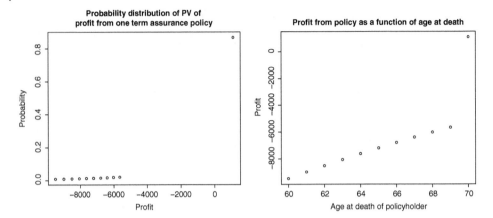

Figure 24.6 Present value of profits: Single 10-year term assurance policies.

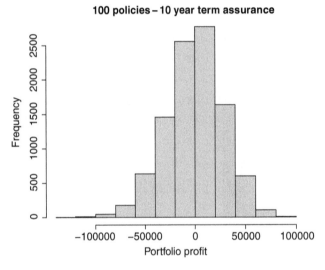

Figure 24.7 Present value of profits: Portfolio of 100 term assurance policies.

The results from Exercise 24.22 (saved in the object term_port) from running 10000 simulations are set out below:

```
var(term_port)^0.5; term_sd_single_profit*10   # sd from sims and theory (portfolio)
[1] 28345.22
[1] 28240.69

quantile(term_port,0.01); qnorm(0.01)*(10*term_sd_single_profit) #99% sims and Normal

      1%
-72649.65
[1] -65697.66
```

As noted earlier, the standard deviation of profits is proportionately significantly higher for the term assurance compared to the whole life. Also, the Normal approximation appears not to be appropriate here (not surprising given the skewed nature from the single policy).

Remark 24.9 Reminder – the profit calculations shown above have been carried out without setting up any reserves. These reserves, or policy values, are discussed later in the chapter.

Exercise 24.23 *Extended exercise – insurer sells annuities and life assurance business*

This exercise illustrates the effect of writing both life assurance and annuity business; the partially negatively correlated lines of business help to reduce the uncertainty of the results to some extent.

An insurer has priced and subsequently sold 100 annuities and 100 whole life policies, all to independent lives (i.e. 200 lives in total). Subsequently, there is a national health crisis resulting in significantly higher mortality rates. Analyse the impact on profits of both the individual departments of the insurer, and the aggregate effect, by running simulations from an appropriate model similar to those described in Section 24.7. The reader should calculate the mean and standard deviation of the present value of the net profits from simulations and theory, and calculate the 99% VaR using these simulations and the Normal approximation.

> For life assurance: "whole, age 60, benefit £400,000, $i = 4\%$".
> For annuities: "age 60, benefit £20,000 payable annually in arrears, $i = 4\%$".
> Mortality rates for pricing, q_x: AM92.
> Mortality rates for profit calculations, q_x: 110% of AM92.

Assume the following: (1) Life assurance premium calculated using Equation 24.6 payable annually in advance; (2) the single annuity premium equal to 20000 a_{60}; (3) ignore expenses.

Results from the simulations are shown in Figure 24.8.

Exercise 24.24 *Re-calculate the simulated distribution with only 100 lives, but with each life buying both an annuity and life assurance policy, that is the total number of policies is 200, as above.*

Exercise 24.25 *Study the effect of various alternative changes in mortality rates, for example there is a sudden reduction in mortality rates for lives under age 70 years.*

24.8 Progression of expected profits throughout the lifetime of a policy: no reserves held

Most of the remainder of the chapter will discuss profits. In this section we discuss how profits emerge through the life of a policy with no reserves being set. In Section 24.9 we discuss policy values which, when set appropriately, aim to minimise the likelihood that

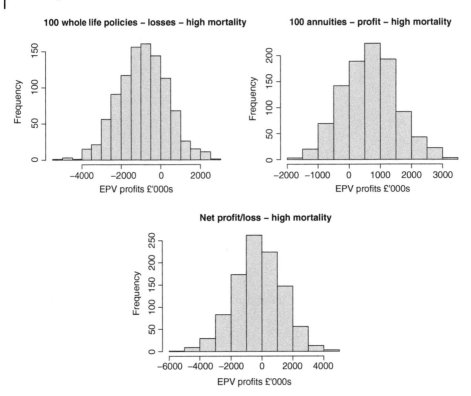

Figure 24.8 Simulated profits from whole life and annuity portfolios policy following an increase in mortality rates.

losses will be made in any particular year. In Section 24.10 we look at the emergence of profits through the life of a policy when such reserves are set up.

Understanding the plots in Figure 24.9 is a key step in appreciating the importance of reserves. They show expected values of profits, at the start of each year, emerging throughout the life of a policy, per policy in force at the start of each year (for a male life, aged 60, with annual benefit payments of £500,000, with no reserves being set). In effect, each single calculation looks at the cashflows over the following year only. Note that these are not probability distributions; compare these with previous plots from Section 24.7 which show probability distributions of profits. The code is set out below:

```
age_stx<-60;        term_x<-20

#Endowment
premium_endw<-AExn(am92_at, age_stx,term_x)/axn(am92_at, age_stx,term_x)
exp_profit_endw<-500000*(premium_endw-c(qx_rates1(60,19),1)/1.04)

#Term assurance
premium_term<-Axn(am92_at, age_stx,term_x)/axn(am92_at, age_stx,term_x)
exp_profit_term<-500000*(premium_term-qx_rates1(60,20)/1.04)

#Whole life
premium_whl<-Axn(am92_at, age_stx)/axn(am92_at, age_stx)
exp_profit_whl<-500000*(premium_whl-qx_rates1(60,40)/1.04)
```

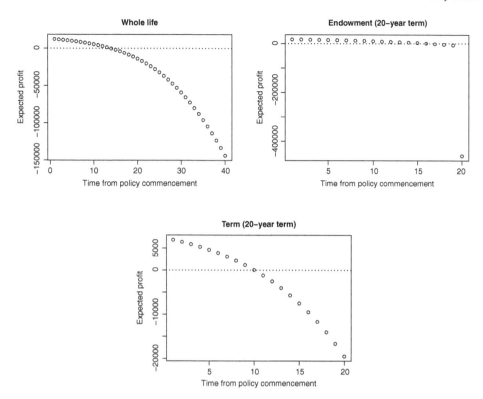

Figure 24.9 Present value of expected profits in each year of a policy.

Note that all calculations ignore expenses, and assume a constant interest rate through-out the term of 4% pa.

Exercise 24.26 *Re-run these calculations using various ages and terms.*

These calculations demonstrate that profits, typically, are expected in the early years of each type of policy, and losses in later years. In the early years of a policy there is, in general, little chance the life will die resulting in a benefit payment, and one would expect a profit in these years.

24.9 Policy Values

24.9.1 Calculating Policy Values

From the previous section we saw that, due to increasing mortality rates with age, the largest profits emerge in the earlier years of a policy, with losses being incurred in the later years. This is unacceptable as the insurer may be unable to make the payments arising later in the policy if profits have previously been distributed to shareholders from the early, profitable years. Thus reserves should be set up in anticipation of these expected cashflows later in the policy.

In effect, by setting up these reserves, profits distributable to shareholders are deferred. The more prudent these reserves the greater is this deferral; indeed, losses will be incurred in early years if reserves are large enough. The extreme alternative, discussed in Section 24.8, is where no reserves are set up and all profits are distributed immediately, resulting in the earliest payment of profits. In Section 24.10 we will incorporate reserves into the profit calculations under various reserving bases, such that the likelihood of losses *in those later years* is reduced. We will see that, by including reserves, the net present value of profits is reduced.

Example 24.7 The same principle applies to annuity contracts. If we sell an annuity policy to a life aged 60 years, we would receive a premium at issue and set up a reserve in respect of the expected payments on the life surviving 1, 2, 3... years; it would be imprudent to take the premium in year one as profit and distribute this to shareholders, knowing that we are likely to be required to make a series of annual payments.

We will determine the amount of reserves by calculating policy values. A general definition of the policy value is the expected present value of future benefits and expenses payable, less the expected present value of future premiums due. The policy value at policy duration t is denoted by $_tV$.

One method used to determine reserves is to calculate the net premium policy value, where expenses are excluded from the calculation. For a whole life policy sold to a life at age x, the net premium policy value at duration t is given by:

$$_tV = S\,A_{x+t} - P\ddot{a}_{x+t} \tag{24.8}$$

where S is the sum assured. P is the net premium (as defined in Equation 24.6) calculated using the reserving basis in place at the valuation date – it is not the premium that is actually paid. This should be compared with the premium used in the gross premium policy value calculation defined below.

Thus the reserve at the end of the first year of a whole life policy sold to a male 60-year-old (per £1 insured), $_1V$, is given by:

```
prem_y<-Axn(am92_at, 60)/axn(am92_at, 60)
t<-1;    (reserve1<-Axn(am92_at, 60+t)-prem_y*axn(am92_at, 60+t))
[1] 0.0257682
```

Remark 24.10 It is important to note the timing assumption of cashflows and the convention used to calculate these reserves – we assume that benefits are paid immediately prior to the policy anniversary, with premiums being received immediately after the anniversary; policy values are calculated between these two cashflows – immediately prior to receiving a premium and immediately after the payment of the benefit.

The following exercise is important, with the results shown in Figure 24.10.

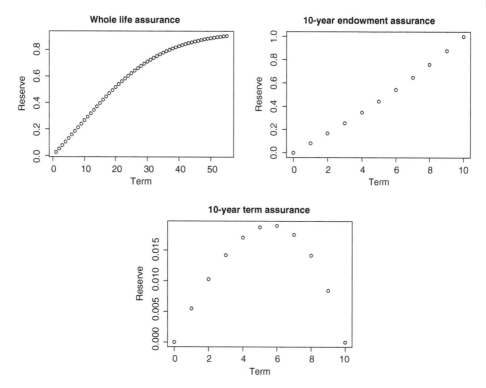

Figure 24.10 Evolution of reserves for policyholder aged 60 years at commencement.

Exercise 24.27 *Write a function to calculate the reserve at time t, $_tV$, for a life aged x at the start of a whole life policy. Repeat for an endowment assurance policy and term assurance policy.*

Proceed to calculate how the reserves for these three types of policies change over the term of each policy for a life aged 60 years, and a term of 10 years where appropriate. Assume i = 4% pa and mortality is in line with the AM92 tables. (See Figure 24.10.)

Note that the reserving basis is likely to change with time; this requires re-calculation of the net premium used in the calculation. For example, at $t = 3$ perhaps the valuation rate of interest has changed from 4% pa to 5% pa. The net premium policy value is now:

```
prem_y3<-Axn(am92_at, 60,i=0.05)/axn(am92_at, 60,i=0.05)
t<-3;    (reserve3<-Axn(am92_at, 60+t,i=0.05)-prem_y3*axn(am92_at, 60+t,i=0.05))
[1] 0.07142421
```

The calculation of reserves is important in estimating the insurer's current liabilities and hence solvency position. It is also important in assessing the future profitability of products; we will see shortly that reserves in general reduce the profitability of contracts, thus it is vital to estimate reserves in advance such that we can analyse the effect on profitability from holding different levels of reserves, and adjust premiums accordingly.

Remark 24.11 Note the technicality resulting in a minor inconsistency of the reserve calculation at expiry of the endowment policy – the reserve is technically zero as there are no further cashflows due at expiry. The term, AExn(am92_at, 70,0), which would in general not be used, calculates a payment immediately, with certainty.

The gross premium policy value is a more intuitive term, more closely reflecting the original definition of the policy value:

$$_tV = S\,A_{x+t} + EXPV - P\ddot{a}_{x+t} \tag{24.9}$$

where, in contrast to Equation 24.8, EXPV is the expected present value of all future expenses (see Equation 24.5 for details of these expenses), and P is now the *actual* premium paid under the policy.

Of course, the actual premium due under a contract may be calculated using whatever assumptions the insurer deems appropriate; they may decide not to include explicit assumptions about expenses, and implicitly allow for expenses by adopting a lower discount rate instead.

An intuitive way to gain an initial understanding of the policy value is to consider it as the value of the policy to the policyholder. We can think of this amount as the amount which the policyholder may be willing to sell it back to the insurer. For example, what would the policy value be one day prior to the end of an endowment assurance policy? Compare a term assurance policy where it is unlikely any payment will be made with one day remaining on the contract. Of course, the calculation depends on the actuarial basis used.

24.9.2 Recursive Formulae – Discrete and Continuous (Thiele)

The following recursive formula, relating $_tV$ and $_{t+1}V$, is informative (where S is paid at age $x + t + 1$, and we ignore expenses):

$$(_tV + P)(1 + i) = S\,q_{x+t} + (1 - q_{x+t})\,_{t+1}V \tag{24.10}$$

This has an intuitive interpretation; the reserve held at the start of the year, $_tV$, plus the premium received together with interest earned over the following year, should be sufficient to meet both the sum insured for all lives which die in that year (paid at the end of the year) and the reserve for all remaining lives.

Exercise 24.28 *Verify the results from Exercise 24.27 using Equation 24.10.*

The recursive approach is particularly useful where the benefits are a function of the policy value; as the policy value at the end of the contract must be zero (as there are no further benefits or premiums due) we can work recursively back from the end of the policy term.

Example 24.8

An insurer has sold a 10-year endowment policy to a 60-year-old male life, with annual premiums of £32,000 (in advance), where the death benefit equals the policy value at the start of the year and the survival benefit is £400,000. Determine the annual policy values under this contract assuming a constant effective interest rate is 3.5% pa.

The code and output are given below:

```
premx<-32000;      int<-0.035;      sss<-400000;      v1<-NULL;      v1[11]<-sss
qa<-qx_rates1(60,10)
for (i in 10:1) {v1[i]<- (v1[i+1]*(1-qa[i])-premx*(1+int))/(1-qa[i]+int)  }
round(v1,0)          #policy values at each year end

 [1]    3500  37011  71739 107734 145048 183737 223859 265477 308659 353474
[11] 400000
```

The continuous-time version of Equation 24.10 is Thiele's differential equation:

$$\frac{d}{dt}\,_tV = \delta_t\,_tV + P_t - \mu_{x+t}(S_t - {}_tV) \tag{24.11}$$

where P_t is the annual rate of premiums paid continuously, μ_x is the annual force of mortality at age x, interest is accrued at the force of interest δ_t per annum, and S_t is the benefit payable immediately on death at time t. Thiele's expression is useful where we may be interested in policy values at times other than at the year-end, where premium or benefit payments are non-annual or irregular, and where we have a non-constant rate of interest (which will be the case in practice). Of course, Equation 24.11 can easily be adjusted to incorporate other cashflows, such as initial and regular expenses. The following is a discretised version of Equation 24.11, using time steps equal to h (also compare Equation 24.10):

$$({}_tV + hP_t)(1 + \delta_t h) \approx S_t\,\mu_{x+t}h + (1 - \mu_{x+t}h)\,_{t+h}V \tag{24.12}$$

Exercise 24.29 *Repeat Exercise 24.27 by applying Equation 24.12, but using the following mortality rates:* $\mu_{age} = e^{-10.2+0.092\times age}$.

The following example includes non-constant interest rates, premium rates, mortality rates, and sum assured.

Example 24.9

Using Equation 24.12, calculate policy values for a 10-year endowment for a life aged 60 years at commencement. A benefit of £5,000 is payable on survival, with the single death benefit decreasing linearly from £15,000 to £5,000 over the term of the policy, with continuously paid premiums decreasing linearly from an initial annual rate of £700 to £300 at $t = 10$.

Assume a linear force of interest where $\delta_0 = 3\%$pa and $\delta_{10} = 4\%$pa, and force of mortality, $\mu_x = e^{-10.2+0.092x}$.

Also calculate the policy values for three such policies, as at today, with 4, 6, and 8 years left on the policy, that is where the policyholders are aged, today, 66, 64, and 62 years respectively.

```
agestart<-60;  nnn<-10;  assured_end<-5000;  assured_start<-15000
h<- 0.01;       steps_no<-nnn/h

prem1<-seq(700,300,length.out=steps_no)               #premium
SSS<-seq(assured_start,assured_end,length.out=steps_no)  #sum assured
muxx<-function(agexx)exp(-10.2+0.092*agexx)           #mortality
deltaxx<-function(timexx) 0.03+0.001*timexx           #interest
value1<-NULL;    value1[steps_no+1]<-assured_end      #policy values

for (i in (steps_no+1):1) {qqq<-(i-1)*h
       value1[i-1]<-value1[i]*(1-deltaxx(qqq)*h)-prem1[i-1]*h+
       h*muxx(agestart+qqq)*(SSS[i-1]-value1[i])}
```

The solution is shown in Figure 24.11.

It would be a useful additional exercise for the reader to simulate paths of future policy values, allowing for uncertainty in interest rates and mortality rates. This can then be combined with a stochastic projection of asset values. In this way the probability that the value of assets will at any time fall below that of the liabilities can be assessed; this could be used to determine the initial risk capital allocation required, given a particular risk appetite.

24.9.3 Recursive Equation with 3 States – HSD Model

The above examples deal with typical life assurance policies discussed in this chapter, where lives can exist either in the alive or dead state. However, Thiele's equation can be generalised to allow for as many states as are required to model a particular scenario. The following example involves three states – healthy, sick, and dead – often referred to as the *HSD* model, or disability model (also discussed in Chapter 22).

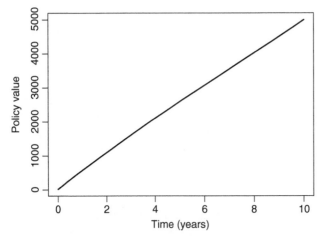

Figure 24.11 Evolution of reserves using Thiele's equation.

Example 24.10

The policy provides sickness and death benefits to a male life, aged 50 years, over a term of 20 years. There is no limit on the number of transitions between H and S states. Whilst sick, the life receives payments at the rate of £150,000 each year. On death a single payment of £400,000 is paid. Premiums are paid at the rate of £20,000 each year whilst the life is healthy. All payments are assumed to be paid weekly.

For the 3-state HSD model, the relevant discretised Thiele equations are as follows:

$$_tV^H = {}_{t+h}V^H(1 - \delta_t h) - hP_t + \mu_x^{HS}h({}_{t+h}V^S - {}_{t+h}V^H) + \mu_x^{HD}h(S - {}_{t+h}V^H) \quad (24.13)$$

$$_tV^S = {}_{t+h}V^S(1 - \delta_t h) + hB_t + \mu_x^{SH}h({}_{t+h}V^H - {}_{t+h}V^S) + \mu_x^{SD}h(S - {}_{t+h}V^S) \quad (24.14)$$

The transition rates we shall adopt are as follows:

$$\mu_x^{HS} = 0.0004 + 0.0000035e^{0.14x}$$

$$\mu_x^{HD} = 0.0005 + 0.000075e^{0.0875x}$$

$$\mu_x^{SH} = 0.1\mu_x^{HS}$$

$$\mu_x^{SD} = \mu_x^{HD}$$

Calculate the policy values, $_tV^H$ and $_tV^S$, throughout the term of the policy, assuming $\delta = 4\%$ pa, and using $h = \dfrac{1}{52}$ years.

```
deltax<-0.04;    agestart<-50;      termz<-20
h<-1/52;         steps_no<-termz/h
prem1<-20000;    sickz<-150000;     dth<-400000

a1=0.0004;  b1=0.0000035;  c1=0.14
a3=0.0005;  b3=0.000075;   c3=0.0875
val1<-val2<-NULL;   val1[steps_no+1]<-val2[steps_no+1]<-0

for (i in (steps_no+1):1) {
qqq<-(i-1)*h
mu12=a1+b1*exp(c1*(agestart+qqq))
mu21=0.1*mu12
mu13=a3+b3*exp(c3*(agestart+qqq))
mu23=mu13
val1[i-1]<-val1[i]*(1-deltax*h)-prem1*h+h*mu12*(val2[i]-val1[i])+h*mu13*
(dth-val1[i])
val2[i-1]<-val2[i]*(1-deltax*h)+sickz*h+h*mu21*(val1[i]-val2[i])+h*mu23*
(dth-val2[i])}
```

The solution is plotted in Figure 24.12. Thus if a valuation was due one year into this particular policy, the liability would equal £10,000 if the life was healthy, and £1.86m if the life was sick.

Exercise 24.30 *Calculate the premium such that $_0V = 0$.*

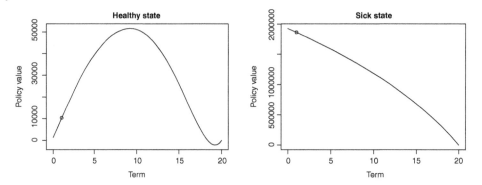

Figure 24.12   Evolution of Thiele reserves for HSD model.

24.10   Profits from Policies where Reserves Are Held

24.10.1   Calculating the Profit Vector

Earlier in this chapter we looked at the pattern of profits which typically emerge through-out the term of various assurance products; we saw that profits are expected in the early years, with losses incurred in later years (see Figure 24.9). Clearly this is unacceptable – some money must be kept back in the profitable, early years such that there is sufficient money to pay benefits later.

We proceeded to calculate policy values for various products. These represented the amount of money that it was estimated, on some basis, would be sufficient, together with future premiums, to pay all future benefits and expenses. Rather than distributing all the profits from those early years we retain some as reserves and invest them.

The profit earned in year t of a policy is given by:

$$\text{Profit}_t = \Delta A_t - \Delta L_t \tag{24.15}$$

where ΔA_t and ΔL_t are the changes in the value of assets and liabilities respectively, over year t – compare this with the definition for profit discussed in Section 24.7 where no reserves were set and L_t was zero. Before proceeding we set out a number of definitions (some of which have been defined earlier in the chapter):

P is the premium payable at the start of each year;

i is the effective rate of interest earned on the invested assets (these assets may typically be invested in a bond fund);

S is the benefit payable on death, here assuming the payment is made at the end of year of death;

q is the probability the life dies during the year.

(Note that each of the above could be time-dependent; for simplicity we are assuming each parameter is constant.)

At the start of each year (time $= t$) we will hold assets equal to the reserves (or liabilities), and the liabilities are simply $_tV$. So $A_t = {}_tV$. Any excess assets can be distributed as profits.

At the end of the year, $A_{t+1} = (A_t + P)(1 + i) - qS$. In respect of the liabilities we need to hold $_{t+1}V$ for each policy still in force at the end of the year – we don't need any money in respect of policies already paid out from deaths that year, thus $L_{t+1} = (1 - q)_{t+1}V$.

Hence, the profit in year t, as at the year end, is given by the important equation:

$$\text{Profit}_t = (A_t + P)(1 + i) - qS - (1 - q)_{t+1}V \tag{24.16}$$

and the profit vector is defined as the vector which contains the profits from each year of the n-year policy. We carry out the profit calculation in respect of each year of the policy, each year re-setting the assets held at the start of the year to the policy value, that is $A_t = {}_tV$.

As we decide to hold reserves and not distribute all of the profits immediately, we instead invest these reserves in assets such as bonds which would be expected to earn a lower return than our insurance business would be expected to (otherwise why bother running an insurance company – we could just invest money in bonds). By holding reserves, the money invested achieves this lower return, thus diluting returns; the payment of profits is deferred and discounted to a greater extent. Thus shareholders will only want insurers to hold a level of reserves which is considered reasonable. The decision as to the amount of reserves to hold is therefore a balance between holding sufficient to ensure benefits can be paid, and minimising reserves in order to receive profits earlier and maximise returns.

Remark 24.12 There are a number of interest rates required in the following calculations. A reserving basis must be set to calculate reserves and obtain the profit vector – this includes an interest rate usually referred to as the valuation rate of interest. In addition, a pricing basis may be specified which is used to calculate the policy premium (alternatively the premium may simply be stated). An investment return assumption is required, often referred to as the "effective rate earned on insurer's assets" – this is the return earned on the assets, A_t, and will typically reflect a bond yield. We shall see shortly that a "risk discount rate" is also likely to be required – this is used to assess the profitable of the contract, reflecting the return required by shareholders.

Example 24.11

The aim of this example is to calculate the profit vectors from an endowment policy on a number of different bases, and to illustrate the effect which various reserving bases have on profits. We shall use discount rates of 4% to 8% pa for the various reserving bases (we also calculate profits using a discount rate of 40% pa, effectively setting initially extremely low reserves).

The policy details are as follows: 20-year endowment, £100,000 benefit, male 60-year-old. Calculate the premium, and profit vectors, under the various reserving bases below:

- Reserving bases: a range of valuation rates of interest, $i = 4\%$ to 8% pa, and 40% pa ; reserves are calculated using net premium policy values
- Pricing basis: $i = 6\%$ pa
- Mortality for pricing and reserving: $AM92$
- Investment return on insurer's assets, an effective rate of 7.5% pa

- Risk discount rate: 10% pa
- No expenses are assumed in the initial calculations; however please see Exercise 24.31 where we include £800 set-up costs.

```
ben_endw<-100000;     age_0<-60;        term_endx<-20;    npv_various<-NULL
inv_x<-0.075      # return earned on reserves; effective rate

pricing_i<-0.06
(prem_act<-ben_endw*AExn(am92_at, age_0,term_endx,i=pricing_i)/
axn(am92_at,age_0,term_endx,i=pricing_i))

[1] 3614.092

q_rate<- qx_rates1(age_0,term_endx)
q_rate[term_endx]<-1   #needed for endowment - payment is made at end
range_val_i<-c(seq(0.04,0.08,by=0.01),0.4); rng_lng<-length(range_val_i)
reserves_rec<-matrix(0,rng_lng,21)
counter<-1;         prof_table<-matrix(0,rng_lng,20)

#START OF LOOP INCORPORATING THE RANGE OF RESERVING BASES
for (acc_i in range_val_i) {(prem_end<-ben_endw*AExn(am92_at, age_0,term_endx,
i=acc_i)/
       axn(am92_at,age_0,term_endx,i=acc_i)) # required for reserving basis

pol_val_endw<-NULL
for (aa1 in 1:term_endx) {pol_val_endw[aa1]<-ben_endw*AExn(am92_at, age_0+aa1-1,
term_endx-aa1+1,i=acc_i)-prem_end*axn(am92_at,age_0+aa1-1,term_endx-aa1+1,i=acc_i)}

pol_val_endw<-c(pol_val_endw,0)  #reserves set
reserves_rec[counter,]<-pol_val_endw

ass_y<-liab_y<-profit_y<-npv_vec<-NULL

for (h1 in 1:term_endx) {
       ass_y[h1]<- pol_val_endw[h1]*(1+inv_x)+(prem_act)*(1+inv_x)
       -ben_endw*q_rate[h1]
       liab_y[h1]<- (1-q_rate[h1])*pol_val_endw[h1+1]
       profit_y[h1]<-ass_y[h1]-liab_y[h1]
       }
prof_table[counter,]<-round(profit_y,0)
counter<-counter+1
}                       #END OF LOOP INCORPORATING THE RANGE OF RESERVING BASES
```

`prof_table` contains the profit vectors from each of the reserving bases used above (4% to 8%, and 40%); they are plotted in Figure 24.13. Thus a prudent reserving basis results in profits being recognised much later in the policy, for example $i = 4\%$ pa. Using a discount rate of 8% pa results in fairly level projected profits. Note that these profits are calculated as at the year end.

In addition to the profits calculated at each year end, we must also include any cashflows required before $t = 0$ in respect of the contract, for example acquisition expenses, or the setting of prudent reserves which may require reserves to be set up at $t = 0$, both of which are particularly likely. (We include such an expense in Exercise 24.31, and set up prudent gross reserves as part of Exercise 24.34.)

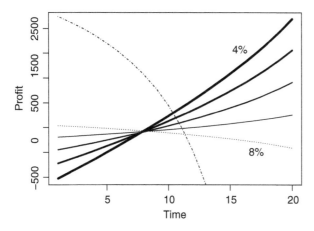

Figure 24.13 Profit vectors with various reserving bases.

The reader is encouraged to experiment with various combinations of reserving, pricing, and effective rates of interest.

As seen above, we must set up reserves such that we can be confident that benefits will be paid in the future. At the start of each year we determine the policy value and hold assets equal to this – this is the value of the assets and liabilities at the start of the year. Both these then change during the year as we receive premiums, pay benefits, earn interest on the assets, and recalculate policy values at the year end; the profit is determined at the end of each year, not just based on the net cashflow but also on the change in policy value. Each year end we take our profit such that we are left with assets which equal the policy values.

24.10.2 Measures of Profit and Profit Testing

Now we have estimated a pattern of profits (i.e. profit vectors) which incorporate potentially appropriate reserves, we can assess the size of those profits. Ultimately we can determine a premium which results in an acceptable profit to the insurer. We proceed using the example in the previous section.

It is important to note that the profit vector contains the expected profit at the end of each year of the policy, *per policy in force at the start of each year*. Effectively, we look at each year in isolation.

The profit signature is calculated directly from the profit vector; it is defined as the expected profit in each year of the policy, *per policy in force at the date of issue*. Thus to obtain the signature we simply multiply the vector by the probability of the contract remaining in force to the start of each year; this should allow for all types of decrements, not just deaths; for example a life may surrender a policy. We shall look at the simpler scenario of where there are no surrenders.

To assess the profitability of a policy design it is preferable to calculate a suitable profit measure. There are various such measures, including Net Present Value, Internal Rate of Return, Discounted Payback Period, and Profit Margin – we shall look at the first two of these measures.

Net Present Value ("NPV") is simply the total discounted present value of each element of the profit signature, applying the risk discount rate (which reflects the return required by shareholders). The Internal Rate of Return ("IRR") was first discussed in Chapter 4 – it is the discount rate (here using the annual effective rate) such that the NPV equals zero.

Starting with the array of profit vectors calculated in Example 24.11 under various reserving bases (called prof_table), we calculate the profit signatures and NPVs using a risk discount rate equal to 10% pa:

```
prof_sig1<-matrix(0,rng_lng,20);       npv_vec<-matrix(0,rng_lng,20)
disc_x<-0.10                           #for npv calc - risk discount rate

for (counter in 1:rng_lng) {
        prof_sig1[counter,]<-prof_table[counter,]*(c(1,surv_fn1(age_0,term_endx-1)))
        npv_vec[counter,]<-prof_sig1[counter,]*(1+disc_x)^-(1:term_endx)
        npv_various[counter]<-sum(npv_vec[counter,])}
```

See the results below (Table 24.1). Note that the NPV (in column "NPV_ign") is highest with the least prudent reserving basis, as was noted earlier; by setting low reserves more profit is distributed early in the contract.

Proceeding to calculate the Internal Rate of Return. These highlight a potential problem with IRR; there may not exist a solution, for example, where all entries are positive:

```
IRR_x<-0;    IRR_y<-NULL
for(xzz in 1:rng_lng){
minz<-0;        maxz<-0.8      #reset ranges for each run
price_est<-100                 #set initial estimate
cc<-1                          #counter to break while loop
while(abs(price_est)>0.1){
if(cc==200){break}
IRR_x<-(minz+maxz)/2
price_est<-presentValue(prof_sig1[xzz,],1:term_endx,IRR_x)
if(price_est < 0){maxz<-IRR_x}else{minz<-IRR_x}
cc<-cc+1
}
IRR_y[xzz]<-IRR_x
if(IRR_y[xzz]==0.8){IRR_y[xzz]=999}}  #999 error code
```

To ensure any non-calculation in the output is clear we include an error code, "999".

Exercise 24.31 *Repeat the above calculations but assuming an initial actual expense equal to £800 at t = 0 is incurred in respect of policy set-up costs.*

The full summary of results, showing NPVs and IRRs, both ignoring ("ign") and including ("inc") the £800 initial costs from Exercise 24.31 are shown below in Table 1. Note that the adjusted NPVs are all £800 lower than the original NPVs:

Table 24.1

	res_i	NPV_ign	IRR1_ign	NPV_inc	IRR_inc
[1,]	0.04	2559	0.228	1759	0.161
[2,]	0.05	2779	0.381	1979	0.197
[3,]	0.06	2989	999.000	2189	0.269
[4,]	0.07	3189	999.000	2389	0.417
[5,]	0.08	3380	999.000	2580	0.636
[6,]	0.40	6111	999.000	5311	999.000

Exercise 24.32 *Calculate the NPV of profits if no reserves are set up.*

The insurer can now determine a premium which may be charged to satisfy particular profit criteria, given a particular reserving basis. For example, the insurer may choose the reserving basis above with $i = 6\%$ pa, and set premiums such that the IRR $= 20\%$ pa.

Exercise 24.33 *Re-run the test with various premiums to assess the effect on profits, and determine the minimum premium which meets the above profit criteria, i.e. IRR > 20% pa. The reader should also be aware of the sensitivity of the IRR from changes to the premium.*

Exercise 24.34 *Re-run the calculations using reserves calculated on a gross premium policy value basis, i.e. reserves based on actual premiums paid.*

The following exercise is similar to Example 24.11. However, it contains a number of extra elements, including the calculation of dependent decrement rates (see Section 24.15 – Appendix) and the use of different bases.

Exercise 24.35 *An insurer sells a 20-year endowment policy to a male, 40-year-old, with sum assured of £100,000 and annual premiums of £4,200. Surrender value of the policy is equal to the net premium policy value at the beginning of each year.*
Profit testing basis uses 90% AM92 mortality. Assume independent surrender decrements of 20% in year 1, and 2.5% each year thereafter. Annual expenses equal to 75% of the initial premium plus £1,000 at the start of year 1, followed by 5% of the premium plus £50 each subsequent year. Reserves are calculated on a net premium policy value basis, with valuation rate of interest equal to 5.5% pa, and 110% AM92. 4.5% pa interest on reserves, and a risk discount rate of 12% pa.
The objective of the exercise is to determine the net present value of profits and IRR.
Also re-run the test with various premiums to assess the effect on profits. Also try various reserving bases, and vary parameters to undertake sensitivity testing, for example increase mortality rates, reduce investment returns, increase surrenders to 50% in year 1.

Exercise 24.36 *Zeroised reserves*

Recalculate the reserves and profits from Example 24.11 and Exercise 24.35 but using "zeroised reserves". This requires setting the minimum possible reserves subject to the constraint that expected profits in each year are non-negative, working backwards from the final year of the policy.

24.11 Profit Uncertainty: Interest Rate and Mortality Risk

Example 24.12

It is important to understand the potential uncertainty in the financial position of a port-folio of policies. Here we look at a portfolio consisting of 1,000 identical whole life policies each with sum assured of £100,000, sold to male 60-year-old lives, and assess the uncertainty of the relative value of assets and liabilities at the end of the first year of the policies (at $t = 1$), allowing for (1) potential changes in interest rates and (2) the number of deaths in the first year. For example, interest rates may fall, or there are more deaths than expected – both of which may lead to a deterioration of the financial position. In addition, we adopt a mis-matched investment strategy by investing in a 1-year zero-coupon gilt such that the asset return over the first year is guaranteed. (The reader is encouraged to incorporate additional levels of uncertainty, for example expenses, investment returns, surrenders – see Exercise 24.39.) Our aim is to determine the distribution of profits earned at the end of year 1, using Equation 24.16. Possible code is included below.

Assumptions:

Mortality – *AM*92 table throughout (pricing and reserving bases);
Assume the number of deaths between exact ages x and $x + 1$, $D_x \sim Bin(1000, q_x)$;
Reserves are set equal to gross premium policy values, with $i = 3.5\%$ pa at $t = 0$;
Policy value at $t = 1$ year, $_1V$, calculated using a stochastic rate of interest; we use a simplistic interest rate model: $i \sim N(0.035, 0.0025^2)$, and assume a flat yield curve;
The initial allocated capital per policy equals $_0V$;
Premium charged is calculated using the equivalence method, using $i = 4\%$;
Assets are invested in a zero-coupon 1-year gilt, such that the return over year 1 equals 5%, and is risk-free (compare Exercise 24.39);
Ignore expenses (to simplify the scenario).

```
set.seed(100);   simskkk<-1000;  popnx1<-1000;  ben_ab<-100000
profit1x<-res_pol<-NULL

inv_ret<-0.05; i_price<-0.04; i_val_mean<-0.035;  i_val_sd<-0.0025
qqqx<-qx_rates1(60,1)        #mortality rate for year 1
deathx1<-rbinom(simskkk,popnx1,qqqx);  popnx2<-popnx1-deathx1   # stochastic
premium_whlx<-ben_ab*Axn(am92_at, 60,i=i_price)/axn(am92_at, 60,i=i_price)  #prem
res_pol_zero<-ben_ab*Axn(am92_at, 60,i=i_val_mean)-
premium_whlx*axn(am92_at, 60,i=i_val_mean)  #initial policy value and assets
asset_zero<-(premium_whlx+res_pol_zero)*popnx1    #includes premiums rec'd at t=0

intxz<-rnorm(simskkk,i_val_mean,i_val_sd)  # stochastic reserve rate
#loop...
for (ttyy in 1:simskkk) {res_pol[ttyy]<-ben_ab*Axn(am92_at, 61,i=intxz[ttyy])-
premium_whlx*axn(am92_at, 61,i=intxz[ttyy])}

reserveat1term<-res_pol*popnx2
assetsat1<-asset_zero*(1+inv_ret)-deathx1*ben_ab
profit1x<-assetsat1-reserveat1term        #see histogram
```

Stochastic interest rates and deaths: Profit in Year 1

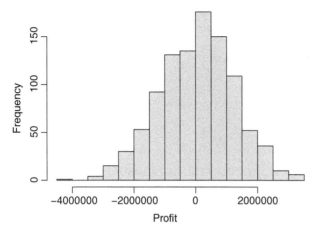

Figure 24.14   Distribution of profits at the end of Year 1.

The simulations obtained from the above code are shown in Figure 24.14. From these results there is a 46% probability that a loss will be made in year 1. Of course, this is a simplistic example; we may usually expect losses in year 1 due to initial expenses (which we have not allowed for in this example).

In Exercise 24.42, we carry out similar calculations to the above, but instead, run through the entire policy term.

Exercise 24.37 *Identify the 10 simulations which result in the worst results, and comment on the number of deaths and change in interest rates included in these simulations.*

Exercise 24.38 *Assess the effect on the uncertainty of year-1 profits if different levels of uncertainty in respect of interest rates and mortality are incorporated in the assumptions, and quantify the relative levels of mortality and interest rate risk.*

Exercise 24.39 **Matched investment strategy.**
Repeat the above calculations, but with all monies invested in 10-year zero-coupon gilts, assuming a flat yield curve and the same interest rate distribution as above, thus incorporating stochastic investment returns over year 1.

Exercise 24.40 *Repeat the above calculations but with different pricing, reserving, and asset return bases.*

Exercise 24.41 *Repeat for term assurance and endowment polices, and compare the level of risk.*

The following exercise is a little more time-consuming than typical exercises.

Exercise 24.42 *Re-run the calculations in Example 24.12 but complete them for the entire policy term, that is run the calculations across each year of the policies. Estimate the probability that the portfolio results in a loss, and also the probability of losses being made in each year.*

The key conclusion from the above exercises is that, although reserves have been set up (equal to the policy values), it is possible that additional reserves would be required following adverse experience such as a larger than expected number deaths occurring, and/or a fall in interest rates.

24.12 Risk Capital and Risk-adjusted Return Measures

Ultimately, calculations similar to those in Section 24.11 can be used to assess the amount of initial risk capital which should be held in respect of a portfolio of policies such that there is an acceptable probability that further capital is not required at a future point, for example as a result of falling interest rates or excess deaths. From Example 24.12 we estimated there was a 46% probability that losses would be made in Year 1 alone – we would therefore require additional initial capital to reduce the risk of the policies not being self-financing. The calculations of various profit measures can incorporate this required risk capital, which can then be used to choose between alternative products to sell.

The reader may wish to explore other measures of profit and returns which incorporate the inherent risk of such products, such as "RAROC" (risk-adjusted return on capital). RAROC calculated from various lines of insurance can be compared to determine which parts of the business should be developed and expanded, and which are less profitable.

This idea can be developed further by assessing correlations between various lines of business and the effect of diversification, impacting on the levels of risk capital required both by individual business lines and by the insurer as a whole. This was touched upon in Exercise 24.23.

24.13 Unit-linked Policies

24.13.1 Introduction

Under a typical unit-linked endowment policy, the policyholder receives a payment at the end of the policy term which reflects investment returns on a particular portfolio of investments, typically held in a collective investment vehicle such as a unit trust (which may invest in bonds and company shares); this is in contrast to a traditional endowment policy where the benefit payment is equal to a pre-determined amount. In addition, a death benefit is incorporated under the policy usually based on the fund value at the date of death (in the example below this is equal to 110% of the fund value).

Premiums are paid by the policyholder into the "policyholder's fund" (alternatively known as the "unit-fund") which, after a small deduction/charge, are invested in the chosen investment vehicle, for example a FTSE100 Index unit trust (for our purposes we will assume the premiums are annual); the amount deducted is allocated to the "insurer's fund" (see below). An Annual Management Charge (AMC), usually a percentage of the value of the fund, is also deducted each year and paid across to the insurer's fund. Such policies generally include a Guaranteed Minimum Maturity Benefit (GMMB); in the example below this guarantee is set equal to the total amount of gross premiums paid by the policyholder over the term. Thus, in this example, the policyholder is guaranteed to receive at least their total contributions back following a fall in the fund value at maturity.

The other fund account mentioned above is the insurer's fund (alternatively known as the "non-unit" fund); a typical schedule is shown below in Table 1.2. This sets out a projection of potential cashflows in the insurer's fund. (A similar schedule will be kept throughout the life of the policy setting out the actual cashflows, which will be different to those estimated.)

At the start of each policy year, the deduction applied to each policyholder premium (usually referred to as the "unallocated premium") is transferred to the insurer's fund (note that this is not an estimate). A deduction is also made to the insurer's fund in respect of estimated expenses which the insurer will incur; note that these are estimates – the actual profits made by the insurer will depend on actual expenses which will turn out to be higher, or lower, than those allowed for above. This net cashflow is rolled up to the year end at the expected rate of return from the assets backing these insurer's funds (which are typically invested in short term bonds). To this is added the AMC (see earlier). An allowance may be required in respect of any additional contractual death benefit which exceeds the value of the policyholder's fund. Taking all these items into account, the net value is the insurer's profit for that year. The process starts again at $t = 1$.

Note that policyholders who surrender the policy mid-term will generally receive the value of the fund at that point in time, therefore having no impact on the insurer's fund. A common alternative is to return only a proportion of the fund or contributions. In the example below we adopt the prior approach.

Remark 24.13 We do not include any calculations in this section relating to reserves. The position here is generally fundamentally different to that from conventional policies – here a separate fund is held in respect of the policyholder's money. However, some reserves are usually required in respect of, for example, investment mis-matching between policyholder liabilities and those assets actually held, a policy design where significant negative cashflows are expected in later years, or AMC's may be inadequate to meet actual expenses incurred by the insurer. For example, there will always exist some tracking error when trying to match an index, liabilities in respect of an onerous GMMB following a market fall, or inadequate AMC's to meet actual expenses.

The reader may wish to incorporate these issues into the exercises below.

The reader should refer to Dickson et al. (2013) Chapter 14, and Booth et al. (2005) Chapter 7.2 for further details on unit-linked policies.

24.13.2 Example with Deterministic and Stochastic Projections

In the example which follows, we run both deterministic and stochastic calculations; the focus of this section/example are the stochastic projections.

Example 24.13

Policy design/ assumptions
The details of the contract and assumptions used in this example are set out below:

- The annual premium, payable in advance, is £6,000. The policy is for 10 years.
- 4% is deducted from the initial premium, with 1% deductions from subsequent premiums. The AMC equals 0.90% of the fund value at the year-end.
- Assume 10% of policyholders surrender in year 1, and 5% in Year 2 (see comment later about cancellation risk).
- We will assume, first, that the policyholder's funds increase at the fixed rate of 8% pa throughout the term. In the exercise which follows we assume a stochastic rate of return with distribution $N(0.08, 0.15^2)$.
- An estimate of actual expenses incurred by the insurer equals 10% of the initial premium plus £200 at $t = 0$, and 0.5% of subsequent premiums (note that the initial expense at $t = 0$ is included in the calculation towards the end of the code below).
- On death before maturity a payment equal to 120% of the fund value is payable. We assume a simple 0.4% probability of death each year.
- Assume 5.5% pa interest is earned on the insurer's funds.
- To assess net present value, use a risk discount rate of 18% pa.

Deterministic calculations
For the deterministic model we simply run one simulation with a small standard deviation.

```
fund_starty<-amc_y<-fund_end<-npv_stoch<-NULL
record_fund_end<-final_year_profit<-NULL

GMMB_prop<- 1 # reader should vary this (see later)
stochq<-0    #0=det or 1-stoch?    #determinstic/stochastic switch
sims_stochxx<-ifelse(stochq==0,1,10000)

termy<-10;    premy<-6000; dth_ben<-0.2    #policy details
expy1<-c(0.04,rep(0.01,9))*premy;    amc1<-0.009 #exp dedns applied
prem_allocy<-premy-expy1
setup_cost<-0.1*premy+200    #initial costs
exp_t<-c(0,rep(0.005*premy,9))

fund_starty[1]<-0
qqy<-0.004                #simplified mortality rates
lapsey<-c(0.1,0.05,rep(0,8))    #lapse rate
ins_int<-0.055            #interest rate earned on insurer's funds
risk_i<-0.18             #NPV calcs - risk discount rate
inv_ret_py<-0.08;    sd_py<-0.15    #stochastic inv ret
#end of assumptions

#policyholder fund calculations..............
set.seed(1000)
```

```
sigmaX<-ifelse(stochq==0,0.0000001,sd_py)    # sd of returns

for (gggjj in 1:sims_stochxx) {
stoch_xx<- (1+rnorm(termy,inv_ret_py,sigmaX))
for (rr in 1:termy) {
fund_end[rr]<-(fund_starty[rr]+prem_allocy[rr])*stoch_xx[rr]
amc_y[rr]<-fund_end[rr]*amc1
fund_end[rr]<-fund_end[rr]-amc_y[rr]
fund_starty[rr+1]<-fund_end[rr]
}
#now insurer's fund calculations...............

inty1<-ins_int*(expy1-exp_t)
fundins_end<-NULL      #profit each year
deathy_exp<-qqy*fund_end*dth_ben  #excess death benefit

fundins_end<-(expy1)-exp_t+inty1+amc_y-deathy_exp   #just rying - NEW NEW NEW
fundins_end[termy]<- fundins_end[termy] - max(termy*premy*GMMB_prop-fund_end[termy],0)

survxx<-NULL;    survxx[1]<-1
      for (kr in 1:(termy-1)) {
      survxx[kr+1]<-(1-qqy)*(1-lapsey[kr])*survxx[kr]
      }
p_sigxx<-fundins_end*survxx
npv_qw<-p_sigxx/(1+risk_i)^(1:termy)
}
```

From the above code we have the following policyholder's year end fund values:

```
round(fund_end,0)    #policyholder fund value at each year end

 [1]   6165 12956 20224 28002 36328 45238 54775 64982 75907 87599
```

Table 1.2 sets out the insurer's fund calculations (noting that the initial costs of £800 are included below the table):

Table 1.2 (insurer's fund)

```
round(cbind(expy1,exp_t,inty1,amc_y,deathy_exp,fundins_end,p_sigxx,npv_qw),1)
```

	expy1	exp_t	inty1	amc_y	deathy_exp	fundins_end	p_sigxx	npv_qw
[1,]	240	0	13.2	56.0	4.9	304.3	304.3	257.8
[2,]	60	30	1.6	117.7	10.4	138.9	124.5	89.4
[3,]	60	30	1.6	183.7	16.2	199.1	168.9	102.8
[4,]	60	30	1.6	254.3	22.4	263.6	222.6	114.8
[5,]	60	30	1.6	329.9	29.1	332.5	279.8	122.3
[6,]	60	30	1.6	410.8	36.2	406.3	340.5	126.1
[7,]	60	30	1.6	497.5	43.8	485.3	405.1	127.2
[8,]	60	30	1.6	590.2	52.0	569.8	473.7	126.0
[9,]	60	30	1.6	689.4	60.7	660.3	546.7	123.3
[10,]	60	30	1.6	795.5	70.1	757.1	624.4	119.3

```
(npv_tot<-sum(npv_qw)-setup_cost)

[1] 509.1018
```

The policyholder's fund value at the end of the contract is calculated to be £87,599, and the NPV of insurer's profits, after allowing for the significant set-up costs of £800, is £509. Note that the insurer faces a risk of losses from early surrenders in the first few years of policies, due to these setup costs.

Example 24.14

Stochastic calculations
The above calculations are now re-run but with the "stochastic switch" in the above code set to "YES" (`stochq<- 1`). The key assumption change is a stochastic investment return; instead of a fixed return of 8% pa we assume the return is Normally distributed: $\sim N(8\%, 15\%^2)$. (The reader may wish to subsequently adopt a more appropriate log-normal distribution.) The results from the simulations are set out in Figure 24.15 (the code is not repeated here).

Figure 24.15 (left) shows the policyholder's fund value at expiry before the payment from the insurer in respect of the GMMB. Compare the mean value from these simulations of £87604, consistent with the deterministic value above (£87,599). Note that the lowest returned fund value the policyholder will actually receive is £60,000 due to the GMMB.

From Figure 24.15 (left) we can see there is a range of policyholder fund values, due to the range of investment returns achieved. Where the fund is less than the GMMB the insurer is required to make up the shortfall. Using the above `set.seed`, 12.5% of simulations resulted in this GMMB biting (when the policyholder's fund was less than £60,000 at $t = 10$), incurring additional costs for the insurer. The reader should analyse those simulations which result in the biggest losses.

An important point to note is the "stochastic" mean NPV of the insurer's profits was £350, compared to the "deterministic" value of £509, and the 5% and 95% percentiles of NPVs of £-1,026 and £828 respectively – as can be seen from Figure 24.15 (right) the distribution of NPVs is highly skewed. *Only looking at expected values from deterministic calculations is highly misleading.*

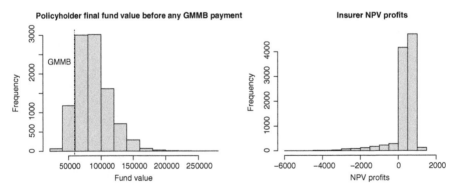

Figure 24.15 Unit-linked fund: Stochastic projections.

Note also that the insurer is particularly at risk from concentration risk if a large number of policies are sold around the same time – the profitability of such policies will be correlated.

Exercise 24.43 *Investigate the impact on NPVs by changing a number of parameters. In particular, analyse the effect of lowering the GMMB level, for example to 50%, or 0%. Similarly, vary the GMMB such that the probability that the NPV is negative is less than 5%. Also change the AMC and risk discount rate. Which parameters have the most influence on the results?*

Exercise 24.44 *Develop calculations which incorporate uncertainty in the number of surrenders and amount of expenses incurred. Also include in the investment return distribution a 1% chance of extreme losses, for example a 50% fall in the fund value.*

Including a GMMB clearly involves considerable risk and the potential for significant losses, and measures should be taken to analyse and manage this risk. There have been several cases of providers incurring significant losses where this risk has been poorly managed.

Remark 24.14 Unit-linked policies are often referred to as equity-linked policies.

24.14 Additional Exercises

The reader may wish to consider re-running a number of the examples and exercises in this chapter using a Gompertz model (or similar) in place of recognised life tables.

As noted at the start of this chapter, the reader may wish to carry out several of these calculations in Excel and compare results.

24.15 Appendix: Dependent and Independent Rates

(Please also see Chapter 22.)

Exercise 24.35 involves a multiple decrement model with three states – alive, dead, surrender – with only two permitted decrements from the alive state. The assumptions in Exercise 24.35 include independent annual probabilities of surrender, q^s, and death, q^d. These need to be translated into forces of transition and dependent rates. We can estimate (constant) forces of transition μ^s and μ^d as follows:

$$q^d = 1 - e^{-\mu^d} \quad \text{and thus} \quad \mu^d = -\log(1 - q^d)$$

$$q^s = 1 - e^{-\mu^s} \quad \text{and thus} \quad \mu^s = -\log(1 - q^s)$$

From μ^d and μ^s, we can obtain the dependent decrement probabilities, aq^d and aq^s:

$$aq^d = (1 - e^{-(\mu^d + \mu^s)}) \frac{\mu^d}{\mu^d + \mu^s}$$

$$aq^s = (1 - e^{-(\mu^d + \mu^s)}) \frac{\mu^s}{\mu^d + \mu^s}$$

For a more detailed discussion on this topic see Macdonald et al. (2018) Chapter 16, and Dickson et al. (2013).

24.16   Recommended Reading

- Benjamin, B. (1968). *Health and Vital Statistics*, 1e. London: Routledge.
- Booth, et al. (2005). *Modern Actuarial Theory and Practice*, 2e. Boca Raton, Florida, US: Chapman and Hall/CRC.
- Charpentier, A. et al. (2016). *Computational Actuarial Science with R*. Boca Raton, Florida, US: Chapman and Hall/CRC.
- Dickson, D.C.M., Hardy, M.R., and Waters, H.R. (2013). *Actuarial Mathematics for Life Contingent Risks*. Cambridge, UK: Cambridge University Press.
- Gerber, H.U. (2010). *Life Insurance Mathematics*, 3e. Berlin Heidelberg: Springer-Verlag.
- McCutcheon, J. (1969). A method for constructing an abridged life table. *Transactions of the Faculty of Actuaries* 32: 404–411.
- MacDonald, A.S., Richards, S.J., and Currie, I.D. (2018). *Modelling Mortality with Actuarial Applications*. Cambridge, UK: Cambridge University Press.
- Spedicato, G.A. https://cran.unimelb.edu.au/web/packages/lifecontingencies/vignettes/an-introductiontolifecontingenciespackage.pdf.
- Spedicato, G.A. https://cran.r-project.org/web/packages/lifecontingencies/lifecontingencies.pdf.
- Tetley, H., M.A., F.I.A. (1946). Actuarial statistics, Volume I, Statistics and Graduation. (Pp. xvi 285. Cambridge University Press, 1946. 21s.). *Journal of the Institute of Actuaries* 73 (1): 173–174.

25

Actuarial Risk Theory – An Introduction: Collective and Individual Risk Models

Peter McQuire

25.1   Introduction

```
library(MASS);  library(moments);  library(copula)
```

Remark 25.1 This chapter includes several calculations which involve simulations. The number of simulations run were often restricted in the interest of computational efficiency, although they have been tested by running a greater number of simulations than included in the text. The reader should ensure sufficient simulations are run which give results which have an acceptable level of stability.

In this chapter we will study models used to assess risks associated with short-term insurance contracts. Examples of such contracts, which typically provide insurance cover for one year or less, include car insurance policies, house insurance policies, and one-year term assurance policies. In contrast with longer-term policies such as whole life assurance (see Chapter 23), these short-term policies may either lapse at the end of the annual contract with no further premium payable, or be renewed, at which point another premium is paid in respect of the risk insured for the next period. Each annual premium is therefore in respect of the risk taken on for the following year. Contrast this with typical life assurance policies where the mortality risk increases throughout the policy as the policyholder gets older; generally these longer-term contracts require a fixed, level premium to be paid throughout the contract such that the earlier premiums are higher than the risk covered for those earlier years. A reserve is therefore built up to meet the expected net outgo in the future; however, no such reserves are required for the short-term products covered in this chapter.

One of the main tasks facing an insurer which writes such short-term contracts is to estimate the total claim amounts that will be paid, or more correctly, incurred over the next year; given that Y policies have been sold the insurer will estimate the distribution of

R Programming for Actuarial Science, First Edition. Peter McQuire and Alfred Kume.
© 2024 John Wiley & Sons Ltd. Published 2024 by John Wiley & Sons Ltd.
Companion Website: www.wiley.com/go/rprogramming.com.

total claim amounts from these policies and hence the distribution of potential profits (and losses) from these policies. Another key task is to calculate the premium which should be charged for each policy to be confident of making sufficient profits.

To help us do this we will look at two key models: the 'Collective Risk Model' (in Sections 25.2 to 25.8), and the 'Individual Risk Model' (Section 25.9).

25.2 Collective Risk Model

Various Collective Risk Models are commonly used in practice. In this chapter we will concentrate on the Poisson Compound Collective Risk Model (Section 25.3), but will also compare, briefly, the Binomial and Negative Binomial variants in Sections 25.5 and 25.6.

25.3 Poisson Compound Collective Risk Model

We can start by analysing historic data from an homogeneous group of policies, and estimate the claim rate, λ, from these policies.

Remark 25.2 Note that this discussion is generally not considered to be part of the recognised Collective Risk Model – most texts assume the total number of claims the insurer will incur to have a $Poisson(\lambda)$ distribution, with no justification for λ. Here we are first justifying how we arrive at an estimate for λ.

This can easily be done using the same Markov theory covered in Chapter 16:

$$\hat{\Lambda} = N_{past}/E \tag{25.1}$$

where N_{past} is the total number of claims incurred from a group of policies, E is the total time for which these policies were exposed to risk, and $\hat{\Lambda}$ is the estimated rate at which claims were incurred *per policy year*. If the number of claims from one policy follows a Poisson process with rate, $\hat{\Lambda}$, and all our policies are assumed to be independent, then the total number of claims from our current portfolio of Y policies may be assumed to follow a Poisson process with parameter $Y\hat{\Lambda}$. Defining $\hat{\lambda} = Y\hat{\Lambda}$, where $\hat{\lambda}$ is the total number of claims expected from our portfolio of Y policies, we can model the total number of claims, N, from the portfolio using a Poisson distribution:

$$N \sim Poisson(\hat{\lambda}). \tag{25.2}$$

For example, if we currently have 1,000 annual policies ($Y = 1,000$) with $\hat{\Lambda} = 0.1$ then $N \sim Poisson(100)$. As referred to above, most texts ignore this preamble regarding the number of policies (Y) and justification for λ, instead simply looking at the total number of claims and the sizes of each of those claims.

We now proceed with the main part of the model. To estimate the total claim amount, S, from this portfolio of policies, we also require a distribution for the size of each individual claim. We'll define the size of the ith claim as X_i and assume all X_i's are identically distributed and independent.

The total claim amount, S, is given by:

$$S = X_1 + X_2 + X_3 + X_4 + ... + X_N. \tag{25.3}$$

It is straightforward to show that the expected total claim amount is:

$$E(S) = E(X)E(N) \tag{25.4}$$

and the variance of the total claim amount is:

$$Var(S) = (E(X))^2 Var(N) + E(N)Var(X). \tag{25.5}$$

It is important to note that, with the Collective Risk Model, we do not concern ourselves with the number, or size, of claims incurred *under each policy*. If we think of a typical portfolio of car insurance policies, we will have no claims from most policies, receive one claim from some policies, with a few policies incurring more than one claim; this model does not identify from which policies the claims were incurred, for example three claims from the same policy would simply be recorded as three claims.

The above expressions apply to any distribution of claim numbers and amounts. Where the number of claims, N, has a Poisson distribution with parameter λ (as discussed earlier in this Section), Equations 25.4 and 25.5 can be simplified as follows:

$$E(S) = \lambda E(X) \tag{25.6}$$

and the variance of the total claim amount is:

$$Var(S) = \lambda E(X^2). \tag{25.7}$$

S is said to have a *Compound Poisson distribution*. We illustrate the above by running a number of simulations.

Example 25.1

Let's say we've reviewed claim payments from the last five years and noted/estimated the following (these parameter values will be used in several examples and exercises throughout this chapter):

- the annual number of claims follows a Poisson distribution;
- the annual rate of claims per policy, Λ, is 0.2;
- the total current number of policies in force, Y, is 1,000;
- thus the total annual number of claims has a Poisson(200) distribution (as previously noted this is usually the starting point in many actuarial texts on Risk Models);
- individual claim amounts have a Gamma(10,2) distribution;
- all amounts are in £'000s.

Setting out these values:

```
lambda_x<-0.2;    YY<-1000
(exp_claim_no<-YY*lambda_x)

[1] 200

gamma1<-10;      gamma2<-2
```

Remark 25.3 Reminder of parameterisation in R: a Gamma(10,2) has a mean equal to 5 and variance equal to 2.5; for Example 25.1 the mean and standard deviation of the individual claim amounts is £5,000 and £1,581.139, respectively.

Simulating 100,000 values from a Poisson(200) distribution:

```
sims_x<-100000;         set.seed(555)
no_of_claims<-rpois(sims_x,exp_claim_no)    #simulate Poisson variables
mean(no_of_claims);     var(no_of_claims)

[1]  200.014
[1]  199.5298

head(no_of_claims)

[1] 195 201 174 174 189 219
```

Next simulate claim amounts for each of the simulated N's, and sum them to obtain the total claim amounts and key statistics (plotted in Figure 25.1, left):

```
total_claim_amount<-NULL;       set.seed(556)
for (i in 1:sims_x){
total_claim_amount[i]<-sum(rgamma(no_of_claims[i],gamma1,gamma2))}
c(mean(total_claim_amount),var(total_claim_amount),sd(total_claim_amount))

[1] 1000.05571 5490.47664   74.09775
```

Comparing these simulated results with the results from Equations 25.4 and 25.5:

$$E(X) = 10/2 = 5 \qquad E(N) = 200$$
$$Var(X) = 10/2^2 = 2.5 \qquad Var(N) = 200$$

$$E(S) = 5 \times 200 = 1,000$$
$$Var(S) = 200 \times 2.5 + 200 \times 5^2 = 5,500$$

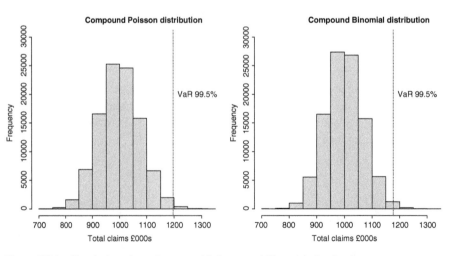

Figure 25.1 Simulations from Compound Poisson and Binomial distributions.

This is consistent with the results simulated above. The reader should experiment with various distributions and parameter values. (The Compound Binomial and Compound Negative Binomial distributions are discussed in Sections 25.5 and 25.6.)

Appropriate use of the Compound Poisson model will depend on a number of factors, including the level of heterogeneity in our data and the degree of dependency between policies; alternative, more detailed models such as the Individual Risk Model may be preferable (see Section 25.9).

Exercise 25.1 *In general, all claims amounts in this chapter have been shown in £'000s, with $X \sim Gamma(10, 2)$ in the example above. Note that the Gamma distribution is easily scalable, that is if $X \sim Gamma(\alpha, \beta)$ then $kX \sim Gamma(\alpha, \frac{\beta}{k})$. Hence we could alternatively have used $X \sim Gamma(10, \frac{2}{1000})$ with all claim amounts in £'s. The reader may wish to re-run the relevant examples and exercises in this chapter as such.*

We will return to these calculations throughout this chapter when we compare alternative models. We now proceed to look at how this model, and indeed most models discussed in this chapter, can be used in practice.

25.4 Applications of the Model

Remark 25.4 If the reader wishes to concentrate on the specific various risk models discussed in this chapter, please move onto Sections 25.5, 25.6, and 25.9.

25.4.1 Setting Appropriate Reserves and Premium Pricing

Insurers will want to hold sufficient reserves to ensure, with a particular level of certainty, that next year's claims can be met. One way to manage this risk is to charge sufficient premium rates such that the premiums received will exceed the claims incurred with a required probability (we should also allow for expenses, but these are ignored here for our purposes). The mean and variance calculations are important; however, it is information regarding the tails of the distribution which is particularly important when assessing solvency, that is 'with what probability will total claims exceed a particular level?'.

Using the results from the simulations in Section 25.3, the 99.5% Value at Risk ('VaR') is estimated as follows:

```
VaR1<-0.995    #used throughout this Section
quantile(total_claim_amount,VaR1)

   99.5%
1196.489
```

Based on these simulations, there is 0.5% chance that our total claims will exceed £1,196,489 next year. If we charged £1,196.49 for each policy we would receive a total premium of £1,196,489 from our 1,000 policies, and would be 99.5% confident all claims will be paid and a profit will be made (see Figure 25.1 (left)).

Remark 25.5 To obtain reliable estimates, usually more than 10^5 simulations would be run; the above result is, however, close to results obtained from running significantly more simulations. Two alternatives methods of estimating tail risk are covered later in the chapter: (1) Panjar's method and (2) using the Normal distribution as an approximation.

However market forces will play a significant part in determining the premium that can be charged, together with a profit requirement. For example, if the premium charged by the typical insurer is only £1,100, we may be unlikely to sell many policies for £1,196, and thus be required to reduce our premium in line with the market rate. If we charge only £1,100, we must then set up an additional reserve to make up for the difference between the agreed risk level (here 99.5% VaR) and expected premiums. The additional reserve is given by:

$$99.5\% VaR - (premium \times number\ of\ policies), \tag{25.8}$$

In this case the difference amounts to £96,489. As noted above the insurer will need to take several other factors into account when setting the premiums, such as fixed and variable expenses, and profit requirements.

25.4.2 Increasing the Number of Independent Policies

If the insurer can increase the number of policies it sells the uncertainty of the total claims will fall (as a proportion of the expected claims).

Exercise 25.2 *By running an appropriate number of simulations, calculate the premium the insurer could charge, using the same parameters as above, if the insurer could sell twice as many policies, that is 2,000 policies and expect to incur 400 claims. (The code used to obtain the following results is similar to that shown above, and applies the same* set.seed.*)*

Results from Exercise 25.2: The 99.5% VaR from these simulations for total claim amounts is now £2,275,467. Following the same logic as above, we can now charge £1,138 for each of our 2,000 policies (receiving a total premium of £2,275,467), and would be 99.5% confident of making a profit. This compares to a premium of £1,196 if only 1,000 policies are sold.

Of course, as noted above, the insurer may not sell many policies at this price. However, by charging the market premium rate of £1,100 (as above) we now require a reserve of only £75,467 (compared to £96,489 if only 1,000 policies had been sold). This leads to a higher expected return on those smaller reserves which, in turn, may allow the insurer to potentially charge lower premium rates.

Exercise 25.3 *By running an appropriate number of simulations, write code to estimate the required 99.5% VaR additional reserve (using Equation 25.8) as the number of policies increases from 100 to 3,000. Assume that a premium of £1,100 is charged, and that all other variables remain unchanged from above. (See Figure 25.2 in which we compare the results from this Exercise with calculations set out in the next section, which uses the Normal distribution approximation. We have run 10^4 simulations to obtain Figure 25.2, but the reader may decide further simulations are required to reduce noise.)*

The above reasoning can be a major factor in explaining why companies may decide to merge; by writing more policies the uncertainty of future claims reduces (as a proportion of the size of the business) thus requiring less capital and improving shareholder returns.

It is easy to forget one overriding principle when setting these rates; the underlying claim risk must be assessed properly and priced appropriately. If we charge a premium which does not reflect the amount of risk we will, in the long run, incur losses and eventually face insolvency.

25.4.3   Adopting a Normal Distribution Approximation

Using a Normal distribution to estimate the probabilities and risk measures discussed above can be computationally efficient. It can also provide a useful check on results obtained from running simulations and other calculations. (The shifted Gamma distribution is an alternative often used.)

Exercise 25.4   *For both scenarios above where (1) 1,000 policies and (2) 2,000 policies are sold, compare the 95%VaR and 99.5%VaR from simulations with those assuming a Normal distribution as an approximation for the compound distributions. Also carry out a Jarque-Bera test and produce appropriate Q-Q plots as a visual test for the appropriateness of a Normal distribution.*

The 99.5%VaRs from Exercise 25.4 for 1,000 policies are calculated below (also shown is the result from running simulations):

```
1000+qnorm(VaR1)*(2.5*200+200*5^2)^0.5        #theory, assuming Normal

[1] 1191.029

qnorm(VaR1,mean(total_claim_amount),var(total_claim_amount)^0.5) #Normal (sims)

[1] 1190.919

quantile(total_claim_amount,VaR1)             #simulations (copied from above)

   99.5%
1196.489
```

The Normal distribution approximation becomes more appropriate when the expected number of claims is sufficiently high; as the reader may have encountered from other chapters, this approximation should be exercised with caution.

Exercise 25.5   *Re-run the above calculations with just 50 policies, that is $E(N) = 10$ and comment on the results.*

Using a Normal distribution approximation we can derive a simple expression, using Equation 25.8, to determine the capital required (this was calculated in Exercise 25.3 by running simulations). It is straightforward to show that, for a given premium, the required capital is given by (based on 99.5% VaR):

$$\text{Capital required} = 2.58\sigma_s - \theta E(S) \tag{25.9}$$

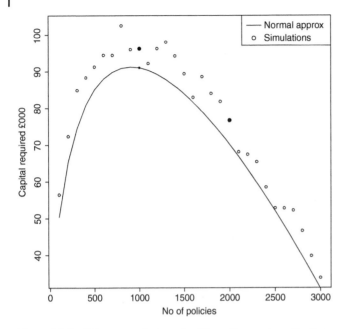

Figure 25.2 Effect on capital required from the number of policies sold – Normal vs simulations (99.5 VaR).

where $E(S)$ and σ_s are the expected claims and standard deviation of claims respectively, and $(1 + \theta)$ is the ratio of the total premium charged to $E(S)$. (In the above example with 1,000 policies, $\sigma_s = 74162$, $E(S) = 1,000,000$ and the per policy premium was 1,100, thus $\theta = 0.1$, and the capital required is 91,338.)

Calculating the capital required as a function of the number of policies sold, with $\theta = 0.1$, and using the Normal distribution approximation (Equation 25.9):

```
no_claims<-NULL; no_pol<-NULL; capital<-NULL; var_S<-NULL; mean_S<-NULL
mean_claim<-5000;    var_claim<-2500000    #claim amounts
claim_freq<-0.2;    excess_prem<-0.1

for (i in seq(1,30)) {
no_pol[i]<-i*100
no_claims[i]<-no_pol[i]*claim_freq
mean_S[i]<-mean_claim*no_claims[i]
var_S[i]<-no_claims[i]*var_claim+no_claims[i]*mean_claim^2
capital[i]<-qnorm(VaR1)*var_S[i]^0.5-excess_prem*mean_S[i]
}
```

The results are plotted in Figure 25.2. This demonstrates that less capital may be required as more policies are written – larger companies can benefit from this. The critical point to note is that σ_s is proportional to $\sqrt{N}$, whereas the total amount of expected claims, $E(S)$, is proportional to N (see Equation 25.9).

It is clear from Figure 25.2 that the Normal distribution approximation is underestimating the capital required relative to the simulations. The reader should also recalculate the capital required using these two methods at various VaR levels – as we move further into the tail of the distribution the Normal distribution approximation becomes less valid.

25.4.4 Return on Capital

Turning briefly to look at the return on capital ('ROC').

$$ROC = \frac{net\ income}{capital} = \frac{TP - S}{VaR - TP} \tag{25.10}$$

where TP and S can either be the expected or actual total premiums and claims.

The insurer will calculate premiums based, as least partly, on the return on capital required, number of policies they expect to sell, and the acceptable probability of insolvency; the relationships between these are complex. Reducing the premium may increase expected futures sales resulting in lower required capital (as shown above), potentially increasing returns for shareholders; this would depend on the elasticity of demand of the policies. The following type of analysis may be carried out:

1. Estimate the expected claim amount per policy.
2. Determine the premium to be charged, taking account of the market.
3. Estimate the number of policies that will be sold at this price (which will depend on the market premium).
4. Calculate the 'Claims-VaR' and capital required based on this number of policies sold and premium.
5. Calculate the expected return on capital.
6. Based on this capital and premium, run simulations allowing for a potential range of actual policies sold, and claims made, obtaining a distribution of ROCs.

   The above can be re-run using different premiums, subsequent policies sold, and capital employed, that is from points 2 and 3.

We could then analyse the potential range of profits and ROC resulting from a range of potential policies actually sold, and potential total claims. For example, we may only sell 800 policies and incur claims at a higher rate than expected; it would be informative to understand the likely range of the ROCs. This could be repeated using different premium rates and resulting range of policies sold until an acceptable outcome was achieved.

Thus a pricing strategy can be developed by analysing the demand elasticity of products, and the effect on capital, profit, and return on capital.

25.4.5 Skewness of the Compound Poisson Model

It is useful to note the skewness of the Compound Poisson distribution, and in particular, the effect which the expected number of claims may have on the shape of the distribution.

From Example 25.1 and Figure 25.1 (left), where $E(N) = 200$, the sample skewness can be calculated as follows (with a skewness of zero indicating no asymmetry):

```
skewness(total_claim_amount)

[1] 0.09040418

#or
mean((total_claim_amount-mean(total_claim_amount))^3)/
sd(total_claim_amount)^3*(sims_x*(sims_x-1))^0.5/(sims_x-2)

[1] 0.09040418
```

This is consistent with Figure 25.1; the simulations appear symmetrical around a mean of 1,000, exhibiting very little skewness.

Exercise 25.6 *Repeat the calculations from Example 25.1, but with $\lambda = 5$ instead of 200, and recalculate the skewness.*

Repeating these simulations with $E(N) = \lambda = 5$, we calculate a skewness equal to 0.5051 (Figure 25.3), considerably higher compared to when $E(N) = 200$ (Figure 25.1 (left)). Clearly the Normal distribution approximation, discussed in Section 25.4.3, may be considered inappropriate where the expected number of claims is small. Similar considerations should be made regarding many of the models discussed in this chapter.

Remark 25.6 There are various alternative measures of sample skewness, for example $\frac{\lambda m_3}{(\lambda m_2)^{1.5}}$ (the methods of moments estimator, where m_2 and m_3 are the second and third central moments) which we do not concern ourselves with here. Our focus in this section is to compare the skewness of various distributions.

25.4.6 Sum of Compound Poisson Distributions

An insurer is likely to have several lines of business, or simply significant heterogeneity, such that the preferred model is to sum various compound distributions from distinct, more homogeneous parts of the business. The total claims amount incurred by the insurer, S_t, is the sum of the total claims from each separate line, department, or portfolio of business the insurer writes:

$$S_t = \sum_i S_i$$

where S_i is the total claim amount from line or department i. A key result is that this sum of independent Compound Poisson distributions is itself a Compound Poisson distribution. Thus where the number of claims from each department has a Poisson distribution, the mean total claims amount is given by (from Equation 25.4):

$$E(S_t) = \sum_i E(S_i) = \sum_i \lambda_i E(X_i) \tag{25.11}$$

If all lines are independent, the variance of the total claims is given by (from Equation 25.5):

$$Var(S_t) = \sum_i Var(S_i) = \sum_i \lambda_i(Var(X_i) + E(X_i)^2) = \sum_i \lambda_i E(X_i^2). \tag{25.12}$$

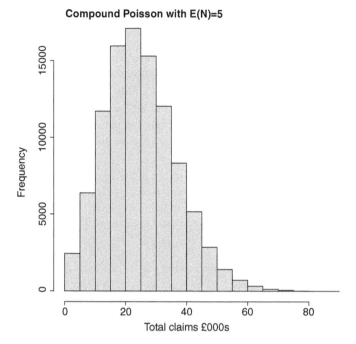

Figure 25.3 Skewed claims distribution.

We can show that:

$$E(S_t) = \lambda_T E(X) \text{ and } Var(S_t) = \lambda_T E(X^2)$$

where

$$\lambda_T = \sum_i \lambda_i \quad ; \quad E(X) = \frac{1}{\sum_i \lambda_i} \sum_i \lambda_i E(X_i) \quad ; \quad E(X^2) = \frac{1}{\sum_i \lambda_i} \sum_i \lambda_i E(X_i^2)$$

thus demonstrating that the sum of independent Compound Poisson distributions is itself a Compound Poisson Distribution.

In the following exercise our aim is to analyse the distribution of total claim amounts an insurer may incur from all its lines of insurance over the following year; clearly this is likely to be one of the key tasks of the actuary.

Exercise 25.7 *An insurer has three separate, independent, lines of insurance (A, B, C). Their claims distributions can be represented by the following:*

- *The annual claim frequency from each of the three lines, (A, B, C), all have a Poisson distribution with $\lambda_A = 10, \lambda_B = 30, \lambda_C = 60$ claims per annum.*
- *The claim amounts have the following distributions: $X_A \sim Normal(100, 10^2)$, $X_B \sim Gamma(25, 0.125)$, and $X_C \sim Exp(1/300)$.*

Calculate the mean and standard deviation of the total claim amounts which the insurer may incur over a year, both by running simulations and using Equations 25.11 and 25.12.

Also estimate the 99.5% VaR of total claims, and draw a Q-Q plot to demonstrate the appropriateness of a Normal distribution in modelling total claim amounts.

The results from Exercise 25.7 are set out in Figure 25.4, and also discussed below.

From Equations 25.11 and 25.12, the mean and standard deviation of the insurer's total claims are, respectively, 25,000 and 3,486; assuming a Normal distribution, the 99.5 %VaR is therefore 33,978. From the simulations run, the mean claims, standard deviation of claims and 99.5 %VaR were 24,993, 3,481, and 34,739, respectively. Indeed, the Q-Q plot is consistent with the skewed nature of the data when compared to a Normal distribution. (Clearly these results depend on the seed used.)

Thus the Normal distribution does not do particularly well in analysing the extremes of the total claims distribution (the mean and standard deviation agree well with theory). When analysing the risk position of the company it may be preferable, therefore, to model future claims using simulations, only using the simplified methods as useful checks. The reader is encouraged to expand on this exercise.

Exercise 25.8 *Re-run the above exercise with respective Poisson parameters of 1, 3, and 6, that is one-tenth the number of claims, commenting on the appropriateness of the Normal distribution approximation.*

Exercise 25.9 *Calculate, by running simulations, $\lambda_T E(X^2)$ (using the above example) and compare with the theoretical value, hence demonstrating that the sum of Compound Poisson distributions is itself a compound Poisson distribution. As noted above, the usefulness of this model is somewhat limited as we will usually be interested in the probabilities of large total claims; thus only knowing the variance of the distribution is often of limited value.*

Thus the sum of Compound Poisson model may be a useful, and parsimonious model in estimating an insurer's aggregate claims distributions resulting from multiple lines of business. However care should be taken analysing the tail of the distribution when claim numbers are small.

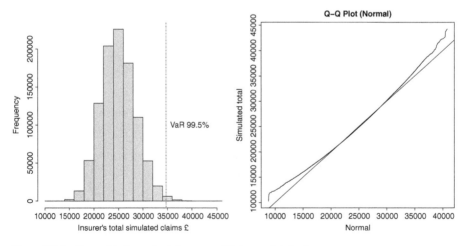

Figure 25.4 Sum of independent Compound Poisson distributions.

Exercise 25.10 *An insurer has 10 separate claims departments, each with different claims distributions. The 10 departments have claim number distributions from Poisson(10), Poisson(20), Poisson(30),..., Poisson(100), and Gamma distributions for the claim amounts, with means ranging from 10,000 to 100,000, and sd's ranging from 2,000 to 20,000, as follows:*

```
departments<-10
lambdax<-seq(10,100,by=10)    #10 Poisson parameters
claim_amn_mean<-seq(10000,100000,by=10000)    #10 claim amount means
claim_amn_sd<-seq(2000,20000,by=2000)         #10 claim amount sds
```

Analyse the insurer's total claim amounts using simulations and Equations 25.11 and 25.12. Re-run the calculations with alternative parameter values.

Remark 25.7 The reader is reminded about the important, simplifying assumption of independence of policies (and departments) used in these calculations. This important issue is discussed briefly in Section 25.11.

Exercise 25.11 *(Challenging – for readers familiar with the material in Chapters 13 and 14.) Re-calculate the variance and 99.5% VaR from Exercise 25.7, but assuming the three lines of business have a degree of correlation which can be modelled with a Gaussian copula, where each of the three correlation coefficients are equal to 0.5. (See Figure 25.5.) Compare this with the results from Exercise 25.7, and also with applying a Normal distribution approximation.*

This concludes our discussion of our first model – the 'Compound Poisson Collective Risk Model'.

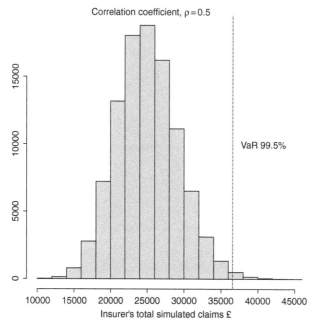

Figure 25.5 Sum of Compound Poisson distributions with dependence (using copulas).

Remark 25.8 Much of the material covered in the above sections should be re-applied using the models covered in Sections 25.5, 25.6, and 25.9.

25.5 Compound Binomial Collective Risk Model

An alternative model, not dissimilar to the Compound Poisson Collective Risk model, is the 'Compound Binomial Collective Risk model', where the total number of claims follows a Binomial distribution. (There are also some similarities with the Individual Risk Model – see Section 25.9.) This model is likely to be more applicable where a maximum of one claim per policy could be incurred; for example, an annual term assurance policy, or where only one claim is permitted under the terms of a non-life policy. It also assumes that policies are independent and identically distributed.

In comparison with the Compound Poisson model, we must stipulate the total number of policies, together with the probability that a policy will incur a claim; unlike the Poisson equivalent, the Binomial model imposes a ceiling on the total number of claims possible, equal to the number of policies.

We proceed to re-run the calculations in Section 25.3 using consistent parameter values. Let's say we've reviewed recent claim payment data and estimated that the probability of a claim from each policy is 0.2. If the total number of policies in force is 1,000, the total number of claims may be assumed to have a *Bin*(1,000, 0.2) distribution.

Exercise 25.12 *Run 100,000 simulations similar to those under the Poisson equivalent model, but using the Compound Binomial distribution (adopt the same Gamma distribution for claim amounts). Use* set.seed(556).

The results from Exercise 25.12 are plotted in Figure 25.1 (right), earlier in the chapter. The simulated Compound Binomial distribution values are as follows:

```
mean(tot_claim_cbin);    var(tot_claim_cbin)
[1]  1000.039
[1]  4506.736
```

Checking these results with theory (as we did for the Poisson model):

$$E(S) = 200 \times 5 = 1,000$$
$$Var(S) = 200 \times 2.5 + 160 \times 5^2 = 4,500$$

This is consistent with our simulated results. Calculating the 99.5% VaR:

```
quantile(tot_claim_cbin,VaR1)

   99.5%
1176.805
```

From these simulation results, there is a 0.5% chance that our total claims will exceed £1,176,805 next year. Compare this with the 99.5% VaR of £1,196,489 from the Compound Poisson model. The longer tail of the Compound Poisson can be observed visually by comparing the plots in Figure 25.1; the higher variance and 99.5% VaR is due to the possibility of more than one claim arising from each policy under the Poisson model.

25.6 Compound Negative Binomial Distribution

The third distribution from the family of compound distributions covered in this chapter is the Compound Negative Binomial distribution. This is often used in place of the simpler Poisson model due to its additional flexibility; the Poisson distribution is often too restrictive in statistical modelling, under which $E(N) = Var(N) = \lambda$. The Compound Negative Binomial model may be more suited where $Var(N) > E(N)$, a characteristic claims data often exhibits as a result of dependencies between policies and/or significant heterogeneity of the policies.

Remark 25.9 The formulation underlying the rnbinom function in the stats package is as follows: $N \sim NegBin(r, p)$, $E(N) = r\frac{1-p}{p}$, and $Var(N) = r\frac{1-p}{p^2}$. The reader should note that alternative formulations exist in various texts.

As we did for both the Compound Poisson and Binomial models, we choose the mean number of claims equal to 200. We adopt $N \sim NegBin(800, 0.8)$, with the same claim amounts distribution, $X \sim Gamma(10, 2)$, and proceed with simulations:

```
sims_nb<-100000;        set.seed(555)
r<-800 ;    p<-0.8        # Negative Binomial parameters
rb<-rnbinom(sims_nb,r,p)
mean(rb);   var(rb)      # note the higher variance

[1] 199.9972
[1] 250.3796

tot_claim_cnegbin<-NULL     #create vector of length x
set.seed(556)
for (i in 1:sims_nb){tot_claim_cnegbin[i]<-sum(rgamma(rb[i],10,2))} # run sims
mean(tot_claim_cnegbin);    var(tot_claim_cnegbin);    sd(tot_claim_cnegbin)

[1] 999.9719
[1] 6761.972
[1] 82.23121

quantile(tot_claim_cnegbin,VaR1)

  99.5%
1219.996
```

Thus the Compound Negative Binomial model allows us to model claims numbers where $Var(N) > E(N)$. The reader should compare the key results from the Compound Poisson, Binomial, and Negative Binomial models. Please also see Section 25.12 where the Negative Binomial model is further discussed.

25.7 Panjer's Recursion Formula

We have seen that the Normal distribution approximation to our risk models may, in some cases, be inappropriate, with results obtained from running simulations often being preferable. Rather than rely on simulations which can be computationally inefficient, it would be useful to have another tool available to model such distributions – Panjer's recursion formula provides us with such an alternative.

Below we compare the 99% VaR from a Compound Poisson(8) distribution with claim amounts $X \sim N(40, 10^2)$, obtained from three methods:

1. Simulations;
2. Approximation using the Normal distribution;
3. Panjer's recursion formulae.

The simulations (to obtain the object `tot_claim_amount_pan_sims` used below) have been run using similar code to that in Section 25.3, and are not repeated here. Panjer's recursion formula is given by:

$$f_S(r) = \sum_{j=1}^{r} \frac{\lambda j}{r} f_X(j) f_S(r - j) \qquad \text{for } r = 1, 2, 3..... \tag{25.13}$$

The method is appropriate only where the number of claims follows either a Poisson, Binomial, or Negative Binomial distribution. (Please see Boland Chapter 3.2, Dickson Chapter 4.5, or Kaas et al. Section 3.5 for a derivation and further discussion on Panjer's method.)

Note that Panjer's formula requires a discrete distribution for claim amounts; the following code discretises the Normal distribution by calculating an approximate probability of claim amounts for 1, 2, 3, 4,..., 80.

```
#panjer
pois_par1<-8                        #Poisson mean - claim number distribution
norm_mean<-40;    norm_sd<-10    #claim amount distribution

max_claim_am<-norm_mean+4*norm_sd   #suitable max ind claim amount
S_max<-pois_par1*max_claim_am*1.5   #suitable range for r

#Normal discretisation
probj<-NULL
for (pp in 1:max_claim_am) {
probj[pp]<-pnorm(pp+0.5,norm_mean,norm_sd)-pnorm(pp-0.5,norm_mean,norm_sd)}

output_pan<-matrix(0,S_max,max_claim_am)
output_pan[1,1]<-exp(-pois_par1)    #set probability of no claims

#now run the Panjer recursion...
for (r in 2:S_max) {
for (jjj in 1:min(r-1,max_claim_am)) {
        output_pan[r,jjj]<-pois_par1*jjj/(r-1)*probj[jjj]*sum(output_pan[r-jjj,])}}

tot_pan<-rowSums(output_pan)
```

```
pan_VaR<-0.99;      match(pan_VaR,round(cumsum(tot_pan),3))    #99VaR Panjer

[1] 623

quantile(tot_claim_amount_pan_sims,pan_VaR)                  #99VaR from simulations

      99%
622.2966
```

See Figure 25.6. Calculating the mean and variance of the Compound Poisson as per Equation 25.5, then applying a Normal distribution approximation:

```
mean_compound1<-pois_par1*norm_mean;  var_compound1<-pois_par1*(norm_mean^2+norm_sd^2)
mean_compound1 + var_compound1^0.5*qnorm(pan_VaR)

[1] 591.2965
```

The Normal distribution approximation significantly underestimates the 99% VaR in this case, mainly due to the small number of claims ($\lambda = 8$). The reader may wish to run more simulations to verify the reliability of these estimates.

Exercise 25.13 *Repeat the above calculations, but with $\lambda = 100$.*

Exercise 25.14 *As noted above, Panjer's recursion formula requires a discrete distribution for claim amounts. For the calculations above, discretisation of a continuous distribution was required. Repeat the first calculation above (with $\lambda = 8$) but with a discrete claim amount distribution, X, where there is an equal chance of a claim equalling 100, 200,..., 5,000.*

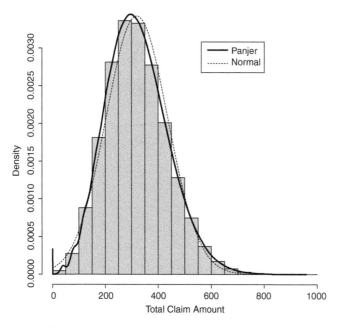

Figure 25.6 Panjer vs Normal vs simulations.

25.8   Closing Thoughts on Collective Risks Models

The main advantage of the Collective Models discussed in Sections 25.2 to 25.7 is their computational efficiency; it is straightforward to estimate the mean and variance of future total claims amounts and the model can be extremely useful if we are satisfied the model reflects reality.

However, a significant issue with these models is the assumption that all policies are independent, and sufficiently homogeneous; as a consequence the variance of claims may be underestimated. A number of approaches can be taken to address this to some extent; one such approach is the Individual Risk Model, discussed in the next section, which, although more computationally onerous, may provide an improved analysis.

25.9   Individual Risk Model

25.9.1   Standard Individual Risk Model

Rather than modelling claims collectively from a portfolio of policies as we did under the Collective Risk Model, we now model the number and amount of claims from each individual policy. An important difference between the Collective and Individual models is the latter specifies both the probability of a claim from policy i, and the claim amount distribution from policy i, for each policy. Thus any heterogeneity identified within the policy portfolio can be modelled appropriately.

In addition, the standard Individual Risk Model permits a maximum of only one claim per policy; the number of claims under policy i therefore has a Binomial$(1, q_i)$ distribution, where q_i is the probability of a claim under policy i. Although appropriate for many life assurance policies, this is a major restriction where policies may incur more than one claim, for example private motor insurance. (We shall look at an obvious variant in Section 25.9.2.)

Independence of policies is usually assumed under this model. Thus risks are independent, but not necessarily identical.

Set out below are the expressions for the mean and variance of the total claims, S, under the standard Individual Risk model:

$$E(S) = \sum_{i=1}^{n} q_i \mu_i \tag{25.14}$$

$$Var(S) = \sum_{i=1}^{n} q_i \sigma_i^2 + q_i(1 - q_i)\mu_i^2 \tag{25.15}$$

where μ_i is the mean of the claim amount if a claim is incurred under policy i, and σ_i^2 is the variance of the claim amount incurred under policy i.

Exercise 25.15   *Derive Equations 25.14 and 25.15.*

We proceed to calculate these statistics with an example which includes non-identical policies.

Example 25.2

An insurer has written 10 policies under a particular line of business which exhibits a significant amount of heterogeneity. The probability of a claim from policy P_i is $0.1 + 0.02i$ where $i = 1, 2, 3...10$. A maximum of one claim can occur for each policy. If a claim occurs for policy P_i, the size of the claim has a Gamma$(10 + i, 2)$ distribution. All policies can be assumed to be independent.

Calculate the mean, variance, and 99%VaR of the total claim amount from this portfolio of 10 policies, using Equations 25.14 and 25.15, and also by running simulations.

First, running just one simulation for all policies to demonstrate the model:

```
b<-NULL;              c<-NULL
set.seed(10555);      policies<-10

for(tt in 1:policies){
b[tt]<-rbinom(1,1,0.1+0.02*tt)
c[tt]<-rgamma(1,10+tt,2)
}
(ddd<-b*c)
```

```
[1] 0.000000 8.750066 0.000000 0.000000 0.000000 7.793794 0.000000 6.949915
[9] 0.000000 0.000000
```

From this single simulation, only policies 2, 6, and 8 incurred a claim. Proceeding with running the model with many simulations:

```
ind_VaR<-0.99
set.seed(500);    sims_ind<-100000;    policies<-10
b<-matrix(0,nrow=policies,ncol=sims_ind); c<-matrix(0,nrow=policies,ncol=sims_ind)

for(tt in 1:policies){
b[tt,]<-rbinom(sims_ind,1,0.1+0.02*tt)
c[tt,]<-rgamma(sims_ind,10+tt,2)
}
ddd<-b*c
sim_sum<-apply(ddd,2,sum)
mean((sim_sum));    sd(sim_sum);    quantile(sim_sum,ind_VaR)
```

```
[1] 17.07348
[1] 10.81892
     99%
45.51141
```

The mean and variance calculations are set out below, together with the 99% VaR assuming a Normal distribution for S:

```
tt<-1:10;  prob_i<-0.1+0.02*tt;  mean_j<-(10+tt)/2;  var_j<-(10+tt)/2^2
mean_theory<-sum(prob_i *mean_j)
var_theory<-sum(prob_i*var_j+prob_i*(1-prob_i)*mean_j^2)
VaR_in_risk1<- qnorm(ind_VaR)*var_theory^0.5+mean_theory   #assuming Normal

mean_theory;    var_theory^0.5;    VaR_in_risk1
```

```
[1] 17.1
[1] 10.81338
[1] 42.25568
```

Here it is clear that the Normal distribution approximation is not appropriate when assessing the tail of the distribution.

Note that if each policy is identical this model is similar to the Binomial Collective Model (although the claim allocation per policy is treated differently).

Exercise 25.16 *A global consultancy firm provides a life assurance benefit for its staff, which consists of 100 actuaries, 400 auditors, 1,000 administrators, and 20 senior managers. All staff are entitled to a life assurance benefit equal to four times their salary, except senior managers who are entitled to eight times. The salary distributions for each category of staff are as follows:*

- *actuaries' salaries ~ Gamma(100,0.001)*
- *auditors' salaries ~ Gamma(12.25,0.000175)*
- *administrators' salaries ~ Gamma(4,0.000133)*
- *senior managers' salaries ~ Gamma(44.44,0.000222)*

The assumed probability of death next year for each life is 1%, except for senior managers who are assumed to have a probability of death equal to 2%. Calculate using simulations, the probability that total life assurance benefit payments next year will exceed £6 m. Compare your answer using Equations 25.14 and 25.15, and the Normal distribution approximation. State your assumptions.

25.9.2 Alternative Model – 'The Poisson Individual Risk Model'

The above model is likely to be appropriate for most types of life assurance products where the number of claims is limited to one per policy. Alternatively, we can adopt a Poisson distribution, with parameter λ_i, to describe the rate of claims per policy, such that each policy may have more than one claim. This is likely to be more suited to non-life products.

Equations 25.14 and 25.15 become:

$$E(S) = \sum_{i=1}^{n} \lambda_i \mu_i \tag{25.16}$$

$$Var(S) = \sum_{i=1}^{n} \lambda_i \sigma_i^2 + \lambda_i \mu_i^2. \tag{25.17}$$

Thus, if $q = \lambda$, the Poisson model has the same mean as the standard 'Bernoulli' individual risk model, but a higher variance due to the possibility of more than one claim from each policy. Note that where λ_i are the same for all policies, these equations are the same as Equations 25.6 and 25.7 from the 'Poisson' Collective Risk Model.

Exercise 25.17 *Re-calculate the mean, variance, and 99%VaR of the total claim amount from the policy portfolio described in Example 25.2 replacing the above standard 'Bernoulli' Individual Risk model, with a Poisson distributed number of claims, with rate parameter $\lambda_i = 0.1 + 0.02 * i$ for policy P_i.*

The key results from Exercise 25.17 are as follows:

```
mean(sim_sum2);      sd(sim_sum2);    quantile(sim_sum2,ind_VaR)

[1] 17.08446
[1] 12.33099
        99%
51.98113
```

The theoretical results from Equations 25.16 and 25.17 are consistent with the simulations:

```
tt<-1:10
var_j<-(10+tt)/2^2    #claim amount var
mean_j<-(10+tt)/2     #claim amount mean
(mean_theoryP<-sum(lambda_i *mean_j))

[1] 17.1

(sd_theoryP<-(sum(lambda_i *var_j+lambda_i *mean_j^2))^0.5)

[1] 12.32071

qnorm(ind_VaR)*sd_theoryP+mean_theoryP

[1] 45.76227
```

The standard deviation and 99% VaR are larger under the 'Poisson' model, due to the greater variability in claim numbers under the Poisson model, for example one simulation above has 5 claims under just one policy. Again, note the importance of stochastic modelling when considering the shape of the distribution and in particular, its tails.

It may be seen that the Individual Risk Model is closest to reality in many cases, particularly if any dependencies which exist can be modelled with copulas.

25.10   Issues with Heterogeneity

It is important to analyse the potential effect of policy heterogeneity when deciding on an appropriate risk model. One of the key problems in adopting a suitable Collective Risk model is where significant underlying heterogeneity exists – the analyst may encounter problems fitting a suitable claims amount distribution, and use of the computationally slower Individual Risk Model may be required. It may prove to be the case that initial application of the more accurate Individual Risk Model, followed by adoption of a suitably adjusted Collective Model, is the most efficient modelling process, resulting in faster computations; however, with modern-day computers and faster computing power these may no longer be concerns. We re-examine Example 25.2 below (which includes a group of policies which have some level of heterogeneity).

Example 25.3

Here we replace the Individual Risk Model applied in Example 25.2 with a broadly equivalent, simpler Collective Risk Model (described in Section 25.5). The Individual Risk Model incorporates details of the heterogeneous policies, whereas the Collective Model ignores this detail.

To determine a Collective Model we fit a Gamma distribution to the claim amounts simulated from Example 25.2 – the reader may wish to fit alternative, potentially suitable distributions. We also determine an 'average' probability of a claim to be applied in the Binomial Collective model (which should be close to 2.1, i.e. 10 policies with an average claim frequency between 0.12 and 0.3):

```
non_zero_claim<-ddd[(ddd>0)]    #from IndRiskBin model above
y2<-fitdistr(non_zero_claim,"gamma",lower=0.0001)    #fitted
(gamma1<-y2$estimate[1]);    (gamma2<-y2$estimate[2])

   shape
10.73062
    rate
1.31831

YY<-10;    (lambda_x<-mean(b))    #Binomial parameter estimate from earlier

[1] 0.209756

sims_x<-100000;    set.seed(555)
no_of_claims<-rbinom(sims_x,YY,lambda_x)#simulate Binomial claims

total_claim_amount_het<-NULL;    set.seed(556)
for (i in 1:sims_x){
total_claim_amount_het[i]<-sum(rgamma(no_of_claims[i],gamma1,gamma2))}
c(mean(total_claim_amount_het),sd(total_claim_amount_het))

[1] 17.04103 11.04246

quantile(total_claim_amount_het,ind_VaR)

    99%
46.8213
```

These calculations should be compared to the standard deviation and 99% VaR from Example 25.2 of 10.8 and 45.5, respectively. Given its simplicity, this collective model may be chosen by the insurer to model claims from this portfolio of heterogeneous policies, if it is deemed to be sufficiently accurate. (The reader is encouraged to repeat this example but with varying degrees of heterogeneity.)

Exercise 25.18 *Similarly, run a suitable Compound Poisson Model in place of the Compound Binomial model in Example 25.3, and compare results.*

The other major issue is that of dependent policies.

25.11   Policies Which Are Not Independent

The reader should now be equipped with the basic models with which to analyse most scenarios involving claim amounts and claim numbers from short-term insurance products. However, it will generally be the case in practice that individual policies will not be independent; the assumption that claims from various policies are independent has been adopted throughout this chapter. There are various ways we can allow for some degree of dependence. An explicit allowance for dependence can be included by incorporating copulas in the Individual Risk Model – the reader is directed towards the relevant chapters in this book.

Alternatively we may adopt the Negative Binomial Model which allows for distributions which exhibit a variance greater than the mean, a characteristic of scenarios where policies are not independent (please see the earlier section on this distribution).

A further option is to apply an approximate adjustment in respect of the aggregate level of dependence; the variance of a sum of n correlated variables, σ^2_{dep}, where each pair has the same Pearson's correlation coefficient, ρ, can be calculated from the following:

$$\sigma_{dep} = \sigma_{ind}\sqrt{1 + (n-1) * \rho} \qquad (25.18)$$

where σ^2_{ind} is the variance of a sum of n uncorrelated variables.

Exercise 25.19 *Derive Equation 25.18.*

Exercise 25.20 *Compare the variance of total claim amounts from an insurer which has 10 departments; first assume claims from departments are independent, then assume a correlation coefficient between each pair of departments equal to 0.2. The following gives such an example where we run simulations, assuming a multivariate Normal distrbution, to demonstrate the results of Equation 25.18.*

```
no_dptms<-10;   sims_dep<-100000

cov_nn1<-matrix(0,no_dptms,no_dptms);   diag(cov_nn1)<-1
sim_data1<-mvrnorm(sims_dep,rep(0,no_dptms),cov_nn1)
sim_sum_1<-rowSums(sim_data1)
var(sim_sum_1)          #independent

[1] 9.973747

rhomm<-0.2
cov_nn2<-matrix(rhomm,no_dptms,no_dptms);   diag(cov_nn2)<-1
sim_data2<-mvrnorm(sims_dep,rep(0,no_dptms),cov_nn2)
sim_sum_2<-rowSums(sim_data2)
var(sim_sum_2)          #some dependency

[1] 28.1429

(var(sim_sum_2)/var(sim_sum_1))^0.5    #ratio from simulations

[1] 1.679791

(1+(no_dptms-1)*rhomm)^0.5             #ratio from above equation

[1] 1.67332
```

Hence if the standard deviation of total claims was estimated to be £10 m assuming independence, then we may estimate that if we assume a level of dependence suitably described by an average correlation between pairs of $\rho = 0.2$, then the standard deviation would be approximately £16.7 m.

Exercise 25.21 *Develop the model from Exercise 25.10 to allow for the pairwise correlations of the total claims between the 10 insurance departments.*

25.12   Incorporating Parameter Uncertainty in the Models

As discussed elsewhere in this book, parameter error is an additional source of uncertainty which should be assessed and incorporated in our models where appropriate. Ignoring such uncertainty in these models may significantly underestimate the extent of future total claim amounts, potentially leading to unforeseen losses and insolvency.

Example 25.4

Here we repeat the calculations from Section 25.3 (the Collective Risk Model) using the same expected parameter values for the Poisson (λ) and Gamma (γ_1, γ_2) distributions (i.e. $\lambda = 200$; $\gamma_1 = 10$; $\gamma_2 = 2$), but incorporating uncertainty in two of them *viz.* λ and γ_1; in the simulation of claim amounts and claim numbers the parameter values are simulated from suitable, independent Normal distributions as follows:

$$\lambda \sim N(200, 10^2) \qquad \gamma_1 \sim N(10, 0.5^2).$$

Each simulation run applies a set of three parameter values chosen from these distributions (with γ_2 fixed). Effectively we are running the calculations described in Section 25.3 100,000 times using parameter values from the two distributions above.

```
poisson_best<-200;    gamma1_best<-10;    gamma2_best<-2

poisson_x<-NULL;  gamma1_x<-NULL;  gamma2_x<-NULL;
no_of_claims<-NULL;    total_claim_amount_unc<-NULL

sims_x<-100000;        set.seed(555)
poisson_x<-rnorm(sims_x,poisson_best,poisson_best*0.05)
no_of_claims<-rpois(sims_x,poisson_x)
gamma1_x<-rnorm(sims_x,gamma1_best,gamma1_best*0.05)

for (i in 1:sims_x){
total_claim_amount_unc[i]<-sum(rgamma(no_of_claims[i],gamma1_x[i],gamma2_best))
}
c(mean(total_claim_amount_unc),sd(total_claim_amount_unc))

[1] 1000.0103   102.6982

quantile(total_claim_amount_unc,VaR1)

   99.5%
1281.617
```

The standard deviation of claims and 99.5% Value at Risk calculated above are £102,698 and £1.282 m respectively, whereas in Section 25.3 where no parameter uncertainty was allowed for, these quantities are significantly lower (£74,098 and £1.196 m, respectively).

Incorporating uncertainty in the Poisson parameter using the Negative Binomial Distribution.

The reader may wish to refer back to Section 25.6 where we briefly discussed the Negative Binomial distribution; this distribution can be used to simulate Poisson realisations where the Poisson parameter itself is a random variable from a Gamma distribution. So, as an alternative to simulating Poisson values by first generating Poisson parameters from a Gamma distribution with parameters α and β, we can simulate from Negative Binomial $(\alpha, \frac{\beta}{1+\beta})$, with mean $\frac{\alpha}{\beta}$ and variance $\frac{\alpha(1+\beta)}{\beta^2}$. For example:

```
set.seed(10000);          alpha1<-10;         beta1<-0.2
simsppp<-100000;          set.seed(1000)
g_p<-rgamma(simsppp,alpha1,beta1)
pois_unc<-rpois(simsppp,g_p)
mean(pois_unc);          var(pois_unc)

[1] 49.99686
[1] 298.7683

neg_bin1<-rnbinom(simsppp,alpha1,beta1/(1+beta1))
mean(neg_bin1);          var(neg_bin1)

[1] 49.85599
[1] 297.3458
```

The Q-Q plot in Figure 25.7 visually demonstrates the similarity in the two distributions.

Remark 25.10 The analysis of parameter uncertainty in such models in the actuarial literature typically concentrates solely on the claim number distribution (λ). The reader may therefore wish to repeat this analysis allowing only for claim number parameter uncertainty.

It is also important to note that we have assumed zero correlation between the parameter values in these simulations; this may not be the case in practice. In particular, when there is a large number of claims, there is often an increase in the proportion of smaller claim amounts; this is often observed in car insurance claims data following periods of bad weather.

Figure 25.7 Q-Q Plot: Comparing Negative Binomial and Poisson-Gamma.

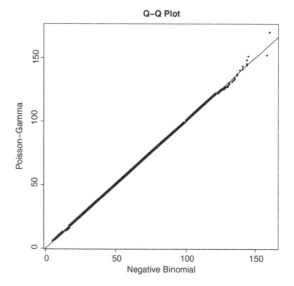

25.13   Claim Amount Distributions: Alternatives to the Gamma Distribution

Throughout this chapter we have adopted the Gamma distribution to model claim amounts; the Gamma distribution is perhaps the first choice when assessing claim amounts which do not have a particularly fat tail. Two alternative distributions commonly employed to model claim amounts are the LogNormal and Pareto distributions. The Pareto distribution in particular allows for modelling of fat-tailed claim amounts, typical of insurance classes such as liability and fire insurance.

The reader is encouraged to re-run the examples and exercises in this chapter using a variety of these alternative claim severity distributions.

25.14   Conclusions

Given the significant computer power now available, stochastic models will often be viewed as the most appropriate modelling approach. In particular, where claim numbers are not sufficiently high, percentile estimates using the Normal distribution are likely to be misleading, thus requiring stochastic modelling using an Individual Risk Model.

The models covered in this chapter allow straightforward verification of the mean and variance of claims; in particular, the Collective Risk Model can provide a simple estimate of total claims if policies can reasonably be considered to be independent and identical. Following appropriate analysis (e.g. of percentile results) the Collective Model may be considered sufficiently accurate to use as the principal model where the number of claims is sufficiently high.

Important issues regarding heterogeneity, potential correlations between policies, and parameter uncertainty should be carefully addressed with the use of suitably detailed models and tools. In many cases it would be recommended to use a number of models, each with their own advantages and disadvantages, to obtain a better overall view on reality.

25.15   Recommended Reading

- Boland, P.J. (2007). *Statistical and Probabilistic Methods in Actuarial Science*. Boca Raton, Florida, US: Chapman and Hall/CRC.
- Booth. et al. (2005). *Modern Actuarial Theory and Practice*, 2e. Boca Raton, Florida, US: Chapman and Hall/CRC.
- Charpentier, A. et al. (2016). *Computational Actuarial Science with R*. Boca Raton, Florida, US: Chapman and Hall/CRC.
- Kaas, R., Goovaerts, M., Dhaene, J., and Denuit, M. (2008). *Modern Actuarial Risk Theory*, 2e. Berlin Heidelberg: Springer-Verlag.
- Sundt, B. (1984). *An Introduction to Non-Life Insurance Mathematics*. Veröffentlichungen des Instituts für Versicherungswissenschaft der Universität Mannheim, Vol. 28, Verlag Versicherungswirtschaft, Karlsruhe.

26

Collective Risk Models: Exercise

Peter McQuire

26.1   Introduction

This chapter takes the form of a guided exercise, based on material in the previous chapter. We present the reader with data and proceed to determine a suitable, relatively simple, model on which the reader can expand.

26.2   Analysis of Claims Data

Our task is to analyse a set of claims data and determine appropriate distributions for both the total number of claims each month, N, the individual claim amount, X_i, and ultimately the total monthly claim amounts.

The dataset includes the number of claims made each month together with the size of each claim made. All claim amounts are in £000's.

```
data<-read.csv("~/loss distn data2.csv")
amounts<-data$claim.size
claimnumbers<-data$monthly.claim.numbers[1:48]
```

There are 48 monthly claims data items and 4829 claim amounts data items. The claims amounts are shown in date order such that the first 95 claims are in respect of month 1, the next 109 claims are in respect of month 2 etc.

Exercise 26.1   *Plot histograms of the data.*

Our objective is to determine the most appropriate model for our data, using the Collective Risk Model discussed in the previous chapter. We will fit distributions to both the claim amounts, X_i, and the monthly claim numbers, N, using the method of maximum likelihood estimation ("MLE") and the `fitdistr` function.

R Programming for Actuarial Science, First Edition. Peter McQuire and Alfred Kume.
© 2024 John Wiley & Sons Ltd. Published 2024 by John Wiley & Sons Ltd.
Companion Website: www.wiley.com/go/rprogramming.com.

Fitting various distributions for the claim amounts, X_i, trying Normal, Gamma, Lognormal, Cauchy, and Weibull distributions (further distributions should be tested to ensure the best fit):

```
a<-NULL
y1<-fitdistr(amounts,"normal");              a[1]<-y1$loglik
names(y1)

[1] "estimate" "sd"        "vcov"     "n"       "loglik"

y2<-fitdistr(amounts,"gamma",lower=0.0001);  a[2]<-y2$loglik
y3<-fitdistr(amounts,"lognormal");           a[3]<-y3$loglik
y4<-fitdistr(amounts,"cauchy");              a[4]<-y4$loglik
y5<-fitdistr(amounts,"weibull");             a[5]<-y5$loglik
```

Which distribution, and parameters, are best?

```
a    #loglikelihoods

[1] -19705.45 -19274.25 -19388.76 -20314.92 -19366.39

which.max(a);    max(a)

[1] 2
[1] -19274.25

y2$estimate[1];  y2$estimate[2]    #parameter MLEs

   shape
3.918835
     rate
0.1380487
```

The 2nd distribution tried, Gamma (3.9188347, 0.1380487), has the highest loglikelihood. The reader should try various tests as described in Chapter 5, such as the Anderson-Darling test, Chi-squared test, or Wald test (also see Boland Chapter 2.3, or Verzani Chapter 10.3). Shown below are the results from the Kolmogorov–Smirnov ("K-S") test:

```
ks.test(amounts,"pgamma",y2$estimate[1],y2$estimate[2])


One-sample Kolmogorov-Smirnov test

data:  amounts
D = 0.0078725, p-value = 0.9257
alternative hypothesis: two-sided
```

The p-scores for the other distributions are all very close to 0.

The Gamma distribution clearly appears to be the best fit for the data. Of course, we should always visually check the fit with the data for example, histograms, QQ tests (Figure 26.1).

Exercise 26.2 *Carry out QQ tests on the above distributions and data.*

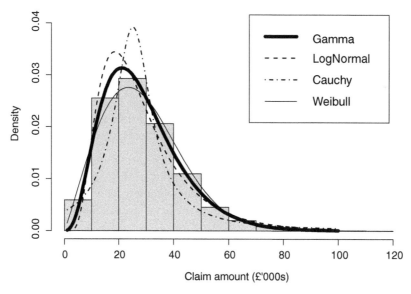

Figure 26.1 Claim amounts data and fitted distributions.

Exercise 26.3 *There are a number of standard distributions for which ready-made func-tions do not exist in R; the Pareto is one such example. Write code to determine its best fitting parameters using MLE.*

We must also fit a suitable distribution to the total monthly claim numbers, N. Apply-ing a similar procedure to that set out above, using the Gamma, Negative Binomial, Lognormal, and Poisson distributions:

```
b<-NULL
y6<-fitdistr(claimnumbers,"gamma");              b[1]<-y6$loglik
y7<-fitdistr(claimnumbers,"negative binomial");  b[2]<-y7$loglik
y8<-fitdistr(claimnumbers,"lognormal");          b[3]<-y8$loglik
y9<-fitdistr(claimnumbers,"poisson");            b[4]<-y9$loglik

#which distn is best?
b

[1] -187.7966 -187.5678 -188.2281 -189.3148

y7$estimate[1];  y7$estimate[2]  #parameters

    size
219.5104
      mu
100.6042
```

Here the Negative Binomial (100.6041667, 219.510436) fits best. (Note the Negative Bino-mial distribution is usually represented as *NegBin (μ, size)*; hence the swapping of the order of the parameters.) The claims in this example exhibit some evidence of dependency; the variance of the claim numbers is significantly higher than the mean. As a result, the Compound Poisson is inappropriate here, with the Compound Negative Binomial being preferred.

Additional checks, as described above, should be carried out. We may not be satisfied that we have tried a sufficient number of distributions; for example we may wish to try distributions which are negatively skewed.

Of course, it may be the case that an alternative model design would be more appropriate. For example, a visual inspection of the data suggests that the claim numbers may have a bi-modal distribution, possibly arising from "summer" and "winter" claims; the analyst may wish to investigate such seasonal claims. We will proceed with the simpler model described above; however this illustrates that great care must be taken in choosing the overall model design.

Remark 26.1 (See Section 26.5.) The estimated standard error of the *size* parameter, 139.36, is particularly large in this case, implying conderable uncertainty regarding the variance of the number of monthly claims, and requires further consideration in this case.

26.3 Running Simulations

Now we have our proposed model (a Collective Risk Model with $X_i \sim$ Gamma (3.919, 0.138) and $N \sim$ Negative Binomial (100.604, 219.51) we proceed to simulate claims:

```
sims<-100000     #no of sims
set.seed(555)
sim_no<-rnegbin(sims,y7$estimate[2],y7$estimate[1])   #note par order
```

These are our simulated claim numbers, N. Proceeding to simulate claim amounts for each of the simulated N's, and summing them to get the total claim amount from each simulation:

```
sim_monthly<-NULL;     set.seed(556)
for (i in 1:sims){sim_monthly[i]<-sum(rgamma(sim_no[i],y2$estimate[1],
y2$estimate[2]))}
mean(sim_monthly);          sd(sim_monthly)

[1] 2854.91
[1] 373.5479
```

From these simulations, the mean total monthly claim amount is £2854910, and the standard deviation of the total monthly claims is £373548.

Of course, it is imperative to check our model with the original data; the simulated total monthly claim amounts from the chosen compound distribution are shown in Figure 26.2, together with the actual monthly claims data for comparison.

We should calculate the mean and standard deviation of the actual monthly claims from the original data. Unfortunately, the data which we were provided did not split the claims by month - it was simply a chronological list of claims. Thus we are required to split the claims up by months for example, the first 95 claims refer to the first month. This is left as an exercise for the reader.

The mean and standard deviation of the total monthly claims from the original data are calculated to be £2855884 and £357570 respectively. It appears that our model may be overestimating the variance of the data. This may be due to the limited number of data points that is, 48 months. The higher variance builds in a level of prudence to our model - whether this is appropriate depends on the purpose of the model for example, will it be used to assess profitability or for solvency purposes?

It may therefore be decided that we need to re-visit our chosen distributions and final model decisions. We leave this as an exercise for the reader.

Exercise 26.4 *Rerun the simulations using the best-fitting Compound Poisson distribution.*

Exercise 26.5 *By calculating appropriate quantile values from the above simulations, draw a Q-Q plot comparing the 48 monthly claim data points with the simulations from the chosen compound distribution. (The QQ plot is shown in Figure 26.3.)*

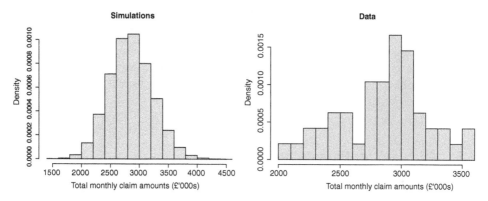

Figure 26.2 Comparing total monthly claim amounts from simulations and data.

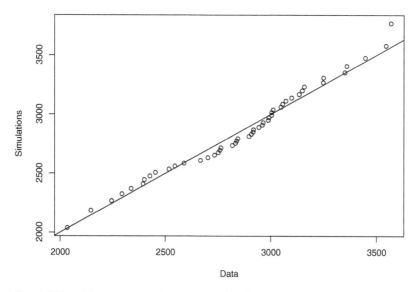

Figure 26.3 QQ plot: comparing total monthly claim amounts from simulations and data.

26.4   Tails of the Distribution

We are likely to be particularly interested in estimating probabilities of incurring large monthly claim amounts. For example, what is the probability that the total claims amount next month will exceed £3.5m? From our proposed model:

```
sum(sim_monthly>3500)/sims
[1] 0.04714
```

This compares with 2 months out of 48 months in the data (months 34 and 37) which exceeded £3.5m that is, 4.17%. Similarly, the 99% VaR calculated from the simulations run using our model is £3.78m.

Exercise 26.6   *Verify the above analysis and calculations.*

The reader may recall from the fitting exercise that the Lognormal distribution was also a potential candidate for the claim amounts distribution. It is left as an exercise for the reader to compare the aggregate claims distribution if the Lognormal distribution was used in place of the Gamma distribution. In addition, the reader may wish to apply Extreme Value Theory (Chapter 28) to the above analysis.

26.5   Allowing for Parameter Uncertainty

The uncertainty of, and correlation between, the parameter estimates determined in Section 26.2 are obtained from:

```
y2$vcov; y7$vcov

                shape              rate
shape 0.005866049 0.000206634027
rate   0.000206634 0.000008285766
                 size               mu
size 19420.1032641691 0.0002108807
mu        0.0002108807 3.0565044778

#95% intervals (only y2 shown)
y2$estimate+y2$sd*1.96    #upper

   shape     rate
4.0689514 0.1436906

y2$estimate-y2$sd*1.96    #lower

   shape     rate
3.7687181 0.1324069
```

In addition, the correlation between parameters values should be incorporated in any simulations which allow for parameter uncertainty.

The reader is encouraged to re-run simulations from Section 26.3, incorporating this parameter uncertainty. One way to do this is to simulate a set of parameter values using the above estimate of the covariance matrix and the mvrnorm function for each simulation (based on the asymptotic normality property of MLEs). For example:

```
set.seed(100); sims_parx<-5
mean_vec1<-c(y2$estimate[1],y2$estimate[2]); sigma_vec1<-y2$vcov
sim_par1<-mvrnorm(sims_parx,mean_vec1,sigma_vec1)
mean_vec2<-c(y7$estimate[1],y7$estimate[2]); sigma_vec2<-y7$vcov
sim_par2<-mvrnorm(sims_parx,mean_vec2,sigma_vec2)
(sim_sum_par_val<-cbind(sim_par1,sim_par2))

        shape      rate      size      mu
[1,] 3.957309 0.1390845 206.9843 100.65542
[2,] 3.908740 0.1382769 206.0940 101.28400
[3,] 3.924904 0.1375456 247.6093  99.71104
[4,] 3.850887 0.1364830 116.4092 102.20177
[5,] 3.909863 0.1380938 202.3168  96.56511
```

We can then proceed to simulate claims data from the model using these parameter values, rather than only the MLEs. As noted in Remark 26.1, further attention is required regarding the uncertainty of the *size* parameter from the Negative Binomial distribution; the estimated standard error is $\sqrt{19420.1} = 139.36$.

26.6 Conclusions

The main objective of this chapter was to fit a compound distribution model to claims data. In addition, a secondary objective was to highlight the importance of assessing whether a proposed model is appropriate, which often depends on the purpose of the modelling exercise. It is certainly not generally acceptable simply to follow modelling algorithms to obtain model parameters; judgement is required in determining whether or not a proposed model is a reasonable reflection of reality. As noted above, the chosen compound distribution may not reflect possible seasonality of claims, nor the variance of claims. Other issues such as heterogeneity, dependence of claims between policies, and parameter uncertainty should be analysed.

26.7 Recommended Reading

- Boland, P.J. (2007). *Statistical and Probabilistic Methods in Actuarial Science*. Boca Raton, Florida, US: Chapman and Hall/CRC.
- Verzani, J. (2014). *Using R for Introductory Statistics*, 2e. Boca Raton, Florida, US: Chapman and Hall/CRC.

27

Generalised Linear Models: Poisson Regression

Peter McQuire

```
library(dplyr)
```

27.1 Introduction

The aim of this chapter is to provide an introduction to this important topic by considering:

- what is a Generalised Linear Model ("GLM");
- the aims of GLMs and what sort of problems can be solved using GLMs;
- the issues encountered when using GLMs.

Rather than cover several versions of GLMs in what is a relatively short chapter, we concentrate on one particular type – Poisson Regression. This is often the model of choice of insurance companies when modelling claim frequencies. We will look at several examples by analysing various datasets; in this way the reader should develop an understanding of the fundamentals of GLMs, and their benefits.

The datasets used in this chapter to which we fit GLMs incorporate data which has been simulated using specific equations, the code for which is included on the website; the reader then has the facility to alter the data to understand the effect such changes have on the model output.

There are numerous excellent books on Multiple Regression and GLMs; please see the Recommended Reading. It is beyond the scope of this book to cover the range of models in any great detail. In particular, the reader is encouraged to study various other types of GLMs such as logistic and probit models which are not covered in this chapter.

27.2 Examples/Exercises/Data

There are four datasets utilised in this chapter. *Data1* is artificially simple in that all lives are exposed to risk for the same period of time; the data includes details of the age and

R Programming for Actuarial Science, First Edition. Peter McQuire and Alfred Kume.
© 2024 John Wiley & Sons Ltd. Published 2024 by John Wiley & Sons Ltd.
Companion Website: www.wiley.com/go/rprogramming.com.

sex of each life, together with the number of claims each life made. This data contains significantly less data in one particular age band. We also summarise this data and re-run our model.

Data2 is more realistic, including lives which are exposed to risk for various periods of time. Data3 demonstrates the importance of considering whether interactions exist between variables, whilst Data4 includes an extra categorical variable – postcode.

As noted above, the format of the examples/exercises includes datasets generated from an appropriate equation; we proceed to fit a GLM to the data and compare the regression coefficients with those used to simulate the data. The reader can proceed to change parameter values, change the volume of data etc. and observe changes to the model output.

The reader who wishes to move straight onto fitting GLMs to data, please go to Section 27.6.2.

27.3 Brief Recap on Multiple Linear Regression

The reader may wish to review Multiple Linear Regression from Chapter 5; below is a brief recap.

The aim of Multiple Linear Regression ("MLR") is to understand the relationship between one variable (the dependent, or response, variable) and various explanatory variables (the independent variables). For example, how many chocolate bars may a company be expected to sell next month if its price is increased by 10p and £1m is spent on advertising. The MLR model is given by:

$$Y_i = \beta_0 + \beta_1 X_{i1} + \beta_2 X_{i2} + ... + \beta_n X_{in} + \epsilon_i \tag{27.1}$$

where $\epsilon_i \sim N(0, \sigma^2)$, and $X_{i1}, X_{i2}, ... X_{in}$ are the n values describing the ith individual or entity. $\beta_0, \beta_1, ..., \beta_n$ represent the regression coefficients to be determined from analysing the observed data. From Equation 27.1:

$$E(Y|X) = \beta_0 + \beta_1 X_1 + \beta_2 X_2 + ... + \beta_n X_n \tag{27.2}$$

Thus if we increase X_1 (or any of the X_i) by one unit then we expect Y to increase by β_1 (or β_2, etc.). However, clearly this model is inappropriate for a number of types of dependent variables e.g. if Y can only take discrete, positive values.

27.4 Generalised Linear Models ("GLMs")

GLMs consist of the following key parts:

- an error distribution (e.g. Poisson, Binomial, Gamma); compare MLR which is restricted to using the Normal distribution;
- a link function, f, which links how the linear function of the explanatory variables is related to the expected value of the dependent variable. In Poisson regression, which is the focus of this chapter, the link function is the natural logarithm.

- the variance function, which determines the relationship between the mean and variance of the dependent variable. We will concentrate solely on Poisson regression and, as such, will not explore this in any detail.

All these options provide significant flexibility when deciding on which model to use in any particular situation.

Remark 27.1 Other commonly used GLMs, particularly in the actuarial field, include logistic regression and probit regression (or model). These will not be covered in this chapter, but the reader may wish to explore these models on completion of this chapter.

Note that the link function used is dependent on the distribution used to model the errors e.g. the logarithmic link function is used where Poisson errors are adopted, the logit link function is used where Binomial errors are adopted.

Thus the most general expression for GLMs is:

$$f(E(Y|X)) = \beta_0 + \beta_1 X_1 + \beta_2 X_2 + \dots + \beta_n X_n \tag{27.3}$$

Note that usually we will ultimately be interested in $E(Y|X)$, not $f(E(Y|X))$, requiring us to apply f^{-1} to predict the final output.

A full discussion of link functions is beyond the scope of this chapter; as noted above we will concentrate solely on Poisson regression.

27.5 Goodness of Fit of GLMs

Before moving on to discuss Poisson regression, we must first briefly discuss how the quality of the various models we will try should be compared and assessed.

As with all models we look at in this book, we must assess how well it predicts the data. For example, we may want to understand whether introducing an extra variable improves our GLM and hence predictions. To measure the goodness of fit of GLMs we calculate a measure called the deviance. Deviance (or residual deviance as defined in the glm output) is defined as follows:

$$deviance = 2(\log L_0 - \log L_1)$$

where $\log L_0$ is the loglikelihood in respect of the full, or saturated, model (the best fit), and $\log L_1$ is the loglikelihood in respect of the proposed model. The higher the $\log L_1$ value, and thus the closer it is to $\log L_0$, the better fit from the model. The null deviance is simply where the proposed model is equivalent to that which includes only one parameter.

The change in deviance in two nested models can be compared with the relevant values from a χ^2 distribution with a number of degrees of freedom equal to the difference in the number of parameters used in the two models. We analyse the change in deviance in several examples which follow.

Ultimately our objective may be considered to be to develop a model which sits somewhere between the null and saturated models, where the goodness of fit is deemed acceptable, but also where the number of parameters is deemed acceptable. Of course, this approach is not limited to GLMs.

27.6 Poisson Regression

27.6.1 Introduction

Poisson regression may be adopted in situations where the number of observed events follows a Poisson distribution (as compared to a Normal distribution in the case of MLR). Examples where Poisson regression may be suitable in an actuarial context include modelling the number of deaths in a large human population, and modelling the number of claims expected under a particular non-life policy. In both examples, we may include variables such as age, sex, location, and occupation.

In the MLR model above, we may be modelling the salaries of actuarial students, using number of exam passes, years of experience, and age as explanatory variables. Poisson regression is similar – we may be modelling the number of claims a policyholder may make given their age, sex, claims history etc. The principal differences with Poisson regression are two-fold; we are interested in proportionate changes in the explanatory variables, and the values of our dependent variables must be positive integers. More specifically, the distribution of the dependent variable should follow a Poisson distribution.

Under Poisson regression, f is the natural logarithm, and $Y_i \sim Poisson(\lambda_i)$. Compare this with MLR where f is simply 1, and $Y_i \sim N(\mu_i, \sigma_i)$. Thus MLR is a particularly simple example of a GLM.

27.6.2 Using Poisson Regression to Model Claim Numbers

Throughout this chapter we will use the example of modelling the number of insurance claims under a policy, for example, a private motor car policy. Our objective is to model the number of claims made by policyholder, i, ultimately to calculate suitable premium rates for policyholders exhibiting various characteristics.

We assume that the number of claims made by policyholder, i, has a Poisson distribution with an expected number of claims equal to λ_i. That is:

$$Y_i \sim Poisson(\lambda_i) \qquad (27.4)$$

and therefore the probability of incurring y claims under this policy is:

$$P(Y_i = y) = \frac{e^{-\lambda_i}(\lambda_i)^y}{y!} \qquad (27.5)$$

However the number of expected claims, λ, will not be the same for everyone; it would be reasonable to assume it may depend on, amongst other things, the age, X_1, and sex, X_2, of the policyholder. So perhaps:

$$\lambda_i = E(Y_i) = \beta_0 + \beta_1 X_{1,i} + \beta_2 X_{2,i} \qquad (27.6)$$

This is similar to our linear regression model (Equation 27.1) with the key difference that the number of claims now has a Poisson distribution rather than a Normal distribution. This is our first Poisson regression model.

However, if the insured's perceived level of risk changed following an event (e.g. they had an accident, or eyesight deteriorated) such that the premium should increase, we would expect the premium to increase by a *proportion* of the existing premium, not a fixed amount. If the insured's current premium was £500, following an accident the premium may be reviewed upwards by 20% to £600 as the insured's estimated level of risk has increased by 20%. If, alternatively, the premium was originally £1,000 it would be reasonable to expect the premium to increase to £1,200 following such an event.

Also the expected claims cannot be negative; $E(Y_i)$ in Equation 27.6 can clearly give a negative value. For these reasons it would seem sensible to use the following model instead.

$$\lambda_i = E(Y_i) = e^{\beta_0 + \beta_1 X_{1,i} + \beta_2 X_{2,i}} \qquad (27.7)$$

The standard Poisson regression model is given by Equations 27.5 and 27.7. Under Equation 27.6 β_i would represent the increase in the expected number of claims following a change in X_i of 1. However, under Equation 27.7, it now represents a proportionate change in the expected number of claims.

Exercise 27.1 *If $\beta_1 = 0.3$ by how much will $E(Y_i)$ increase if $X_{1,i}$ changes from 46 to 47 (with no other changes)?*

In the above setting, λ is simply a number and its value will correspond to the length of time of observations. We will change our approach shortly (where we have non-unit periods of time in our data) but for now will continue where all our data is for the same unit period of time; in our first example we observe every policyholder for exactly one year. The R code for the glm function is as follows:

```
glm_model = glm(Y ~ X1 + X2, family = poisson)
```

By including "`family = poisson`" the function automatically assumes the form in Equation 27.7 (otherwise, Equation 27.6 is assumed). To find the regression coefficients, β, we maximise the Poisson loglikelihood of our data, which, from Equation 27.5 is:

$$\log L = \sum_i (-\lambda_i + y_i \log \lambda_i) \qquad (27.8)$$

where y_i is the number of defined events of interest which occurred to life i, and the sum is over all lives, or policies.

Exercise 27.2 *Write down the loglikelihood of the first 2 items in our data set (see "GLM-Data1 simple.csv").*

Example 27.1

Let's look at a particular example of modelling the number of claims, Y, from a car insurance policy using age and sex as predictor variables (X_1 and X_2). The idea is that by estimating values for the regression coefficients, β, we can estimate the expected number of claims based on the policyholder's age and sex, and hence charge appropriate premiums to a range of policyholders.

The data analysed in this first example is particularly simple and somewhat artificial (so as not to detract from the key learning points). We have observed each policyholder for exactly one year, with the same number of lives at each age (except for ages 61–63 years, for reasons we shall see shortly).

We shall choose $\beta_0 = -0.5, \beta_1 = 0.3, \beta_2 = -0.01, (\beta_{3,4...} = 0)$, and arbitrarily set $male = 1$ and $female = 0$. The age range is 20 to 64 years. We have proceeded to simulate claims data using these parameter values, with the data saved in "*GLMData1 simple.csv*".

```
rate_glm<-function(age,sex){EY<-exp(-0.5+(if(sex==1){0.3}else{0})-0.01*age);
return(EY)}
male_rate<-rate_glm(20:64,1);    female_rate<-rate_glm(20:64,0)   #"true rates"
```

These rates are plotted in Figure 27.1 as $(\times, +)$, where they are com-pared with the fitted rates using the GLM. The data should be downloaded using `claimdata1<-read.csv("~/GLMData1_simple.csv")`:

```
head(claimdata1,3)
   X age sex claims
1 1  20   1      0
2 2  21   1      1
3 3  22   1      0
```

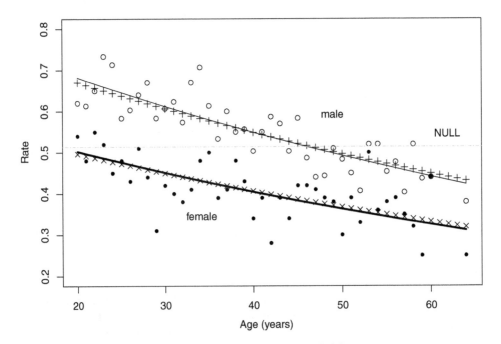

Figure 27.1 Fitting of GLM, compared with crude rates and underlying rates.

We can run the following code to regress on age and sex, using the `glm` function:

```
glm_model1 = glm(claims ~ age + sex, family = poisson, data = claimdata1)
summary(glm_model1)#parameter results


Call:
glm(formula = claims ~ age + sex, family = poisson, data = claimdata1)

Deviance Residuals:
    Min       1Q   Median       3Q      Max
-1.1677  -1.0197  -0.9141   0.5900   3.8792

Coefficients:
              Estimate Std. Error z value Pr(>|z|)
(Intercept) -0.4715109  0.0421423  -11.19   <2e-16 ***
age         -0.0108445  0.0008851  -12.25   <2e-16 ***
sex          0.3052889  0.0270435   11.29   <2e-16 ***
---
Signif. codes:  0 '***' 0.001 '**' 0.01 '*' 0.05 '.' 0.1 ' ' 1

(Dispersion parameter for poisson family taken to be 1)

    Null deviance: 17115  on 16799  degrees of freedom
Residual deviance: 16828  on 16797  degrees of freedom
AIC: 31385

Number of Fisher Scoring iterations: 5
```

We have our first model:

$$E(Y_i) = \mu_i = e^{-0.4715109 - 0.0108445X_{1,i} + 0.3052889X_{2,i}} \tag{27.9}$$

where $X_{1,i}$ is age, and $X_{2,i}$ is the indicator for sex, both for life i. The coefficient estimates obtained from this GLM are consistent with the values used to simulate the data.

The intuition behind Equation 27.9 is as follows: the number of expected claims decreases by around 1% for each increasing year of age (X_1), with male rates approximately 36% higher ($e^{0.31} - 1$) than female rates. We can obtain the logarithm of the expected number of claims for each life in our data using the `predict` function, and subsequently the expected number of claims (by taking the exponential):

```
head(exp(predict(glm_model1)))

        1         2         3         4         5         6
0.6817370 0.6743838 0.6671100 0.6599146 0.6527969 0.6457559

male_rate[1:6]    #compare underlying rates

[1] 0.6703200 0.6636503 0.6570468 0.6505091 0.6440364 0.6376282
```

Remark 27.2 If you include `data= mydataset` in the `glm` argument (as we have done) it will look for appropriately named variables first in "mydataset"; if it cannot be found it will look elsewhere. It's preferable to have all the data in this single data object, but you must ensure the columns are named accordingly.

The model is plotted in Figure 27.1, with lines denoting the modelled claim numbers. As noted earlier, the crosses indicate the true, underlying claim numbers from `male_rate` and `female_rate` objects calculated earlier.

Example 27.2 Comparing models: deviance calculations

We re-run the model with (1) only `sex` and (2) only `age`, and compare deviance results. Also run below is the NULL model.

```
glm_model1_noage = glm(claims ~ sex, family = poisson, data = claimdata1)
glm_model1_nosex = glm(claims ~ age, family = poisson, data = claimdata1)
summary(glm_model1_noage)$deviance;    summary(glm_model1_nosex)$deviance

[1] 16979.33
[1] 16963.51
```

The deviance results clearly show the model is improved with both age and sex being included, with a residual deviance equal to only 16828 (the χ^2 (1) 95% level is only 3.84). Turning to the null model:

```
glm_model1_NULL = glm(claims ~ 1, family = poisson,data = claimdata1)
summary(glm_model1_NULL)$deviance;    exp(summary(glm_model1_NULL)$coef[1])

[1] 17114.62
[1] 0.5140476
```

The null model effectively gives the best possible model when constrained to using only one rate. (This value is also plotted in Figure 27.1.)

27.7 Data with Varying Exposure Periods

27.7.1 Claim Rates and the *Offset*

It may indeed be the case that we wish to model the number of events, for example, the number of puppies in a litter, number of gold medals a country wins at the Olympics; this was the method used in Section 27.6.2.

More typically, however, we will be interested in *rates*, as is the case with claims – it is important to distinguish whether a policyholder has made *n* claims in one year or ten years.

The data in the example above contained exactly one year of exposure for each policyholder. This is clearly not realistic; at any point in time when a valuation is undertaken, policies will be in force, and the term will not have been completed.

Fortunately, our model can easily be adjusted where we hold policy exposure data by modelling claim *rates*; our model is now given by:

$$\mu_i = e^{\beta_0 + \beta_1 X_{1,i} + \beta_2 X_{2,i}} \tag{27.10}$$

where μ_i is the annual rate of claims for life *i* for example, 0.5 claims per year. Note that $\lambda_i = \mu_i t$. As we will still be dealing with claim numbers and exposure periods in the data, Equation 27.7 becomes:

$$E(Y_i) = \lambda_i = t\, e^{\beta_0 + \beta_1 X_{1,i} + \beta_2 X_{2,i}} \tag{27.11}$$

As before, λ is the expected number of claims in a period of time, t, over which we observe a risk. Equation 27.5 becomes:

$$P(Y_i = y) = \frac{e^{-\mu_i t}(\mu_i t)^y}{y!} \tag{27.12}$$

So, in general, for each policy, i, we will have the following loglikelihood:

$$logL_i = -\mu_i t_i + y_i \log \mu_i \tag{27.13}$$

where μ_i is the annual rate of claims. And from Equation 27.11:

$$\log E(Y|X) = \beta_0 + \beta_1 X_{1,i} + \beta_2 X_{2,i} + \log t \tag{27.14}$$

The glm function includes an $offset$ argument, with the following required code:

```
glm_model = glm(Y ~ X1 + X2, offset= log (exposure time), family = poisson)
```

The offset is simply the log of the exposure data ($\log t$). The above code is consistent with Equation 27.14.

Remark 27.3 It was noted in Chapter 1 that $dplyr$ is an excellent data manipulation package, allowing an advanced level of manipulating large datasets. This package is highly recommended to the reader whose work includes large-scale data manipulation.

Given that most readers will have covered several chapters of this book by now, it is likely that some experience of this package may have been accrued. There is a small amount of coding in this chapter which uses functions from this package to manipulate data (without any detailed explanation).

27.7.2 Application to Aggregated Data in Section 27.1

The data in Example 27.1 includes details of each individual policy, with each policy exposed to one year. Below we aggregate the data, grouping all policy exposure time by age and sex, using functions from the $dplyr$ package (an extract of the summarised data is included below):

```
gf<-claimdata1 %>% group_by(age,sex) %>% summarize(sum(claims))

ds<-claimdata1 %>% count(age,sex)
data_agg<-cbind(ds,gf[,3]);  colnames(data_agg)<-c("age","sex","exps","sum_claims")
head(data_agg,4)

  age sex exps sum_claims
1  20   0  100         54
2  20   1  300        186
3  21   0  100         48
4  21   1  300        184
```

For example, there were 54 claims from female lives from 100 years of exposure, and 186 claims from male policies from 300 years of exposure (it's the same data as above). This is the format in which data is often presented.

Applying the `glm` function to this summarised data, `data_agg`, now including the `offset` argument:

```
glm_model2 = glm(sum_claims~age+sex,offset=log(exps),family=poisson,data=data_agg)
summary(glm_model2)$coefficients   #summary of results (not full)

              Estimate   Std. Error   z value     Pr(>|z|)
(Intercept) -0.47151089 0.0421440528 -11.18808 4.664422e-29
age         -0.01084447 0.0008851704 -12.25128 1.653652e-34
sex          0.30528893 0.0270450876  11.28815 1.501688e-29

expclaimcheck2<-function(age,sex){exp(summary(glm_model2)$coef[1]+
summary(glm_model2)$coef[2]*age+summary(glm_model2)$coef[3]*sex)} #check
expclaimcheck2(c(20,21),c(1,1))   #as above

[1] 0.6817370 0.6743838
```

The results are exactly the same as above – the GLM model is the same. Figure 27.1 shows the fitted GLM together with the crude rates at each age (i.e. the number of claims divided by the total exposure time at each age), with males rates o, and female rates • (see code below).

```
#crude rates - used in plot
crudeqq_m<-filter(data_agg %>% mutate(rateopop=data_agg[,4]/data_agg[,3]),sex==1)
crudeqq_f<-filter(data_agg %>% mutate(rateopop=data_agg[,4]/data_agg[,3]),sex==0)
```

Ignoring the offset gives erroneous results – we are just looking at claim numbers and ignoring over what period they were incurred:

```
trial_wrong = glm(sum_claims ~ age + sex, family = poisson,data=data_agg)
summary(trial_wrong)$coefficients[,1:2]   #highly misleading!!!

              Estimate   Std. Error
(Intercept)  4.13365930 0.0421440528
age         -0.01084447 0.0008851704
sex          1.40390122 0.0270450876
```

Example 27.3 Below we run the simple Normal regression model to fit the data, as described by Equation 27.1, using both the `glm` and `lm` functions (see Chapter 5). The reader may wish to plot these over the underlying rates and compare Figure 27.1.

```
glm_mod_norm = glm(claims ~ age + sex, family = gaussian, data = claimdata1)
summary(glm_mod_norm)$coefficients[,1:2]   #summary of results (not full)

              Estimate    Std. Error
(Intercept)  0.630306152 0.021251171
age         -0.005541584 0.000448387
sex          0.144761905 0.012684600

summary(glm_mod_norm)$aic;   summary(glm_model1)$aic   #better AIC from Poisson
```

```
[1] 36264.47
[1] 31385.27

lm(claims ~ age + sex, data = claimdata1)$coef #this is just the simple linear model

  (Intercept)            age            sex
 0.630306152 -0.005541584   0.144761905
```

The AIC values clearly indicate that the Poisson regression model is superior.

Example 27.4 Random exposure periods for a portfolio of polices – Data2

In this example, each life in the portfolio of insurance policies is exposed to random amounts of time between 0 and 12 months (see `timerandxx` below), and exhibit the same underlying age/sex-dependent claim rates as previously used. Proceeding to fit the GLM to the downloaded data; `claimsdata2<-read.csv("~/GLMData2_times.csv")`:

```
head(claimsdata2,3)

    X age_y sexx claimsxx    t_randx
1 1    20    1        0 0.6239546
2 2    21    1        1 0.9722192
3 3    22    1        1 0.6600340

glm_exp_var = glm(claimsxx~age_y+sexx,offset=log(t_randx),
family=poisson,data=claimsdata2)
summary(glm_exp_var)$coefficients    #summary of results (not full)

                Estimate    Std. Error   z value        Pr(>|z|)
(Intercept) -0.48903646 0.0237643318 -20.57859   4.269123e-94
age_y       -0.01018447 0.0005296033 -19.23038   2.061235e-82
sexx         0.30245753 0.0138258917  21.87617  4.381470e-106
```

Again, the model estimates regression coefficients which are consistent with the underlying values used to produce the data.

Exercise 27.3 *Verify that lives which have been exposed for less than 0.1 years exposure have significantly fewer total claims compared with those lives with more than 0.9 years exposure.*

Exercise 27.4 *Plot a graph of the fitted rates against the underlying rates (similar to Figure 27.1).*

27.8 Categorical and Continuous Variables

27.8.1 Problem with Continuous Variables

The reader may have noticed a problem with using age as a continuous variable, implicitly assuming that claim frequency will change exponentially with age – in our examples the claim frequency falls by approximately 1% with each additional year of age, a fairly restrictive assumption. (This approach actually does represent a reasonably accurate depiction of human mortality rates, particularly between the ages of 30 and 80 years, as discussed in Chapter 17.) Note that further flexibility is afforded by GLMs by using the log of the explanatory variable, from Equation 27.14 e.g. $\log \mu = \beta_0 + \beta_1 \log X_{1,i} + \log t$.

A more robust model is clearly needed where this constraint is not appropriate, as will generally be the case. For example, claim rates may perhaps increase over particular age ranges and then decrease over others, and claim rates will vary by post code. We need a model which predicts the relative level of these claims.

27.8.2 Categorical Variables

Much of the data analysed in this chapter contains information on the age and sex of each life. However, we have not yet discussed how these two variables fundamentally differ. Age is an example of a continuous variable, whereas sex is a categorical variable. Continuous variables are more straightforward and require little explanation; other examples include weight, blood pressure, salary, revenue, and share price. It is important to note that continuous variables can be treated as categorical factors – we shall see an example shortly.

The treatment of categorical factors requires a little more explanation. Further examples of categorical variables include the make and model of a car, the degree of a patient's medical symptoms relating to a particular illness (e.g. no symptoms/difficulty walking/most of the time spent in bed/bedridden), or post code, for which there may be hundreds of categories. Often the number of categories will be chosen by the analyst designing the investigation; other times there may be no choice. Note that categorical variables can be either numeric or non-numeric; continuous variables are always numeric.

The various categories available are referred to as "levels" e.g. for sex there are two levels. Categorical factors require one of the levels to be designated as the "base level" by the analyst/modeller. A number of indicators for each categorical variable, one less than the number of levels, is also required which take the values 0 or 1. In this way each life or entity can be fully described.

For example, we may use level of expertise (e.g. beginner/competent/expert) and height (cms) as two variables to include in a model to predict the number of points scored per game by a basketball player. We may set beginner as the base level for the expertise variable, requiring two indicators to categorise each life. For example, a competent player who is 180cm tall would be labelled (1,0,180), and an expert who is 170cm tall labelled (0,1,170). A beginner (height = 165) is neither competent or expert and would be labelled (0,0,165).

It is important to note that categorical variables should, in general, be factor type vectors in R (see Example 27.6).

Remark 27.4 Please refer to Goldburd M, et al. (2020) for a more detailed discussion of categorical and continuous variables.

The dataset analysed in this section is the same as in our first example *"GLMData1 simple.csv"*, downloaded as follows: `getdata<-read.csv("~/GLMData1_simple.csv")`.

In this example, however, we set up nine levels, each covering a 5-year range, for the age factor, setting age range 20–25 years as the baseline. The following code does this:

```
age_band1<-(age<30)*(age>=25)
age_band2<-(age<35)*(age>=30)
.......
age_band7<-(age<60)*(age>=55)
age_band8<-(age<65)*(age>=60)
```

The sex variable is also required; we set female $= 0$; male $= 1$ (this is equivalent to setting female as the baseline). Thus each level has either a value 0 or 1. For example, life 1 in the data is described as: (0,0,0,0,0,0,0,0,1) – male life aged between 20 and 25; the last life included in the data can be described as: (0,0,0,0,0,0,0,1,0) – female life aged between 60 and 65. The first eight entries represent age levels and the final value identifies the sex. The baseline category defined here is therefore female lives between the ages of 20–25 years. However the following code is somewhat quicker; for this we need the cut function from the dplyr package, which automatically creates a factor object:

```
age_cat_fac<-cut(getdata$age,seq(19,64,5)) #easiest - already a factor
levels(age_cat_fac)   #confirm the levels of the factor object

[1] "(19,24]" "(24,29]" "(29,34]" "(34,39]" "(39,44]" "(44,49]" "(49,54]"
[8] "(54,59]" "(59,64]"
```

Note that the glm function takes, by default, the first level as the baseline, in this case age band 19–24 years. A 31 year-old female life would be identified as (0,1,0,0,0,0,0,0,0). Adding our new variable to the data:

```
getdata[,5]<-age_cat_fac;  colnames(getdata)[5]<-c("age_cat_fac");  head(getdata)

  X age sex claims age_cat_fac
1 1  20   1      0     (19,24]
2 2  21   1      1     (19,24]
3 3  22   1      0     (19,24]
4 4  23   1      0     (19,24]
5 5  24   1      1     (19,24]
6 6  25   1      0     (24,29]
```

Now run the glm function (note that all the data is held in the "getdata" dataframe):

```
#Note: default - missing 19-24
trial3 = glm(claims ~ age_cat_fac + sex, family = poisson, data=getdata)
summary(trial3)$coefficients

                     Estimate Std. Error     z value      Pr(>|z|)
(Intercept)        -0.7048602 0.03562855 -19.783577 4.123733e-87
age_cat_fac(24,29] -0.0936356 0.04091984  -2.288269 2.212185e-02
age_cat_fac(29,34] -0.0745378 0.04071685  -1.830638 6.715463e-02
age_cat_fac(34,39] -0.1513613 0.04155118  -3.642767 2.697228e-04
age_cat_fac(39,44] -0.2376133 0.04254552  -5.584918 2.338102e-08
age_cat_fac(44,49] -0.2863528 0.04313507  -6.638514 3.168604e-11
age_cat_fac(49,54] -0.3286814 0.04366433  -7.527458 5.173762e-14
age_cat_fac(54,59] -0.3810256 0.04434121  -8.593035 8.470149e-18
age_cat_fac(59,64] -0.4644326 0.06302622  -7.368879 1.720683e-13
sex                 0.3052889 0.02704350  11.288811 1.490389e-29
```

Choosing the base level

The above code uses, as the default, the first category as the base level (or intercept); however, care should be taking in simply using the default. In general the base level should be chosen which contains a relatively significant amount of data (often the most populated category is defined as the base level).

In this example the default happens to be suitable – indeed, most claims are incurred by younger ages in the dataset. The data simulated actually has less data in the last age category; we re-run the above calculations using the `relevel` function to define a new baseline:

```
getdata$age_cat_fac<-relevel(getdata$age_cat_fac, ref = 9)
trial4 = glm(claims ~ age_cat_fac + sex, family = poisson,data=getdata) #key
summary(trial4)$coefficients

                     Estimate Std. Error    z value     Pr(>|z|)
(Intercept)       -1.16929283 0.06037867 -19.365993 1.494459e-83
age_cat_fac(19,24] 0.46443258 0.06302622   7.368879 1.720683e-13
age_cat_fac(24,29] 0.37079697 0.06364450   5.826065 5.674967e-09
age_cat_fac(29,34] 0.38989478 0.06351418   6.138704 8.319735e-10
age_cat_fac(34,39] 0.31307130 0.06405225   4.887749 1.019954e-06
age_cat_fac(39,44] 0.22681932 0.06470171   3.505616 4.555523e-04
age_cat_fac(44,49] 0.17807977 0.06509089   2.735863 6.221690e-03
age_cat_fac(49,54] 0.13575115 0.06544283   2.074347 3.804707e-02
age_cat_fac(54,59] 0.08340701 0.06589638   1.265730 2.056098e-01
sex                0.30528892 0.02704350  11.288811 1.490389e-29
```

By choosing a base level with less data we obtain regression coefficients which appear to have more uncertainty; we can in fact have more confidence in these values. The reader should note (and verify) that the regression coefficients values obtained under both models are identical.

Returning to our first model (with baseline 20–24 years) we can easily get the modelled rates for each life in our data; for example, for the 1st male life, aged 20 years, using $(\text{exp}(\text{predict}(\text{trial3})[1])) = 0.6706075$ – this is the modelled claim rate for a male life in the 20–24 year age range.

There is the potential for material discontinuities between ages at each range boundary. Care is therefore required when choosing ranges for the categories; the same rate may now be used for all ages in the same range, which may lead to some degree of selection, for example, 20-year-olds may view the premiums as cheap, and 24-year-olds as expensive, potentially resulting in selling lots of loss-making policies to ages at the lower age of each range. Clearly, larger age ranges should be adopted with care. A major issue with categorical factors is often choosing the number and size of ranges to use.

So we have a model using categorical variables:

$$\mu_i = e^{-0.7048602-0.0936356X_{2,i}-0.0745378X_{3,i}+.....-0.4644326X_{9,i}+0.3052889X_{10,i}} \tag{27.15}$$

We list all the rates from Equation 27.15 below (also plotted in Figure 27.2, with male rates: o; female rates: •):

```
expclaims3<-function(x,sex){zxzx<-exp(summary(trial3)$coef[1]
+summary(trial3)$coef[x]*(if(x==1){0}else{1})+summary(trial3)$coef[10]*sex)
return(zxzx)}

rate_cat<-matrix(0,9,2)
for (i in 1:2) {
for(j in 1:9){rate_cat[j,i]<-expclaims3(j,i-1)}}
colnames(rate_cat)<-c("female","male")
(rate_cat<-cbind(seq(20,60,by=5),rate_cat)) #5 year age bands

            female      male
 [1,] 20 0.4941776 0.6706075
 [2,] 25 0.4500053 0.6106649
 [3,] 30 0.4586820 0.6224393
 [4,] 35 0.4247640 0.5764120
 [5,] 40 0.3896628 0.5287791
 [6,] 45 0.3711262 0.5036246
 [7,] 50 0.3557448 0.4827517
 [8,] 55 0.3376026 0.4581325
 [9,] 60 0.3105865 0.4214712
```

A decision would now be required whether we should use the continuous or the categorical variables (or indeed whether different categorical variables should be tried). Further analysis is beyond the scope of this chapter.

Exercise 27.5 *The regression above used nine 5-year age ranges. Re-calculate the GLM with (a) forty-five 1-year ranges (b) three 15-year ranges. Plot the rates obtained on Figure 27.2.*

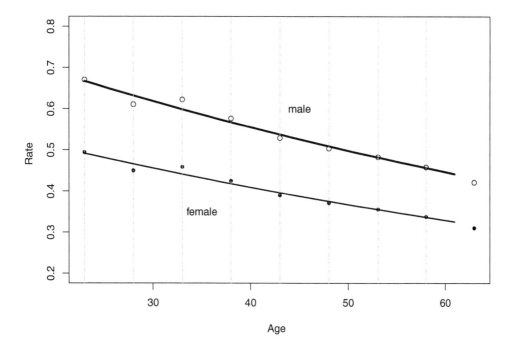

Figure 27.2 Comparing rates using categorical factors with a single function.

27.9   Interaction between Variables

Perhaps a policyholder's age has a different impact on the claims experience of males to that of females. We can model such effects by incorporating an interaction term. This important subject of interactions between variables is also discussed in Chapters 5 and 21.

Example 27.5

The objective is to fit a GLM which includes interaction terms to model claims data. The function `rate_inter` (see below) calculates the fictitious underlying rates which reflects the characteristic described above. These underlying rates (together with the fitted model) are shown in Figure 27.3. Claims data simulated from `rate_inter` is included in the file *"GLMData3 intact.csv"*. The rates are such that male rates at younger ages are higher than female rates, but lower at older ages.

By using an interaction term, analyse this possible effect contained within this simulated data.

The underlying rates used to simulate the data in this example are as follows:

```
rate_inter<-function(age,sex) {aa<- exp((if(sex==1){-0.6-0.01*age}
else{-0.2-0.02*age})); return(aa)}
```

The following code can be used to analyse relevant interactions; note that `age*sex` is equivalent to `age+sex+age*sex`.

```
glm_model = glm(claims ~ age * sex, family = poisson, data = ourdata)
```

Downloading the data for this example: `claimsdata3<-read.csv("~/GLMData3_in tact.csv")`, and proceeding to fit the model, first without an interaction term, then with this term:

```
####   1 -  simple model without interaction term
trial_exprand_int0 = glm(claimsxx ~ sexx + age_coly ,offset = log(time_randxx),
family = poisson,data=claimsdata3)

summary(trial_exprand_int0)$coefficients    #summary of results

              Estimate    Std. Error     z value       Pr(>|z|)
(Intercept) -0.374941139 0.0263177441 -14.2467051   4.699359e-46
sexx         -0.009451324 0.0155646355  -0.6072307   5.436978e-01
age_coly     -0.015357916 0.0006068762 -25.3065060  2.708694e-141

summary(trial_exprand_int0)$deviance; summary(trial_exprand_int0)$aic

[1] 54661.43
[1] 84864.35

######  2 - now with interaction term
trial_exprand_int2 = glm(claimsxx ~ sexx * age_coly ,offset = log(time_randxx),
family = poisson,data=claimsdata3)
```

```
summary(trial_exprand_int2)$coefficients    #summary of results

                 Estimate    Std. Error    z value       Pr(>|z|)
(Intercept)    -0.17220120  0.0350597931  -4.911643   9.031615e-07
sexx           -0.41897053  0.0503942986  -8.313848   9.264794e-17
age_coly       -0.02055732  0.0008633327 -23.811588   2.533353e-125
sexx:age_coly   0.01038551  0.0012153461   8.545309   1.281926e-17

summary(trial_exprand_int2)$deviance;    summary(trial_exprand_int2)$aic

[1] 54588.31
[1] 84793.23
```

Visually from Figure 27.3, the "interaction model" provides a good fit; the simpler model without the interaction term may be considered to be less appropriate. These observations are confirmed from the AIC and deviance values, both of which are lower for the interaction model, and we may decide to reject the hypothesis that the interaction term should not be included in the model. By not including appropriate interaction terms in the model, sub-optimal models may be adopted.

The decision as to which model to use will depend on the purpose of the investigation, with further analysis required before deciding on a final model.

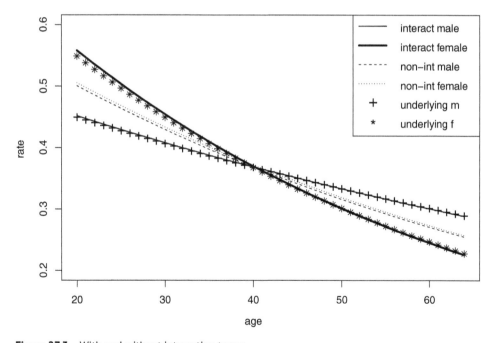

Figure 27.3 With and without interaction terms.

27.10 Over-dispersion

The issues relating to over-dispersion are not analysed in any of the datasets in this chapter. Over-dispersion relates to the phenomenon where the variance of the data exceeds that of the mean; this can be a product of heterogeneity existing in data we assumed was homogenous, and/or dependence between policies. Indeed the existence of such a characteristic is highly probable given the likelihood of heterogeneity and dependence in real policy data. This would imply that a Poisson distribution, and therefore our Poisson regression model, may not be entirely appropriate for scenarios where this characteristic is more present. Adoption of the Poisson distribution in such cases is likely to result in underestimating the probability of large total claims being incurred; underestimating the likelihood of insolvency events is clearly undesirable. To address this the overdispersed Poisson or Negative Binomial distributions are generally used in place of the Poisson. See McCullagh and Nelder (1989) for alternative approaches.

An example of over-dispersion was seen in Chapter 25.

27.11 Miscellaneous Exercises

Exercise 27.6 *Using "GLMData4 pcode.csv", determine a GLM for the data using three variables: age, sex and postcode (where postcode can be 1, 2, 3, or 4).*

The code used to simulate the claims data is as follows:

```
rate_glm_postcode<-function(age,sex,codex){aa<-exp(-0.5+(if(sex==1){0.3}else{0})-
0.01*age+if(codex==2){0.05}else if(codex==3){0.1}else if(codex==4){0.35}else{0});
return(aa)}
```

The following is a possible model obtained from an analysis of the data from Exercise 27.6:

$$E(Y|X) = \mu_i = e^{-0.49+0.3X_{1,i}-0.01X_{2,i}+0.04X_{3,i}+0.09X_{4,i}+0.35X_{5,i}} \tag{27.16}$$

This is consistent with the formula used to simulate the data.

Exercise 27.7 *Re-run the examples and exercises in this chapter but with significantly larger, and smaller datasets. The code is included on the website for such simulations. The reader should, of course, vary the values of the regression coefficients.*

Exercise 27.8 *Re-run the model in Exercise 27.1 with an interation term i.e.* age * sex, *and comment on the results.*

Exercise 27.9 *Simulate data similar to that in Exercise 27.6, but include a relatively high claim rate for a chosen age group for example, under 25 years of age. Fit a GLM to the data you have simulated.*

Exercise 27.10 *Simulate claim data for 10,000 policyholders incorporating variables for five occupations, from three countries, as well as age, sex, and number of past claims, choosing* *your own parameter values. Fit the model using Poisson regression and compare the estimated regression coefficients.*

27.12 Further Study / Next Steps

This chapter has provided an introduction to the important topic of Generalised Linear Models, and in particular Poisson regression. The interested reader may wish to make the following topics a priority for further study in this area:

- the study of variance functions (i.e. how the variance of the dependent variable varies with its mean) and "non-Poisson GLMs" for example, Gamma distributions, Negative Binomial and inverse Gaussian,
- all our examples modelled claim frequencies; this should be extended to model total claim amounts by introducing distributions in respect of amounts per claim made e.g. Compound Poisson distributions,
- it is also recommended that the reader considers studying two alternative GLMs commonly used by actuaries: logistic, and probit models.

27.13 Recommended Reading

- Crawley, MJ. (2013). *The R Book*, 2e. Chichester, West Sussex, UK: John Wiley and Sons.
- Dobson, A.J. and Barnett, A.G. (2018). *An Introduction to Generalized Linear Models*, 4e. Chapman and Hall/CRC.
- Everitt, B.S. and Hothorn, T. (2006). *A Handbook of Statistical Analyses using R*. Boca Raton, Florida, US: Chapman and Hall/CRC.
- Goldburd, M., Khare, A., Tevet, D., and Guller, D. (2020). *Generalised Linear Models for Insurance Rating*. 2e. Casualty Actuarial Society. CAS Monograph Series Number 5 Available at. www.casact.org.
- Kaas, R., Goovaerts, M., Dhaene, J., and Denuit, M. (2008). *Modern Actuarial Risk Theory*. Berlin Heidelberg: Springer-Verlag.
- McCullagh, P. and Nelder, J.A. (1989). *Generalized Linear Models*. 2e. Volume 37 of Monographs on Statistics and Applied Probability. Chapman and Hall/CRC.
- Nelder, J.A. and Wedderburn, R.W.M. (1972). *Generalized linear models. Journal of the Royal Statistical Society*: Series A (General) 135: 370–384.
- Seeber, G.U.H. (2005). Poisson regression. In: *Encyclopedia of Biostatistics* (ed. P. Armitage and T. Colton). John Wiley and Sons (online).

28

Extreme Value Theory

Peter McQuire

```
library(evir)
library(stats4)
library(tseries)   #only for exercise
```

28.1   Introduction

Risk actuaries and analysts allocate much of their time to assessing the likelihood of particular events occurring which could result in severe financial consequences to an institution, and then implementing procedures to manage these potentially catastrophic risks. Extreme Value Theory ("EVT") is a branch of mathematics which can help with analysing these risks.

Risks can broadly be assessed in two ways – qualitatively and quantitatively. This is particularly true of extreme risks. In situations where there is little available data, a quantitative analysis is unlikely to yield results which are statistically credible, thus requiring a qualitative approach to be adopted. Development of quantitative models to analyse such extreme, infrequent events is clearly problematic. For example we may want to assess the risk of bank insolvency resulting from internal fraud; as there is relatively little data on such events a qualitative approach is likely to be required.

However, when we have sufficient historic data at our disposal, a quantitative approach may be possible; the total volume of data may be such that, even for extreme events, we have an adequate amount of data. For example:

- an equity manager may wish to estimate maximum daily losses that could reasonably be expected from the equity portfolios she is managing; the example included later in the chapter requires us to estimate the probability that an equity fund may suffer a significant fall in value in one day;
- an insurance company will be required to estimate the distribution of future total claims which may be incurred next year. This includes both what is most likely to happen, and also the probability and size of catastrophic events which may result in insolvency (which sit in the tail of the distribution).

R Programming for Actuarial Science, First Edition. Peter McQuire and Alfred Kume.
© 2024 John Wiley & Sons Ltd. Published 2024 by John Wiley & Sons Ltd.
Companion Website: www.wiley.com/go/rprogramming.com.

Extreme Value Theory is a quantitative approach which can help us assess the likelihood of extreme, potentially catastrophic events.

Remark 28.1 The results obtained throughout this chapter are presented in a variety of graphical forms, from comparing cumulative distribution plots with empirical plots, QQ-plots, and to plots showing only the tails of distributions. The reader should experiment with various plots to determine which communicate the results most effectively.

28.2 Why Use EVT?

Why should we use this additional theory – can we not simply use one of the many standard distributions which we are already familiar with (e.g. Normal, Gamma, Lognormal, Cauchy, Pareto etc.)?

With respect to the claims example above, we may have determined a best-fit distribution for the claims data, with this single distribution used to model all claims data from the smallest to the largest claims. Fitting techniques will be weighted by the parts of the distribution where there is most data – the chosen distribution is likely to fit the data particularly well around these expected values. Critically, however, it may be a poor fit where there is little data e.g. for extreme events which occur rarely (see Figure 28.5 (right) and the related discussion). Thus estimates regarding the probability of insolvency may be made based on our chosen model which is appropriate in the normal course of events, but fits poorly when dangerous events occur e.g. severe flooding, global equity markets crashing, pandemics. This is a recipe for disaster.

Thus it may well be the case that no single statistical distribution can accurately model the real world in all scenarios (the everyday and the exceptional).

This is where EVT can play its part – the analysis looks only at extreme data and filters out other data, thus fitting a distribution only to the extreme data, and aiming to model this as accurately as possible. For risk managers this is key – to accurately assess the likelihood of these extreme, unlikely events happening which may result in the insolvency of companies, entire industries, and significant harm to global economies.

There are broadly two groups of models which focus on analysing extreme values: the "block maxima" approach, and the "threshold exceedances" approach. In our discussions which follow we will consider only a threshold exceedances approach – the Generalised Pareto Distribution or "GPD" (often referred to as "Points over Threshold"). The GPD models are widely considered to be a more powerful and useful approach to analysing extreme events in practice. The reader should of course review the block maxima group of EVT models.

28.3 Generalised Pareto Distribution – "GPD"

This method requires a threshold amount, u, to be set, below which all data is removed from our analysis, leaving only data which is considered to be "extreme." Given the data, z, we consider only the data $z - u$ (conditional on $z > u$) for our analysis.

The key idea of the GPD approach to analysing extreme values is that the conditional distribution of $z - u$, irrespective of the underlying distribution of the full unconditional data, converges to a Generalised Pareto Distribution as we increase u. Thus GPD can be applied to extreme data without concerning ourselves with the distribution of the underlying data. Clearly therefore, when applying this methodology, we must be careful that the conditional data we do use can reasonably be considered to be extreme i.e. the value of u used must be sufficiently large such that GPD is appropriate. Thus using a large value of u is preferable; however this comes at a cost of reducing the quantity of data in our analysis.

The GPD cumulative distribution function, G, is central to our understanding and is given by the following expression:

$$G_u(x) = 1 - (1 + \frac{\xi x}{\beta})^{-\frac{1}{\xi}}$$

(28.1)

where x is the value *in excess of our chosen threshold, u*. The shape and scale parameters are given by ξ and β respectively.

Remark 28.2 Note that $G_u(x)$ is often shown in a slightly different form, using the reciprocal of the shape parameter, ξ:

$$G_u(x) = 1 - (1 + \frac{x}{\xi \beta})^{-\xi}$$

(28.2)

We shall apply the prior usage, Equation 28.1, which is consistent with the usage in R.

The pdf of the GPD distribution is as follows:

$$g_u(x) = \frac{1}{\beta}(1 + \frac{\xi x}{\beta})^{-\frac{1+\xi}{\xi}}$$

(28.3)

We will require Equation 28.3 shortly when estimating parameters for our model.

Exercise 28.1 *Show that Equation 28.3 is the pdf by taking the derivative of $G_u(x)$ with respect to x.*

A plot of $G_u(x)$ can be obtained with the following example code (shown in Figure 28.1 together with results from Example 28.2).

```
cdf_shape_8<- pgpd(seq(0,20,0.1),xi=0.8,beta=2.653)
plot(seq(0,20,0.1),cdf_shape_8)
```

Thus ξ, the shape parameter, determines the significance of the tail; the larger the value of ξ the more significant is the tail. When interpreting these formulae it is critical to understand that $G_u(x)$ is a *conditional* cumulative distribution function. For example, in the context of claim amounts, $G_u(x)$ is the probability that the claim amount in excess of u is less than x, given that the claim amount is greater than u.

Our objective is usually to determine the probability that the total claims (or whatever it is we are estimating) exceed a particular level, say y. If p is the probability that the total

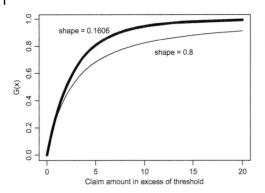

Figure 28.1 Cumulative distribution function of the Generalised Pareto Distribution.

claims amount exceeds our chosen threshold, u, then the probability that the total claims exceed y is given by: $p \times (1 - G_u(y - u))$.

Example 28.1

With a threshold, $u = 100$, the probability of incurring a claim amount less than 120, given that it exceeds 100, is calculated from $G_{100}(20)$.

Let's say we have analysed a set of data, determined the most appropriate values for the parameters ξ and β (see later in this chapter), and using Equation 28.1, calculated $G_{100}(20)$ = 60%. Therefore, the probability that a claim exceeds 120, given that it exceeds 100, is 40%. If the estimated probability that claims exceed our threshold is 5% (which may be obtained empirically or from another fitted distribution), the probability that a claim exceeds 120 is 5% x 40% = 2%. As noted above the conditional probability of 40% has been calculated using only extreme data and should therefore be more accurate.

We will proceed to analyse a set of insurance claims, determine parameter values, and estimate the claims distribution.

Example 28.2 : Insurance claims data

The following dataset represents individual property claims (in £'000s) from an insurer's property book. Downloading the data:

```
evtdata<-as.matrix(read.csv("~/GPD_evt.csv"))
```

(Note that the gpd function used below requires the data to be held in a vector object, thus it is easiest if we extract the data directly into that form.)

We need to select our extreme data by choosing a suitable threshold; here we extract data above 40. Note that the choice of threshold is somewhat arbitrary; often 5-10% of data is retained in an extreme value analysis. If we choose a threshold which is too low the data may no longer be considered extreme and the theory is less applicable; if the threshold is too high the standard errors in the parameter values will be unacceptably large. Any analysis should therefore be repeated using various threshold levels.

```
thresholdx <-40
(no_extreme<-sum(evtdata>thresholdx ))        #data points exceeding threshold

[1] 100

(no_data<-length(evtdata))                    #data points in total

[1] 2000
```

We will use the gpd function in the `evir` package to calculate the MLEs of the parameters, ξ and β. Note that we are not required to specifically extract the extreme data – the function does this for us.

```
gpd_model <- gpd(evtdata, thresholdx )
```

We can obtain all the relevant output using `gpd_model` (we have not printed the full output here as it is quite extensive). The parameter values and standard errors obtained were:

```
gpd_model$par.ests

      xi       beta
0.1606419 2.6528743

gpd_model$par.ses

      xi       beta
0.1371132 0.4480582
```

In summary, the original data consisted of 2000 points; with a chosen threshold of 40 our EVT analysis was carried out on the 100 largest data points. The MLE parameter values determined were $\xi = 0.1606419$ and $\beta = 2.6528743$. It should be noted that the uncertainty of the ξ parameter value is large and should be allowed for in any further analysis which the insurer may undertake. This is mainly as a result of using only 100 data points in the extreme value analysis.

Exercise 28.2 *Verify the parameter values obtained from the* gpd *function by writing your own function using Equation 28.3, together with the* mle *function.*

We can obtain a plot of $G_{40}(x)$ using the pgpd function (shown in bold in Figure 28.1):

```
data_for_plot<- pgpd(seq(0,20,0.1),gpd_model$par.ests[1],0,gpd_model$par.ests[2])
```

Thus the probability of a claim exceeding 45, given it is greater than 40, is given by 1-$G_{40}(5)$:

```
(cond_prob1<-1-pgpd(5,gpd_model$par.ests[1],0,gpd_model$par.ests[2]))#pgpd is the G(x)

      xi
0.1927286
```

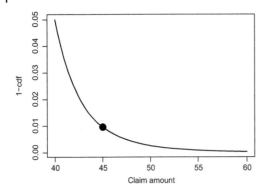

Figure 28.2 Probability of exceeding claim amount (in the tail).

As we know that only 5% of the data exceeded the threshold of 40, the probability that our claim exceeds 45 can be estimated from: $0.05 \times 0.1927286 = 0.0096364$. That is, the estimated probability that a claim will exceed 45 is 0.96%.

The plot of similar calculations over a range of values is shown in Figure 28.2 – with this plot we have reached our objective (the point calculated above is highlighted in the plot).

Exercise 28.3 *Calculate the 99.9%VaR. Repeat this calculation using $\xi = 0.3$ and comment on the results.*

Another useful function, not specific to extreme value theory, is ecdf. This function produces the unconditional *empirical* cdf that is, the proportion of data which is less than a specified value (the plot is obtained using plot(ecdf(evtdata))). Figure 28.3 (left) shows the plot of the empirical cdf together with the fitted GPD parametric plot – our EVT model appears to be an excellent fit for the data in this case.

Exercise 28.4 *Write your own function which plots the empirical cdf.*

The "QQ-plot" is straightforward to obtain under the package (Figure 28.3 right); again the plot visually confirms a reasonable fit of the model to the extreme data:

```
qplot(evtdata, xi = gpd_model$par.ests[1], threshold = thresholdx, line = TRUE)
```

Exercise 28.5 *Repeat the above analysis using threshold values between 35 and 45, recalculating the estimated probability that claims will exceed 45 with the new parameter values. Note the effect on the standard errors of the parameter values calculated.*

Remark 28.3 Further analysis relating to the choice of threshold is beyond the scope of this book. The reader is advised to study this important topic further using the recommended reading (see McNeil, Frey, Embrechts (2005) Sec 7.2).

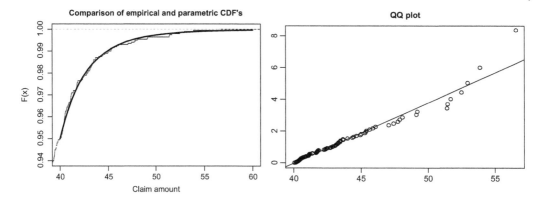

Figure 28.3 Visual checks on the fitted model.

28.4 EVT Analysis of Historic Daily Equity Market Returns (S&P 500)

In this section we apply EVT to analyse the risk of extreme equity market losses, addressing questions such as "how bad could investment returns on equity markets be tomorrow?" We will apply the same methods as those set out in Section 28.3. We then proceed to compare the model with a simple Normal distribution.

Remark 28.4 It is worth noting at this point that applying an extended EVT model may be preferable in this particular context given the dependency between daily returns e.g. GARCH-EVT models (see Christoffersen (2016) Chapter 6.8); please see Exercise 28.11 below. For our purposes we will continue to model the data using the standard GPD model.

28.4.1 Basic EVT Analysis

Uploading the equity data (a plot of which is shown in Figure 28.4):

```
evtdatainrxx<-read.csv("~/fin_crisis_GARCH2.csv")
evtdatainrx<-evtdatainrxx[,2]
```

The data contains the daily returns on the S&P500 between 2001 and 2020. In this example we are interested in large negative values i.e. investment losses; our analysis is made simpler if we multiply all the data by -1.

```
evtdatainr<- (-evtdatainrx)                    #changes losses to positives
```

Setting a loss threshold equal to 2.145% leaves us with around 4% of the data to use in our extreme value analysis. Proceeding with estimating the parameters:

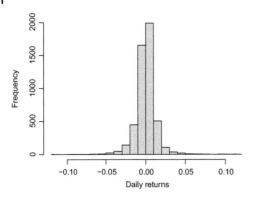

Figure 28.4 Daily equity returns.

```
threshold2<-0.02145
no_extreme2<-sum(evtdatainr>threshold2)      #data points exceeding threshold
no_data2<-length(evtdatainr)
(prop_extreme<- no_extreme2/no_data2)        #proportion of data used

[1] 0.03884662

gpd_model2 <- gpd(evtdatainr, threshold2) #the model
(shape_par_eq<-gpd_model2$par.ests[1]);       (scale_par_eq<-gpd_model2$par.ests[2])

        xi
0.2082175
       beta
0.009453003
```

Rather than use the built-in function for $G(x)$ let's write our own function (we have used the alternative format of the cdf – Equation 28.2):

```
gpdcdf<-function(xx,shape_par,scale_par){1-(1+xx/shape_par/scale_par)^-shape_par}
shape_par_eq<-shape_par_eq^-1
```

The results of the proposed GPD model, together with the data, are plotted in Figure 28.5. (Also included in the plot are the best fitting Normal and Cauchy distributions – this is discussed in Section 28.4.2.)

Our objective is to estimate the probability of incurring an extreme daily loss, however we may wish to define this. Let's say our client has a £100 m equity fund and wants to estimate the probability of losing at least 4% of its value tomorrow:

```
(gpd_cond<-gpdcdf(0.04-threshold2,shape_par_eq,scale_par_eq))   #conditional prob

        xi
0.8070559

(actualcdf4<-(1-gpd_cond)*prop_extreme)

        xi
0.007495227
```

Using this model we estimate there is a 0.75% probability that our client's fund will fall below £96 m tomorrow.

Exercise 28.6 *Verify your calculations using the* pgpd *package function.*

28.4.2 Will a Normal Distribution (and Other Alternatives) Do Just as Well?

The purpose of using Extreme Value Theory is that we believe it may model the *tails* of the data better than using one distribution for the entire data. If a simple distribution works equally well (or better) we should be using that simpler model (this is the principle of parsimony). Imagine that an initial modelling of the data has been undertaken which involved fitting various standard distributions, the conclusion of which was that the Normal distribution provided the best fit.

Exercise 28.7 *Fit a Normal distribution to the equity data and compare this with the GPD model.*

The results are plotted in Figure 28.5. Calculating the probability that losses will exceed 4% using the Normal model:

```
mean(evtdatainr);    sd(evtdatainr)
[1] -0.000279135
[1] 0.01248031

(normal_cdf4<-1-pnorm(0.04,mean(evtdatainr),sd(evtdatainr)))
[1] 0.0006245693
```

Applying a Normal distribution, we would estimate the probability of a daily loss exceeding 4% to be 0.06%, whereas the GPD model predicted a 0.75% probability, thus underestimating the true risk by more than a factor of 12 (assuming that we have verified our GPD model).

It is clear that using the best fitting Normal distribution in this case would significantly under-estimate the true level of risk. This reiterates a common theme of this book – approximations using the Normal distribution should be treated with caution.

Exercise 28.8 *Carry out further investigations of the appropriateness of a Normal distribution to model equity returns by plotting the empirical cdf together with the cdf of the best fitting Normal distribution.*

Exercise 28.9 *Try fitting a Cauchy distribution to the data, and compare with the GPD and Normal models.*

The results to Exercises 28.8 and 28.9 are included in Figure 28.5 (right).

From Exercise 28.9 and Figure 28.5 (right), it is worth noting that, although the Cauchy achieves a higher $logL$ (1.5388×10^4) than the Normal distribution (1.4806×10^4) and can be considered to be more appropriate if modelling the entire data, it is clearly less satisfactory than the Normal distribution for modelling the tails. As discussed in the Introduction, analysing the entire data set is not a successful recipe for understanding tail

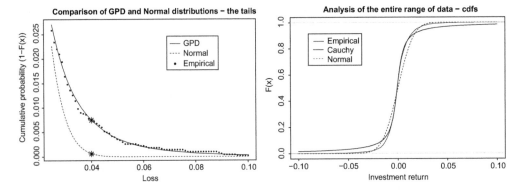

Figure 28.5 Analysis of daily equity returns.

risk – the Cauchy achieves the better *logL* by achieving a better fit in the main body of the distribution where most of the data lies; however, it vastly over-estimates the amount of tail-risk.

From a presentational viewpoint, the two plots in Figure 28.5 should be compared. From Figure 28.5 (left) it is clear that the Normal distribution does not model the tail well; however Figure 28.5 (right) could be misleading and lead to the incorrect adoption of the Normal distribution as a reasonable model.

Exercise 28.10 *Calculate the 99%VaR using the various models discussed above.*

Remark 28.5 This analysis of equity return data comes with one major health warning – it is unlikely that the daily data is independent. (The GARCH chapter discusses this characteristic in more detail.) This can potentially be dealt with by analysing the extreme standardised data by first applying of an appropriate GARCH model.

Exercise 28.11 *(For readers familiar with GARCH models – Chapter 31.) Develop a GPD model by first applying a GARCH model to the data extracted in Section 28.4.1.*

There are several other useful functions available within the `evir` package which we encourage the reader to review.

28.5 Data for Further EVT Analysis

There are numerous data sets available in R which lend themselves to EVT analysis.

The univariate dataset comprising of 2,167 fire losses incurred in Denmark over the period 1980 to 1990 was collected at Copenhagen Reinsurance. They have been adjusted for inflation to reflect 1985 values and are expressed in millions of Danish Krone (see McNeil (1997)). The data from this study is available in R from the `evir` package:

```
data(danish) # loads data into object danish
length(danish)
```

```
[1] 2167
```

```
head(danish)
```

```
[1] 1.683748 2.093704 1.732581 1.779754 4.612006 8.725274
```

Exercise 28.12 *Carry out GPD analysis on this Danish fire data, starting with a threshold equal to 6.*

The extRemes package contains several data sets, such as damage which contains data on hurricane damage between 1926 and 1995 (not run here):

```
data(damage)
damage$Year
damage$Dam
```

28.6   Recommended Reading

- Balkema A, and de Haan, L (1974). *Residual life time at great age.* Annals of Probability, 2, 792-804 10.1214/aop/1176996548.
- Christoffersen PF. (2016). *Elements of Financial Risk Management (2nd edition).* Oxford, UK: Elsevier.
- Danielsson, J. (2011). *Financial Risk Forecasting.* Chichester, West Sussex, UK: Wiley.
- Dowd, K. (2005) *Measuring Market Risk (2nd Edition).* Chichester, West Sussex, UK: Wiley.
- Embrechts, P., Klüppelberg, C., Mikosch, T. (1997). *Modelling Extremal Events for Insurance and Finance.* Berlin Heidelberg: Springer-Verlag.
- Hull J. (2015) *Risk Management and Financial Institutions (4th edition).* Hoboken, New Jersey, US: Wiley.
- McNeil, A.J. (1997). *Estimating the Tails of Loss Severity Distributions Using Extreme Value Theory.* Dept Mathematics, ETH Zentrum, CH-8092 Zurich.
- McNeil, Frey, Embrechts. (2005) *Quantitative Risk Management* Princeton, NJ: Princeton University Press.
- Pielke, R.A. Jr. and Landsea, C.W. (1998). *Normalized hurricane damages in the United States: 1925-95.* Weather Forecast, 13(3), 621–631.
- Resnick, S. (1997). *Discussion of the Danish Data on Large Fire Insurance Losses.* ASTIN Bulletin, 27(1), 139-151. doi:10.2143/AST.27.1.563211.
- Sweeting, P. (2017) *Financial Enterprise Risk Management (2nd edition).* Cambridge, UK: Cambridge University Press.

29

Introduction to Machine Learning: k-Nearest Neighbours (kNN)

Peter McQuire

```
library(class)
library(gmodels)
```

29.1 Introduction

Machine learning is a vast, diverse, and rapidly evolving subject. The objective of this chapter is to provide an introduction to machine learning by discussing one particular example, k-nearest neighbours, or "kNN". We will look at two examples of kNN – an artificially simple one followed by a more realistic example.

Given the complexity of the subject, it is beyond the scope of this book to cover the important area of machine learning in further depth; the reader is encouraged to refer to the list of Recommended Reading at the end of the chapter.

kNN is a relatively simple machine learning tool, thus an excellent one with which to start your journey. Having said that, it remains one of the most frequently used machine learning algorithms. The application of kNN is widespread, used in the fields of investments, insurance, healthcare, and image recognition, to name but a few. kNN is a supervised machine learning algorithm (as opposed to an unsupervised machine learning algorithm, such as "k-means"). By this we mean that we start with data which is assigned to a known particular class, and develop an algorithm (which we "supervise") that can be used to allocate new data to the appropriate class.

For example, we may be interested in predicting how individuals may vote in a political election, given some information about that person e.g. age, salary. By starting with historic data containing this information together with which party they voted for historically, we may be able to develop an algorithm which is successful in predicting how particular people may vote in the future.

Other examples of situations in which kNN has been used are:

- diagnosis of medical illnesses;
- whether insurance should be provided to an individual;
- the decision to invest in particular start-up companies.

It is best understood by way of a simple example.

R Programming for Actuarial Science, First Edition. Peter McQuire and Alfred Kume.
© 2024 John Wiley & Sons Ltd. Published 2024 by John Wiley & Sons Ltd.
Companion Website: www.wiley.com/go/rprogramming.com.

29.2   Example 1 – Identifying a Fruit Type

29.2.1   Data

Let's get our data:

```
fruit <- read.csv("~/KNNpmacsimple.csv", stringsAsFactors = FALSE)
head(fruit,4)
```

```
  type radius mass
1   A   3.612  199
2   M   5.364  299
3   A   2.781  233
4   M   3.098  303
```

Our data consists of the size and mass of 100 fruits, all of which are one of two "classes" – apples ("A") or melons ("M").

29.2.2   Overview of the Process

Our task is to develop an algorithm, or model, which will help us identify whether a fruit is either an apple or a melon, given information only about its mass and size.

To do this we start with data consisting of the type of fruit (i.e. apple or melon), and the size and mass of each fruit. We then develop a kNN algorithm such that when we are presented with this information we can determine, with a certain level of confidence, whether it is an apple or a melon. If our proposed model correctly predicts all the fruit in the initial data sample, we may be confident in our new model; however if, say, 10% of predictions the model makes are incorrect we may want to develop our algorithm further or choose an alternative type of model. Once we are satisfied our model meets certain criteria we may proceed to use it on unknown fruit.

We first split the data into a "training data" set and a "test data" set – here we have allocated 80 fruit to the training data and 20 to the test data.

Using kNN terminology, our objective is to determine which "class" the subject is in given a number of "covariates". The classes in this example are apple and melon. The covariates are the radius and mass.

The choice of the relative sizes of the training and test data is somewhat arbitrary – it's generally regarded sensible to have around 80% in the training data.

29.2.3   How does the kNN Algorithm Work?

The distance from each fruit in the testing data to each fruit in the training data is measured. For each of the 20 fruit in the testing data the function records the k nearest fruit out of the 80 fruit in the training data eg A-M-A-A-A. Thus 1,600 measurements are made in this example. For each test fruit the function returns its most popular class of fruit. We will initially set $k = 5$.

For example, see Figure 29.1, left. The last test data point (data number 100) is shown in large point size, together with all 80 training points. All 80 distances from this single test data point are calculated, from which the 5 shortest distances are identified. In this case all 5 nearest fruit are apples resulting in the model specifying this fruit as an apple.

The classes within the data will often be words, as is the case here; these need to be held as integers. Thus we first need to replace the "class" (or type) data with a factor-type object by using the factor function:

```
fruit$type <- factor(fruit$type, levels = c("A", "M"),
labels = c("apple", "melon"))
```

This over-writes the 1st column in the data. Although the data continues to be shown as the fruit, the underlying data is now held as factors – it is important to note that the data held has been changed.

```
head(fruit,4)

   type radius mass
1 apple  3.612  199
2 melon  5.364  299
3 apple  2.781  233
4 melon  3.098  303
```

It is straightforward to run the kNN calculations using the knn function from the class package. The kNN function requires as inputs the training set covariates, the test set covariates, and the class of fruit in the training data set.

Collating each of these three groups of data (and also the class of fruit in the test data set, which is required later):

```
fruit_all <- as.data.frame(fruit[,2:3]) #covariates
fruit_train <- fruit_all[1:80, ] #covariates
fruit_test <- fruit_all[81:100, ] #covariates
fruit_train_class <- fruit[1:80, 1] #class
fruit_test_class <- fruit[81:100, 1] #class
```

The kNN function returns the predicted class for each of the 20 fruit in the test data:

```
est_fruit<- knn(train = fruit_train, test = fruit_test,
cl = fruit_train_class, k = 5)
est_fruit

 [1] apple melon apple melon apple melon apple melon apple melon apple melon
[13] apple apple apple apple melon apple melon apple
Levels: apple melon
```

This is the output from the model.

Exercise 29.1 *Compare the estimated fruit type with the actual fruit type in the test data, noting any discrepancies.*

From Exercise 29.1, the only discrepancy is the 14th fruit in the test data, the melon circled in Figure 29.1 (right), which the kNN algorithm has returned as an apple. This melon is particularly light.

```
fruit[94,]

        type radius mass
94 melon    4.5   250
```

Running the CrossTable function compares the actual fruit type, *x*, with the predicted fruit type, *y* (the full results table is not shown here):

```
results_1<-CrossTable(x = fruit_test_class, y = est_fruit,
prop.chisq = FALSE)
```

The key results can be obtained from:

```
results_1$t

        y
x        apple melon
   apple    11     0
   melon     1     8
```

The table tells us the following:

- Of the 11 apples in the testing data, the model correctly identified all of them as apples.
- Of the 9 melons in the testing data, the model identified 8 as melons and 1 as an apple.

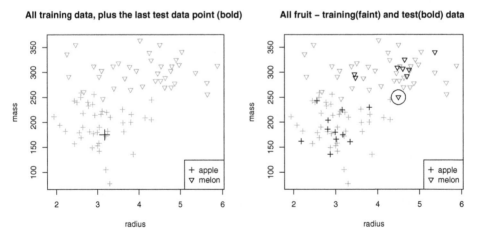

Figure 29.1   kNN analysis.

Whether this is sufficiently accurate is a matter of judgement, and depends on the consequences of particular errors in the model output.

29.2.4 Normalising Our Data

There is, however, a significant problem with the above method – the mass values (y-axis) are significantly larger than the size values. This is simply a consequence of the units used. It may have surprised the reader that fruit 14 was identified by the model as an apple, given its proximity to several melons in Figure 29.1 (right). However, when the distances between the data points were calculated, all that really mattered was the mass of each fruit – we may almost measure only the distance along the y-axis (see Figure 29.2, left, where both axes have the same scale).

Therefore we need to standardise the values in some way. There are various methods which can be used to achieve this, among the most common are:

- `normalise` which compares the distance of a data point from the minimum value with the entire range of values.
- `scale` which uses the number of standard deviations away from the mean
- using quantiles of the data based on a particular distribution

We will use the "normalise" method in this chapter (the normalised data is plotted in Figure 29.2, right).

```
normalize <- function(x) {
return ((x - min(x)) / (max(x) - min(x)))
}
fruit_all_norm <- as.data.frame(lapply(fruit[,2:3], normalize))
fruit_train_norm <- fruit_all_norm[1:80, ]
```

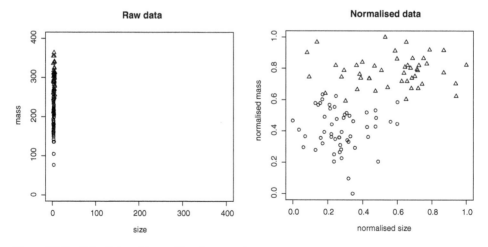

Figure 29.2 Potential problem with using raw data.

```
fruit_test_norm <- fruit_all_norm[81:100, ]
head(fruit_test_norm,4)

        radius       mass
81  0.4720954  0.5331010
82  0.7133435  0.7944251
83  0.3084729  0.5156794
84  0.6872146  0.8641115
```

Exercise 29.2 *Run the* kNN *function as above on the normalised data. (Only the results table is shown below.)*

```
          y
x       apple melon
  apple    11     0
  melon     0     9
```

Now equal weight is given to the size and mass of the fruit. So when the mass is the dominant covariate (as is the case with the raw data) the lightest melon looks more like an apple, but when both sizes and masses are used appropriately (as is the case with the normalised data) it looks more like a melon. Fruit 14 was a light melon, but it was quite large in size.

Exercise 29.3 *Carry out the calculations using alternative methods of normalising the data, and compare the results.*

29.2.5 Varying *k*

We can also vary the number of k-nearest neighbours. Let's try $k = 1$:

```
p_normalised_k1<- knn(train = fruit_train_norm, test = fruit_test_norm,
cl = fruit_train_class, k = 1)
table1<-CrossTable(x = fruit_test_class, y = p_normalised_k1,
prop.chisq = FALSE)
```

```
          y
x       apple melon
  apple    11     0
  melon     1     8
```

It appears that using $k = 5$ may be superior to $k = 1$. The reader may wish to check this visually from Figure 29.1.

29.2.6 Using Our Model

We can now apply our model; say we have two fruit with normalised parameters $(0.7, 0.8)$, and $(0.3, 0.4)$ – we would expect these to be a melon and an apple respectively:

```
fruit1 <- knn(train = fruit_train_norm, test = c(.7,.8),
cl = fruit_train_class, k = 5)
fruit2 <- knn(train = fruit_train_norm, test = c(.3,.4),
cl = fruit_train_class, k = 5)
fruit1;  fruit2

[1] melon
Levels: apple melon
[1] apple
Levels: apple melon
```

The model predicts the unknown fruit to be (1)a melon and (2)an apple.

We defer discussion of the output from the kNN algorithm for now; first we need to understand how best to analyse the results.

29.3 Analysis of Our Model – the Confusion Matrix

We must assess the quality of the proposed model. How many times did the model predict the outcome correctly? What level of accuracy is required given the potential consequences of a particular situation? We would be unlikely to accept a model which produces a high number of false predictions. In the case of identifying patients with life threatening illnesses, as in the next section, wrongly predicting that a patient is disease-free is clearly extremely undesirable (incorrectly giving the "all clear"); this is know as a "false negative". Also misdiagnosing a patient with an illness (a "false positive") is undesirable as this may result in unnecessary treatment, e.g. surgery, or chemotherapy which could have major side effects.

Thus several measures of model performance have been introduced within machine learning; we will cover a few of them below.

To illustrate this we will use a simple example where we are interested in developing a model to remove spam email from employees inboxes; we return to the cancer diagnosis model below. A model has been built from which the following results were obtained:

	pred work	pred spam
actual work	850	50
actual spam	25	75

We received 1,000 emails in total, 100 of which turned out to be spam. The model would have resulted in 50 of our work emails being deleted (bad news!) and keeping 25 of the 100 spams (a bit annoying but better than nothing).

Four categories are defined: TP (true positive), TN(true negative), FP(false positive) and FN (false negative). From the above results: TP = 75, TN= 850, FP= 50, and FN= 25.

The following are some of the most commonly used measures to asses the suitability of a machine learning model (the results from the above table are also shown):

- Recall (or sensitivity) – of all the actual spam emails how many did the model predict correctly?

$$\frac{TP}{TP+FN} = 75\%$$

- Precision – of all the spam predicted how many were actually spam?

$$\frac{TP}{TP+FP} = 60\%$$

- Specificity – of all the actual work emails how many were correctly identified?

$$\frac{TN}{TN+FP} = 94.4\%$$

- Accuracy – what proportion of emails were correctly identified?

$$\frac{TP+TN}{all\ entries} = 92.5\%$$

- Error – what proportion was incorrectly identified?

$$\frac{FP+FN}{all\ entries} = 7.5\%$$

A possible conclusion from this model analysis may be that incorrectly identifying 5.6% of work emails as spam is unacceptable – we may require a higher specificity figure. We may be willing to accept a poorer spam identification rate if would could remove any risk of deleting actual work emails.

We now look at an example which could have a important practical application.

29.4   Example 2 – Cancer Diagnoses

In our second example we look at a more typical set of data relating to the diagnosis of potentially cancerous tumours (WDBC dataset). Instead of trying to predict the type of fruit based on two covariates (size and mass), our objective now is to predict whether a patient has a cancerous tumour based on 31 covariates relating to the tumour.

At the time of writing the data was available at the following address:

```
https://archive.ics.uci.edu/ml/datasets/Breast+Cancer+
Wisconsin+%28Diagnostic%29
```

The data file is called wdbc.data; the data file is also saved in the book's website.

The aim is to identify which tumours are likely to be cancerous (malignant) based on a number of covariates. The classes here are malignant and benign. Preparing the data ready for kNN modelling:

```
cancer_raw <- read.csv("~/wdbc1.data", stringsAsFactors = FALSE,header = FALSE)
```

```
number_patients<-dim(cancer_raw)[1]
cancer<-cancer_raw[,-1]   #removing id
dim(cancer) #we have data on 569 patients with 31 covariates

[1] 569   31

cancer[,1] <- factor(cancer[,1], levels = c("B", "M"),
labels = c("benign", "malignant"))#replace diagnosis column data with data
as a factor
cancer_norm <- as.data.frame(lapply(cancer[,2:31], normalize))
```

Exercise 29.4 *Using the methods set out earlier in the chapter, develop the kNN algorithm, by using the* kNN *and* CrossTable *functions, and produce the kNN table of results using the following:*

- *Use a testing data set of the last 100 patients (and 469 in the training set).*
- *Select an appropriate value for k, say k = 21.*

The results from Exercise 29.4 are as follows:

```
table_2

$t
            y
x              benign malignant
  benign        77        0
  malignant      2       21

$prop.row
            y
x              benign    malignant
  benign     1.00000000 0.00000000
  malignant  0.08695652 0.91304348

$prop.col
            y
x              benign    malignant
  benign     0.97468354 0.00000000
  malignant  0.02531646 1.00000000

$prop.tbl
            y
x            benign malignant
  benign      0.77    0.00
  malignant   0.02    0.21
```

The key conclusions from the results are as follows:

- 91.3 % of malignant tumours were correctly identified ("sensitivity").
- However, 8.7% of patients with cancer were incorrectly given the "all clear". This is a major concern with our current algorithm.
- No patients were incorrectly diagnosed with malignant tumours which may have resulted in unnecessary treatment.
- The overall accuracy was 98% (77 benign and 21 malignant cases were correctly identified out of 100, with 2 incorrect).

We can obtain these measurements using the `confusionMatrix` function from the `caret` package (not run here):

```
library(caret);  library(e1071)
confusionMatrix(p,cancer_test_class,positive="malignant")
```

29.5 Conclusion

This chapter has aimed to provide a flavour of the application of machine learning. It is likely that, in the modern world, it will be important to develop a knowledge and understanding of a range of such techniques.

Machine learning methods which the reader may wish to explore next, following studying this chapter, include:

- decision trees
- neural networks
- dimension reduction techniques
- clustering methods
- regression methods

29.6 Recommended Reading

- Aggarwal, C.C. (2015). *Data mining : the textbook.* Boston, Mass. : Springer US.
- Dua, D. and Graff, C. (2019). *UCI Machine Learning Repository* [http://archive.ics.uci.edu/ml]. Irvine, CA: University of California, School of Information and Computer Science.
- Hastie, T, Tibshirani, R, and Friedman, J (2009). *The Elements of Statistical Learning: Data Mining, Inference, and Prediction.* Second Edition New York: Springer NY.
- Lantz, B. (2015). *Machine Learning with R,* 2e. Birmingham, UK: Packt Publishing.
- Wisconsin Diagnostic Breast Cancer (WDBC) dataset obtained by the University of Wisconsin Hospital.

Sources:

1. Dr. William H. Wolberg, General Surgery Dept. University of Wisconsin, Clinical Sciences Center Madison, WI 53792
2. W. Nick Street and Olvi L. Mangasarian, Computer Sciences Dept. University of Wisconsin, 1210 West Dayton St., Madison, WI 53706

30

Time Series Modelling in R

Dr Alfred Kume

```
library(fGarch)
library(forecast)
library(tseries)
library(latex2exp)
```

30.1   Introduction

Time series data are generally considered as a set of observations ordered in time, typically at equally spaced intervals such as days, hours, minutes, and months. The main aim of the time series analysis is to understand the driving dynamics of the data generating process and use this for forecasting future values.

The key departure from the standard statistical theory is that the data do not satisfy the independent and identically distributed (i.i.d.) assumption. This generates some technical challenges in the modelling process that are found difficult by some students. Based on our experience by teaching this material, once some of these difficulties are overcome, the modelling is then considerably simplified to the level of a regression modelling. In this chapter, we will go through some of these challenging points while covering the basic theory and the relevant computations in R/Rstudio. This joint approach will provide further insight into some potential issues of working with real time series data.

We normally fit to the real (time series) data some specific *stochastic processes*, defined as a collection of infinitely many random variables y_i, indexed by the discrete time points $i = 0, \pm 1, \pm 2, \cdots$. The time series observations are then considered as a finite sequence (of length T say), of single observations from the finite set of random variables $y_1, y_2, \cdots, y_T$. Therefore, any plot of time series observations displayed *should be considered as a single path of realisations from this set, where the i-th observation is a single realisation from the random variable y_i*. From these T data points, we need to make inference on the corresponding set of T random variables. This is only possible if some efficient parametrisation is imposed. In order to achieve this, the time series models are constructed in an incremental fashion,

R Programming for Actuarial Science, First Edition. Peter McQuire and Alfred Kume.
© 2024 John Wiley & Sons Ltd. Published 2024 by John Wiley & Sons Ltd.
Companion Website: www.wiley.com/go/rprogramming.com.

starting from the simplest model like that of the White Noise. This process is performed by exploring linear dependence among observations.

30.2 Linear Regression Versus Autoregressive Model

The main difference between linear regression and basic time series models is based on the way the dependence among observations is addressed. We will illustrate this based on the data in Figure 30.1. This data can be loaded as:

```
PoundvsEuro<-read.csv("BOE_Euro_Pound_2010_20.csv")[,2]
```

The Figure 30.1 reports some analysis of this time series data. In particular, if one is to adopt a simple linear regression model here (see the fitted line in the first plot) they could initially conclude that Sterling's exchange rate is in the long term increasing. Using this linear trend for predicting the future is not realistic however, as the regression line is way above the last 500 observations. Also if we are to believe this trend we could conclude that sometime in the past the exchange rate could have been negative! *Note that the t-values associated to the intercept and slope parameter in regression model are 387.25924 and 10.64407 (highly significant).*

This is certainly not the right model though, as the exchange rates evolve with a more complicated dynamics than that of a simple regression line. For example, the observations in the range of time points 0-1000 and 2000-2500 (see the left plot of Figure 30.1), are below the corresponding fitted line and therefore there is some local time dependence which is

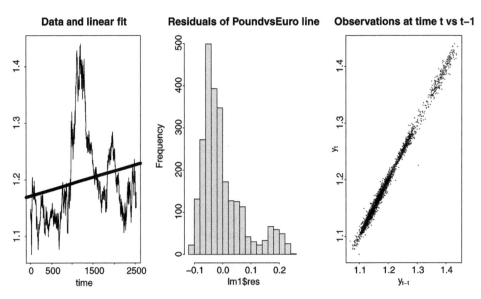

Figure 30.1 Daily exchange rate for Pound versus Euro with linear regression fitted (first), histogram of residuals (middle) and lag 1 dependence (third).

not captured by the linear model. More specifically, a simple diagnostic check of this linear regression model will fail see e.g. the middle plot of residuals in Figure 30.1.

Exercise 30.1 *Perform a standard diagnostic check for the fitted regression model.*

One sensible alternative is to explore the time dependence of subsequent observation values. For example the plot of values of y_t vs y_{t-1} (shown in the third plot of Figure 30.1) suggests presence of a strong correlation.

Exercise 30.2 *Perform a similar analysis to the above for the exchange rate data of Pound versus Dollar:*

```
PoundvsDollar<-read.csv("BOE_Dollar_pound_2010_2020.csv")[,2]
```

It is clear now that such a strong dependence between subsequent values will have to be taken into account in the modelling strategy. A natural and simple form of such dependence could be explained by a linear relationship between y_t and y_{t-1} such as

$$y_t = a_0 + a_1 y_{t-1} + \varepsilon_t. \tag{30.1}$$

This equation represents the standard parametrisation of AR(1), the Auto-Regressive model of order 1 (see later). Alternatively

$$(y_t - \mu) = a_1(y_{t-1} - \mu) + \varepsilon_t \quad a_0 = \mu(1 - a_1).$$

This model assumes a linear dependence of subsequent observations after the common mean μ is removed. Note that ε_t can be regarded as zero-mean error components for varying t. One could consider more complicated models for such dependence structure either by taking higher time (lag) dependency like a linear function of involving lag two as in AR(2) model

$$(y_t - \mu) = a_1(y_{t-1} - \mu) + a_2(y_{t-2} - \mu) + \varepsilon_t$$

or even taking non-linear functions such as $(y_t - \mu)^2 = \alpha(y_{t-1} - \mu)^2 + \varepsilon_t$. The basic models in time series analysis consider similar linear (auto)regressive approach among the observations with some lag difference. These are sufficiently easy to implement while special care in their parametrisation is required (see later).

30.3 Three Components for Time Series Modelling

In Figure 30.2 we show the collection of $T = 140$ observations displayed using the command ts.plot in R. We could load and plot this data set and its logs y (as in the code below) to produce Figure 30.2:

```
Airline_data <-read.csv("Airline_data.csv")[,2];par(mfrow=c(1,2));
y=log(Airline_data);#We apply the log transform to stabilise for variance.
ts.plot(Airline_data, main="Original Airline data", ylab="");
ts.plot(y,main="log of Airline data ",ylab="")
```

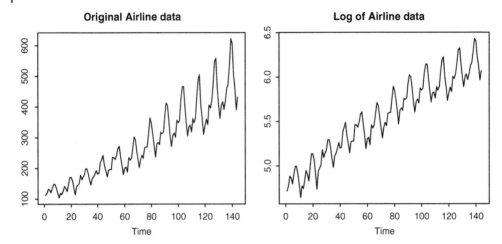

Figure 30.2 Plots of the original data and their log.

The data represent the monthly passenger numbers for some Airline company and is used in various books and R packages for illustrative purposes. We carry out a detailed analysis of this particular dataset later in Section 30.7 at page 545. We adopt the log transformation here as a special case of the Box-Cox transformations seen in the Chapter 5. In a simple visual inspection, each plot in Figure 30.2 shows some periodic/seasonal behaviour together with some upward going trend. These components, together with some additional stochastic, or irregular, terms are addressed either separately or jointly in the time series theory, depending on the level of expertise required. The basic set of components essential in describing the main features in time series data are as follows:

- **trend** A deterministic function of time
- **seasonal** A deterministic periodic function of some known period.
- **irregular-random** A collection of random time-dependent quantities.

They can be assumed to be collectively contributing to the time series in additive fashion as:

$$y_t = T_t + S_t + I_t \tag{30.2}$$

An example of such a decomposition could be:

$$y_t = \underbrace{2 + 0.5t}_{T_t:\ \text{linear trend}} + \underbrace{2\sin(t\pi/3.5)}_{S_t:\ \text{period 7 amplitude 2}} + \underbrace{0.75I_{t-1} + \varepsilon_t}_{I_t:\ AR(1)}$$

In Section 30.7.2 we introduce an alternative model which combines the seasonal and the irregular terms in a multiplicative fashion. Decompositions of the form (30.2) could be performed in an ad-hoc way in R by simply using the command decompose (see output of `plot(decompose(ts(y , frequency = 12)))` in Figure 30.16 in page 545. *Note that the period needs to be pre-specified initially.* We set it to 12 in this example.

30.4 Stationarity

In general, after an appropriate transformation of the original data such as the decomposition above (or other transformations which we will see later), the irregular part, I_t, needs to be studied further. In many situations it is assumed that the irregular part satisfies the weak or second-order **stationary** assumptions. Intuitively, stationarity is present in the time series data if the key statistical properties – the mean, variance, and pairwise dependence of data points – are invariant of the time window of observations. The formal definition is based on the reference to the stochastic process such that each component y_t satisfies:

$$E(y_t) = \mu, \text{ for all t} \tag{30.3}$$

$$var(y_t) = E[(y_t - \mu)^2] = \sigma^2 < \infty, \text{ for all t} \tag{30.4}$$

$$cov(y_t, y_{t-s}) = E[(y_t - \mu)(y_{t-s} - \mu)] = \gamma_s, \text{ for all t, s} \tag{30.5}$$

with μ, σ^2 and γ_s constant parameters. Clearly, $\gamma_0 = \sigma^2$ and $\gamma_s = \gamma_{-s}$ and γ_s: **auto-covariance** for lag s is a discrete symmetric function defined only at $s = 0, \pm 1, \pm 2 \cdots$. The rescaled version of γ_s called the **Auto-correlation function** (**ACF**) is defined as:

$$\rho_s = \frac{\gamma_s}{\gamma_0}$$

ρ_s takes values in $[-1, 1]$ and is essential in describing the main properties of the time series processes. We will use this function for identifying certain models. Suppose that we observed some $y_1, y_2, \cdots y_T$ data values from a stationary process, the corresponding sample estimates of μ, γ_s and ρ_s are:

$$\hat{\mu} = \bar{y} = \frac{1}{T} \sum_{t=1}^{T} y_t \quad \hat{\gamma}_s = \frac{1}{T} \sum_{t=s+1}^{T} (y_t - \bar{y})(y_{t-s} - \bar{y})$$

and

$$\hat{\rho}_s = \frac{\hat{\gamma}_s}{\hat{\gamma}_0} = \frac{\sum_{t=s+1}^{T} (y_t - \bar{y})(y_{t-s} - \bar{y})}{\sum_{t=1}^{T} (y_t - \bar{y})^2} \tag{30.6}$$

Note that even though the observations y_i are NOT assumed to be independent from each other, their sample mean and variance estimates can still be used.

In Figure 30.3 we display four time series realisations, three of which, loosely, illustrate scenarios where stationary may be violated. In the top left plot (1), the data indicates similar patterns across the whole observations' window hence likely to be coming from some stationary model; in the top right plot (2), the variance however seems to change from $t = 1000$; in the bottom left plot (3), the variance seems to be constant but the mean is changing following a linear trend; in the bottom right plot (4), we notice too much variation to assume a common mean.

The simplest model which is then used to construct more advanced ones is that of **White Noise** process ε_t. It is defined as a zero mean process, that is, $E(\varepsilon_t) = 0$ such that it does not have any apparent correlation among the data points:

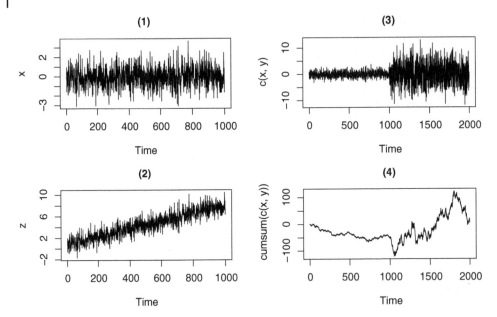

Figure 30.3 Some time series examples with different visual patterns.

$$\gamma_s = E(\varepsilon_t \varepsilon_{t-s}) = \begin{cases} \sigma^2 & s = 0 \\ 0 & s \neq 0 \end{cases} \Rightarrow \rho_s = \begin{cases} 1 & s = 0 \\ 0 & s \neq 0 \end{cases}$$

A realisation from White Noise is the plot (1) in Figure 30.3 in page 528. A more general set of models is constructed as linear combination of some lagged White Noise terms up to lag q and/or a linear combination of some lagged y_t terms up to lag p. These are the well known $ARMA(p,q)$, **Auto-regressive and Moving-Average models** of orders p and q:

$$y_t = a_0 + a_1 y_{t-1} + a_2 y_{t-2} + \cdots + a_p y_{t-p} + \varepsilon_t + \beta_1 \varepsilon_{t-1} + \beta_2 \varepsilon_{t-2} + \cdots + \beta_q \varepsilon_{t-q}.$$

The parameters a_i and β_i are estimated during the model fitting process based on the given data. These models are very flexible for practical data modelling. If either $q = 0$ or $p = 0$ these models reduce to the pure Auto-regerssive AR(p)=ARMA(p,0) and Moving-Average MA(q)=ARMA(0,q) processes respectively, while White Noise is simply ARMA(0,0).

Whilst stationarity is always guaranteed for a pure MA processes, that is not the case for those with Autoregressive parts unless some additional constraints on the parameters $a_1, a_2, \cdots a_p$ are imposed. We will now illustrate the type of parametric constraints required and their implications in the parameter estimation.

For example, for the AR(1) case: $y_t = a_0 + a_1 y_{t-1} + \varepsilon_t$ we require $-1 < a_1 < 1$ for the stationarity condition to be met; otherwise the overall variance of the process will explode to infinity. In order to see this based on the sample paths of such processes, we construct initially a simple function for generating some data from AR(1) model 30.1.

```
ar1simulate=function(a0=0,a1=0.5, n=100,sigma=1){y=1:(n+1);y[1]=a0
for (i in 2:length(y)) y[i]=a1*y[i-1]+rnorm(1,sd=sigma);return(y[-1])}
```

Note that we will use the `arima.sim` function later in the chapter to perform simulations for more general models. We now plot four different time series paths for various values of a_1 and show them in Figure 30.4 each with the same `set.seed`.

The first two examples related to stationary AR(1) indicate that observations have some mean reversion property (in this example the mean is 0); each subsequent data point tends to be "pulled back" or regresses to the mean, such that the data varies around some common value; hence the stationarity holds. Also note the greater range of values with the larger a_1 value. However, in the last two (non-stationary) series that is not the case as the observations are varying too much to be considered as stable. In fact, for the $a_1 = 1.1$ case, the observations are exploding as their magnitude is increasing exponentially. It is clear from the model equation the closer the value of $|a_1|$ is to 0 the less each subsequent data point is influenced by the previous term and the lower the overall variability of the data.

Example 30.1 [Sample path variability] Figure 30.4 shows a single realisation path from four different time series models. For each of these models one can generate a set of five realisation paths and plot them jointly in the same plot:

```
set.seed(12115);ts.plot(ar1simulate(n=100,a1=0),ylab="")
for (i in 1:4) lines(ar1simulate(n=100,a1=0))
```

and re-run with our other choices for a_1 (see Figure 30.5).

What can you notice? Are the paths from the first two models which are stationary more alike to each other (and hard to distinguish) as opposed to the non-stationary ones?

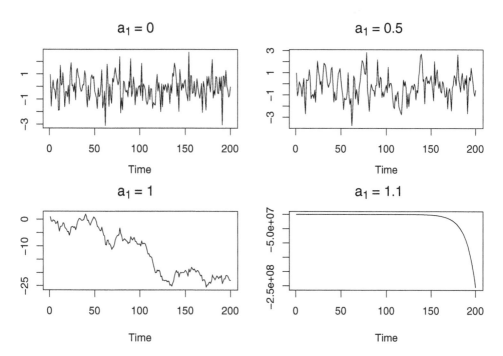

Figure 30.4 Realisations of AR1 series with different a_1 values.

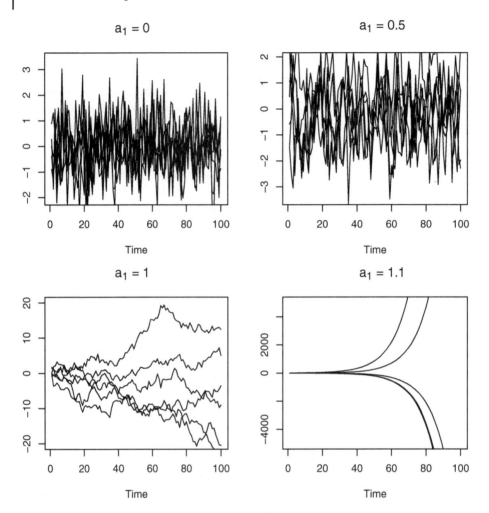

Figure 30.5 A few more runs of AR1 series with different a_1 values.

Although we can't see each path separately, for the first two models which are stationary, we can see that each path is constrained to within a range of values, that is, it regresses back to 0, and tend to never get too far from 0. For these models the parameter estimation is possible. The reader may wish to separately run, plot, and observe additional individual runs.

Each of the five paths of the last two models which are non-stationary (sampled as in the bottom row of Figure 30.5), are very much different from each other. For example, the bottom right paths are varying so much and reaching the infinity exponentially fast, even with varying signs! The bottom left ones however, belong to a particular model called **Random Walk**, since $a_1 = 1$. In this case the process is called **Integrated** such that differencing of observations as $\Delta y_t = y_t - y_{t-1}$ will result to a stationary case (see the output in Figure 30.6 where the command `diff` is used in the following for differencing).

```
par(mfrow=c(1,2),cex.main=2);set.seed(12115);
ts.plot(ar1simulate(n=100,a1=1),main="Random Walk paths",ylim=c(-15,15),ylab="")
for (i in 1:5) lines(ar1simulate(n=100,a1=1),ylab="")
set.seed(12115);ts.plot(diff(ar1simulate(n=100,a1=1)),main="The differenced paths",
ylab="")
for (i in 1:5)lines(diff(ar1simulate(n=100,a1=1)),ylab="")
```

In general, the non-stationary AR model $y_t = a_0 + y_{t-1} + \varepsilon_t$ is called **Random Walk with Drift** and if the starting value y_0 is given one could see that:

$$y_t = \underbrace{y_0 + ta_0}_{Trend} + \sum_{i=1}^{t} \varepsilon_i \qquad (30.7)$$

If $a_0 = 0$ then the slope term above disappears and the process is simply the ordinary **Random Walk without Drift**. Equation 30.7 confirms that the variance of y_t is now $t\sigma^2$ (increasing with t, see that confirmed in the left plot of Figure 30.6, for $a_0 = 0 = y_0$). However, $\Delta y_t = y_t - y_{t-1} = a_0 + \varepsilon_t$ is stationary (see paths after differencing at Figure 30.6). Note that despite the linear drift term of (30.7), the ordinary linear regression between y_t and time t is not appropriate since the error structure there has non constant variance $\sum_{i=1}^{t} \varepsilon_i$ (see the Exercise 30.4 below for a discussion regarding regression modelling of the trend). In fact for such models the three assumptions of stationarity are violated.

In general, one can difference the data more than once, say d times as $\Delta\Delta \cdots \Delta y_t = \Delta^d y_t$, but in practice $d = 1$ or $d = 2$. Note that a process y_t is called $ARIMA(p, d, q)$ **of orders (p,d,q)** if $\Delta^d y_t$ is stationary $ARMA(p, q)$:

$$\Delta^d y_t = a_0 + a_1\Delta^d y_{t-1} + a_2\Delta^d y_{t-2} + \cdots \Delta^d y_{t-p} + \varepsilon_t + \beta_1\varepsilon_{t-1} + \beta_2\varepsilon_{t-2} + \cdots + \beta_q\varepsilon_{t-q}$$

Exercise 30.3 *Derive theoretical expressions for the mean and variance of y_t AR(1) model with $-1 < a_1 < 1$. Run simulations of an AR(1) model and compare the sample mean and sample variance of the output with the theoretical values.*

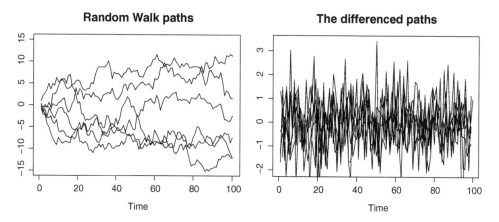

Figure 30.6 Some Random Walk paths and their differences.

An example of an AR(1) model used in finance is Vasicek's interest rate model.

Exercise 30.4 *Comment on the following lines of code and convince yourselves as to why the linear regression is not appropriate for estimating the trend term as in Equation 30.7 with $y_0 = 10$ and $a_0 = 2$.*

```
par(mfrow=c(1,2)); set.seed(1234); RW=10+cumsum(rnorm(100,sd=3));
WNDrift=RW+2*(1:100);ts.plot(RW);ts.plot(WNDrift);abline(a=10,b=2);
tt=1:100; abline(lm( WNDrift~ tt),lty=2)
```

30.5 Main Tools in R for ARIMA Modelling

The key commands that you will need to know for modelling time series data are:

- `acf`, `pacf` which calculate these functions from the data (see the definition of pacf later.)
- `arma.acf` which calculates the theoretical ACF and PACF functions.
- `arima.sim`, `arima` which simulate and fit the arima models. Understanding their use and output is very important.
- `tsdisplay` is another function (of the package `forecast`) which could be used for displaying the time series data.
- `auto.arima` this function needs to be carefully implemented as it could occasionally lead to the wrong models.

The first step in modelling stationary time series data is to explore the autocorrelation structure since it provides further insight into the underlying time dependence. For example the first two plots in Figure 30.4 might seem to be visually similar with no noticeable difference. However, the corresponding plots of their ACF and PACF (shown in Figure 30.7) offer more insight as to how they differ:

```
set.seed(12115);par(mfrow=c(2,2))
d1<-ar1simulate(n=1000,a1=0);   acf(d1); pacf(d1)
d2<-ar1simulate(n=1000,a1=0.5); acf(d2); pacf(d2)
```

See the use of `acf` and `pacf` in the R code. For the White Noise data d1, the ACF and PACF, the ***partial autocorrelation function*** entries seem to differ from those of d2 (more about the theoretical justification later). In fact an initial visual inspection of similar plots is used for model building. We showed earlier how calculation of ACF and sample ACF values is performed. The values of "PACF" and their sample counterparts are simply obtained as functional derivations from those of ACF described in the following.

30.5.1 PACF as a Derivation of ACF and Their General Behaviour for ARMA(p,q) Models

We will explain briefly the rationale as to why PACF is a particular transformation of ACF values. In Table 30.1, we show the tabular form of ACF and PACF, where the functions f_s

Table 30.1 The derivation of PACF from ACF.

ACF	ρ_1	ρ_2	ρ_3	...	ρ_s
PACF	$\phi_{11} = f_1(\rho_1) = \rho_1$	$\phi_{22} = f_2(\rho_1,\rho_2) = \frac{\rho_2-\rho_1^2}{1-\rho_1^2}$	$\phi_{33} = f_3(\rho_1,\rho_2,\rho_3)$	...	$\phi_{ss} = f_s(\rho_1,\rho_2,\rho_3...\rho_s)$

are sequentially defined as the solutions for the coefficient a_s in some appropriate set of s linear equations involving $a_1, ...a_s$ and $\rho_1, \rho_2, \rho_3...\rho_s$. In particular:

$$f_1: \rho_1 = a_1 \Rightarrow a_1 = \rho_1 = f_1(\rho_1)$$

$$f_2: \begin{cases} \rho_1 = a_1 + a_2\rho_1 \\ \rho_2 = a_1\rho_1 + a_2 \end{cases} \Rightarrow a_2 = \frac{\rho_2-\rho_1^2}{1-\rho_1^2} = f_2(\rho_1,\rho_2)$$

$$f_3: \begin{cases} \rho_1 = a_1 + a_2\rho_1 + a_3\rho_2 \\ \rho_2 = a_1\rho_1 + a_2 + a_3\rho_1 \\ \rho_3 = a_1\rho_2 + a_2\rho_1 + a_3 \end{cases} \Rightarrow a_3 = f_3(\rho_1,\rho_2,\rho_3) \quad \text{and so on.}$$

These equations for each $s = 1, 2, 3, ..$, are derived from the Yule-Walker equations of the auto-covariance or auto-correlation entries for the pure AR(s) processes. If for example the true process was a pure AR(p) then it can be seen as an AR(∞) process with $0 = a_p = a_{p+1} = \cdots$, and which in turn implies that $f_{p+1}(\rho_1,\rho_2,..\rho_{p+1}) = a_{p+1} = 0$, $f_{p+2}(\rho_1,\rho_2, ..., p+2) = a_{p+2} = 0$ and so on. Therefore if the sample PACF values for some real data are close to zero after some value p then the AR(p) process is a likely model. But the ACF values will be non zero and decay exponentially to 0 for pure AR processes. Such behaviour could indicate an AR(p) process.

In a reflective fashion and as seen in many time series books, for a pure MA(q) process it can be shown that only the first q entries of the ACF: $\rho_1, \rho_2, ..., \rho_q$ are non zero while the PACF values will be non zero and decay exponentially to 0. Hence if the data exhibit a similar pattern an MA process is very likely to be a good fit.

If we have a general ARMA(p,q) process however, the behaviour of ACF and PACF functions is seen in an additive form of the corresponding pure components, see Table 30.2.

In general, the sample data from such ARMA(p,q) models also follow the same theoretical patterns as those described on Table 30.2. For example, the ACF and PACF plots in Figure 30.7 which correspond to the data shown at the top row of Figure 30.3 suggest that there are not significant spikes in the ACF-PACF plots and therefore the data seem to have a close to zero correlation structure just as the White Noise process, AR(0) does. In fact these data were indeed generated from the White Noise. Additionally the second series considered in Figure 30.7 displays a significant spike at the PACF and an exponential decay to the ACF, consistent to the AR(1) with $a_1 = 0.5$ used.

Example 30.2 The following lines of code generate two time series data; one from a AR(1) model and one from a MA(1) model:

```
set.seed(123);dat1=arima.sim(n=200,list(ar=c(0.7)));dat2=arima.sim(n=200,
list(ma=c(0.6)))
```

Table 30.2 The general behaviour of ACF and PACF for ARMA models.

Model	ACF(ρ_s)	PACF(ϕ_{ss})
AR(p)	$\rho_s \neq 0$, $\forall s$ but (exp) decaying to 0	$\phi_{ss} \neq 0$ for $s \leq p$; $\phi_{ss} = 0$ $s > p$
MA(q)	$\rho_s \neq 0$ for $s \leq q$; $\phi_{ss} = 0$ $s > q$	$\phi_{ss} \neq 0 \forall s$; exp-decaying to 0
ARMA(p,q)	$\rho_s \neq 0$, $\forall s$ but (exp) decaying to 0 for $s > q$	$\phi_{ss} \neq 0 \forall s$; but exp-decaying to 0 for $s > p$

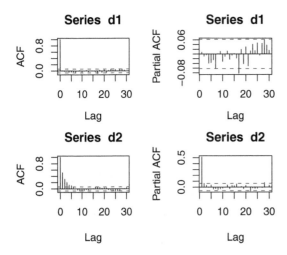

Figure 30.7 ACF and PACF plots of a White Noise and AR(1) generated using ar1simulate.

Generate for each of them the time series plots using ts.plot. Do you notice any difference among them? Explore the differences on the acf and pacf plots.

From the graphical output of this example, one can see that valid models capturing the time dependence of the data can only be inferred after exploring the ACF and PACF plots of the data and not from the original ts.plot output.

30.5.2 How to Simulate and Obtain the Theoretical Values of ACF and PACF for ARMA Models

Further insight in understanding the sample properties of ARIMA models is by simulating data points from them. The function arima.sim is easily applied for pre-specified ARIMA models, as in

```
set.seed(1239);x=arima.sim(n=200,list(ma=c(0.7,0.6)))
```

In the lines above we are asking to generate random data x from an $ARMA(0, 2)$ process with MA part of order 2 with β values c(0.7,0.6), (for simulating from more general models see later or run ?arima.sim). One can generate a set of relevant plots for this data as in Figure 30.8:

```
tsdisplay(x)
```

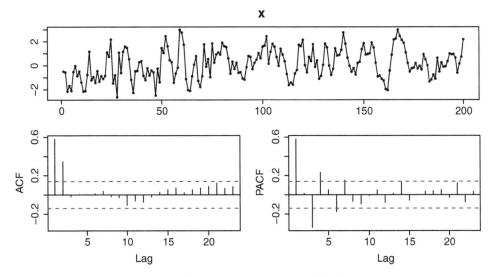

Figure 30.8   Standard output of tsdisplay, data and sample ACF/PACF plots.

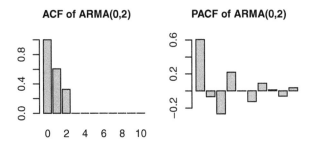

Figure 30.9   Theoretical ACF and PACF values.

Note that the function `tsdisplay` is part of the library `forecast`. The graphical output is apparently a bit more efficient then separate plots. Note that unlike that of `acf`, the output of `tsdisplay` does not show the entry for lag 0 which is always 1. The theoretical values for the simulated ARMA model are obtained using the command `ARMAacf` as in Figure 30.9.

```
par(mfrow=c(1,2))
barplot(ARMAacf(ar = c(0),ma=c(0.7,0.6), lag.max = 10, pacf = FALSE),
main="ACF of ARMA(0,2)" )
barplot(ARMAacf(ar = c(0),ma=c(0.7,0.6), lag.max = 10, pacf = TRUE),
main="PACF of ARMA(0,2)" )
```

Note the use of `barplot` command above. The sample and the theoretical functions are generally not identical but similar, due to the sampling variability from the random data. A similar comparison can be adopted for some ARMA(1,1) model with `ar=c(0.5)` and `ma=c(0.6)` (plots not shown here):

```
y=arima.sim(n=200,list(ar=c(0.7),ma=c(0.6)));tsdisplay(y)
```

and compare the ouput with the corresponding theoretical values (plots not shown here):

```
par(mfrow=c(1,2))
barplot(ARMAacf(ar = c(0.7),ma=c(0.6), lag.max = 10, pacf = FALSE),
main="ACF of ARMA(1,1)" )
barplot(ARMAacf(ar = c(0.7),ma=c(0.6), lag.max = 10, pacf = TRUE),
main="PACF of ARMA(1,1)" )
```

Exercise 30.5 *The following time series are generated from MA(1), MA(2) and ARMA(1,2) processes using the* arima.sim *command. Analyse the ACF and PACF plots of each of those data sets and check whether the theoretical features as in Table 30.2 are present.*

```
set.seed(1234);x1=arima.sim(n = 250, list( ma = c(1.279)),sd = sqrt(0.1796))
x2=arima.sim(n = 250, list(ma = c(1, .5)),sd = sqrt(0.1796));
x3=arima.sim(n = 250, list(ar = c(0.6), ma = c(.3, .5)),sd = sqrt(0.1796))
```

In this exercise, you might prefer to use the command tsdisplay.

30.6 Identifying a Set Possible Models to the Data Including the Order of Differencing

As explained in Table 30.2, sample versions of ACF and PACF of the given data could identify the possible values of p and q. Note that due to the sample variability, there could be more than one possible model identified visually from such plots such as ARMA(1,1) or ARMA(1,2) or ARMA(2,1) might produce very similar ACF-PACF plots. So we might like to consider more than one possible set of values p and q. In fact the possible orders of a possible ARIMA(p,d,q) model (including the level of differencing) is also not clearly known apriori. For determining differencing (or the unit root presence) there is a test called the Augmented Dickey-Fuller (ADF) test and is run as adf.test from the package tseries. The unusual feature of such a test is that the null hypothesis H_0 is that the model is non-stationary and needs to be differenced versus the alternative H_1 stationary or explosive. For example, a sample data from random walk is shown here in Figure 30.10:

```
set.seed(1123);x=ar1simulate(n=100,a1=1);tsdisplay(x)
```

the ADF test produces a p-value 0.6635 and hence it seems to suggest that there is not any evidence against the null, that is, non-stationary model hence differencing is a valid assumption.

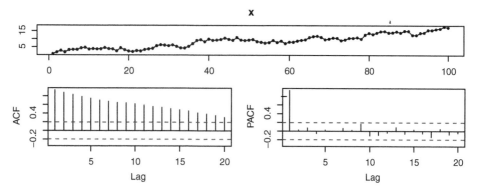

Figure 30.10 A random walk realisation.

```
adf.test(x, alternative = c("stationary", "explosive"))


Augmented Dickey-Fuller Test

data:  x
Dickey-Fuller = -3.1292, Lag order = 4, p-value = 0.1089
alternative hypothesis: stationary
```

We expect this conclusion here as the data were generated from the random walk. Theoretically, the ACF of the random walk exhibits a rather slow decay and that is why you see a linear decay in the ACF plot in Figure 30.10 while PACF could be misleading. However, the differenced data `diff(x)` are stationary as they are simply White Noise observations and `adf.test` is indeed suggesting the alternative hypothesis (see the small p-value this time):

```
adf.test(diff(x), alternative = c("stationary", "explosive"))


Augmented Dickey-Fuller Test

data:  diff(x)
Dickey-Fuller = -4.6624, Lag order = 4, p-value = 0.01
alternative hypothesis: stationary
```

Therefore the differencing order could be just $d = 1$.

Exercise 30.6 *Show that the differencing order $d = 1$ is also reasonable assumption for* PoundvsEuro *data and interpret the output of* tsdisplay(diff(PoundvsEuro)).

30.6.1 Model Fitting to Time Series Data

In general, the model identification is related to the creation of a list of possible order values of some ARIMA model candidates. These could include various values of p, d and q.

When we fit a particular ARMA(p,d,q) model in R, the computer output will display the parameter estimates for that particular fixed model. These estimates are obtained using one of the methods

- conditional least square estimates
- maximum likelihood estimates
- moment estimates (only for the pure AR processes)

By default the command `arima` uses the first method above as it is the quickest and not too different to the second (mle) one which can be sometimes more numerically challenging. We will illustrate the model fitting with the series x3 which is a realisation from ARMA(1,2) as in Exercise 30.5 on page 536. Let us forget for the moment what the true model that generated this dataset is and graphically explore its properties as in:

```
tsdisplay(x3)
```

As it can be seen from the corresponding output of Figure 30.11, there seem to be some significant spikes perhaps in the first 2,3 lags for both ACF and PACF. These suggest that a good model could be adopted for a set of combinations of p and q as

```
fit1=arima(x3,order=c(1,0,1));fit2=arima(x3,order=c(2,0,1));
fit3=arima(x3,order=c(1,0,2));fit4=arima(x3,order=c(2,0,2))
```

You can see R output for each of these models by simply calling them by their name, for example

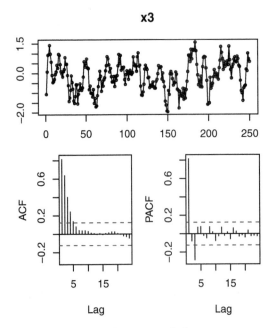

Figure 30.11 tsdisplay output of x3.

```
fit3

Call:
arima(x = x3, order = c(1, 0, 2))

Coefficients:
         ar1      ma1     ma2   intercept
      0.6415   0.2393   0.448    -0.1513
s.e.  0.0636   0.0673   0.071     0.1162

sigma^2 estimated as 0.1553:  log likelihood = -122.74,  aic = 255.49
```

Each of these four models is essentially a list containing all the relevant quantities as

```
names(fit3);

 [1] "coef"       "sigma2"     "var.coef"  "mask"     "loglik"   "aic"
 [7] "arma"       "residuals"  "call"      "series"   "code"     "n.cond"
[13] "nobs"       "model"
```

In a situation when there are a few competing models to consider, a common measure of ranking/comparing them is the AIC (Akaike Information Criterion) score. Note that the alternative BIC is also used (see the notes on regression in Chapter 5 about this). One definition of the AIC score is (or some rescaled version) is

$$AIC(p, q) = \log(\hat{\sigma}^2) + \frac{p + q + 1}{T}$$

where $\hat{\sigma}^2$ is the estimated variance of the residuals, and T is the number of observations. The value of 1 in the $p+q+1$ is present if the intercept a_0 is fitted to the model. *The smaller the AIC score the more parametrically efficient the model is.*

For example, the four models considered above produce the AIC values 283.94, 277.54, 255.49, and 257.11 respectively. The value of the third model is the lowest and therefore this is our *preferred one* out of the four considered. Note that the absolute value of AIC means nothing and the fact that model fit3 is chosen does not mean that this is a good model as this optimal choice depends entirely on the list of competing models considered. Your expertise in producing such a list is important. In fact the command auto.arima is supposed to perform an exhaustive model search but sometimes it *incorrectly* produces a model which is not necessarily the one with a lower AIC value to that of fit3 that we found earlier.

```
(autofit<-auto.arima(x3))

Series: x3
ARIMA(2,0,3) with zero mean

Coefficients:
         ar1      ar2      ma1      ma2     ma3
      0.3475   0.1685   0.5541   0.5475   0.1745
s.e.  0.6890   0.4551   0.6844   0.1895   0.3112

sigma^2 = 0.1592:  log likelihood = -123.36
AIC=258.73   AICc=259.07   BIC=279.86
```

Note in particular the magnitude of the standard errors for the parameters relative to their estimates. Hence a low t-value, suggesting over parametrisation for this model. This is reflected to a higher AIC value than that of our chosen `fit3` which is an ARMA(1,2). Therefore this function should be used with care.

30.6.2 Parameter Estimation for Pure Auto-Regressive Models

As indicated in the standard theory, the method of moments or the Yule-Walker estimator, is applicable only for the pure Auto-regressive models. Similar to `auto.arima`, in R you can fit automatically based on the AIC criterion, the pure AR models using `ar.yw`, `ar.lme` and `ar.ols` for Yule-Walker method, Ordinary Least squares and MLE respectively. Each of these three methods produce similar but different parameter estimates:

```
ar.yw(x3);ar.mle(x3);ar.ols(x3)
```

The ouput for the first line above is

```
Call:
ar.yw.default(x = x3)

Coefficients:
      1        2        3
 0.8560   0.1739  -0.2865

Order selected 3  sigma^2 estimated as  0.1672
```

See also `?ar`, for more information about these functions.

30.6.3 Diagnostic Plots

Similar to the linear regression theory, the most common goodness of fit measure for a particular model relates to the assumption that residuals are statistically close to the White Noise, namely they do not have any correlation structure. Hence, if correct, the chosen model has already captured the time dependence dynamics and residuals are just noise. This analysis is part of the standard theory but is summarised graphically in R using `tsdiag`. It produces three plots: the model residuals, their ACF, and a sequence the p-values of the Ljung-Box tests (the degrees of freedom need some adjustments) which formally test whether the residuals are indeed a realisation path from the White Noise. For example for `fit2`

```
tsdiag(fit2)
```

Standardized Residuals

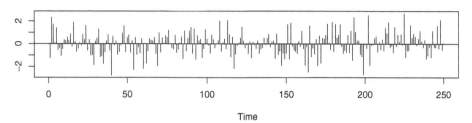

Time

ACF of Residuals

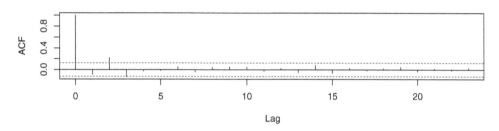

Lag

p values for Ljung–Box statistic

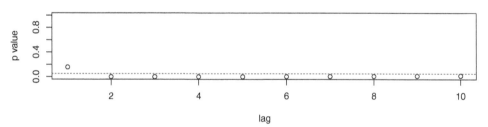

lag

Figure 30.12 tsdiag output of model fit2.

indicates that the ACF plot has some significant spike in the lag 2 and the p-values are extremely low suggesting that the White Noise hypothesis is not supported. See also the p-values in the third plot there as being below the dashed line which represent the critical regions for rejecting the (null) White Noise assumption at 5% significance level. So such diagnostics would have flagged model `fit2` as problematic. However the opposite is observed for our chosen model `fit3`. Strong evidence of White Noise is displayed supporting a good fit to this data.

```
tsdiag(fit3)
```

Standardized Residuals

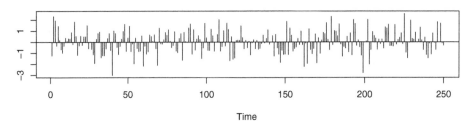

ACF of Residuals

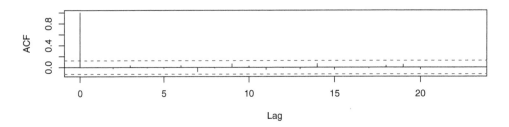

p values for Ljung–Box statistic

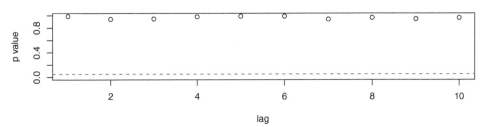

Figure 30.13 tsdiag output of model fit3.

Note however, that the same is observed for tsdiag(fit4) (plot not shown but is very similar to that of tsdiag(fit3)). Note that fit4 is inferior to our chosen model fit3 as it is over-parametrised, see the standard deviation for the ar2 parameter when calling this model in R as fit4. Be aware that, many incorrect models like fit4 could also pass such diagnostic tests!

As a conclusion, the diagnostic tools like that of tsdiag, need to be used ONLY as a diagnosis to check the adequacy of the chosen model from the AIC/BIC score (including considerations of the t-values of parameters), and NOT as a tool for model comparison.

Finally, the Box.test command also generates the individual p-values of the Ljung-Box tests shown graphically on tsdiag. For example, the second entry of the p-values plotted for fit3 can be obtained individually as

```
Box.test(residuals(fit3),lag=2,type="Ljung")

Box-Ljung test

data:  residuals(fit3)
X-squared = 0.12371, df = 2, p-value = 0.94
```

Exercise 30.7 *Show that the p-value of the Ljung-Box statistics for the residuals of* fit2 *at lag 8 is 0.01. Is the White Noise assumption for these residuals realistic?*

Exercise 30.8 *In Excercise 30.6 at page 537, we explored whether* PoundvsEuro *data required differencing. Show how can you run the command* tsdiag *for confirming that random walk is not a bad fit to this data. Explore the presence of a drift in this case.*

30.6.4 Forecasting

Once decided on the most appropriate model, we can also forecast the future values. In our case we have the chosen model fit3, for which we forecast at five time points ahead as

```
forecast(fit3, h=5)

    Point Forecast       Lo 80      Hi 80       Lo 95      Hi 95
251     0.26903595  -0.2360306  0.7741025  -0.5033967   1.041469
252     0.06600399  -0.6070258  0.7390337  -0.9633062   1.095314
253    -0.01190218  -0.8573281  0.8335237  -1.3048695   1.281065
254    -0.06187786  -0.9687774  0.8450216  -1.4488609   1.325105
255    -0.09393653  -1.0249546  0.8370815  -1.5178057   1.329933
```

where in addition to the forecast values, the output provides by default the 80% and 95% confidence intervals. In particular for the future observation at time point 254, the 95% confidence interval is (-1.44887, 1.323251). It is more appropriate to plot these quantities as

```
plot(forecast(fit3, h=15))
```

The confidence intervals of the forecast values are indicated as shaded intervals in Figure 30.14 (see ?forecast for more details). Note that the forecast values are obtained as conditional expectations of the future observations given a particular model. Hence, wrong models forecast badly. Additionally in the case of stationary models, the conditional variance of future forecast values tends to be increasing slowly towards that of the marginal variance; see the shaded area in Figure 30.14 and compare that with that of the following example.

Example 30.3 A plausible model that can be fitted to the data PoundvsEuro is that of ARIMA(0,1,0). We can fit and forecast the next 150 values ahead as

Forecasts from ARIMA(1,0,2) with non-zero mean

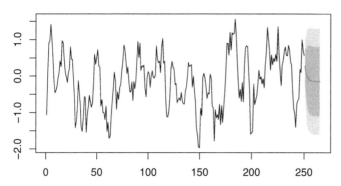

Figure 30.14   Forecasting values of model fit3 and their shaded confidence intervals.

Forecasts from ARIMA(0,1,0)

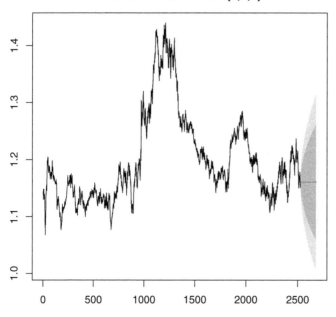

Figure 30.15   Forecasting values and shaded confidence intervals for PoundsvsEuro.

```
plot(forecast(arima(PoundvsEuro,order=c(0,1,0)),h=150))
```

What do you notice in the variance of the forecast values?

Here the shaded regions are increasing linearly since the process is not stationary, that is, $d \geq 1$ and the variance increases linearly with time.

Decomposition of additive time series

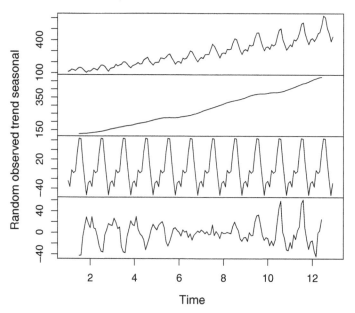

Figure 30.16 Decomposition of the airline data.

30.7 Dealing with Real Data far from Stationary

30.7.1 Non Parametric Approaches

Most of the real data are indeed far from the ideal stationary setting. The original Airline data as in Figure 30.2 is one of them. We can use again the standard command decompose to split the data into three main components: Trend, Seasonal, and Irregular.

```
x=ts(Airline_data, frequency = 12)#generates a time series structure with the
periodicity specified.
dec=decompose(x); plot(dec)
```

The output of this command saved here as dec, is a list of objects:

```
names(dec)

[1] "x"       "seasonal" "trend"    "random"   "figure"   "type"
```

If one wanted to explicitly obtain the data from this decomposition you could call these elements explicitly as: dec$random, dec$seasonal, or dec$trend. In particular the random term which looks close to stationary could be modelled further using the procedures of ARIMA fitting and forecasting as seen before. However, the irregular term still

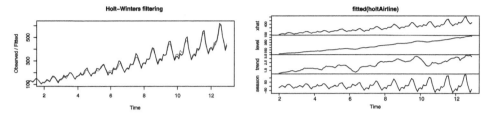

Figure 30.17 Holt-Winters algorithm output to Airline Data.

shows some seasonality behaviour and the variance is not constant as some cycles seem to show more variation than some others.

Exercise 30.9 *Generate the ts.plot of the irregular term in the decomposition for Airline data.*

In fact another ad-hoc non parametric approach to dealing with real and highly irregular data is that of *exponential smoothing* referred to as the Holt-Winters algorithm.

```
holtAirline<- HoltWinters(x);plot(holtAirline);plot(fitted(holtAirline))
```

Some tuning parameters are decided in a rather ad-hoc basis. However, in the current use the default optimal values are chosen. See ?HoltWinters for more information and a rather extreme decomposition with no trend as in plot(fitted (HoltWinters(x,alpha=1))). For a theoretical description see Chapter 9 in the *Introduction to Time Series and Forecasting*.

30.7.2 Airline Data Modelling Using Multiplicative Seasonal Models

As seen in the ad-hoc decompositions (30.16), the random parts are far from statoinary as non constant variance is observed. It is therefore more appropriate to perhaps consider the seasonal and random effects jointly in the ARIMA framework. In fact this can be possible with the help of some variance/trend stabilising transformation like the differencing or log and also modelling the seasonality in terms of the seasonal differencing for a predefined period s like

$$\Delta^s y_t = y_t - y_{t-s}.$$

In our Airline data $s = 12$ as we observe monthly dependence, but s can take any value depending on the individual data at hand. We consider here the log-transform:

```
y=log(Airline_data)
```

and generate various graphical output for seasonal differences for $s = 4$ and $s = 12$ see Figure 30.18.

In particular, these plots confirm that the ordinary difference joint with an additional seasonal difference at $s = 12$ seem to generate ACF and PACF which decay very quickly except some spikes at lags 1 and 12 for $\Delta\Delta_{12}y_t$ series. These in fact give rise to the presence

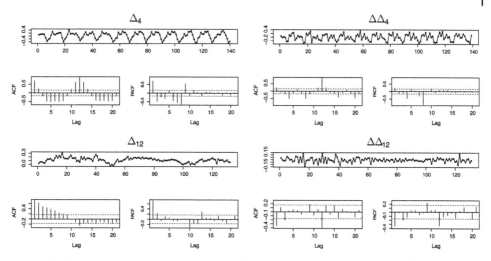

Figure 30.18 A set of graphical output of log Airline data y, for combinations of ordinary Δ and seasonal Δ_4, Δ_{12} differencing (quarterly and monthly).

of β_1 and β_{12} in the following Multiplicative Seasonal ARIMA model of orders ($p = 0, d = 1, q = 1)(P = 0, D = 1, Q = 1)_{12}$ with seasonal difference $D = 1$ and ordinary difference $d = 1$:

$$\Delta\Delta^{12}y_t = (1 + \beta_1 L)(1 + \beta_{12}L^{12})\varepsilon_t$$

where L is the backshift or lag operator. Such models are very easily fitted using arima as:

```
(fit=arima(y, order=c(0,1,1), seasonal=list(order=c(0,1,1), period=12)));tsdiag(fit)


Call:
arima(x = y, order = c(0, 1, 1), seasonal = list(order = c(0, 1, 1), period = 12))

Coefficients:
          ma1       sma1
      -0.4018    -0.5569
s.e.   0.0896     0.0731

sigma^2 estimated as 0.001348:  log likelihood = 244.7,  aic = -483.4
```

As we can see from the diagnostics plots at Figure 30.19, the fitted model seems to confirm a good fit (no significant spikes at the residual ACF). Hence the spikes of the ACF plots of the bottom right time series of Figure 30.18 for $\Delta\Delta_{12}y_t$ are indeed captured by the introduction of $\beta_1 = -0.4018$ and $\beta_{12} = -0.5569$ model parameters. We can very easy generate the forecast function values for the next 24 months using (output shown in the left plot of Figure 30.20):

```
plot(forecast(fit, h=24))
```

Standardized Residuals

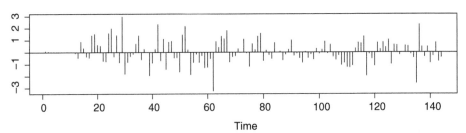

ACF of Residuals

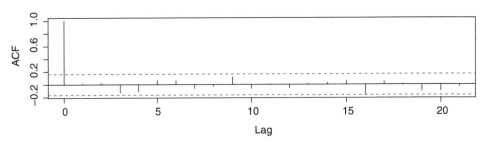

p values for Ljung–Box statistic

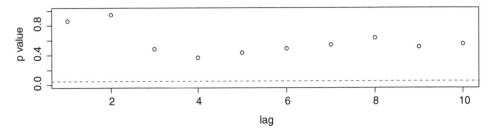

Figure 30.19   tsdiag ouput of fit.

One could also extract and plot the relevant forecast bands in the original scale by applying its inverse (the exp function) as the right plot in Figure 30.20, using this code:

```
plot(exp(c(forecast(fit, h=24)$fit,forecast(fit, h=24)$mean)),type="l",col="blue",
ylim=c(100,860))
ax=1:24+length(x);ax1=sort(ax,decreasing = TRUE);bx=exp(forecast(fit, h=24)
$upper[,1]);
bx1=exp(forecast(fit, h=24)$lower[,1])[24:1];polygon(c(ax,ax1),c(bx,bx1),lty=2)
ax=1:24+length(x);ax1=sort(ax,decreasing = TRUE);bx=exp(forecast(fit, h=24)
$upper[,2]);
bx1=exp(forecast(fit, h=24)$lower[,2])[24:1];polygon(c(ax,ax1),c(bx,bx1),lty=3)
```

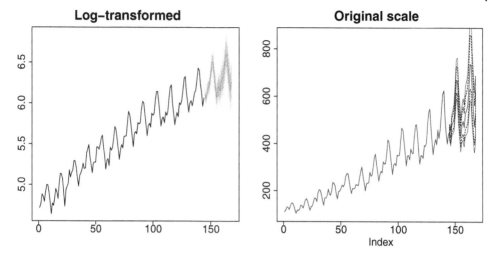

Figure 30.20 Forecast values for the next 24 months and their confidence intervals for the log-transformed and original data.

Finally, one could also consider a range of alternative models here; a plausible choice would be that which ignores the ordinary difference but introduces an AR(1) term instead, and with additional orders $(p = 1, d = 0, q = 1)(P = 0, D = 1, Q = 1)_{12}$

```
(fit1=Arima(y, order=c(1,0,1), seasonal=list(order=c(0,1,1), period=12)));

Series: y
ARIMA(1,0,1)(0,1,1)[12]

Coefficients:
          ar1       ma1      sma1
        0.9959   -0.3994   -0.5526
s.e.    0.0050    0.0898    0.0741

sigma^2 = 0.001379:  log likelihood = 245.61
AIC=-483.22    AICc=-482.91    BIC=-471.69
```

In this model the AIC score is still above that of `fit` and also the parameter `ar1` is 0.9959319 with associated standard deviation 0.0049735. This suggests that the 95% confidence interval contains the value $a_1 = 1$, and hence the ordinary difference is a valid option here as it is more parametrically efficient, that is, more *"parsimonious"* model. Not surprisingly though the output of `tsdiag(fit1)` in this case looks good even though `fit` is more appropriate as it has lower AIC value (see below).

The output of `auto.arima` in this case is far from that with orders $(p = 0, d = 1, q = 1)$ $(P = 0, D = 1, Q = 1)_{12}$. In fact it suggests a model ARIMA(3,1,3) with AIC value -289.483, which is far from the figures -483.399 and -483.223 that we obtained earlier for `fit` and `fit1` respectively. You might like to explore independently the model suggested by `auto.arima(y)`.

As you can see from the examples covered in this chapter, there is a variety of modelling expertise required when dealing with time series data. Many ready-made tools are actually available in R/Rstudio environment. Theoretical principles however largely determine the correct model choice.

30.8   Recommended Reading

- Enders, W. (2009). *Applied Econometric Time Series*, 4e. Wiley Series in Probability and Statistics.
- Brockwell, P.J., and Davis, R.A. (2016). *Introduction to Time Series and Forecasting* – 3d Edition. Springer Texts in Statistics.
- Tsay, R.S. (2014). *Multivariate Time Series Analysis: With R and Financial Applications*. Wiley.

31

Volatility Models – GARCH

Peter McQuire

```
library(moments)
library(tseries)
library(rugarch)
```

31.1   Introduction

In this chapter we will look at a special family of time series models known as GARCH models. *Generalised Auto Regressive Conditional Heteroskedastic* models have been extensively used by financial analysts since their introduction in the 1980s, particularly when studying equity market returns and movements in exchange rates.

We will study the characteristics of GARCH models first by simulating and fitting appropriate models, and comparing them with the Normal distribution. We then proceed, in the form of an exercise, to analyse US daily equity return data over the period from 2001 to 2020 (a period which included the Global Financial Crisis 2008 and the COVID-19 pandemic).

31.2   Why Use GARCH Models?

It is worth first highlighting two key aspects of typical GARCH models to provide motivation for their use in modelling equity markets. After all, many analysts simply use a Normal distribution to model equity market returns; why should we use these more advanced GARCH models?

1. GARCH models exhibit a phenomenon known as "volatility clustering". It may be that, for a period of time, it appears that the data comes from a distribution with a higher variance than is usually the case, with many relatively large values occurring over a small period of time. GARCH models exhibit the characteristic of non-constant variance (or

R Programming for Actuarial Science, First Edition. Peter McQuire and Alfred Kume.
© 2024 John Wiley & Sons Ltd. Published 2024 by John Wiley & Sons Ltd.
Companion Website: www.wiley.com/go/rprogramming.com.

"heteroskedasticity"); we tend to see periods of higher uncertainty, followed by periods of lower uncertainty.

The US equity market showed significant volatility around September 2008 during the Global Financial Crisis and again in March 2020 at the commencement of the COVID-19 pandemic (daily returns from the S&P 500 index are shown in Figure 31.1).

It appears from Figure 31.1 that the daily returns from late 2008 and early 2020 come from a different underlying distribution from other periods of time. Such markets are prime candidates for GARCH models.

2. GARCH models are particularly suited to model time series data which exhibit fat-tails. That is, when compared with a Normal distribution with the same mean and variance, the GARCH model typically provides higher probabilities of obtaining extreme values. Data which exhibits this property of fat-tails is referred to as *leptokurtik*. Equity market return data typically exhibits such leptokurtic behaviour, making GARCH models potentially suitable candidates to analyse equity markets. Figure 31.2 compares two distributions, both with a mean of 100 and standard deviation of 15. Under the Normal distribution $P(X < 50) = 0.04\%$ whereas under the alternative "fat-tailed" distribution $P(X < 50) = 0.8\%$ i.e. 20 times higher.

If the fat-tailed nature of GARCH models was the only advantage then it may be the case that we could simply use fat-tailed distributions such as a Cauchy or t-distribution, or apply Extreme Value Theory to model such behaviour; however, such a marginal distribution will not incorporate any time dependency and the subsequent concept of volatility clustering noted above, the principal reason for applying GARCH models.

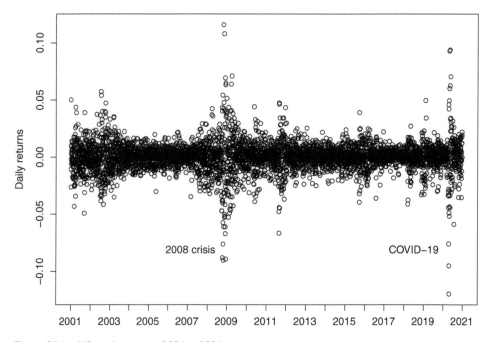

Figure 31.1 US equity returns 2001 – 2021.

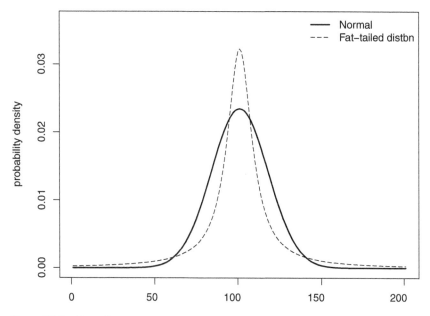

Figure 31.2 Fat tails.

31.3 Outline of the Chapter

The chapter is split into the following sections:

- Simulations from a GARCH model are run, to which a GARCH model is fitted and analysed.
- Demonstrate, partially at least, the stationarity of GARCH models by calculating the long-term variance using simulations.
- Present an exercise requiring the analysis of actual equity-market return data (using the data shown in Figure 31.1).
- Briefly discuss and compare extensions to the basic GARCH model.

Throughout this chapter we will approach our study of GARCH models in the context of modelling daily equity returns. Of course there are many other fields in which GARCH models will be helpful, for example, foreign exchange markets, derivative markets.

31.4 Key Theoretical Concepts with GARCH

As noted above, the key characteristic of GARCH models is their "heteroskedasticity" – the variance of the distribution at time t, σ_t^2, is not constant but depends on t. GARCH models help us estimate σ_t^2 such that we can, for example, estimate probabilities of significant equity market losses more accurately.

Define a random variable, X_t, to represent the innovations – we can think of these as the daily return from an equity index such as the FTSE100:

$$X_t = \sigma_t \epsilon_t \tag{31.1}$$

where ϵ_t is a random variable from a distribution with mean of 0 and variance of 1, and σ_t is the standard deviation at time t. (The most common choices for ϵ_t are Normal innovations, which we will predominantly use in this chapter, or Student-t innovations – see Exercise 31.1.) If we choose $\epsilon_t \sim N(0, 1)$ then:

$$X_t \sim N(0, \sigma_t^2) \tag{31.2}$$

The crucial piece of notation in this equation is the "t" suffix after σ^2; thus the X_t 's come from distributions with different variances. (You may find it informative to read the Appendix to this chapter which describes a distribution consisting of just three different Normal distributions, and how this relates to GARCH models.) However we are still to set out our GARCH model for σ_t^2; the model we will be using to model equity market daily returns is as follows:

$$\sigma_t^2 = \omega + \alpha X_{t-1}^2 + \beta \sigma_{t-1}^2 \tag{31.3}$$

where ω, α and β are the parameters of the model which are to be determined from the data, typically using MLE. Note that $\omega, \alpha, \beta > 0$ is required to ensure positive volatility.

Equation 31.3 is known as a GARCH(1,1) model as there is one X^2 term and one σ^2 term on the right of the equation. In general, you can include as many terms as are required to provide the most appropriate model. For example, a GARCH(2,3) model can be written as follows:

$$\sigma_t^2 = \omega + \alpha_1 X_{t-1}^2 + \beta_1 \sigma_{t-1}^2 + \alpha_2 X_{t-2}^2 + \beta_2 \sigma_{t-2}^2 + \beta_3 \sigma_{t-3}^2 \tag{31.4}$$

Derivation of GARCH – a Brief Description
As noted earlier, a good example of the application of GARCH models is in estimating equity market volatility. A sensible starting point in estimating the variance of tomorrow's return distribution may be to calculate the variance of the last 1,000 days' returns. However this places equal weights on each day's return, when we may think it is more sensible to place greater weight on more recent days, for example, yesterday – GARCH assumes a geometrically decreasing weight. The final, key step is to include an extra term, ω, in respect of the long-term variance value to which the variance should regress (see Section 31.8). When we put these ideas together we arrive at Equation 31.3.

Please see Chatfield and Xing (2019) and other recommended reading for a fuller discussion of the main theoretical concepts and derivations.

31.5   Simulation of Data Using a GARCH Model

Data from a GARCH model can therefore be simulated using Equations 31.1 – 31.3 as follows: σ_2^2 is calculated from X_1^2 and σ_1^2, X_2^2 is then simulated from a Normal distribution with variance σ_2^2; σ_3^2 is calculated from X_2^2 and σ_2^2, and so on.

However, there is one problem we must address first; we do not have a value for σ_1. This can be a significant issue and the value chosen can materially affect our analysis. One approach is simply to use the standard deviation of returns from the data. We will not concern ourselves with any further discussion of this topic here and simply choose a value of $\sigma_1 = 0.015$.

Let's see what a GARCH(1,1) simulation looks like. The parameter values chosen below are fairly typical values obtained from an equity-market analysis:

```
sims<-10000 #number of simulations
garch_sim<-numeric(sims);   zz<-numeric(sims);   sigma<-numeric(sims)

omega<-0.0000007;   alpha<-0.08;   beta<-0.91   #set the parameters

set.seed(1263);    zz<-rnorm(sims,0,1)

sigma[1]<-0.015 #our chosen starting value for sigma
garch_sim[1]<-sigma[1]*zz[1]

#now simulate our GARCH series
for(i in 2:sims){sigma[i]<-(omega+alpha*garch_sim[i-1]^2+beta*sigma[i-1]^2)^0.5
garch_sim[i]<-sigma[i]*zz[i]}
```

Figure 31.3 is a fairly typical plot from a GARCH simulation; the most striking characteristic is the "clustering" effect discussed earlier, for example, around $t = 7,000$, where many large values are simulated over a short period of time. σ_t will be particularly large around these points:

```
which.max(abs(garch_sim))      #time of the maximum value

[1] 6977

garch_sim[which.max(abs(garch_sim))]   #the value at this time

[1] 0.0629772

sigma[which.max(abs(garch_sim))];   sigma[which.max(abs(garch_sim))+1] #sd's

[1] 0.01910253
[1] 0.02549618
```

Exercise 31.1 *Analyse how the values of σ_t vary with time in this simulated data. Also, consider how reasonable it is that such a value at $t = 6977$ was obtained given σ_{6977}.*

It is worth understanding why the GARCH(1,1) equation produces these values. Eventually an unusually large value of X_t will be simulated, say 3 standard deviations from zero. This will result in a significantly larger σ_{t+1} (the extent of this depends on the α value), and this in turn will make it more likely that a large X_{t+1} will subsequently be simulated. Note

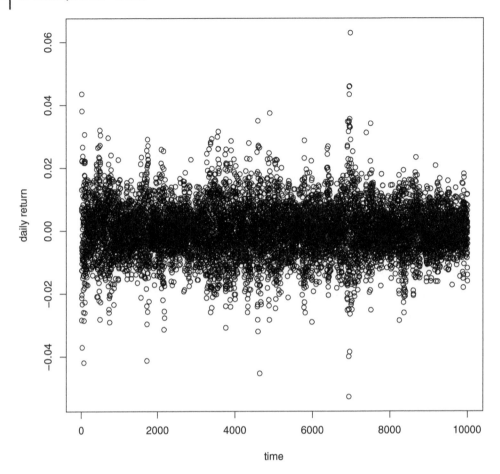

Figure 31.3 Simulated GARCH(1,1)

that the larger the value of β, the slower σ_t will change, and the more likely it is that we continue to obtain large X_t values for some time. Thus β determines how long the shocks may last.

It is also important to note that the GARCH(1,1) model is not stationary if $\alpha + \beta > 1$; the long-term variance is undefined in this case. (The reader may wish to re-run the simulations with $\beta = 0.93$.) This topic of stationarity is discussed further in Section 31.8.

We will carry out further analysis shortly, once we have fitted parameters to the data.

31.6 Fitting a GARCH Model to Data, and Analysis

31.6.1 Fitting a GARCH Model

We now proceed to fit a GARCH(1,1) model to this simulated data to check that our results are consistent with the parameters used in the simulation. (This is a fairly common method used throughout this book.) We remind the reader of the parameters used to simulate the data:

$$\omega = 7 \times 10^{-7} \quad ; \alpha = 0.08 \quad ; \beta = 0.91$$

There are several "GARCH modelling" packages in R, including `rugarch`, `bayesGARCH`, `Rmetrics`, `betategarch`, and `fGarch` which the reader may wish to explore. We will predominantly use the `garch` function from the `tseries` package to determine the parameters using maximum likelihood estimation. We will also briefly look at the `rugarch` package later in the chapter.

Assuming a Normal distribution for each of the daily returns, the likelihood function, L, of the data over the next n days is simply:

$$L = \prod_{t=1}^{n} \frac{1}{\sqrt{2\pi} \, \sigma_t} \, e^{\frac{-x_t^2}{2\sigma_t^2}} \tag{31.5}$$

and the *logL* simplifies to (ignoring constants):

$$logL = \sum_{t=1}^{n} -\log \sigma_t - \frac{x_t^2}{2\sigma_t^2} \tag{31.6}$$

To use the `garch` function we must first create a time series object by using the `ts` function:

```
garch_sim_ts<-ts(garch_sim)
```

And applying the `garch` function:

```
fit_garch<-garch(garch_sim_ts,order=c(1,1))        #try GARCH(1,1)
```

The full output is not shown here given its size (which can be obtained using `summary`), but the key parameters are:

```
fit_garch$coef

        a0            a1            b1
7.807937e-07 9.139638e-02 8.971023e-01

fit_garch$vcov

          a0            a1            b1
a0   1.360664e-14  2.106757e-10 -4.737952e-10
a1   2.106757e-10  3.405725e-05 -3.274333e-05
b1  -4.737952e-10 -3.274333e-05  3.924201e-05
```

An initial comparison confirms a reasonable agreement with the parameters used to simulate the data. We should of course try various models, not just the $GARCH(1, 1)$ model (which we know the data was simulated from). The reader may wish to try the simpler $ARCH(1)$ model, or perhaps an $ARCH(10)$ (ARCH models have $\beta = 0$):

```
fit_arch<-garch(garch_sim_ts,order=c(0,1))#try ARCH(1)
fit_arch10<-garch(garch_sim_ts,order=c(0,10)) #try ARCH(10)
```

We can use AIC to determine the best, most parsimonious model:

```
(AIClist<-c(AIC(fit_garch),AIC(fit_arch),AIC(fit_arch10)))

[1] -70522.98 -69007.46 -70299.84
```

This implies that $GARCH(1, 1)$ provides the best, most parsimonious model for the data from the three models tried (with an AIC of -70523). The reader may also wish to verify that the $GARCH(1, 1)$ model actually results in a higher $\log L$ than the ARCH(10) model, even though it uses eight fewer parameters!

31.6.2 Further Analysis of the Data; Comparison with the Normal Distribution

The histogram of the simulated data (Figure 31.4) is typical of the output from a GARCH model, and looks approximately Normally distributed. We can analyse whether the simulated data is Normally distributed by applying the Jarque-Bera test:

```
jarque.test(garch_sim)


Jarque-Bera Normality Test

data:  garch_sim
JB = 2289.9, p-value < 0.00000000000000022
alternative hypothesis: greater
```

Given the above p-value we can reject the hypothesis that the simulated data is Normally distributed (with a very high level of confidence). As we shall see shortly, the data is highly leptokurtic.

We can gain a better understanding of the data by analysing the most extreme data points. We will proceed to compare the z-scores of all such data points (i.e. how many standard deviations they are away from the mean), first using the variances from our GARCH model, and then using the constant, single variance from a fitted Normal distribution. The times of the largest simulations are:

```
extreme_times<-NULL
(extreme_times<-which(abs(garch_sim)>0.035)) #the times of the largest values

  [1]    6    9   22   69 1722 4640 4901 6938 6941 6946 6949 6958 6961 6962 6977
```

Note the clustering around $t = 6,950$. We can list the 15 most extreme values at all the above times:

```
garch_sim[extreme_times]

 [1]  0.04358406  0.03810041 -0.03705837 -0.04192210 -0.04130942 -0.04537060
 [7]  0.03744538 -0.04003580 -0.05282579  0.04580602  0.04338802  0.04595831
[13] -0.03853513  0.03558077  0.06297720
```

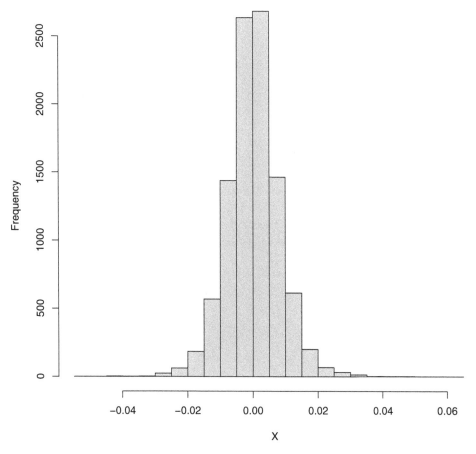

Figure 31.4   Histogram of GARCH(1,1) returns.

The following are the standard deviations of the Normal distributions from which the extreme values were simulated i.e. σ_t:

```
sigma[extreme_times]

 [1] 0.01502345 0.01859150 0.01616440 0.01066193 0.01370105 0.01213086
 [7] 0.01025799 0.01392575 0.01891801 0.02288268 0.02339325 0.02199430
[13] 0.02263545 0.02420222 0.01910253
```

The z-scores of the most extreme points are therefore:

```
(zgarch<-garch_sim[extreme_times]/sigma[extreme_times])

 [1]  2.901069  2.049346 -2.292592 -3.931944 -3.015054 -3.740097  3.650361
 [8] -2.874948 -2.792355  2.001777  1.854724  2.089556 -1.702424  1.470145
[15]  3.296798
```

If we had fitted the single, best-fitting, Normal distribution to all the data we would have the following, much larger, z-scores:

```
(znorm<-garch_sim[extreme_times]/sd(garch_sim))
 [1]   5.483434  4.793521 -4.662418 -5.274338 -5.197255 -5.708203  4.711109
 [8]  -5.037017 -6.646162  5.762985  5.458770  5.782144 -4.848214  4.476517
[15]   7.923339
```

This is the key point. By using a Normal distribution with constant volatility (σ = 0.0079483) we get z-values which are clearly inconsistent with a Normal distribution. The GARCH model copes better with the extreme events than the simple Normal model. (A quick check should convince the reader: one would expect z-values exceeding 5.0 only once every 2.6 million data points – from the data we obtained 10 such data points out of 10,000, clearly far more than we would expect from a Normal distribution.)

Thus if we were using a Normal model in this case we are very likely to underestimate the risks of large, and frequent losses. This is a good example of "model risk" as was seen during the GFC 2008; inappropriate Normal models were applied when assessing financial risk leading to an underestimation of the probability of extreme events occurring which could lead to financial disaster.

Exercise 31.2 *Carry out a Jarque-Bera test on the z-scores from the GARCH model.*

Exercise 31.3 *Draw a QQ-plot of the simulated data,* garch_sim, *against the Normal distribution.*

Finally let's calculate the kurtosis of the data – this gives a measure of the extent of the fat-tails of the distribution. The kurtosis of a $GARCH(1, 1)$ model is given by (see McNeil, Frey, Embrechts (2005)):

$$\kappa_x = \frac{3(1 - (\alpha + \beta)^2)}{(1 - (\alpha + \beta)^2 - 2\alpha^2} = 5.4 \tag{31.7}$$

Calculating the estimated kurtosis from the data:

```
kurtosis(garch_sim_ts)
[1] 5.328983
```

Given that a Normal distribution has a kurtosis of 3, our $GARCH(1, 1)$ model exhibits significant kurtosis. From the above results we can conclude that the simulated data is fat-tailed that is, it is leptokurtic; essentially many more extreme events occur than is predicted from a best fitting Normal distribution.

31.6.3 Further Analysis of the Data; Volatility Clustering

The principal aspect to analyse is that of volatility clustering, characteristic of such models. Notice that there are several short periods where the returns are significantly more uncertain (Figure 31.3). This is an important observation – with GARCH we will tend to have long periods of higher volatility for example, between $t = 6,900$ and $7,000$, and lower volatility for example, between $t = 8,000$ and $t = 10,000$.

To analyse this clustering effect statistically we can look at the autocorrelations of the squared standardised returns – if our GARCH model is a good fit then these terms should appear as white noise and thus show no correlation.

Carrying out a Ljung-Box test:

```
standard_squared<-(garch_sim/sigma)^2
Box.test(standard_squared, type = "Ljung-Box",lag = 10)


Box-Ljung test

data:  standard_squared
X-squared = 13.413, df = 10, p-value = 0.2015
```

Hence there is insufficient evidence to reject the hypothesis that the transformed data is white noise. However, if we look at the raw data squared:

```
squared<-(garch_sim)^2
Box.test(squared, type = "Ljung-Box", lag = 10)


Box-Ljung test

data:  squared
X-squared = 4126.6, df = 10, p-value < 0.00000000000000022
```

Clearly the squared values are correlated and not white noise – large, absolute values tend to cluster together. The results are perhaps most clearly seen by plotting the ACFs of (1) the squared standardised returns and (2) the squared returns.

Exercise 31.4 *Plot (1) the acf of the standardised squared returns and (2) the acf of the squared returns (shown in Figure 31.5).*

Exercise 31.5 *Based on our analysis of the above simulated data, and by running further simulations, estimate the 95% VaR of accumulated losses over the next 10 days. Compare your answer with the 95% VaR if the best fitting Normal distribution was used.*

Also estimate the probability of the initial fund value losing more than 10% of its initial value at any point over these 10 days (again using the two models).

Exercise 31.5 commentary: For example, using set.seed(100), the 95% VaR of 10-day return using the above $GARCH(1, 1)$ model is a 7.5% loss, while the 95% 10-day VaR with the underlying Normal distribution is (from theory) only a 4.1% loss. Using the Normal distribution could significantly underestimate the degree of risk in our position.

Exercise 31.6 *Effect of parameter values on GARCH models*

The reader should experiment running GARCH simulations using various parameter values. You may wish to start with the following:

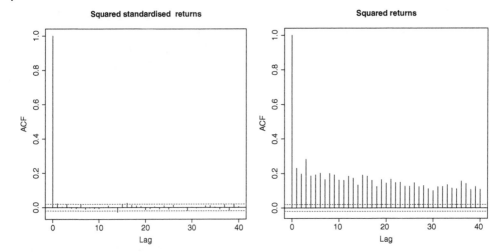

Figure 31.5 ACFs.

1. $\omega = 0.0000007 \quad \alpha = 0.03 \quad \beta = 0.968$

2. $\omega = 0.0000007 \quad \alpha = 0.15 \quad \beta = 0.845$

3. $\omega = 0.00004 \quad \alpha = 0.3 \quad \beta = 0$

The reader should aim to draw conclusions about the effect of the parameters on GARCH models.

31.7 A Note on Correlation and Dependency

It should be clear from the material covered in this chapter that basic GARCH processes exhibit zero correlation; what has happened in the past has no influence on whether the next term in the series is above or below the average. This is easily demonstrated by obtaining the ACF of such a process:

```
acf(garch_sim)    #using the simulations from earlier in the chapter
```

However, such GARCH processes are not independent processes. If the previous error term is sufficiently large this may result in a larger variance for the next term, which in turn results in a higher probability that the next term will be large. Thus "what happened yesterday affects what might happen today" – the terms are not independent of each other (Figure 31.5 demonstrates this).

Exercise 31.7 *Prove that a GARCH process has uncorrelated, but dependent innovations. That is, a GARCH process is a white noise process with dependent variables.*

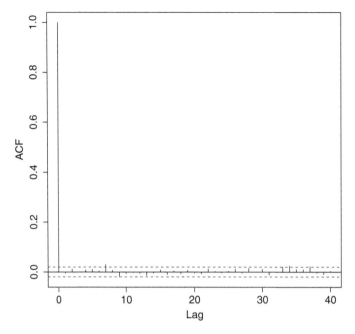

Figure 31.6 ACF of GARCH(1,1) simulations.

31.8 GARCH Long-Term Variance

The long-term variance of a $GARCH(1, 1)$ model, σ^2, mentioned briefly in Section 31.4, is given by the following equation (note here that σ does not include a "t" subscript):

$$\sigma^2 = \frac{\omega}{1 - \alpha - \beta} \tag{31.8}$$

(The reader should ensure they can derive the above expression.) Given the parameter values chosen in Section 31.5 this would result in $\sigma^2 = 0.00007$, or a standard deviation of 0.0083666.

It is important to be clear about the difference between the conditional variance, σ_t^2, and the long-term variance, σ^2. Broadly speaking, the conditional variance tells us how uncertain returns are at time, t, based on the returns over recent history, whereas the long-term variance tells us how uncertain returns are expected to be on a day, say, one or two years from now.

Our objective in this section is to demonstrate that Equation 31.8 holds by simulating many values of the daily return between $t = 0$ and $t = 1000$ and calculating the variance of the simulations.

We have projected forward for 1,000 timesteps using our GARCH model, and run 10,000 simulations of each run. The initial volatility for each simulation has been set equal to 1.5% (this value was chosen so the trend is clearly visible).

```
set.seed(2000); steps<-1000; sims<-10000
x<-matrix(nrow=steps,ncol=sims)
z<-numeric(steps)#set up a vector for N(0,1)
sigma<-matrix(nrow=steps,ncol=sims)
sigma[1,]<-0.015
sdev <-numeric(steps)
omega<-0.0000007; alpha<-0.08; beta<-0.91

for(j in 1:sims){x[1,j]<-0
z<-rnorm(steps,0,1)#simulate 1000 N(0,1)

for(i in 2:steps){sigma[i,j]<-(omega+alpha*x[i-1,j]^2+beta*sigma[i-1,j]^2)^0.5
x[i,j]<-sigma[i-1,j]*z[i]}}
```

These simulated results are plotted in Figure 31.7 (left). The standard deviation of the 10,000 data point sample at $t = 1000$ is:

```
sd(x[1000,])
```

```
[1] 0.008420584
```

This should be compared with the theoretical volatility of 0.0083666 (calculated above).

Exercise 31.8 *There is a material amount of noise, as can be seen in the Figure 31.7 (left). Calculate the average standard deviation between times 500 and 1000, and compare with the theoretical value of σ. (The solution is included below.)*

```
sd_garch_sims<-apply(x,1,sd);
mean(sd_garch_sims[500:1000])
```

```
[1] 0.008395346
```

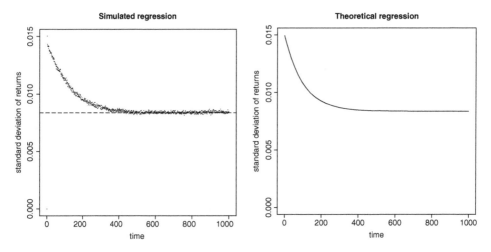

Figure 31.7 Regression to the long-term variance.

Exercise 31.9 *Determine an expression for σ_{n+t} from $t = 0$ to $t = 1000$ in terms of σ_n and σ. A plot of the result is shown in Figure 31.7 (right).*

Exercise 31.10 *Calculate 95% VaR at $t = 200$ and $t = 1,000$ from the simulations carried out above.*

This demonstrates both the regressive nature, and the variance stationary nature, of GARCH models. It is clear from Figure 31.7 that the variance of our random variable, once the system has settled down, is constant. (This is provided that $\alpha + \beta < 1$, otherwise the variance is undefined – the reader may wish to re-run simulations with $\alpha + \beta > 1$.)

31.9 Exercise: Shocks to Global Equity Markets – The Global Financial Crisis 2008, and COVID-19

In the exercise which follows, the reader is required to carry out similar routines and tests as covered earlier, but to real, historic equity return data. The reader should therefore also critique their final model and identify areas in which the model can be improved.

Exercise 31.11 *Download the equity data from the following file: "fin_crisis_GARCH2.csv". The data is plotted in Figure 31.1. The sheet sets out daily equity return data for the period 2001 to 2020; critically it includes both the impact of the financial crisis in 2008 and that of COVID-19 in 2020 on global stock markets.*
The objective is to fit the most appropriate GARCH model and carry out a similar analysis of the data to that discussed earlier.

[Part-answer to Exercise 31.11.]

```
fit_garch$coef[1];fit_garch$coef[2];fit_garch$coef[3]

            a0
0.000002244972
           a1
0.1197502
           b1
0.8630405

Box.test(standard_squared, type = "Ljung-Box",lag = 10) #still doesn't work very well


Box-Ljung test

data:  standard_squared
X-squared = 16.275, df = 10, p-value = 0.09204
```

Perhaps the main conclusion to be reached from this exercise is that the $GARCH(1, 1)$ model, although a significant improvement on a simple Normal distribution model, still has significant flaws in modelling equity data from stressed periods.

In particular, the GARCH model does not tend to adjust sufficiently quickly to periods of significantly higher volatility, and does not cope with extremely bad one-off days for example, 27 February 2007 when the Chinese stock market had the biggest loss in 10 years, affecting global equity markets.

Exercise 31.12 *Assess the model to see how many data points were not adequately predicted by your new model e.g. did around 1% of data points exceed 2.33 standard deviations, and compare this with a simple Normal distribution with constant variance.*

Exercise 31.13 *Refit the model (3 times) using data only up to (i)26 September 2008, (ii)28 November 2008, and (iii)6 March 2020, and calculate the daily 99% VaR which the model predicted for the following day. How did the model perform?*

Exercise 31.14 *Draw a Q-Q plot of the data as a visual test of whether the data is Normally distributed.*

It was noted in the Introduction to this chapter that ϵ_t is a random variable from a distribution with mean of 0 and variance of 1; previously we have used $\epsilon_t \sim N(0, 1)$ in our models. An obvious alternative is the scaled Student-t distribution viz. $\epsilon_t \sim t(v)\sqrt{\frac{v-2}{v}}$. (Note that the option of t innovations is not available under the tseries package; in Example 31.1 we use functions from the rugarch package.)

Example 31.1 Fit a $GARCH(1, 1)$ model with $Student - t$ innovations to the data (in file " fin_crisis_GARCH2.csv"), compare with the Normal innovations model from Exercise 31.11, and determine which is the more appropriate model. (Sample code using rugarch package functions is shown here.)

```
xx_equity<-read.csv("~/fin_crisis_garch2.csv")
```

```
student_spec <- ugarchspec(mean.model=list(armaOrder=c(0,0), include.mean = FALSE),
variance.model = list(garchOrder=c(1,1)), distribution="std")    #std = student t
student_fit <- ugarchfit(student_spec,xx_equity);    coef(student_fit)

         omega          alpha1            beta1           shape
0.000001380549 0.119548721715 0.876422906764 6.402490746699

normal_spec <- ugarchspec(mean.model=list(armaOrder=c(0,0), include.mean = FALSE),
variance.model = list(garchOrder=c(1,1)),distribution ="norm")    #Normal
normal_fit <- ugarchfit(normal_spec,xx_equity);    coef(normal_fit) #similar to
results above

         omega          alpha1            beta1
0.000002238471 0.119544278422 0.863342715983
```

The AICs have been calculated here (alternatively one can obtain AIC/k, where k is the number of data points, using the infocriteria function).

```
-(likelihood(student_fit)-5)*2;    -(likelihood(normal_fit)-4)*2    #AICs

[1] -32579.59
[1] -32368.2

infocriteria(student_fit)[1];    infocriteria(normal_fit)[1]       #AIC/k

[1] -6.524147
[1] -6.481819
```

This suggests that the GARCH model with Student-*t* innovations is superior to Normal innovations. The `rugarch` package is a particularly useful package for those readers wishing to develop an understanding of extensions to the basic GARCH models (see Section 31.10.)

31.10   Extensions to the GARCH Model

The material covered in this chapter provides an introduction to the topic of GARCH models. The reader should, however, be aware of the many proposed variant models, of which some of the most commonly used are:

- Integrated GARCH (IGARCH)
- Exponential GARCH (EGARCH)
- TGARCH
- ARMA-GARCH

We will not discuss any of these in any detail in this book. However we have included sample code and output to illustrate how the modelling, started earlier, on the historic equity data, could be moved forward.

Exercise 31.15 *Leveraged GARCH models have been developed to attempt to reflect the asymmetry which exists in many financial data sets. The idea is that bad news, resulting in market falls, has a bigger impact on future price uncertainty than good news. As company values fall, the financial leveraging of the firm may increase leading to the likelihood of greater uncertainty in the future. Many GARCH-type models have been developed, such as EGARCH, TGARCH and GJR models. The code below illustrates the application of an EGARCH model (with t innovations) to the data in Section 31.9 and compares it with the GARCH(1, 1) models from earlier in the chapter. See Nelson (1991) for details of the EGARCH model.*

```
eg_spec = ugarchspec(variance.model = list(model = "eGARCH", garchOrder = c(1,1)),
mean.model = list(armaOrder = c(0,0), include.mean = FALSE),
distribution.model = "std")

egarch_fit = ugarchfit(data = xx_equity, spec = eg_spec,
solver = "solnp")

(likelihood(egarch_fit)-5)*2   #AIC

[1] 32803.97

coef(egarch_fit)

     omega     alpha1      beta1     gamma1       shape
-0.2177754 -0.1648104  0.9763004  0.1542547  6.9196703
```

The AICs from the four models run in this chapter on the data introduced in Section 31.9 (`fin_crisis_garch2.csv`) are shown below, suggesting the EGARCH with *t* innovations model to be the most suitable:

Model	AIC
Normal	−29608
GARCH(1,1) Normal	−32368
GARCH(1,1) t	−32580
EGARCH(1,1) t	−32804

31.11 Appendix – A Mixture of Normal Distributions

This example may provide some insight into the fat-tailed nature of GARCH models (and provides further coding practice). The reader should note that the following does not represent a GARCH model; however it does exhibit some of the characteristics of the models discussed in this chapter, i.e. it consists of a mixture of various Normal distributions with the same mean but different variances. We will see that such a mixture of Normal distributions is not itself a Normal distribution.

Let's take three Normal distributions, each with the same mean (of zero) but with different variances of 1, 4 and 9. We will take samples from each distribution, throw them all together, and examine what this distribution looks like. This is an example of a mixture distribution.

We will use the Jarque-Bera test as a test for a Normal distribution. (Reminder – the kurtosis of a Normal distribution is 3.)

```
set.seed(1000)
aa<-rnorm(100000,0,1);   bb<-rnorm(100000,0,2);   cc<-rnorm(100000,0,3)
kurtosis(aa);   kurtosis(bb);   kurtosis(cc)   # all close to 3, as expected

[1] 3.008378
[1] 2.986277
[1] 2.984294
```

But when we look at the combined data:

```
data_combine<-c(aa,bb,cc);        kurtosis(data_combine)

[1] 4.491923
```

The kurtosis is significantly greater than 3 which suggests the mixture is not Normally distributed. Let's check more rigorously with the Jarque-Bera test (here only showing the p-value):

```
jarque.test(aa)[2];  jarque.test(bb)[2];  jarque.test(cc)[2]

$p.value
[1] 0.3147259
$p.value
[1] 0.1000348
$p.value
[1] 0.3299359
```

```
jarque.test(data_combine)[2]                    #Ho rejected!!

$p.value
[1] 0
```

With this p-value result we can reject the hypothesis that the combined data is Normally distributed. Thus the mixture distribution, consisting of three Normal distribution each with the same mean but different variances, is not Normally distributed, and has significantly fatter tails than a Normal distribution.

31.12 Recommended Reading

- Charpentier A et al. (2016). Computational Actuarial Science with R. Chapman and Hall/CRC.
- Chatfield and Xing. (2019) The Analysis of Time Series: An Introduction (7th edition) (Chapman and Hall/CRC).
- Danielsson. (2011) Financial Risk Forecasting. Wiley.
- Hull, J. (2015) Risk Management and Financial Institutions (4th edition). Wiley.
- Lütkepohl, H. (2010) Applied Time Series Econometrics (Themes in Modern Econometrics). Cambridge University Press.
- McNeil, Frey, Embrechts. (2005) Quantitative Risk Management. Princeton University Press.
- Sweeting, P. (2017) Financial Enterprise Risk Management (2nd edition). Cambridge.

32

Modelling Future Stock Prices Using Geometric Brownian Motion: An Introduction

Peter McQuire

32.1 Introduction

In this chapter we develop, starting with the simple random walk, one of the most widely used models for projecting future share prices – Geometric Brownian Motion ("GBM"). To see where we are aiming, we have set out below, with no explanation at this point, the discrete time approximation of GBM, where S_t is the stock price at time t:

$$S_{t+\Delta t} \approx S_t + \mu S_t \Delta t + \sigma S_t \sqrt{\Delta t}\, \epsilon \tag{32.1}$$

which is exact as $\Delta t \to 0$. Using the rules of stochastic calculus, we arrive at the exact solution for the price S_t at time t:

$$S_t = S_0(e^{(\mu - \frac{\sigma^2}{2})t + \sigma\sqrt{t}\epsilon}) \tag{32.2}$$

where μ is the expected rate of return per unit time from the stock, σ is the standard deviation of the returns from the stock (often referred to as the volatility), and $\epsilon \sim N(0, 1)$.

Our objective is to develop a model for future stock prices, explaining the importance of each term in these model equations. The chapter is structured to take the reader through various time series models, starting with the basic discrete random walk with Gaussian errors and ending with GBM, discussing at each point how the model is improved. It is therefore really a discussion on model development.

A detailed discussion setting out the mathematics underlying GBM is covered comprehensively in several excellent texts, and is not repeated here. The serious student should ensure a detailed study of this important topic is undertaken. Details of the underlying mathematics are available in several textbooks to varying degrees, such as Baxter et al. (1996), Hull (2005), Löffler(2020), Neftci (1996), Schuss (1980).

In particular, the reader is directed to the exercises at the end of this chapter, which highlight the importance of simulating paths, and to look at applications of model variants.

R Programming for Actuarial Science, First Edition. Peter McQuire and Alfred Kume.
© 2024 John Wiley & Sons Ltd. Published 2024 by John Wiley & Sons Ltd.
Companion Website: www.wiley.com/go/rprogramming.com.

32.1.1 Discrete Gaussian Random Walk

Example 32.1

We start with a discrete random walk with standard Normal (or Gaussian) errors:

$$S_t = S_{t-1} + \epsilon_t \tag{32.3}$$

and

$$S_t = S_0 + \sum_{i=1}^{t} \epsilon_i \tag{32.4}$$

for $t = 1, 2, ...$, where $\epsilon_t \sim N(0,1)$ distribution. (If ϵ_t is any random process with constant mean and variance, then S_t is known as a general random walk.) Running one simulation from $t = 0$ to $t = 1000$ and plotting the results (see Figure 32.1 left):

```
s_price1<-NULL; s_price1[1]<-0; set.seed(1237)
for(i in 1:1001){s_price1[i+1]<-s_price1[i]+rnorm(1,0,1)}
```

Or you may prefer the simpler coding:

```
set.seed(1237)
cumsum(rnorm(1000))
```

Figure 32.1 (left) shows just one realisation of the stochastic process described in Equation 32.3. Indeed there are an infinite number of paths and no single realisation should be viewed as "typical". The same comment applies to all stochastic models which follow in this chapter.

Exercise 32.1 *(Revision of time series.) Is the random walk mean and variance stationary? Write code to run appropriate simulations and hence demonstrate your answer.*

Example 32.2

To start our model development, we include some flexibility in the level of uncertainty. If the error terms have a variance of σ^2, rather than 1, then we have:

$$S_{t+1} = S_t + \sigma \epsilon_{t+1} \quad \text{or} \quad S_{t+1} = S_t + \epsilon_{t+1}^* \tag{32.5}$$

where $\epsilon_t^* \sim N(0, \sigma^2)$. Re-running the simulations, using the same seed as above:

```
s_price2<-NULL; s_price2[1]<-0
sigma<-5                        #larger volatility
set.seed(1237)
for(i in 1:1001){s_price2[i+1]<-s_price2[i]+rnorm(1,0,sigma)}
```

(See Figure 32.1 right). However, equity return data suggests that prices, in general, increase with time.

Example 32.3

The next step in building our model is perhaps to add a term in respect of a general upward trend, or *drift*, μ, to Equation 32.5:

$$S_{t+1} = S_t + \mu + \sigma \epsilon_{t+1} \tag{32.6}$$

Note that the units of μ here are the same as those for S e.g. £. Let's assume that the share price, S (with $S_0 = 100$), increases by £0.3 every time step ($\mu = 0.3$), with $\sigma = 3$ (see Figure 32.2, left).

```
set.seed(1237)              #same as above
s_price3<-NULL;   s_price3[1]<-100
mu_z<-0.3;    sigma_z<-3
for(i in 1:1001){s_price3[i+1]<-s_price3[i]+mu_z+rnorm(1,0,sigma_z)}
```

Simulations using the same seed (set.seed(1237)), but with $\mu = 0.1$ and $\sigma = 6$, are shown in Figure 32.2, right.

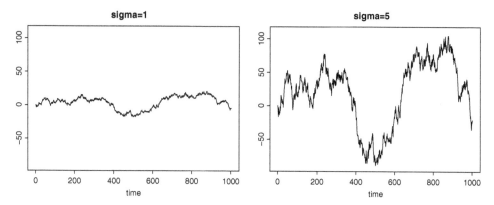

Figure 32.1 Simulated discrete Gaussian random walk (with low and high volatility).

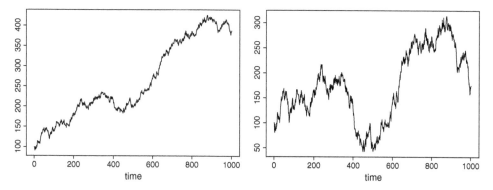

Figure 32.2 Two Simulated discrete Gaussian random walks with drift (different parameter values).

Exercise 32.2 *Estimate the mean and variance of X_{1000} by running several sets of simulations and compare with the theoretical values.*

Example 32.4

We have so far been using time-steps of one unit time. Instead of limiting ourselves to integer values of t we can generalise our expression by allowing any time period of choice, Δt:

$$S_{t+\Delta t} = S_t + \mu \Delta t + \sigma \sqrt{\Delta t}\, \epsilon_t \tag{32.7}$$

Note that the units of μ are now "£ per unit time" e.g. £ per year, or £ per day etc. We also assume, as will usually be the case throughout this chapter, that our parameters are not functions of time.

But why do we include a standard deviation of $\sigma \sqrt{\Delta t}$ in Equation 32.7? This deserves some explanation. Assuming each ϵ_t is independent and identical with each other, $Var(S_{10}) = 10 Var(S_1)$. In general, $Var(S_t) = t\, Var(S_1)$, and thus $sd(S_t) = \sqrt{t}\, sd(S_1)$. In the limit as $\Delta t \to 0$, Equation 32.7 takes the form:

$$S_{t+dt} = S_t + \mu dt + \sigma \sqrt{dt}\, \epsilon_t \tag{32.8}$$

noting from above that $\sqrt{dt}\, \epsilon_t \sim N(0, dt)$. This model is known as *Brownian Motion with drift*, or a *generalised Weiner process* and is usually written in the following form:

$$dS_t = \mu dt + \sigma dB_t \tag{32.9}$$

where dB_t is the Brownian process, and equals $B_{t+dt} - B_t$. Brownian motion has the following properties:

- $dB_t \sim N(0, dt)$
- $B_0 = 0$ (the value of the process at $t = 0$)
- $B_t \sim N(0, t)$ or $B_t \sim \sqrt{t}\, N(0, 1)$ and also
- $B_t - B_s \sim N(0, t - s)$ or $B_t - B_s \sim \sqrt{t - s}\, N(0, 1)$ where $t > s$ (used in Equation 32.9)
- B has independent increments
- B has continuous sample paths.

Note that standard Brownian Motion is defined under Equation 32.8 where $\mu = 0$ and $\sigma = 1$. The reader is encouraged to study further the mathematical approach to Brownian Motion and its variants; good starting points are J. S. Dagpunar (2007), Baxter and Rennie (Ch3) (1996), or Loffler (2020), the latter of which stresses the importance of mathematical rigour to the reader who requires absolute accuracy in this area; in particular Loffler's Introduction chapter is recommended.

Remark 32.1 Note that the terms "Brownian Motion" and "a Weiner process" are equivalent.

Remark 32.2 Note the following, more general version of Equation 32.9, where both μ and σ can vary with time:

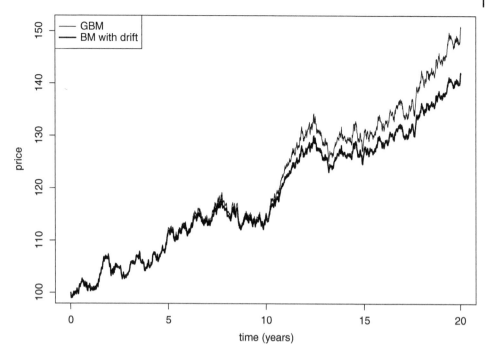

Figure 32.3 Simulated BM with drift, and GBM.

$$dS_t = \mu_t dt + \sigma_t dB$$

Turning back to the matter in hand, we would like to simulate S_t using this BM with drift model, which is possible using the discrete version (Equation 32.7). We assume $\mu = 2$ pa, and $\sigma = 3$ pa. Simulating prices every 0.01 years for 20 years (see Figure 32.3):

```
set.seed(5000)
x_bmdr<-NULL; x_bmdr[1]<-100
mu<-2  ; sigma<-3
timestep<-0.01  #allow smaller time periods
tot_time<-20; no_stepsx<-tot_time/timestep
t<-NULL; t[1] <- timestep    #used for plot

for(i in 1:no_stepsx)
{x_bmdr[i+1]<-x_bmdr[i]+mu*timestep+rnorm(1,0,sigma*timestep^0.5)
t[i+1]<-i*timestep}         # t is useful in plotting the results
```

Given the parameter values used here we would expect a share price of 102 after 1 year, 104 after 2 years, ..., and 140 after 20 years. (The 2nd curve plotted in Figure 32.3 is the subject of Exercise 32.5.)

Exercise 32.3 *Rerun the above with alternative parameter values and terms.*

Exercise 32.4 *By running simulations, demonstrate that the $Var(S_{20}) = 180$, irrespective of which time-step we use eg 20 steps each of 1 year, or 20,000 steps each of 0.001 years. (This* *should demonstrate to the reader that the use of $\sqrt{\Delta t}$ is correct.) Also verify that $E(S_{20}) = 140$. Repeat these calculations at $t = 1, 2, 3 \ldots$*

Is this a reasonable model for share prices? For example, shouldn't the price uncertainty be proportionate to the price? We can further improve the model by using Geometric Brownian Motion.

32.2 Geometric Brownian Motion

The actual price of a share is arbitrary, dependent on the (largely) arbitrary number of shares in which the company has been split into. If a company is worth £1m and split into one million shares, the share price is £1. A share price increase of £1 tomorrow will be viewed very favourably i.e. an increase of 100%. If only 1,000 shares are allocated then the share price is £1,000; if the share price goes up £1 that's only an increase of 0.1%. As we are ultimately modelling the value of the company, we should therefore model the proportionate change in the share price; in our previous model we assumed the share price would continue increasing at a constant monetary amount and was independent of the share price.

Similarly, it would make sense to model uncertainty (i.e. volatility, or standard deviation) to be proportional to the share price. If there are one million shares each of £1 in value, you may expect this to vary by up to 2 or 3 % that day, and be worth between £0.97 and £1.03 at the end of the day. If there were only 1,000 shares with a price of £1,000 you may expect it to be between £970 and £1,030.

Thus it makes sense to be modelling drift rates and volatilities which are proportionate to the share price. A further potential problem with Equation 32.8 is that there is a non-zero probability that $S_t < 0$ (easily observed from running simulations with suitable parameter values).

Therefore, an improvement to Equation 32.8 is given by:

$$S_{t+\Delta t} \approx S_t + \mu S_t \Delta t + \sigma S_t \sqrt{\Delta t}\, \epsilon_t \tag{32.10}$$

This is a discrete, approximate version of Geometric Brownian Motion ("GBM"). The equation may look familiar – it was first shown at the start of this chapter. The equation may be used to simulate a path of future share prices (see Exercise 32.8).

It is important to note the units of μ has changed to "per unit time" eg 0.08 per annum. Note also that this expression is exact only in the limit as $\Delta t \rightarrow 0$ (as S appears on both sides of the equation); Geometric Brownian Motion (i.e. the continuous time version of Equation 32.10) is given by:

$$dS_t = \mu S_t dt + \sigma S_t dB \tag{32.11}$$

(alternatively referred to as Exponential Brownian Motion).

Using the rules of stochastic calculus and Ito's lemma (see Baxter (1996) or Dagpunar (2007)) we can obtain the exact expression for S_t:

$$S_t = S_0 e^{(\mu - \frac{\sigma^2}{2})t + \sigma\sqrt{t}\epsilon} \tag{32.12}$$

where S_0 is the share price at $t = 0$. As noted earlier, both μ and σ are assumed to be constant. This equation allows us to simulate share prices, S_t, at any future time, t, *without any approximations.*

We can use Equation 32.12 to simulate a path of share prices, say daily or annually over the next 10 years (see Figure 32.6), or to simulate share prices in 10 years' time without the need to simulate a path (which will save us time if a path is not required). A number of exercises are included near the end of this chapter to demonstrate the importance of simulating price paths. Note also that for any positive value of S_0, S_t will be positive for any t, as required (see Exercise 32.11).

Further application of Ito's lemma leads to the following important result; given that S_t follows a GBM process with parameters μ and σ, it has the following Lognormal distribution:

$$S_t \sim lnN[lnS_0 + \mu t - 0.5\sigma^2 t, \; \sigma^2 t] \tag{32.13}$$

Thus if returns over a time dt follow $N(\mu dt, \sigma^2 dt)$, the share price at time t, S_t, follows a log-normal distribution as set out in Equation 32.13. Therefore we can simulate asset prices at a future time using either Equations 32.12 or 32.13.

From the properties of the log-normal distribution, the theoretical mean and standard deviation of S_t are given by:

$$E(S_t) = S_0 e^{\mu t} \qquad Var(S_t) = S_0^2 e^{2\mu t}(e^{\sigma^2 t} - 1) \tag{32.14}$$

Plotted curves of the mean and standard deviation of the stock price against time are shown later in this chapter in Figure 32.7 (calculated using Equations 32.14), where they are compared with statistics obtained from simulations.

Exercise 32.5 *Applying the same* set.seed *used in Example 32.4, re-run the "BM with drift" simulations shown in Figure 32.3, but using GBM with parameters values* $\mu = 0.02pa$ *and* $\sigma = 0.03pa$ *i.e. applying Equation 32.12. This second set of simulations is the second plot shown in Figure 32.3.*

32.3 Applications of GBM, and Simulating Prices

The aim of this section is to simulate future asset prices using the models described earlier in this chapter, to demonstrate their equivalence, and to analyse results from these models.

We shall assume the following GBM parameter values throughout this section:

$$\mu = 10\%pa; \qquad \sigma = 30\%pa$$

Remark 32.3 Note that a GBM process requires that both μ and σ are constant, and do not change value with time. We will look at model variants briefly later in this chapter.

Given these values (and $S_0 = 100$), from Equation 32.13 we now have at $t = 1$, and $t = 10$:

$$S_1 \sim lnN[4.6601702 \, ; \, 0.09] \qquad S_{10} \sim lnN[5.1551702 \, ; \, 0.9]$$

It is straightforward to simulate future asset values from these two log-normal distributions (at $t = 1$ and $t = 10$):

```
mu<-0.1;        sigma<-0.3
set.seed(10000)
s_price4<-rlnorm(100000,log(100)+mu*1-0.5 *sigma^2*1,sigma*1^0.5)   #t=1
mean(s_price4);    sd(s_price4)

[1] 110.6406
[1] 33.9083

s_price4_ten<-rlnorm(100000,log(100)+mu*10-0.5 *sigma^2*10,sigma*10^0.5) #t=10
mean(s_price4_ten);    sd(s_price4_ten)

[1] 271.4476
[1] 325.2992
```

The results are plotted in Figure 32.4. Alternatively we can use GBM and apply Equation 32.12, thus demonstrating their equivalence.

Exercise 32.6 *Re-run the simulations, but using Equation 32.12 and* set.seed(1000)*. The results, together with the appropriate log-normal distributions, are plotted in Figure 32.5; key statistics are shown below:*

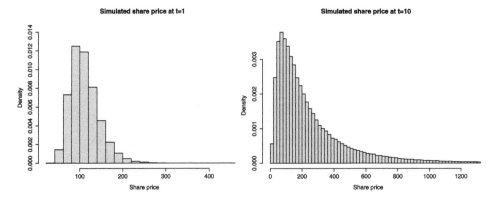

Figure 32.4 Simulated Share prices at t=1 and t=10 years using Eqn 32.13.

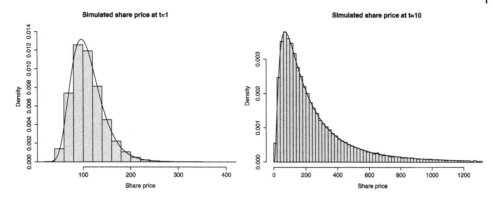

Figure 32.5   Share price distributions at t=1 and t=10 years: histogram using Equation 32.12 and density using Equation 32.13.

```
mean(price_simx1);        sd(price_simx1)

[1] 110.5264
[1] 33.8293

mean(price_simx10);       sd(price_simx10)

[1] 271.5853
[1] 327.4949
```

The theoretical mean and standard deviation at $t = 1$ and $t = 10$ are:

```
100*exp(mu*timezz);        100*exp(mu*timezz)*sqrt(exp(sigma^2*timezz)-1)

[1] 110.5171
[1] 33.9153

100*exp(mu*timez);         100*exp(mu*timez)*sqrt(exp(sigma^2*timez)-1)

[1] 271.8282
[1] 328.4066
```

The 99% VaR's at $t = 1$ and $t = 10$ from both theory and simulations are set out below:

```
(var99_theory_1<-qlnorm(0.01,4.66,0.3)); (var99_sims_1<-quantile(price_simx1,0.01))

[1] 52.56737
      1%
52.73227

(var99_theory_10<-qlnorm(0.01,5.155,0.9^.5));
(var99_sims_10<-quantile(price_simx10,0.01))

[1] 19.0684
      1%
19.25111
```

Exercise 32.7 *Estimate the probability that $S_{10} < 80$, using both the lognormal distrbution, and the simulated values.*

Simulating share price paths

There may be circumstances where we are required to simulate a path of prices e.g. to estimate the probability that the share price will fall below a particular level at any point (see Exercise 32.10); this is most easily achieved using Equation 32.12. The following code records the price at each annual time step from $t = 0$ to $t = 10$ for each of our simulated paths:

```
delta<-1 #time period of simulations (run every year)
no_of_sims<-100000;    no_of_steps<-11
log_price<-matrix(0,nrow=no_of_sims,ncol= no_of_steps)
log_price[,1]<-log(100)

set.seed(10)
for(i in 1:no_of_sims){
y<-sigma*rnorm(no_of_steps,0,delta^.5)
for(j in 2:no_of_steps){log_price[i,j]<-log_price[i,j-1]+(mu-sigma^2/2)*delta+y[j-1]
}}
price<-exp(log_price)
```

Figure 32.6 includes five simulated paths, showing the 0.4th, 9th, 29th, 57th, and 74th percentiles at $t = 10$.

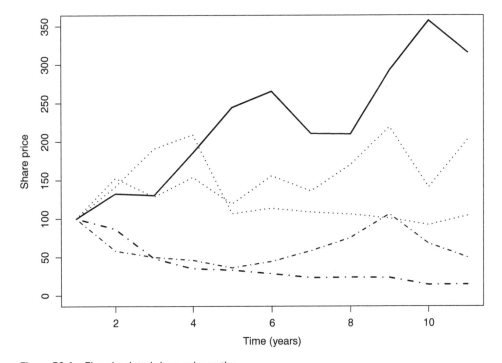

Figure 32.6 Five simulated share price paths.

Exercise 32.8 *The reader may find it improves ones intuitive understanding to simulate future prices using Equation 32.10; however this will be approximate due to the requirement to discretize the time steps. Experiment with $\Delta t = 1$ and $\Delta t = 0.01$.*

It is informative to determine the MLE log-normal parameters from this simulated data. At $t = 1$:

```
mle_gbm_1<-fitdistr(price[,2],"lognormal")
mle_gbm_1$estimate[1];   mle_gbm_1$estimate[2]

 meanlog
4.660012
    sdlog
0.299895
```

At $t = 10$:

```
mle_gbm_10<-fitdistr(price[,11],"lognormal")
mle_gbm_10$estimate[1];   mle_gbm_10$estimate[2]

 meanlog
5.153427
     sdlog
0.9494811
```

The estimated logN parameter values are consistent with the values used in the GBM model.

Exercise 32.9 *Calculate, using theory and the simulations from the share price paths carried out above, the sample mean and standard deviation at all the time steps from our simulations (the results at $t = 0, 1, 2 \ldots 10$ are plotted in Figure 32.7).*

Exercise 32.10 *By simulating prices every hour for one year (252 working days), estimate the probability that the share price falls below 70 **at some point** in the year. Use the following parameter values: $\mu = 0.1pa$, $\sigma = 0.3$, and $S_0 = 100$. Also calculate the expected number of days for which the share price will be below 70 over the next year.*

Exercise 32.11 *Repeat Exercise 32.10, but with $S_0 = 0.01$ and thus verify that $S_t > 0$, $\forall t$.*

Exercise 32.12 *The board of an investment company has set as a risk constraint that there must be less than a 5% probability that their £100m equity fund ($S_0 = 100$) falls below £90m at any point in the next year. Calculate the maximum permissible standard deviation, σ, that can be tolerated such the the board's risk tolerance is met (assume $\mu = 10\%$ pa and that $S(t)$ is a GBM process).*

Exercise 32.13 *(This exercise has some similarities with the Merton Credit Risk Model). An institution's current asset value is 100. It also has liabilities equal to 90, due in 2 years. Assuming that both the asset and liability value are independent GBM processes with parameters $\mu_A = 10\%$, $\sigma_A = 30\%$, and $\mu_L = 10\%$, $\sigma_L = 20\%$, what is the probability that $A < L$ at any time in the next 2 years? What additional level of assets would be required such that there is less than 5% chance over the next two years that $A < L$ at any time?*

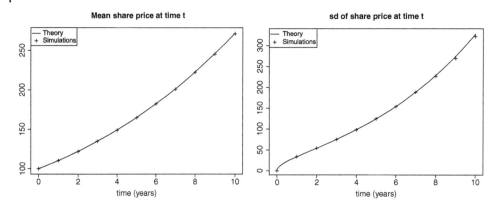

Figure 32.7 Evolution of the mean and standard deviation of the share price; comparing theoretical and simulated values.

Exercise 32.14 *(This exercise requires some knowledge of copulas – see Chapter 13). An equity portfolio consists of equal holdings in 10 different company shares. The current value of the portfolio is £100m. Each of the 45 pairwise correlations has been estimated such that the Pearson correlation coefficient $\rho = 0.4$ for all pairs. Assume that for each company share: GBM is appropriate, $\mu = 10\%$ pa, and $\sigma = 30\%$. Calculate the 1-year 95% Value at Risk of the portfolio value. Re-calculate the 1-year 95% Value at Risk assuming $\rho = 0.1$.*

Exercise 32.15 *(This exercise incorporates rare shocks to the GBM model.) Run an adjusted GBM model, where for α days per year, where $\alpha \sim Poisson(0.5)$ pa, the return is instead either -10% or +10% with equal probability i.e. we expect one extreme day every two years. Apply the same GBM parameters as previously (i.e. $\mu = 0.1$pa and $\sigma = 0.3$pa, and $S_0 = 100$). Run 1m simulations of daily price paths over one year, and plot the distribution of share prices at $t = 1$ year. Also estimate the probability that $S_t < 70$ at some point during the year.*

The following exercises adopt the model described in Remark 32.2, incorporating further flexibility by allowing μ and σ to vary with time.

Exercise 32.16 *Repeat Exercise 32.10 with $S_0 = 100$, but in place of the constant parameter values $\mu = 10\%$pa, $\sigma = 30\%$pa, select daily values from the following distributions: $\mu \sim N(0.1, 0.1^2)$ and $\sigma \sim N(0.3, 0.1^2)$.*

Exercise 32.17 *(For readers familiar with GARCH models – see Chapter 31). GBM assumes that both the drift and volatility are constant, which may not be particularity realistic. Incorporating a GARCH (1,1) model into Equation 32.12 as follows:*

$$S_{t+\Delta t} = S_t e^{(\mu - \frac{\sigma_t^2}{2})\Delta t + \sigma_t \sqrt{\Delta t}\, \epsilon}$$

where $\sigma_t^2 = \omega + \alpha X_{t-1}^2 + \beta \sigma_{t-1}^2$, and $X_t = \ln \frac{S_t}{S_{t-1}}$, and using the following parameter values:

$$\Delta t = \frac{1}{252} \quad S_0 = 100 \quad \sigma_0 = 0.008367 \quad \mu = 0.10 \quad \omega = 0.0000007 \quad \alpha = 0.08 \quad \beta = 0.91$$

simulate 100,000 paths of daily share prices (hence $\Delta t = \dfrac{1}{252}$, assuming there are 252 working days in a year) and calculate the standard deviation and 99% VaR of the share price in one year (at $t = 1$). Perform appropriate tests to determine whether $\ln S_1$ is Normally distributed, and how the model compares with GBM.

32.4 Recommended Reading

- Baxter, M. and Rennie, A. (1996). *Financial Calculus: An Introduction to Derivative Pricing*. Cambridge, UK : Cambridge University Press.
- Brandimarte, P. (2006). *Numerical Methods in Finance and Economics*, 2e. Hoboken, New Jersey, US: Wiley.
- Campbell, J.Y., Lo, A.W., and MacKinlay, A.C. (1997). *The Econometrics of Financial Markets*. Princeton, NJ: Princeton University Press.
- Dagpunar, J.S. (2007). *Simulation and Monte Carlo : With Applications in Finance and MCMC*. Chichester, West Sussex, UK: John Wiley and Sons.
- Glasserman, P. (2004). *Monte Carlo Methods in Financial Engineering*. New York: Springer-Verlag.
- Gulisashvili, A. (2012). *Analytically Tractable Stochastic Stock Price Models*. Berlin Heidelberg: Springer-Verlag.
- Hirsa, A. and Neftci, S.N. (2014). *An Introduction to the Mathematics of Financial Derivatives*, 3e. Academic Press.
- Hull, J. (2012). *Options, Futures and other Derivatives*, 8e. Harlow, Essex, UK: Pearson.
- Andreas Löffler, Lutz Kruschwitz (Contributor) (2020). *The Brownian Motion: A Rigorous but Gentle Introduction for Economists*. Cham, Switzerland: Springer.
- Ross, S. (2011). *An Elementary Introduction to Mathematical Finance*, 3e. Cambridge: Cambridge University Press.
- Schuss, Z. (1980). *Theory and Application of Stochastic Differential Equations*. New York, Chichester: Wiley.
- Wilmott, P. (2007.) *Introduces Quantitative Finance*, 2e. Chichester, West Sussex, UK: Wiley.

33

Financial Options: Pricing, Characteristics, and Strategies

Peter McQuire

```
library(matrixcalc)
library(derivmkts)
```

33.1   Introduction

The purpose of this chapter is to develop an understanding of the fundamental concepts of financial options, and to write code to tackle various problems involving such options.

The chapter starts with a brief discussion of the key components of a financial option contract, and applies the famous Black-Scholes-Merton ("B-S-M") formula to analyse the dependence of the option price on a number of factors. We proceed to discuss how options may be used in the market, that is, speculation, leveraging, and risk management. In particular, we look at a technique used by options traders known as delta hedging.

Please note that only a brief description of financial options is included below; it is assumed that the reader will have alternative sources of learning material in this respect. In particular, we will not formally derive the B-S-M equation, although a sketch is provided in Section 33.11; the full derivation can be found in countless textbooks and its duplication here is unwarranted. Apart from a discussion on the riskless portfolio in Section 33.11, we will cover little theory in this chapter.

33.2   What is a Financial Option?

A financial option is a contract which gives the buyer of the option contract the right (or "option") to either buy or sell an agreed asset at an agreed time (or times) for an agreed

R Programming for Actuarial Science, First Edition. Peter McQuire and Alfred Kume.
© 2024 John Wiley & Sons Ltd. Published 2024 by John Wiley & Sons Ltd.
Companion Website: www.wiley.com/go/rprogramming.com.

price. If the option is to buy the underlying asset it is a call option; if the option is to sell the underlying asset it is a put option.

Remark 33.1 Our discussions will generally involve call options throughout the chapter and leave calculations relating to put options for the reader's self-study.

There are various types and varieties of options. We will consider only European options, where the option to buy or sell exists only at the expiry date specified in the contract. (Compare this with American-style options where the owner of the option contract has the right to exercise the option at any time prior to the expiry date.)

The asset (specified in the contract) to be bought or sold under the contract varies considerably; examples include stocks, bonds, currencies, commodities (such as gold, copper, milk, cattle), and financial indices.

Let us first take a simple example (more realistic examples will be discussed later in the chapter). If you think the price of gold will rise considerably (compared to the average view of the market) you may consider buying a gold call option. This guarantees that you can buy an agreed quantity of gold at an agreed price (the "strike price") at an agreed future date (the "expiry date"). If the current price at which you can buy a specified quantity of gold is £10 (this is known as the spot price) and the strike price under the option contract is £15, by buying this option, say for £0.50, you have the option to buy the gold for £15 at the expiry date of the option. If the price of gold at expiry has risen to £20 you can buy the gold at £15 and immediately sell it for £20, thus making a profit equal to £4.5.

Options can be either exchange-traded or over the counter ("OTC"). Exchange-traded options are bought and sold through exchanges, such as the Chicago Board Options Exchange, Boston Options Exchange and Eurex Exchange, and have standardised terms for example, standardised strike prices, asset quantities, and expiry dates. Standardisation provides traders with reduced costs and greater liquidity (by allowing them to offset initial positions), and simplifies processes making it easier for traders to interact with each other. The exchanges also greatly reduce counterparty risk.

In contrast, OTC options are designed to meet the precise requirements of the parties involved (e.g. setting out the particular quantity, expiry date, and strike price), and are particularly important in the interest rate and foreign exchange markets. OTC contracts will often involve special conditions.

For further details on the mechanics of options markets we refer the reader to Hull (2022).

33.3 What are Financial Options Used for?

There are a number of examples presented in this chapter which go some way to explaining how these options are used in practice. In risk management, options can be used to limit potential losses, effectively providing a form of insurance, for example, to limit equity portfolio losses beyond a certain level by buying an equity index put option (known as a "protective put").

Several speculative trading strategies can be adopted by taking particular positions in various types of options and other assets. For example, if it is thought the market may exhibit considerable volatility, traders can simultaneously buy call and put options (known as a "straddle"). They can also be used to gain a far greater exposure to asset price changes than otherwise would be possible given a trader's access to capital; if a trader believes a stock's price may increase significantly, greater exposure can be obtained to such increases by buying call options on the stock than could otherwise be obtained simply by buying the stock itself (see Section 33.2).

33.4   Black, Scholes and Merton Differential Equation

In 1973 Black and Scholes published their paper "The Pricing of Options and Corporate Liabilities" which developed the famous equation enabling option traders to easily price particular types of financial options (given a particular set of simplifying assumptions). Merton also contributed significant work in this area resulting in Scholes and Merton winning the Nobel Prize in Economics in 1997. The equation explains how the price of a derivative, f, will change both over time, t, and following changes to the price of the underlying asset, S.

The Black-Scholes-Merton ("B-S-M") differential equation is as follows:

$$\frac{\partial f}{dt} + \frac{1}{2}\sigma^2 S^2 \frac{\partial^2 f}{\partial S^2} + rS\frac{\partial f}{\partial S} = rf \tag{33.1}$$

where r is the risk-free interest rate and σ is the standard deviation of returns from the underlying asset.

It is crucial to note that the B-S-M model is concerned with the pricing of derivatives where the price of the underlying asset can be assumed to follow a Geometric Brownian Motion ("GBM") process, described in Equation 33.2:

$$S_{t+dt} = S_t + \mu S_t dt + \sigma S_t\, dB \tag{33.2}$$

where $dB \sim N(0, dt)$, and μ is the expected return on the underlying asset. (Equation 33.2 was discussed in Chapter 32.) A quite different model should be applied where it would be considered inappropriate to model the underlying asset price using GBM, such as a typical bond.

Remark 33.2 A full discussion of the B-S-M differential equation is beyond the scope of this book. A further discussion of the riskless portfolio used to derive it is included in Section 33.11. Many excellent texts exist on this important subject, such as Baxter et al. (1996), Hull (2012), and Wilmott (2007).

33.4.1   Assumptions Underlying B-S-M Formulation

The key assumptions used to derive Equation 33.1 are as follows:

● the stock price, S, follows a Geometric Brownian motion with constant volatility, σ, and constant drift rate, μ

- the risk-free rate of interest, r, is constant
- no dividends are payable from the stock for the lifetime of the derivative
- there are no expenses or taxes.

33.4.2 Solution to B-S-M Equation for European Call Options

Equation 33.1 can be solved for a European call option by stipulating the appropriate boundary conditions. The price of a European call option, c, is given by the following formula:

$$c = S_0 N(d_1) - K e^{-rT} N(d_2) \tag{33.3}$$

where:

$$d_1 = \frac{\ln(S_0/K) + (r + \sigma^2/2)T}{\sigma\sqrt{T}}; \qquad d_2 = \frac{\ln(S_0/K) + (r - \sigma^2/2)T}{\sigma\sqrt{T}}$$

K is the strike price
S_0 is the current share price
T is the time to expiry
σ is the volatility, or standard deviation, of the underlying asset price
r is risk-free interest rate.

The strike price, K, is the price written in the call option contract for which the asset may be bought at expiry.

By buying the option for c and taking a short position in an appropriate number of underlying shares, the investor can , in theory, over a short time interval, guarantee a particular rate of return (the risk-free return). This is discussed further in Section 33.10.

33.4.3 Call Option Price Function

The example option contract we will look at is as follows:

"The option is to buy one share in XYZ Ltd for £90 on 1 December 2023."

Let's assume today's date is 1 October 2023, so there are 2 months till the contract expiry date ($T = \frac{2}{12}$ years). The current price of one share of XYZ Ltd is £100. The risk-free interest rate is equal to 3% pa and the volatility equal to 30% pa.

The option owner can buy one share for £90 under the contract in 2 months, and can buy the share now for £100 (the spot price). As a first approximation, we may estimate that the current value of this contract is £10. This is known as the "intrinsic value" of the option.

However, there will be an additional value known as the "time value" - it allows for the potential for the option to increase in value and the non-linear nature of the payoffs from options; if the spot price increases 20% to £120 at expiry you will receive £30 (as you can buy the shares for £90 under the contract and immediately sell them for £120), but if the share price falls 20% the option will expire worthless. The average of these is £15 - higher than the intrinsic value. Hence you may be willing to pay more than £10 today for this option contract.

Let's define the variables we need to price our option, and assign values:

```
spotprice<-100;   strike<-90;     expy<-2/12
risk_free_rate<-0.03;    vol<-0.3
```

It is straightforward to write a function to calculate the price of the call option (using Equation 33.3):

```
callprice<- function(s_price, strike,risk_free_rate,vol,expry){
d1<-(log(s_price/strike)+(risk_free_rate+vol^2/2)*expry)/vol/expry^0.5
d2<-(log(s_price/strike)+(risk_free_rate+vol^2/2)*expry)/vol/expry^0.5-vol*expry^.5
s_price*pnorm(d1,0,1)-strike*exp(-risk_free_rate*expry)*pnorm( d2,0,1)
}
```

The price of the above European call option with two months to expiry, given the assigned values, is therefore:

```
(y<-callprice(spotprice,strike,risk_free_rate,vol,expy))
```
```
[1] 11.60921
```

Remark 33.3 Of course, there are various packages available in R which include calculations relevant to financial derivatives. The `derivmkts` package includes the greeks function which produces the following output, some of which is discussed in this chapter:

```
greeks(bscall(spotprice, strike, vol, risk_free_rate, expy, 0))

                    bscall
Premium    11.60920940
Delta       0.83205734
Gamma       0.02050101
Vega        0.10250384
Rho         0.11932754
Theta      -0.03115957
Psi        -0.13867622
Elasticity  7.16721791
```

(Note that the Theta value calculated here is a daily, not annual, value.) The `Premium` output agrees with our calculation above. The reader is encouraged to review this package, and others, as they proceed through this chapter.

33.5 Calculating the Option Price Using Simulations

It is informative to calculate the price of the above option using simulations of the price of the underlying share at expiry, and may help the reader gain a better understanding of the pricing process.

First simulate share prices in two months' time using the GBM model developed in Chapter 32:

```
expiry1<-2/12;              set.seed(1000)
rand_z<-rnorm(10000000,0,vol*expiry1^0.5)
sim_price<-spotprice*exp((risk_free_rate-vol^2/2)*expiry1+rand_z)
```

The payment received at expiry under the call option is equal to the share price less the strike price, subject to a minimum of zero. We must also discount this payment back to the valuation date. The price is the average of these simulated values:

```
mean(pmax(0,sim_price-strike)/(1+risk_free_rate)^expiry1)
```

```
[1] 11.61124
```

This is often referred to as "the interest rate discounted expected payoff", and is in agreement (subject to stochastic error) with the price of the option calculated above using Equation 33.3 (i.e. 11.6092094). Note that the risk-free interest rate should be used for both the projection of asset prices and the discounting of the pay-off at expiry. Of course, the B-S-M formulation is quite restrictive; however, by using simulations we can calculate the option price using more flexible assumptions, and also calculate the price of similar, but more complex contracts (such as "exotic options"). For example, an option contract may require a path of share prices to be simulated, rather than just the price at expiry.

Exercise 33.1 *Calculate the price of the above option contract, but where the payoff is max(S^* – K,0), where S^* is the average price of the stock at the end of each day over the 30 days prior to expiry. (This is an example of an Asian option.)*

Exercise 33.2 *Calculate the price of an American put option, using the method set out in Longstaff-Schwartz (2001). Adopt the same parameter values as those in Section 33.4.3 for the European call option.*

33.6 Factors Which Affect the Price of a Call Option

33.6.1 Share Price

We first analyse how the price of the above option varies with the price of the underlying asset at a fixed point in time. Applying our function over a range of share prices, say from £50 to £150, and setting the time to expiry equal to six months:

```
s_price<-100;  rf_rate<-0.03;  vol<-0.3;  strike<-90  ;  exp_t<-0.5
call_fn_share<-callprice(seq(s_price*0.5,s_price*1.5,1),strike,rf_rate,vol,exp_t)
```

The results are shown in Figure 33.1 (left). Also shown is the option price with 3 days to expiry.

The delta of an option, Δ, is defined:

$$\Delta = \frac{\partial f}{\partial S} \tag{33.4}$$

that is the gradient of the curves in Figure 33.1 (left). Δ of this call option is plotted in Figure 33.1(right). Here Δ has been calculated numerically:

```
delta1<-function(shareprice, strike,rf_rate,vol,expiry)
{(callprice(shareprice+.0001,strike,rf_rate,vol,expiry)
-callprice(shareprice,strike,rf_rate,vol,expiry))/.0001}

delta_numer<-delta1(seq(60,120,1), strike,rf_rate,vol,0.5) #at 6 months
delta_numer[seq(1,61,by=10)]

[1] 0.04140582 0.15674463 0.35254634 0.57015913 0.74967052 0.86922807 0.93735283
```

For a European call option we have an exact expression for Δ:

$$\Delta_c = N(d_1) \tag{33.5}$$

Exercise 33.3 *Compare the approximate numerical results calculated above against the accurate values using Equation 33.5.*

When the contract is nearing expiry, Δ changes particularly rapidly when the share price is close to the strike price. This is of particular importance when considering problems relating to the hedging of portfolios which include options (see "Delta hedging" later).

Exercise 33.4 *Payoff diagrams show the value of the option contract at expiry. Draw payoff diagrams for long and short positions on call and put options.*

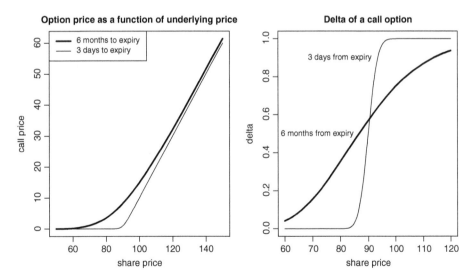

Figure 33.1 Call Option Price and Delta as a function of underlying price.

33.6.2 Time to Expiry

Exercise 33.5 *Plot a graph showing the price of the call option (described in Section 33.4.3) as a function of time to expiry, assuming a constant share price of £100. Start with a time to expiry of 2 years. Repeat the calculations, but with a constant share price equal to £80.*

Figure 33.2 (left) shows that the value of the call option falls as we approach expiry (at $t = 2$), due to the greater certainty about the final share price (assuming a constant share price). From the higher curve in this plot we can see that immediately before expiry (say one minute) the option is worth £10 that is, the difference between the share price and the strike price - it is unlikely the share price will be significantly different to £100 and therefore has little "time value".

33.6.3 Combined Effect of Share Price and Time to Expiry

Exercise 33.6 *Using similar code to that used above, plot the option price against share price at times to expiry of 2 years, 1.5 years, 1 year, 6 months, 3 months, and 0.01 years (see Figure 33.2, right).*

The reader may imagine thousands of curves moving gradually down from the top, thick curve to the bottom curve in the graph, each one representing time moving forward second by second to expiry, with a time to expiry axis coming out of the paper.

Exercise 33.7 *Assuming the share price with 2 years to expiry is £95, and that the share price increases at a force of interest equal to 5% each year, calculate the call option price at each of the times shown in Figure 33.2, right. (This demonstrates one possible path which the option price will take.)*

Combining this information on one 3-d graph, we can visually represent our solution to the B-S-M differential equation (Equation 33.1). We have used the outer function - this

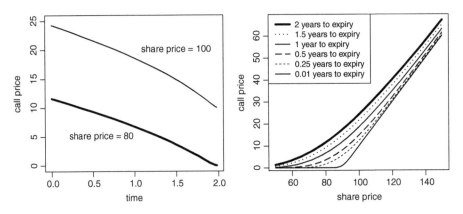

Figure 33.2 Option price as a function of underlying price and time to expiry.

effectively creates a 2-d grid of share prices between £50 and £120, and at every time from 2 years to expiry, in 0.1 year steps:

```
k<-matrix(0,36,21);    shareprice<-seq(50,120,2);    expiry<-seq(0,2,0.1)
k<-outer(shareprice,strike=90,risk_free_rate =0.03,vol=0.3,expiry,callprice)
```

The solution is plotted in Figure 33.3 using the `persp` function:

```
persp(shareprice,expiry,k,xlab="share price",zlab="call price",theta=-35,phi=20,
ticktype="detailed")
```

Thus the call price follows a path on the surface as we move through time to expiry. Simulating a possible path of the share price and corresponding option price:

```
set.seed(20000)
share_price<-NULL;    share_price[1]<-100    #starting share price
mu<-0.03;       sigma<-0.3;     strike<-90
timestep<-0.01;     t<-NULL     ;     time_ex<-2
uncertx<-NULL;     uncertx<-rnorm(200,0,sigma*timestep^0.5)
for(i in 1:(time_ex/timestep)){share_price[i+1]<-share_price[i]+mu*share_price[i]*
timestep+share_price[i]*rnorm(1,0,sigma*timestep^0.5)
share_price[i+1]<-share_price[i]*exp((mu-sigma^2/2)*timestep+uncertx[i])
}
t<-seq(0,2,by=0.01)
option_pricex<-callprice(share_price,strike,mu,sigma,rev(t))#based on sim share
price
```

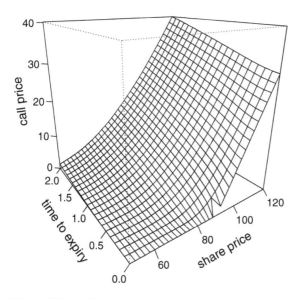

Figure 33.3 3-dimensional plot of Option price.

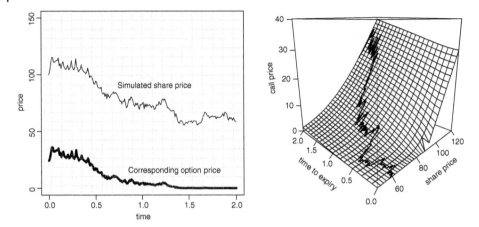

Figure 33.4 Simulated share price and option price.

In this particular simulation the share price performs poorly, resulting in the option having little value for the entire last six months. The simulated price path is plotted in Figure 33.4. Note that all solutions will follow a path on this surface.

Exercise 33.8 *The reader should run many simulations, also experimenting with various parameter values (*set.seed(1000) *provides a better performing simulation).*

Remark 33.4 For more examples of drawing 3-d graphs the reader may benefit from reading the persp demo in R:

```
demo(persp())
```

33.6.4 Other Factors

The option price depends on a number of other factors, such as the volatility of the underlying stock, and the risk-free rate of interest.

Exercise 33.9 *Analyse how (i) the volatility of the underlying share price, and (ii) the risk-free interest rate affects the price of the call option.*

33.7 Greeks

In Section 33.6.1 we briefly discussed the delta, Δ, of an option. We revisit Δ in Section 33.10 where we demonstrate how to manage option price risk by ensuring our portfolio is "delta neutral".

Delta is just one of the measurements used by traders to manage risk; the other "Greeks" commonly used are theta (θ), gamma (γ), vega (ν), and rho (ρ). It is beyond the scope of this book to discuss these in detail (other than delta hedging, which is discussed in

Section 33.10); it is left to the reader to explore the Greeks further (see Hull (2012)). The following exercise provides an introduction to the Greeks.

Exercise 33.10 *Using the* greeks *function from the* derivmkts *package (see Remark 33.3) investigate how* θ, γ, *and* ν *vary as a function of (a)share price, (b)time to expiry, and (c)volatility.*

33.8   Volatility of Call Option Positions

In this section we compare the relative volatility of the option price with that of the underlying share price. We first look at a deep "in-the-money" position on a call option. (In the case of a call option, an in-the-money position refers to the scenario where $S > K$, and it is likely that the option will be exercised.)

Remark 33.5 Note that we use a real, expected return from the asset, μ, when projecting share prices, and the risk-free interest rate when calculating the option price.

Example 33.1 In the money scenario

Consider a company share with a current price of £140. Under a call option contract there are three weeks to expiry, and the strike price is again £90. We assume $r = 3\%$ pa and $\sigma = 30\%$ pa, respectively for the risk-free interest rate and volatility, and the expected return, $\mu = 10\%$ pa. The option price with three weeks to expiry (at $t = 0$) is:

```
share_price_start<-140;     strike<-90      #deep in-the-money
mu_y<-0.1;   r_x1<-0.03;     sigma<-0.3;    expire_x<-3/52
(option_price_start<-callprice(share_price_start,strike,r_x1,sigma,expire_x))

[1] 50.15563
```

We proceed to simulate share prices in two weeks, and hence prices of the call option one week prior to expiry. These are calculated below, and shown in Figure 33.5. An appropriate measure of risk is the 99%VaR loss, which is also calculated below for both the underlying share and option:

```
set.seed(500);   sims<-100000;     share_price_end<-NULL
rndm <-rnorm(sims,0,sigma*(2/52)^0.5)
share_price_end<-share_price_start*exp((mu_y-sigma^2/2)*(2/52)+rndm)  #2 weeks later
option_price_end<-callprice(share_price_end,strike,r_x1,sigma,1/52)
option_profit<-(option_price_end-option_price_start)/option_price_start

1-quantile(option_price_end,0.01)/option_price_start

      1%
0.353724

1-quantile(share_price_end,0.01)/share_price_start

      1%
0.1259823
```

The share price 99%VaR loss is 13% (and gain of 15%), whereas the option price 99% VaR loss is 35% (and a gain of 41%) - we are exposed to the risk of far greater losses (and gains) from this option position compared to simply holding the underlying shares.

Thus, if we invested £1m in the shares we estimate there is a 1% chance that our fund is lower than £0.87m, and higher than £1.15m. However, if we invested £1m in the above options we estimate there is a 1% chance that our fund is lower than £0.65m, and higher than £1.41m (see Figure 33.6).

By investing any limited capital we have in options we can gain a much greater exposure to share price movements than we could by investing directly. This is one of the main reasons for trading in options.

Exercise 33.11 *Repeat the above calculations but with a starting share price of (a) £91 and (b) £85.*

From Exercise 33.11 it is clear that different amounts of gearing and uncertainty can be achieved by using options with various strike prices; the less the position is in-the-money the more exposure to share price changes we can achieve.

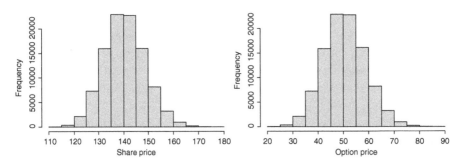

Figure 33.5 Distribution of prices after two weeks.

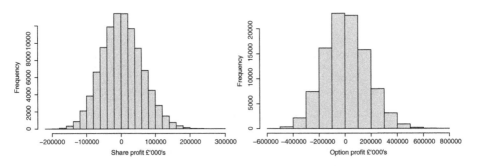

Figure 33.6 Distribution of profit/loss after two weeks, from £1m fund.

Example 33.2 Speculation - deep out the money scenarios

Here we look at the position where $S < K$ - the call option is said to be "out-of-the-money".

A battery research company has been developing car batteries, aimed to be more efficient than the industry average. The current share price of the company is £80. Results of the latest tests are due to be announced next week.

An equity analyst believes there is a much higher chance of successful results than the market does, and speculates that the share price of the company may even exceed £100 next week. Her manager has allocated £1,000 to her "speculation" budget. She is considering either buying company shares or call options on the company shares.

What would the profits be under both strategies if the share price increased over the next week by 25% to £100? If she buys shares the resulting profit would equal £250. Alternatively, if the analyst buys £1,000 of call options with two weeks to expiry and a strike price of £90 (using the same parameter values as in previous examples):

```
strike<-90;     r_x1<-0.03;  sigma<-0.3
(option_price_start<-callprice(80,strike,r_x1,sigma,2/52))

[1] 0.04439868

(option_price_end<-callprice(100,strike,r_x1,sigma,1/52))

[1] 10.05875

option_price_end/option_price_start*1000

[1] 226555.2
```

Following the 25% increase in share price the analyst's fund has increased from £1,000 to £226555. The downside is that there is a much greater chance that most of the initial fund will be lost.

Exercise 33.12 *What is the option fund value if the share price falls or remains the same? Compare this with the strategy of buying shares.*

Exercise 33.13 *Starting with $S = £80$, repeat the simulations from Section 33.8, producing the corresponding histogram to Figure 33.5, and compare the results.*

33.9 Put Options

In contrast with a call option, the buyer of a put option contract has the right (or "option") to *sell* an agreed asset at an agreed time (or times) for an agreed price.

The corresponding price of a European put option, p, is given by the following formula (the price of a call option is given by Equation 33.3):

$$p = Ke^{-rT}N(-d_2) - S_0N(-d_1) \qquad (33.6)$$

Exercise 33.14 *It is left to the reader to write appropriate code and perform calculations for put options, similar to the material discussed above for call options.*

Exercise 33.15 *Plot pay-off diagrams from protective puts and straddles (discussed in Section 33.3).*

33.10 Delta Hedging

Delta hedging is a strategy which can be employed to manage the market risk of a portfolio consisting of financial options. The concept of delta hedging is also crucial in developing the B-S-M pricing model of financial derivatives (see Section 33.11).

Example 33.3

A bank has written a call option with a strike price of £90 on a share with a current share price equal to £80. A key risk to the bank is from an increase to the share price; for example, it may equal £105 at expiry, in which case, to meet our contractual obligation at expiry we would be required to buy the share for £105 and sell it to the option buyer for £90, making a loss of £15 (less the relatively small premium received when the option was sold). To manage this risk we could simply buy the underlying share for £80 on selling the contract, pass over the share at expiry and pocket the premium. However, if the share price falls in value we incur the share price loss, only partially being offset by the option price received earlier. This is known as a covered-call strategy. We can think of this as a poor quality hedge.

Exercise 33.16 *Plot the resultant pay-off diagram from this covered call strategy.*

We can alternatively "delta-hedge" this risk by holding the asset in the correct proportions. Rather than holding no position in the stock, or buying the full quota of stock, a partial exposure to the stock, equal to Δ, is taken. A brief explanation is included in Section 33.11; we shall proceed for now simply to demonstrate this hedging strategy.

Example 33.4

Using the same example and parameter values as those set out in Section 33.4.3, the option price, calculated using B-S-M with 2 months to expiry, is £11.6092, and $\Delta = 0.8321$. If the share price increases by £1 (over a very small time) the option price will increase, approximately, by 0.8321 to 12.4413; therefore, we only need to own 0.8321 shares to mostly offset the loss incurred under the option. The option actually increases to 12.4512.

As with the hedging strategy discussed in Chapter 11, we are looking to invest in an alternative vehicle which is highly correlated with the entity we are trying to hedge; clearly buying the appropriate amount of shares appears a sensible approach.

Remark 33.6 Readers who have read Chapter 11 will recall a similar approach - we were looking for an asset (the hedging asset) which is highly correlated with the underlying asset, and then to determine an appropriate quantity of the hedging asset to buy such that the combined monetary changes of the original investment and our hedging vehicle are close to zero. There the hedging ratio was $\rho \frac{\sigma_1}{\sigma_2}$; here the hedging ratio is Δ.

Perhaps the critical point in understanding delta hedging is that Δ is a function of the underlying asset price, S (see Figure 33.1). We may have bought a number of shares to hedge our options position at time t when the share price was S_t; however if the share price increases, Δ also increases, meaning we are now required to hold more shares to remain hedged. (Note that Δ also changes with *time*.) Thus we are required to continuously buy or sell appropriate amounts of the underlying shares to remain hedged i.e. holding Δ shares. In practice, such rebalancing will occur at discrete points in time which will result in a less than perfect hedge. However, greater hedging frequency will result in better hedging but higher transaction costs, requiring a cost/benefit analysis.

The crux of the issue is that we can determine a portfolio, consisting of the call option and the underlying share, for which we know the return over a short period of time i.e. a guaranteed, riskless return (discussed further in Section 33.11).

Example 33.5

You have sold a European call option on 1,000 shares with one year to expiry. By simulating future share price movements, write appropriate delta-hedging code to determine the distribution of net positions at expiry, and hence demonstrate the effects of delta hedging. Assume the writer of the option starts with a cash account balance equal to £100,000. (*Assumptions*: strike price = share price (at $t = 0$) = 90, risk free rate = 3%pa, volatility = 30%pa, expected real-world return on asset = 8%pa.)

```
strike<-share_p0<-90
r_x1<-0.03;        sigma<-0.3;        mu_x1<-0.08

contract_exp<-1        #time to expiry of call option
share_cont<-1000       #the contract is to buy 1,000 shares
```

```
simsw<-10000
timestep<-1/500    #time interval between calculations

account_y<-share_pricex<-no_shares<-share_buy<-share_buy_pounds<-
callp_t<-z_delta<-matrix(0,simsw,contract_exp/timestep+2)

set.seed(700);  time_exp<-NULL;  time_exp[1]<-contract_exp;  k<-1

#first we calculate everything in the first time period - no sims required here
no_shares[,1]<-0
share_pricex[,1]<-share_p0
option_sell_price<-callprice(share_p0,strike,r_x1,sigma,contract_exp)*share_cont
accounty_start<-100000+option_sell_price      #money held after sale of option
z_delta[,1]<-pnorm((log(share_pricex[1,1]/strike)+(r_x1+sigma^2/2)*time_exp[1])/
sigma/time_exp[1]^.5)

share_buy[,1]<-z_delta[,1]*share_cont
share_buy_pounds[,1]<-share_buy[,1]*share_p0    #delta hedging!!!!
no_shares[,1]<-share_buy[,1]
account_y[,1]<- (accounty_start-share_buy_pounds[,1]) #account value at start

#now run simulations over the term to expiry
norm_sim<-rnorm(simsw*(contract_exp/timestep+1),0,1)

for (j in 1:simsw) {
for(i in 2:(contract_exp/timestep+1)){
time_exp[i]<- round(time_exp[i-1]-timestep,3)  #set time to expiry for next loop
```

```
account_y[j,i]<-account_y[j,i-1]*exp(r_x1*timestep)
share_pricex[j,i]<-share_pricex[j,i-1]*
exp((mu_x1-sigma^2/2)*timestep+norm_sim[k]*sigma*timestep^0.5);   k<-k+1

callp_t[j,i]<-callprice(share_pricex[j,i],strike,r_x1,sigma,time_exp[i])*share_cont
z_delta[j,i]<-pnorm((log(share_pricex[j,i]/strike)+(r_x1+sigma^2/2)*time_exp[i])/
sigma/time_exp[i]^.5)
share_buy[j,i]<-z_delta[j,i]*share_cont-no_shares[j,i-1]
no_shares[j,i]<-sum(share_buy[j,])
share_buy_pounds[j,i]<-share_buy[j,i]*share_pricex[j,i]
account_y[j,i]<- account_y[j,i]-share_buy_pounds[j,i]}}   #at start of each year


value_at_end<-account_y[,1+contract_exp/timestep]*exp(r_x1*timestep)+
no_shares[,1+contract_exp/timestep]*pmin(share_pricex[,1+contract_exp/timestep],
strike)
```

The key results are as follows:

```
mean(value_at_end);   sd(value_at_end)

[1] 103047.1
[1] 418.427

(exp_val_at_end<-100000*exp(r_x1*contract_exp))   #compare the risk-free return

[1] 103045.5
```

If we simply held our risk-free cash account we would have £103045 at expiry (with certainty). With the hedged "call-shares" portfolio, the mean value from the simulations is £103047, with a standard deviation of only £418, and estimated 1% probability that the account will be less than £101950. We have removed much of the uncertainty compared to if we had only held the option, or applied the covered call approach. The option writer can now proceed to sell these option contracts largely risk-free, whilst charging a commission for each sale.

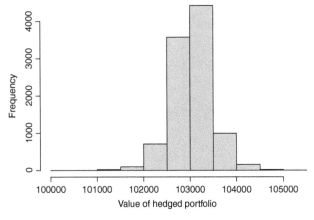

Figure 33.7   Delta hedging: Portfolio value at expiry.

Exercise 33.17 *Re-run this example but starting with a share price equal to 110 instead of 90.*

Exercise 33.18 *Re-run the example but starting with 5 days to expiry, and rebalancing the portfolio every hour that is, 120 buy/sells.*

Exercise 33.19 *Re-run the example but rebalance only once at 6 months. Repeat these calculations with various time steps, with the aim of understanding the effect which less frequent rebalancing has of the quality of the hedge. (Indeed re-run the calculations with no delta hedging.)*

33.11 Sketch of the B-S-M Derivation

Remark 33.7 For a full discussion of this important topic please refer to one of the texts in the Recommended Reading list.

Our portfolio in Example 33.5 consists of a short position in one call option (i.e. we have sold this contract), with value c, and α underlying shares (with share price, S). The value of this portfolio, π, is therefore:

$$\pi = -c + \alpha S \tag{33.7}$$

We want to understand how the portfolio value, π, may change in the future. Over a short time, dt, the share price will change by dS, resulting in a change to the option price, dc. Therefore, the portfolio value changes by:

$$d\pi = -dc + \alpha dS \tag{33.8}$$

From Equation 33.2, we have $dS = \mu S dt + \sigma S dB$. Thus, in general, the portfolio value will change by a random amount, due to the dB term. Applying a famous result from Ito's lemma: $dc = \frac{\partial c}{\partial t}dt + \frac{\partial c}{\partial S}dS + \frac{1}{2}\sigma^2 S^2 \frac{\partial^2 c}{\partial S^2}dt$, we have:

$$d\pi = -\frac{\partial c}{\partial t}dt - \frac{\partial c}{\partial S}dS - \frac{1}{2}\sigma^2 S^2 \frac{\partial^2 c}{\partial S^2}dt + \alpha\, dS \tag{33.9}$$

Looking at the dS terms, if we make $\alpha = \Delta$, where $\Delta = \frac{\partial c}{\partial S}$, then the above becomes:

$$d\pi = -\frac{\partial c}{\partial t}dt - \frac{1}{2}\sigma^2 S^2 \frac{\partial^2 c}{\partial S^2}dt \tag{33.10}$$

Equation 33.10 does not contain any dS terms, and hence, crucially, any dB terms i.e. there is now no random element; the return is guaranteed. The increase in the fund, $d\pi$, must equal $\pi r dt$, where r is this risk-free rate of return (demonstrated in Example 33.20 below). It is then straightforward to write down the B-S-M differential equation (see Equation 33.1).

From Equations 33.9 and 33.10 we can see that if we buy $\alpha = \Delta$ shares the return over this small time interval, dt, is guaranteed. For each subsequent dt we should recalculate Δ based on the new share price, and buy or sell the appropriate number of shares. Thus, by holding a portfolio of shares and a short position in call options, we can, in theory at least, guarantee our future return over some short time period. This is what we can see from Example 33.5; we start with a fund of £100,000 and end with a fund value distribution with mean equal to £103047, and a relatively low standard deviation of £418.

Exercise 33.20 *By adjusting the code used in Example 33.5, analyse movements over the first short time period i.e. from time to expiry of 1 year to 0.998 years, and hence demonstrate Equation 33.10 and $d\pi = 0.002\pi r$. Also identify scenarios which result in the hedge working poorly.*

33.12 Further Tasks

As noted in the Introduction, this chapter provides an introduction to financial options, covering the basics of pricing, leveraging, and hedging. The reader is encouraged, using the recommended Reading below, to study the following areas:

- exotic options such as barrier, lookback, compound, Asian, and digital options;
- apply code to put options;
- gamma hedging.

33.13 Appendix

Revisiting Equation 33.1

The reader may find it informative to note the impact of each term in Equation 33.1 by calculating the value of each term over a discrete set of times to expiry, whilst experimenting with various parameter values. Note that the price calculations are approximate given the discrete nature of the calculations. Equation 33.1 becomes:

$$\delta c = (rc - \frac{1}{2}\sigma^2 S^2 \frac{\partial^2 c}{\partial S^2} - rS\frac{\partial c}{\partial S})\,\delta t \qquad (33.11)$$

Exercise 33.21 *Calculate the call option price at discrete points in time using a constant share price of £90 (equal to the strike price) and $\delta t = 0.01$, and repeat the exercise with a constant share price of £70 and £110. Use the following parameter values:*

```
dt<-0.01; timezz<-0.2; rate<-0.03; vol<-0.3; strike<-90;
```

Set out below is an extract of the results with a share price of £90:

```
      time gamma delta start 1st term 2nd term 3rd term change    end
[1,] 0.20 0.033 0.545 5.073    0.002   -0.120   -0.015 -0.133 4.940
[2,] 0.19 0.034 0.543 4.939    0.001   -0.123   -0.015 -0.136 4.803
[3,] 0.18 0.035 0.542 4.801    0.001   -0.126   -0.015 -0.139 4.662
```

The "start" column shows the theoretical option price from B-S-M, with the "end" column showing the approximate option price at the end of δt. The 1st term, 2nd term, and

3rd term refer to the 3 terms on the RHS of Equation 33.11 which sum to give the approximate change in the option price over δt; we can see in this scenario the fall in price is dominated by the 2nd term (which includes gamma, Γ) - this is not surprising given the share price equals the strike price resulting in a large Γ value.

Exercise 33.22 *Comment on the level of approximation in these calculation as we approach the expiry time.*

33.14   Recommended Reading

- Baxter, M. and Rennie, A. (1996). *Financial Calculus: An Introduction to Derivative Pricing.* Cambridge, UK : Cambridge University Press.
- Gottesman, A. (2016). *Derivatives Essentials: An Introduction to Forwards, Futures, Options and Swaps.* (Wiley Finance).
- Hull, J. (2012). *Options, Futures and Other Derivatives*, 8e. Harlow, Essex, UK: Pearson.
- Hull, J. (2022). *Fundamentals of Futures and Options Markets*, 9e. Harlow, Essex, UK: Pearson.
- Longstaff, F. A. and Schwartz, E. S. (2001). Valuing American options by simulation: a simple least-squares approach. *The Review of Financial Studies* 14 (1): 113–147.
- Wilmott, P. (2007). *Introduces Quantitative Finance*, 2e. Chichester, West Sussex, UK: Wiley.

Index

R Programming for Actuarial Science, First Edition. Peter McQuire and Alfred Kume.
© 2024 John Wiley & Sons Ltd. Published 2024 by John Wiley & Sons Ltd.
Companion Website: www.wiley.com/go/rprogramming.com.